The Geology of Mars
Evidence from Earth-Based Analogs

With the prospect of a manned mission to Mars still a long way in the future, research into the geological processes operating there continues to rely on interpretation of images and other data returned by unmanned orbiters, probes, and landers. Such interpretations are necessarily based on our knowledge of processes occurring on Earth. Terrestrial analog studies therefore play an important role in understanding the origin of geological features observed on Mars.

This book presents contributions from leading planetary geologists to demonstrate the parallels and differences between these two neighboring planets, and to provide a deeper understanding of the evolution of the Solar System. Mars is characterized by a wide range of geological phenomena that also occur on Earth, including tectonic, volcanic, impact cratering, aeolian, fluvial, glacial, and possibly lacustrine and marine processes. This is the first book to present direct comparisons between locales on Earth and Mars and to provide terrestrial analogs for newly acquired data sets from Mars Global Surveyor, Mars Odyssey, Mars Exploration Rovers, and Mars Express.

The results of these analog studies provide new insights into the role of different processes in the geological evolution of Mars. This book will therefore be a key reference for students and researchers of planetary science.

MARY CHAPMAN is a research geologist with the Astrogeology Team at the United States Geological Survey in Flagstaff, Arizona. She is also the Director and Science Advisor for the NASA Regional Planetary Image Facility there. Her research interests center on volcanism and its interactions with ice and other fluids, and she has a keen interest in the development of future robotic and human exploration of the Solar System.

Cambridge Planetary Science

Series Editors: F. Bagenal, F. Nimmo, C. Murray, D. Jewitt, R. Lorenz and S. Russell

F. Bagenal, T. E. Dowling and W. B. McKinnon *Jupiter: The Planet, Satellites and Magnetosphere*
L. Esposito *Planetary Rings*
R. Hutchinson *Meteorites: A Petrologic, Chemical and Isotopic Synthesis*
D. W. G. Sears *The Origin of Chondrules and Chondrites*
M. G. Chapman *The Geology of Mars: Evidence from Earth-based Analogs*

THE GEOLOGY OF MARS

Evidence from Earth-Based Analogs

Edited by

M. G. CHAPMAN

United States Geological Survey

CAMBRIDGE
UNIVERSITY PRESS

CAMBRIDGE UNIVERSITY PRESS
Cambridge, New York, Melbourne, Madrid, Cape Town,
Singapore, São Paulo, Delhi, Tokyo, Mexico City

Cambridge University Press
The Edinburgh Building, Cambridge CB2 8RU, UK

Published in the United States of America by Cambridge University Press, New York

www.cambridge.org
Information on this title: www.cambridge.org/9780521206594

First published 2007
First paperback edition 2011

A catalogue record for this publication is available from the British Library

Library of Congress Cataloging in Publication data
The geology of Mars : evidence from earth-based analogs / edited by M. G. Chapman.
p. cm. (Cambridge planetary science series)
Includes bibliographical references and index.
ISBN-13: 978-0-521-83292-2 (hardback)
1. Mars (Planet)Geology. 2. Mars (Planet)Surface. 3. Mars (Planet)Volcanoes.
I. Chapman, Mary G. II. Title. III.
Series.
QB641.G435
559.9'23—dc22

2006021246

ISBN 978-0-521-83292-2 Hardback
ISBN 978-0-521-20659-4 Paperback

Additional resources for this publication at www.cambridge.org/9780521206594

Contents

Color plates are located between pages 210 and 211
(These plates are also available for download from www.cambridge.org/9780521206594)

Preface: the rationale for planetary analog studies

Just before I left to attend the June 2001 Geologic Society of London/Geologic Society of America Meeting in Edinburgh, Scotland, I received two e-mail messages. The first was from a UK-based freelance science writer, who was producing a proposal for a six-part television series on various ways that studies of the Earth produce clues about Mars. He requested locations where he might film, other than Hawaii. I was amazed that he seemed not to be aware of all of the locations on Earth where planetary researchers have been studying geologic processes and surfaces that they believe are analogous to those on Mars. In retrospect, his lack of knowledge is understandable, as no books were in existence on the topic of collective Earth locales for Martian studies and no planetary field guides had been published that included terrestrial analogs of the newly acquired data sets: Mars Global Surveyor, Mars Odyssey, Mars Exploration Rovers, and Mars Express. [Historically, NASA published a series of four Comparative Planetary Geology Field Guides with four locales having analog features for comparison with Mars, each book on a different subject and area (volcanic features of Hawaii, volcanism of the eastern Snake River Plain, aeolian features of southern California, and sapping features of the Colorado Plateau). However, all of these books were based on Viking data, intended for researchers in the field, were not widely distributed, and are now out of print (NASA has not published any more field guides).] The second e-mail was from Science Editor Susan Francis of Cambridge University Press, requesting that I stop by their booth at the Edinburgh meeting to discuss a possible topic for a new book on the geology of Mars. Following this e-mail correspondence, I came up with a topic that highlights the current research of geologists who study various environments on Mars using Earth-based analogs.

Planetary geologists commonly perform terrestrial analog studies in order to better understand the geology of extraterrestrial worlds, in order to know more about our solar system. Especially Mars, because although the radius of Mars is about half that of the Earth, its gravity is about a third of our own, and the current Martian atmosphere is very thin, dry, and cold – it is the one planet in the solar system whose surface is most similar to our own. The geology of Mars is characterized by a wide range of geological processes including tectonic, volcanic, impact cratering, aeolian, fluvial, glacial and possibly lacustrine and marine. However, other than the ongoing processes of wind, annual carbon dioxide frosts, and impact cratering, most active geologic processes on Mars shut down millennia ago, leaving a red planet frozen in time. Many of the almost perfectly preserved surface features and deposits of Mars appear visually very similar to analogous terrestrial locales, leading researchers to propose similar processes and origins for deposits on both planets. In order to test their hypotheses, logically researchers visit and study these analog areas on Earth to determine characteristics that (1) provide evidence for the origin of surfaces on Mars and (2) can be detected by instruments and astronauts on current and future missions. Currently, the Mars Global Surveyor, Mars Odyssey, and Mars Express spacecraft and onboard instruments continue to orbit the planet and acquire data, while the active Mars Exploration Rovers explore the surface of Gusev Crater and the Meridiani plains. Recent data from these missions show that our earlier interpretations of Mars geology need to undergo expansion and revision. In this book, examples of new insights into these processes on Mars underline the need for study of Earth processes and analogs and the application of these results to a better understanding of the geological evolution of Mars. In addition, future rover and spacecraft missions are also being planned for upcoming launch opportunities. Within the next 20 years, perhaps astronauts may be sent to Mars. Missions to Mars are expensive. It is necessary and cost effective to attempt to be certain that our mission instruments and personnel are equipped and trained to detect and discern the nature of Martian terrains before they are deployed on that planet. Therefore, research geologists investigate terrestrial analog environments to develop criteria to better identify the nature of planetary deposits from remote surface measurements and orbiting spacecraft data.

The first chapter in this book by Jim Head discusses how our Viking-based view of Mars has changed based on the new data we are receiving from the current Mars missions. The rest of the chapters detail how specific rocks and environments on Earth are studied in order to better interpret data from Mars. I would like to thank all the authors that participated in this long-overdue book. The chapters in this book were improved by helpful comments and suggestions from our peer reviewers and I appreciate and want to thank for their time and efforts Devon Burr, Nathalie Cabrol, Bill Cassidy, Dean Eppler, Sarah Fagents, Paul Geissler, Trent Hare, Jeff Kargel, Lazlo Keszthelyi, Goro Komatsu, Nick Lancaster, John McHone, Dan Milton, Bill Muehlburger, Kevin Mullins, Horton Newsom, Tom Pierson, Jeff Plescia, Sue Priest, Susan Sakimoto, Ian Skilling, Jim Skinner, Ken Tanaka, Tim Titus, Wes Ward, Lionel Wilson and Jim Zimbelman.

Mary Chapman

Contributors

E. Ács, *Hungarian Danube Research Station of Institute of Ecology and Botany of the Hungarian Academy of Sciences, Göd, Hungary*
Jayne C. Aubele, *New Mexico Museum of Natural History and Science, Albuquerque, NM, USA*
Victor R. Baker, *Department of Hydrology and Water Resources, University of Arizona, Tucson, AZ, USA*
Nadine G. Barlow, *Department of Physics and Astronomy, Northern Arizona University, Flagstaff, AZ, USA*
Geoffrey Briggs, *NASA Ames Research Center, Moffett Field, CA, USA*
D. Brunstein, *CNRS UMR 8591, Laboratoire de Géographie Physique, Meudon, France*
N. A. Cabrol, *Space Science Division, MS 245-3, NASA Ames Research Center, Moffett Field, CA, USA; and SETI Institute, 515 N. Whisman Road - Mountain View, CA 94043, USA*
Mary Chapman, *US Geological Survey, Flagstaff, AZ, USA*
G. Chong, *Departamento de Geología, Universidad Católica del Norte, Avda., Antofagasta, Chile*
François Costard, *UMR 8148 IDES, Université Paris-Sud, Orsay, France*
Larry S. Crumpler, *New Mexico Museum of Natural History and Science, Albuquerque, NM, USA*
C. Demergasso, *Laboratorio de Microbiología Técnica, Departamento de Química, Universidad Católica del Norte, Avda., Antofagasta, Chile*
L. Escudero, *Laboratorio de Microbiología Técnica, Departamento de Química, Universidad Católica del Norte, Avda., Antofagasta, Chile*
Sarah A. Fagents, *University of Hawaii at Manoa, Honolulu, HI, USA*
D. A. Fike, *Massachusetts Institute of Technology, Cambridge, MA, USA*
François Forget, *Laboratory for Dynamic Meteorology, CNRS, Paris, France*

P. Galleguillos, *Laboratorio de Microbiología Técnica, Departamento de Química, Universidad Católica del Norte, Avda., Antofagasta, Chile*
Emmanuele Gautier, *CNRS UMR 8591, Laboratoire de Géographie Physique, Meudon, France*
Brian Glass, *NASA Ames Research Center, Moffett Field, CA, USA*
Tracy K. P. Gregg, *The University at Buffalo, Buffalo, NY, USA*
I. Grigorszky, *Debrecen University, Botanical Department, Debrecen, Hungary*
B. H. Grigsby, *Schreder Planetarium/ARISE, Redding, CA 96001, USA*
E. A. Grin, *Space Science Division, MS 245-3, NASA Ames Research Center, Moffett Field, CA, USA; and SETI Institute, 515 N. Whisman Road - Mountain View, CA 94043, USA*
E. B. Grosfils, *Department of Geology, Pomona College, Claremont, CA, USA*
Andrew J. L. Harris, *Hawaii Institute of Geophysics and Planetology, University of Hawaii at Manoa, Honolulu, HI, USA*
James W. Head, *Department of Geological Sciences, Brown University, Providence, RI 02912, USA*
A. N. Hock, *University of California Los Angeles, Los Angeles, CA, USA*
Jennifer Jasper, *NASA Ames Research Center, Moffett Field, CA, USA*
Vincent Jomelli, *CNRS UMR 8591, Laboratoire de Géographie Physique, Meudon, France*
Lazlo Keszthelyi, *US Geological Survey, Flagstaff, AZ, USA*
K. T. Kiss, *Hungarian Danube Research Station of Institute of Ecology and Botany of the Hungarian Academy of Sciences, Göd, Hungary*
Goro Komatsu, *International Research School of Planetary Sciences, Universita' d'Annunzio, Pescara, Italy*
Ruslan O. Kuzmin, *Vernadsky Institute, Russian Academy of Sciences, Moscow, Russia*
Nicolas Mangold, *UMR 8148 IDES, Université Paris-Sud, Orsay, France*
Lucia Maninangeli, *International Research School of Planetary Sciences, Universita' d'Annunzio, Pescara, Italy*
Alfred McEwen, *University of Arizona, Tucson, AZ, USA*
C. P. McKay, *Space Science Division, MS 245-3, NASA Ames Research Center, Moffett Field, CA, USA*
D. Mège, *Laboratoire de planétologie et géodynamique, Université de Nantes, Nantes cedex, France*
Jeffrey E. Moersch, *Department of Geological Sciences, University of Tennessee, Knoxville, TN, USA*
Jason M. Moore, *William Cotton & Associates, Los Gatos, CA, USA*
Peter J. Mouginis-Mark, *Hawaii Institute of Geophysics and Planetology, University of Hawaii at Manoa, Honolulu, HI, USA*

Horton E. Newsom, *Institute of Meteoritics and Department of Earth and Planetary Sciences, University of New Mexico, Albuquerque, NM, USA*
Gian Gabriel Ori, *International Research School of Planetary Sciences, Universita' d'Annunzio, Pescara, Italy*
Jean-Pierre Peulvast, *UMR 8148 IDES, Université Paris-Sud, Orsay, France*
Scott K. Rowland, *Hawaii Institute of Geophysics and Planetology, University of Hawaii at Manoa, Honolulu, HI, USA*
R. A. Schultz, *Department of Geological Sciences, University of Nevada, Reno, NV, USA*
Virgil Sharpton, *Geophysical Institute, University of Alaska, Fairbanks, AK, USA*
John L. Smellie, *British Antarctic Survey, Cambridge, UK*
Kelly Snook, *NASA Goddard Space Flight Center, Greenbelt, MD 20771, USA*
K. Szabò, *Eötvös L. University, Microbiological Department, Budapest, Hungary*
C. Tambley, *Department of Astrophysics, Universidad Católica del Norte, Avda., Antofagasta, Chile*
K. L. Tanaka, *US Geological Survey, Flagstaff, AZ, USA*
Thorvaldur Thordarson, *University of Hawaii at Manoa, Honolulu, HI, USA*
B. Tóth, *Hungarian Danube Research Station of Institute of Ecology and Botany of the Hungarian Academy of Sciences, Göd, Hungary*
Steven H. Williams, *National Air and Space Museum, Smithsonian Institution, Washington, DC, USA*
J. Zambrana Román, *Servicio Nacional de Geología y Minería (SERGEOMIN), La Paz, Bolivia*
James R. Zimbelman, *Center for Earth and Planetary Studies, National Air and Space Museum, Smithsonian Institution, Washington, DC, USA*

1

The geology of Mars: new insights and outstanding questions

James W. Head
Department of Geological Sciences, Brown University

1.1 Introduction

The major dynamic forces shaping the surfaces, crusts, and lithospheres of planets are represented by geological processes (Figures 1.1–1.6) which are linked to interaction with the atmosphere (e.g., eolian, polar), with the hydrosphere (e.g., fluvial, lacustrine), with the cryosphere (e.g., glacial and periglacial), or with the crust, lithosphere, and interior (e.g., tectonism and volcanism). Interaction with the planetary external environment also occurs, as in the case of impact cratering processes. Geological processes vary in relative importance in space and time; for example, impact cratering was a key process in forming and shaping planetary crusts in the first one-quarter of Solar System history, but its global influence has waned considerably since that time. Volcanic activity is a reflection of the thermal evolution of the planet, and varies accordingly in abundance and style.

The stratigraphic record of a planet represents the products or deposits of these geological processes and how they are arranged relative to one another. The geological history of a planet can be reconstructed from an understanding of the details of this stratigraphic record. On Mars, the geological history has been reconstructed using the global Viking image data set to delineate geological units (e.g., Greeley and Guest, 1987; Tanaka and Scott, 1987; Tanaka *et al.*, 1992), and superposition and cross-cutting relationships to establish their relative ages, with superposed impact crater abundance tied to an absolute chronology (e.g., Hartmann and Neukum, 2001). These data have permitted reconstruction of the geological history and the relative importance of processes as a function of time, and determination of the main themes in the evolution of Mars. Three major time periods are defined: Noachian, Hesperian, and Amazonian. Although absolute ages have been

The Geology of Mars: Evidence from Earth-based Analogs, ed. Mary Chapman. Published by Cambridge University Press.

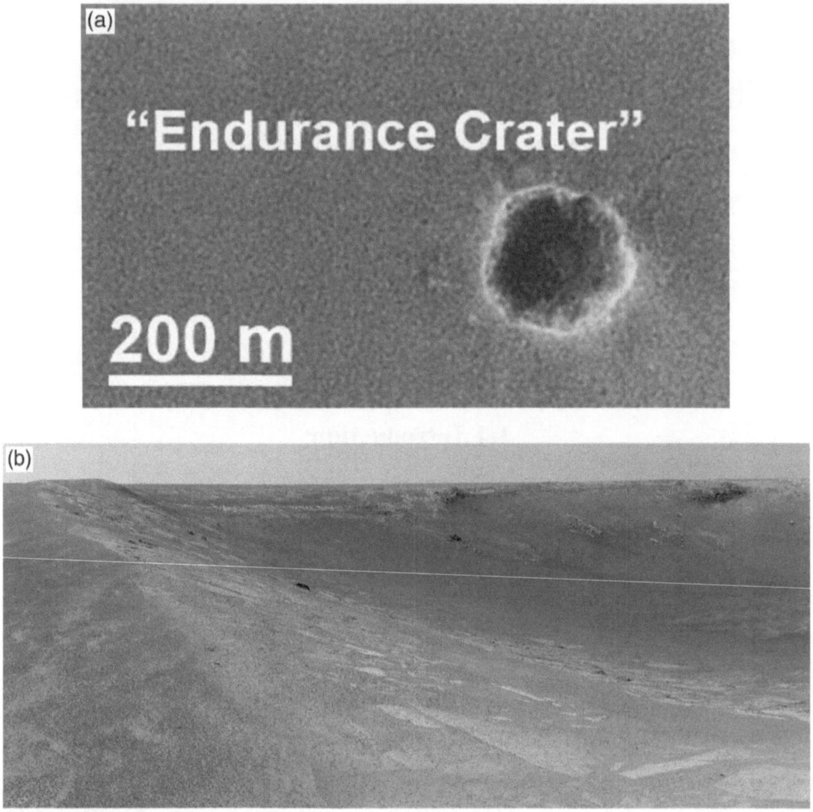

Figure 1.1. Impact crater landforms and processes. (a) NASA's Mars Exploration Rover Opportunity landed on Jan. 24 on a small bowl crater within the Meridiani Planum region later nicknamed "Eagle Crater." After about two months of examining rocks and soils within that crater, the rover set out toward a larger crater informally named "Endurance." During an extended mission following its three-month prime mission, Opportunity finished examining Endurance (1b), and explored a type of landscape to the southeast called "etched terrain" where additional deposits of layered bedrock are exposed. The underlying image for the map was taken from orbit by the Mars Orbiter Camera (MOC) on NASA's Mars Global Surveyor. (NASA/JPL/MSSS). (b) This image taken by the panoramic camera on the Mars Exploration Rover Opportunity shows the interior of the impact crater known as "Endurance." The exposed walls provide a window to what lies beneath the surface of Mars and thus what geologic processes occurred there in the past. While recent studies of the smaller crater nicknamed "Eagle" revealed evidence for an ancient evaporating body of salty water, that crater was not deep enough to indicate what came before the water. Endurance explored this question in the rocks embedded in vertical cliffs. Endurance is ~130 m across. Images such as these bridge the gap between orbital views and sample analysis and provide an important scale perspective when using terrestrial analogs. (NASA/JPL/Cornell).

Figure 1.1. (cont.) (c) Nighttime THEMIS IR image of a ~90 km diameter impact crater along the northeastern margin of Hellas Basin. Bright areas on the surface are warmer than dark areas. Bright areas along the rim of the crater (and along the rim of the smaller superposed crater in the center of the image) are likely to be exposed bedrock that show a higher thermal inertia than the surrounding soil. Image: I07269009 (ASU). (d) Daytime THEMIS IR image of the same crater in 1c. Surface temperature readings are largely dependant on solar reflectance during the day, so small-scale variations in surface composition are not as easily detected, but morphology is enhanced. This combination provides important additional information in interpreting the surface process and geologic history. Image: I07987004 (ASU).

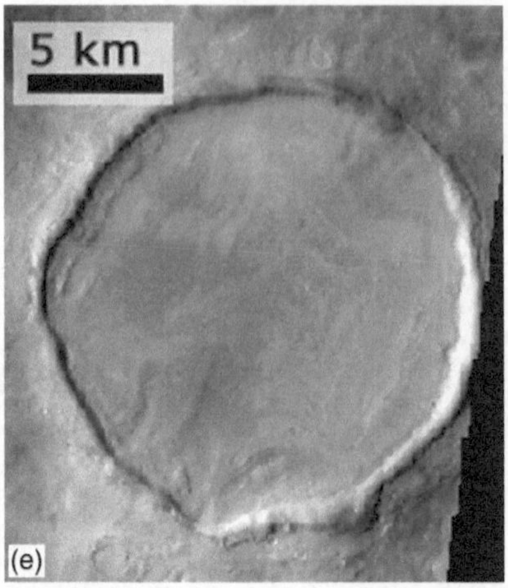

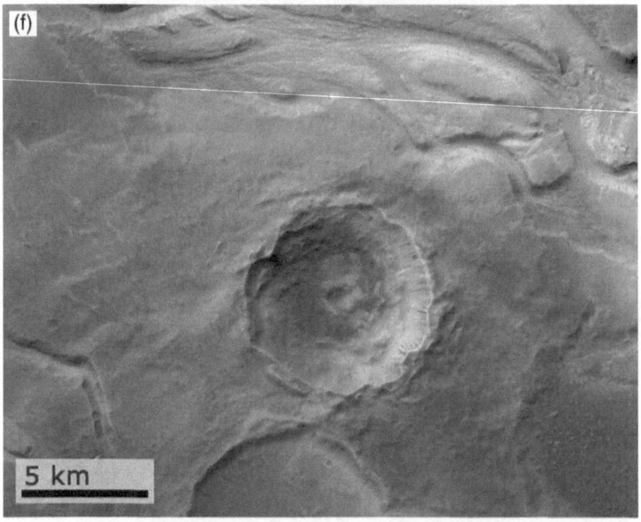

Figure 1.1. (cont.) (e) THEMIS Visible image V03679003 of a highly modified impact crater in the Adamas Labyrinthus region, within Utopia Planitia, at 43.9° N, 101.7° E. (ASU). (f) High Resolution Stereo Camera on board the Mars Express spacecraft took this image of an impact crater to the west of Mangala Valles and just south of its northern reaches (top of image), at 15° S, 205° E. (ESA). (g) The Haughton meteorite impact crater, on Devon Island, Nunavut, in the Canadian high arctic, is 20 km in diameter and formed 23 million years ago. It is one of the highest-latitude terrestrial impact craters known on land (75°22' N, 89°41' W) and is the only crater on Earth known to lie in a polar desert environment similar to that of Mars. Terrestrial analogs such as these provide important information on the nature of impact cratering and modification processes on Mars (see marsonearth.org; Image: obtained via GSFC by Landsat 7, bands 4, 3, and 2).

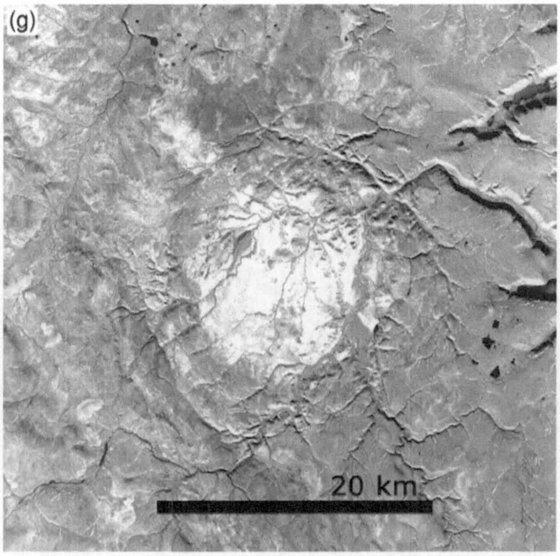

Figure 1.1. (cont.)

assigned to these periods (e.g., Hartmann and Neukum, 2001) (Noachian, ~4.65–3.7 Gyr; Hesperian, ~3.7–3.0 Gyr; Amazonian, ~3.0 Gyr to present), lack of samples from Mars whose context and provenance are known means that these assignments based on crater densities are dependent on estimates of cratering rates and thus are model dependent. Further confidence in these assignments must await a better understanding of the flux in the vicinity of Mars and radiometric dating of returned samples from known units on the surface of Mars.

Confidence in understanding the nature of the geological processes shaping planetary surfaces is derived from: (1) data: the amount and diversity of planetary data at hand, (2) terrestrial analogs: the level of understanding of these processes on Earth and their applicability, and (3) physical modeling: the manner in which planetary variables modulate and modify the processes (e.g., position in the Solar System, which influences initial state, composition, and solar insolation with time; size, which influences gravity and thermal evolution; and presence and nature of an atmosphere, which influences dynamic processes such as magmatic explosive disruption, ejecta emplacement, lava flow cooling, eolian modification, and chemical weathering). On Mars, our understanding of the geological history at the turn of the century was derived largely from the framework provided by the comprehensive coverage of the Mariner and Viking imaging systems (e.g., Mutch *et al.*, 1976; Carr, 1981; Scott and Tanaka, 1986; Greeley and Guest, 1987; Tanaka and Scott, 1987; Tanaka *et al.*, 1992).

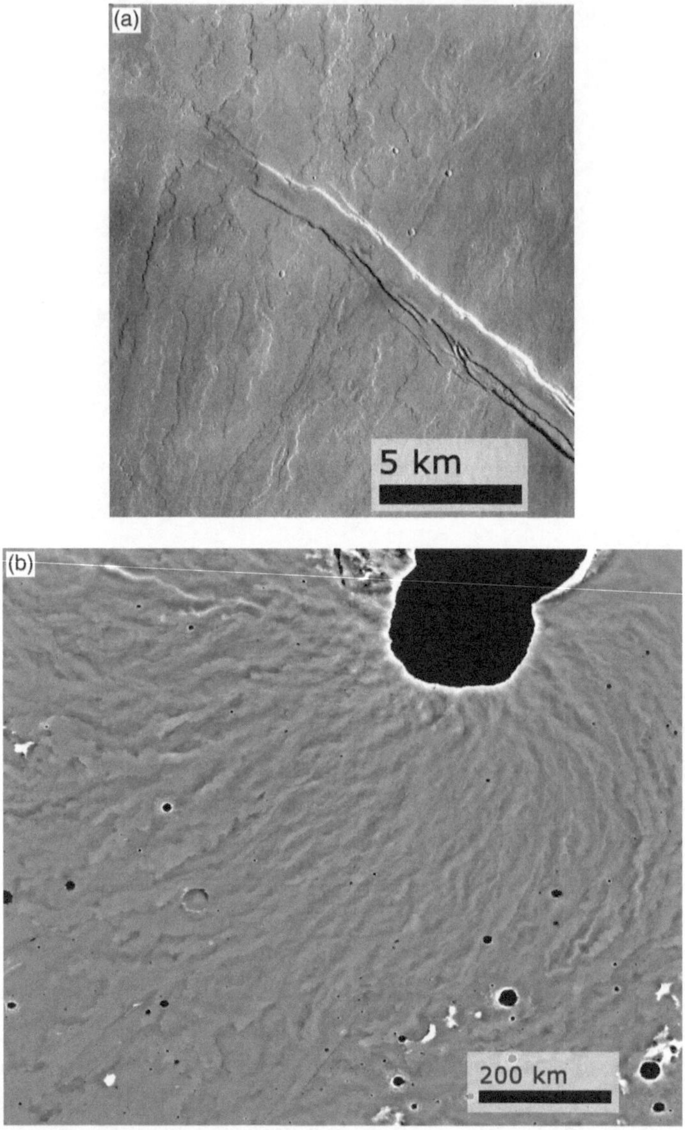

Figure 1.2. Volcanic landforms and processes. (a) Lobate lava flows from Olympus Mons. The relative timing of these volcanic flows and the formation of the structural feature can be deduced by which flows are cut by the fracture and which flows fill and cross the fracture. (THEMIS V02064003; ASU) (b) Lava flows of Arsia Mons, the southernmost of the Tharsis Montes. In this MOLA detrended altimetry data image, the regional topographic slope has been removed and individual lava flows become highlighted. The blacked out area represents the flanking rift zone (lower lobe) and the summit edifice and caldera (upper portion of blacked out area). These new data and modes of presentation provide important tools in the mapping and comparison of lava flows to terrestrial analogs.

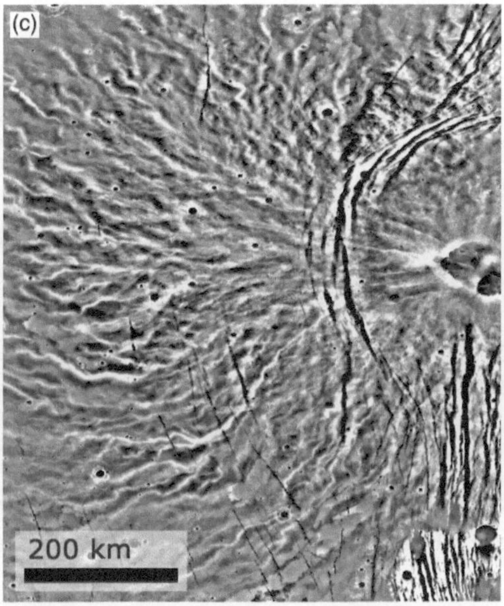

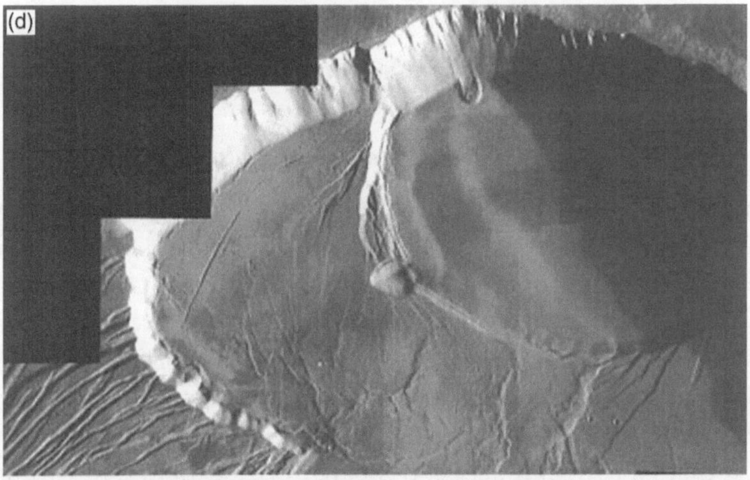

Figure 1.2. (cont.) (c) The western part of the summit and flank of Alba Patera, a massive shield volcano in the northern part of Tharsis. The MOLA detrended topographic representation shows the western part of the summit caldera and edifice, concentric faults, and the extensive western lava flow complex. (d) Multiple calderas on the summit of Olympus Mons, the largest volcano on Mars. Sequential collapse of the calderas can be assessed from the cross cutting relationship, with the youngest being in the top right. The surfaces of the caldera floors are flooded by lavas and then further deformed by wrinkle ridges and graben. Width of the caldera in the upper right is ~30 km. (THEMIS Visible image I04848014) (ASU)

Newly acquired data sets (Mars Global Surveyor, Mars Odyssey, Mars Exploration Rovers, and Mars Express) and increased understanding of terrestrial analogs and their application are fundamentally and irrevocably changing our view of Mars and its geologic history. Global high-resolution topography, comprehensive high-resolution images, thermal mapping of rock and soils types and abundance, enhanced spectral range and resolution, mapping of surface and near-surface water and ice, probing of shallow crustal structure, mapping of gravity and magnetic anomalies, roving determination of surface geology, physical properties, geochemistry and mineralogy, astrobiological investigations, and sounding of the subsurface are some of the ways our understanding is changing. In this contribution, the current view of the geology of Mars is summarized, some key outstanding questions are outlined, and an assessment is made as to where changes from new data and a better understanding of terrestrial analogs is likely to take us in the near future.

1.2 Geological processes and their importance in understanding the history of Mars

1.2.1 Impact crater landforms and processes

Impact craters (Figure 1.1) occur on virtually all geological units and in the cases of older units, such as the heavily cratered uplands, basically characterize and shape the terrain (Figure 1.1c,d), forming the first-order topographic roughness of the Martian uplands (Smith *et al.*, 1999; Kreslavsky and Head, 2000). Several large basins (Hellas, Argyre, Isidis, Utopia) dominate regional topography and crustal thickness. Impact craters cause vertical excavation and lateral transport of crustal material, and future sample return strategies will call on this fact to gain access to deeper crustal material. Ejecta deposit morphologies in younger craters (e.g., Barlow *et al.*, 2000; Barlow and Perez, 2003) provide important clues to the nature of the substrate and also reveal the nature of the impact cratering process, particularly in reference to Martian gravity conditions, presence of an atmosphere, and icy substrates. Impact melts and ejected glasses are also likely to be important (Schultz and Mustard, 2004). Older impact craters provide clues to the types of modification processes operating on landforms (e.g., Pelkey and Jakosky, 2002; Pelkey *et al.*, 2003; Forsberg-Taylor *et al.*, 2004) (Figure 1.1c–f). Impact craters can also be sites of long-term geothermal activity due to heating and impact melt emplacement, and can serve as sinks for ponded surface water (e.g., Carr, 1996; Rathbun and Squyres, 2002).

The number of impact craters forming as a function of time, the flux, is a critical aspect of impact crater studies as it provides a link to absolute chronology provided by radiometrically dated samples returned from well-characterized lunar surfaces. Tanaka (1986) described the crater density of a range of stratigraphic units on Mars, and Ivanov and Head (2001) discussed a conversion from lunar to Martian cratering rates, which set the stage for correlation of crater density with absolute age on Mars. Hartmann and Neukum (2001) show that, in agreement with Martian meteorite ages, significant areas of late Amazonian volcanic and other units have ages in the range of a few hundred million years, while most of the Noachian probably occurred before 3.7 Gyr ago. In the less reliably dated intermediate periods of the history of Mars, Hartmann and Neukum (2001) use the Tanaka *et al.* (1987) tabulation of areas (km^2) resurfaced by different geological processes in different epochs, to show that many processes, including volcanic, fluvial, and periglacial resurfacing, show much stronger activity before ~3 Gyr ago, and decline, perhaps sharply, to a lower level after that time.

Future sample return missions must focus on the acquisition and return for radiometric dating of key geologic units that can be characterized in terms of the impact cratering flux. This step is of the utmost importance in establishing the geologic and thermal evolution of Mars, and the confident interplanetary correlation that will reveal the fundamental themes in planetary evolution. Characterization of impact craters at all scales on Mars is important to obtain a much more firm understanding of the cratering process. Currently there are uncertainties in the nature of the excavation process that influence the size frequency distribution and thus the dating of surfaces. The role of volatiles in the process of excavation, ejecta emplacement, and immediate landform modification is poorly understood. New high-resolution data on the topographic, physical properties, and mineralogic characteristics of impact craters and their deposits are beginning to revolutionize our understanding of the cratering process on Mars (Malin and Edgett, 2001), and radar sounding and surface rovers will add significantly to this picture. Until this improved picture emerges, the full potential of impact cratering as a "drilling" and redistribution process cannot be realized. Terrestrial analogs (Figure 1.1) must play a critical role in contributing to this new understanding and the documentation of Earth impact craters in a host of different geological and climate

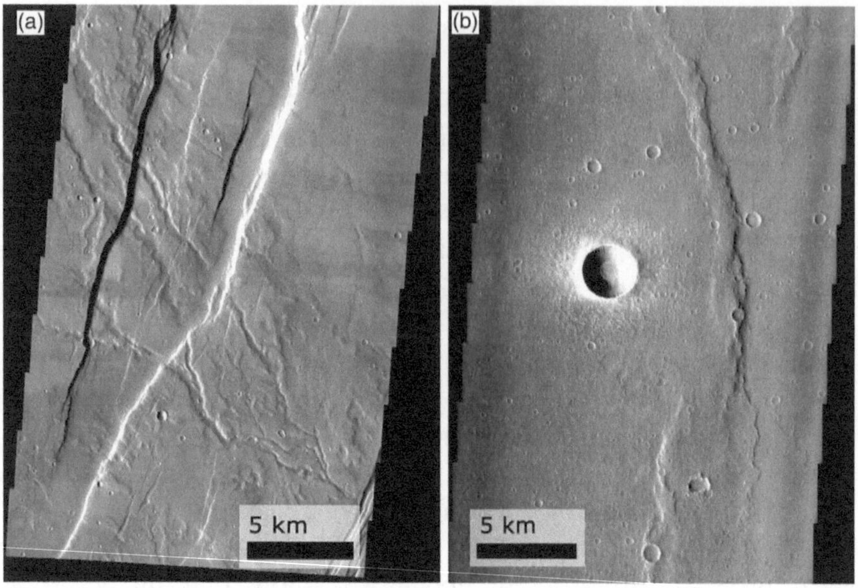

Figure 1.3. Tectonic landforms and processes. (a) Tantalus Fossae, a graben system, along the eastern flank of Alba Patera. Note that the lava flows and channels are cut by the graben. (THEMIS Visible image V02625006) (ASU) (b) Ridged plains of Lunae Planum located between Kasei Valles and Valles Marineris in the northern hemisphere of Mars. Wrinkle ridges are seen along the eastern side of the image. The broadest wrinkle ridges in this image are up to 2 km wide. A 3 km diameter young fresh impact crater is also seen; the sharp well-defined crater rim and the ejecta blanket contrasts with older more degraded impacts (Figure 1.1) and is indicative of a very young crater that has not been subjected to significant erosional processes. (THEMIS image V01388007) (ASU)

environments on Earth (submarine, desert, polar, temperate) is beginning to provide new insight (e.g., Barlow *et al.*, Chapter 2 in this volume).

1.2.2 Volcanic landforms and processes

Early Mars space missions (Mariner 9, Viking) showed clearly the importance of volcanic processes in the history of Mars (Figure 1.2). The huge shield volcanoes of the Tharsis and Elysium regions, extensive lava plains (Figure 1.2a–c), and low-profile constructs (paterae), permitted mapping and characterization of the extent, timing, and styles of volcanism on Mars (Greeley and Spudis, 1978; Mouginis-Mark *et al.*, 1992; Greeley *et al.*, 2000a). Currently and in the near future, new high-resolution images

(Malin *et al.*, 1998), information on surface compositions (McSween *et al.*, 1999; Christensen *et al.*, 2000a, b), and topographic data (Smith *et al.*, 1998, 1999, 2001) provided by the Mars Global Surveyor are providing new insight into Martian volcanism and permit comparison to theoretical analysis of the ascent and eruption of magma on Mars (e.g., Wilson and Head, 1994).

The majority of the ancient crust of Mars is certainly of magmatic (volcanic and plutonic) origin but due to the role of heavy impact bombardment, primary landforms such as flows and structures which might be vents, are not observed in images of the oldest terrains. Thermal Emission Spectrometer (TES) data suggest basaltic compositions for the Martian highlands where most of the ancient crust is found (Christensen *et al.*, 2000a, b). Thus, impact, together with eolian and aqueous geological processes have modified the highlands so extensively that morphological traces related to early putative volcanism are not readily found with currently available data. Paterae are some of the earliest recognizable volcanic features, including Tyrhenna, Hadriaca, Amphitrites, and Peneus Paterae in the Hellas region (Greeley and Spudis, 1978) and several in Syrtis Major (e.g., Hiesinger and Head, 2004). Early eruptions are thought to have involved magmas rising through water-rich megaregolith, leading to extensive phreatic-magmatic activity and the emplacement of ash-rich shields (Crown and Greeley, 1993). Eruptions were apparently followed by effusive activity, emplacing complex sequences of flows which radiate from central calderas. Evolved lavas (Warner and Gregg, 2003) and explosive volcanism may also have occurred in Tharsis (Hynek *et al.*, 2003).

One of the largest volcanoes in the Solar System, Alba Patera (Figure 1.2c), is a central vent structure covering more than 4.4×10^6 km^2. Cattermole (1987) described tube-and-channel fed flows and flows emplaced as massive sheets on the edifice. It contains a caldera some 100 km across, the floor of which includes small cones of probable spatter and pyroclastic origin. Mars Observer Laser Altimeter (MOLA) data have enabled the detailed morphology of the edifice to be understood and it is now clear that Alba Patera shares some of the major characteristics of the younger Tharsis Montes, with flanking rift zones and a complex summit (Ivanov and Head, 2001).

Tharsis Montes, together with Olympus Mons, represent the most impressive volcanoes on Mars. These and related volcanoes on Tharsis and those found in Elysium regions consist of more than a dozen major constructs. High-resolution images show that most of these volcanoes were built from countless individual flows, many of which were emplaced through

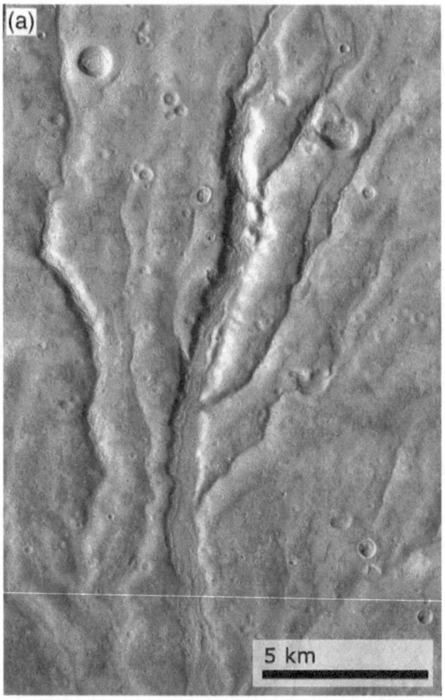

Figure 1.4. Fluvial landforms and processes. (a) Dendritic valley networks to the south of Tharsis in the Aonia Terra region. (THEMIS image V06907004) (b) Outflow channels in southern Chryse Planitia. Sinuous channels are seen to cut through cratered plains, sometimes sourcing in extensive collapse areas. Downstream, hydrodynamically shaped islands are common and the margins of the channels have undergone collapse in several places. MOLA gradient map. (c) MOC high-resolution image of the north wall of a 7 km diameter crater on the floor of the much larger Newton Crater (~287 km across). This crater is only about seven times larger than Meteor Crater in northern Arizona and these types of high-resolution images illustrate how the scale gap between planetary images and terrestrial analogs is closing. The north wall of the crater has many narrow gullies eroded into it. These are hypothesized to have been formed by flowing water and debris flows. Debris transported with the water created lobed and finger-like deposits at the base of the crater wall where it intersects the floor. Many of the finger-like deposits have small channels indicating that a liquid – most likely water – flowed in these areas. The scene is illuminated from the left; north is up. (MOC mosaic of three different images; near 41.1° S, 159.8° W; NASA/JPL/MSSS.)

channels and lava tubes, and representing a style of volcanism analogous to Hawaiian eruptions (Greeley, 1973). Multiple stages of magma ascent and withdrawal are suggested by complex summit calderas (Wilson *et al.*, 2001) (Figure 1.2d). Some of the deposits in the Elysium region have a morphology suggestive of lahars, i.e., emplaced as water-rich slurries (Christensen, 1989;

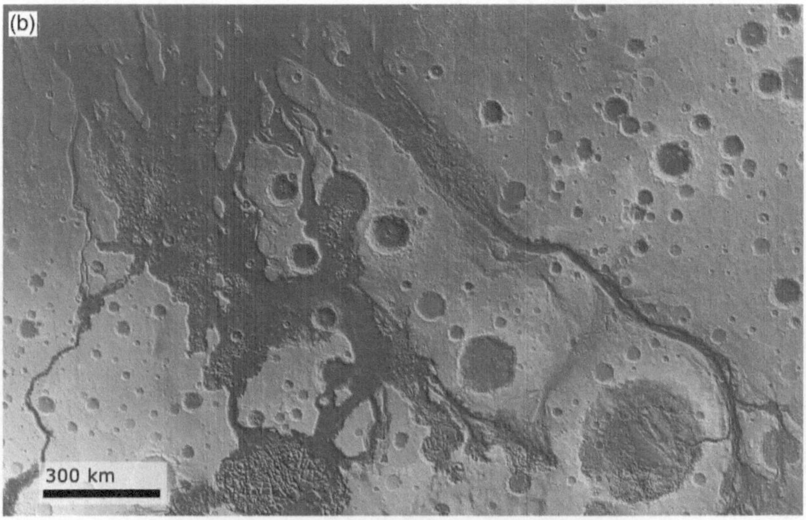

Figure 1.4. (cont.)

Russell and Head, 2003). The subdued appearance of some of the Elysium volcanic summits has been proposed as pyroclastic material, suggesting plinian styles of eruption (Mouginis-Mark *et al.*, 1982; Wilson and Mouginis-Mark, 2001, 2003a, b; also on Tharsis; Scott and Wilson, 2002). The flanks of Olympus Mons are marked by terrain which appears to have

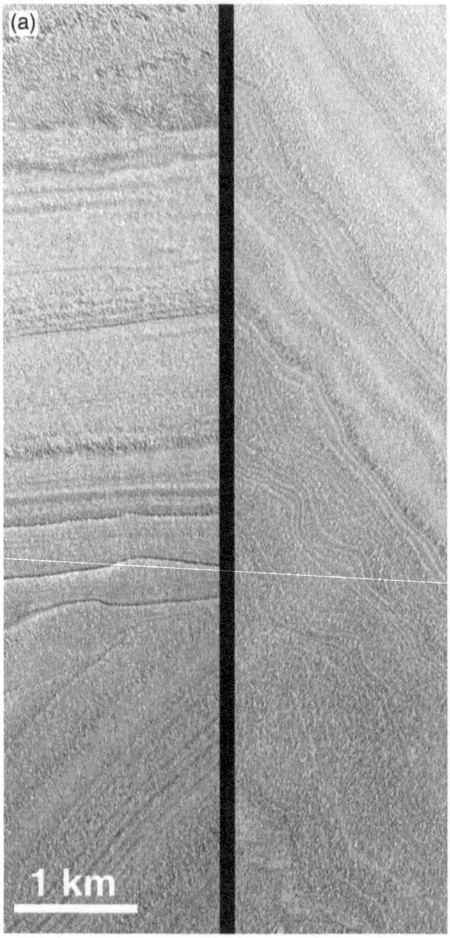

Figure 1.5. Polar, circumpolar, glacial, periglacial, and mass-wasting landforms and processes. (a) Polar layered terrain in the north polar cap, interpreted to consist of alternating layers of ice-rich (brighter) and dust-rich (darker) material. Alternating layers are interpreted to be related to climate change. The left image shows an angular unconformity in a layered outcrop suggesting abrupt changes in the sequence of the layers caused by an interval of erosion followed by resumed deposition. The unconformity represents a break in the sequence of deposition, and thus a gap in the record; the amount of material removed is unknown. The presence of the angular unconformity indicates that before the upper, horizontal layers were deposited, there was a period in which the lower layers were tilted and eroded. The image on the right shows some of the north polar cap layers that that appear to have been folded or deformed which can occur when layers flow as they adjust to the added weight of subsequent deposition. (Image NASA/JPL/MSSS) (b) The north polar cap of Mars overlies a basal unit represented by a series of layered materials that have characteristics that contrast with the overlying bright layers, which are thought to be dominated by ice and dust (a). The lower layers are interpreted to be less icy and contain some amount of dark sand, which

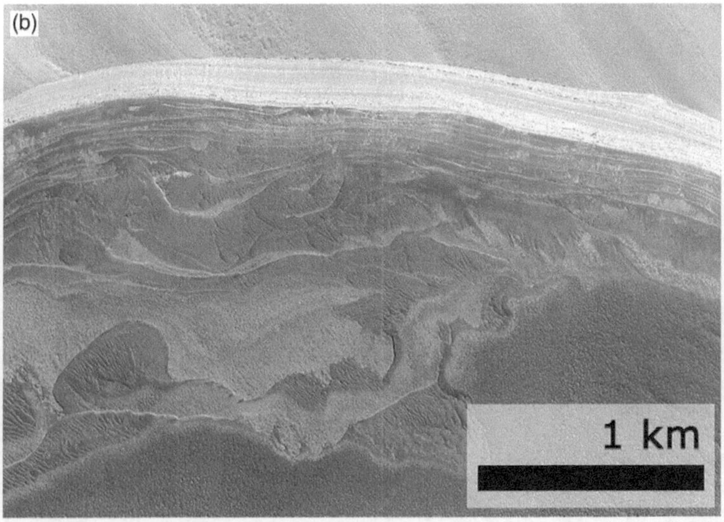

Figure 1.5. (cont.) can be seen to erode much differently than the predominantly material. (83.9° N, 237.9° W; illumination is from the lower left.) (MOC image: E0201209; NASA/JPL/MSSS) (c) Left and middle. Sketch map and image of lobate viscous flow feature on a crater wall on Mars (247° W/38.6° S) (MOC image M18/00898). North is at the top of the image, and illumination is from the northwest. Right. Image mosaic of Mullins Valley, within the Dry Valleys of Antarctica (77°54' S, 160°35' W). This debris-covered alpine glacier moves at very slow rates and is composed of almost pure glacial ice lying below less than a meter of sublimation till in which polygons have formed. (Images acquired by the CAMBOT camera on the Airborne Topographic Mapper project; NASA/GSFC) (d) Geological sketch map of the western Arsia Mons fan-shaped deposit (modified from Zimbelman and Edgett, 1992) superposed on a MOLA topographic gradient map (fan-shaped deposits: R, ridged; K, knobby; S, smooth) (other adjacent deposits: SA, shield; SB, degraded western flank; SC, smooth lower western flank; CF, caldera floor; CW, caldera wall; PF, flank vent flows from Arsia Mons; P, undivided Tharsis plains). (e) Facies of the fan-shaped deposit and possible terrestrial analogs. (a) Ridged facies; (b) Knobby facies; (c) Smooth facies; (a–c are Viking Orbiter images); (d) Drop moraines in the Antarctic Dry Valleys; (e) Sublimation tills in the Antarctic Dry Valleys; (f) Rock glacier in the Antarctic Dry Valleys (portions of USGS aerial photographs; TMA 3079/303, (d); 3078/006, (e); 3080/275, f; all 11-21-93). (f) Possible periglacial features in Utopia Planitia (near 48.0° N, 225.7° W). Although the Martian northern plains are often considered to be flat and featureless, this MOC image shows pitted and fractured plains unlike anything found by MOC elsewhere on Mars. A suite of sharply oulined pits and fractures indicate that the upper surface materials are strong and indurated. The parallel and polygonal alignments of fractures and pits indicate that this area has been subjected to directional stress. The pits are interpreted to mean that ground ice has been removed from beneath the rigid, upper crusted material. (Illumination is from the lower left; MOC image E02-00880, NASA/JPL/MSSS.) (g) Debris aprons surrounding massifs in the area on the eastern rim

(c)

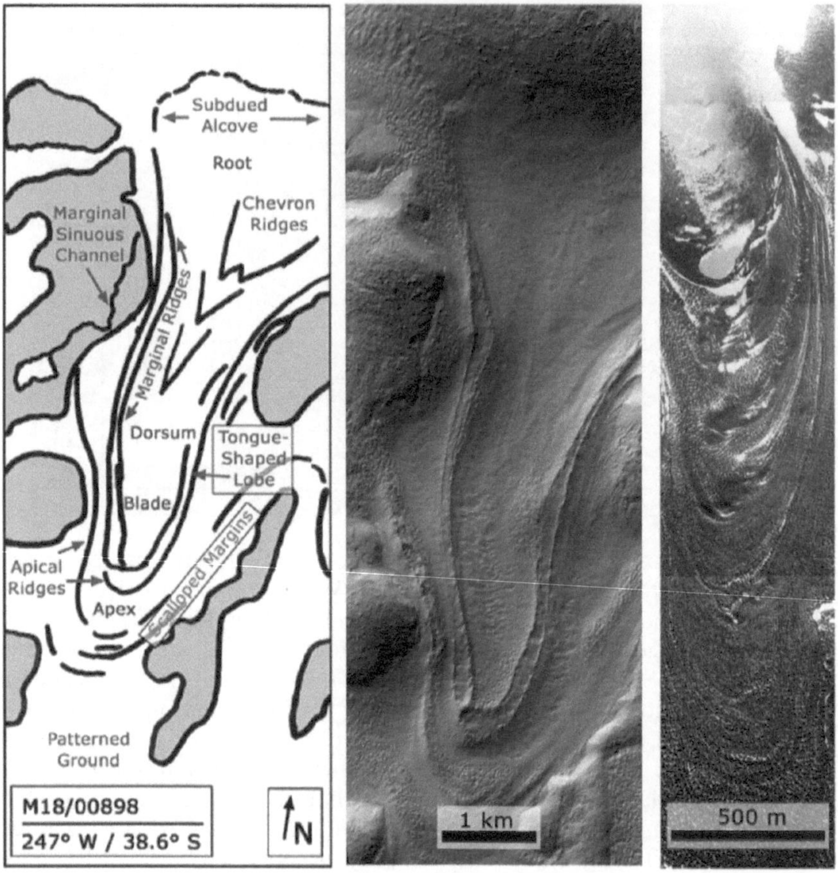

Figure 1.5. (cont.) of the Hellas Basin. Note the multiple debris aprons and their convergence into multiple lobes flowing downslope. THEMIS images superposed on MOLA topography. (h) Large wall slump and landslide on the interior of Ganges Chasma. Such features are common in the interior walls of Valles Marineris. Subframe of THEMIS image I01001001. (ASU) (i) Multiple wall slumps and landslides in THEMIS image I01699006. (ASU)

failed by gravitational collapse, leaving mass wasted deposits covering hundreds of thousands of square kilometers.

Volcanic plains, the youngest of which show flow fronts and embayment into older terrains, represent by far the greatest areal extent and inferred volume of volcanic materials on Mars and are thought to represent flood eruptions. More problematic are the extensive ridged plains (e.g., Hesperia Planum) characterized by wrinkle ridges (Figure 1.3b), and thought to be analogous to mare basalts on the Moon. A wide variety of smaller volcanoes (e.g., Plescia, 2000; Stewart and Head, 2001) and volcanic features are

(d)

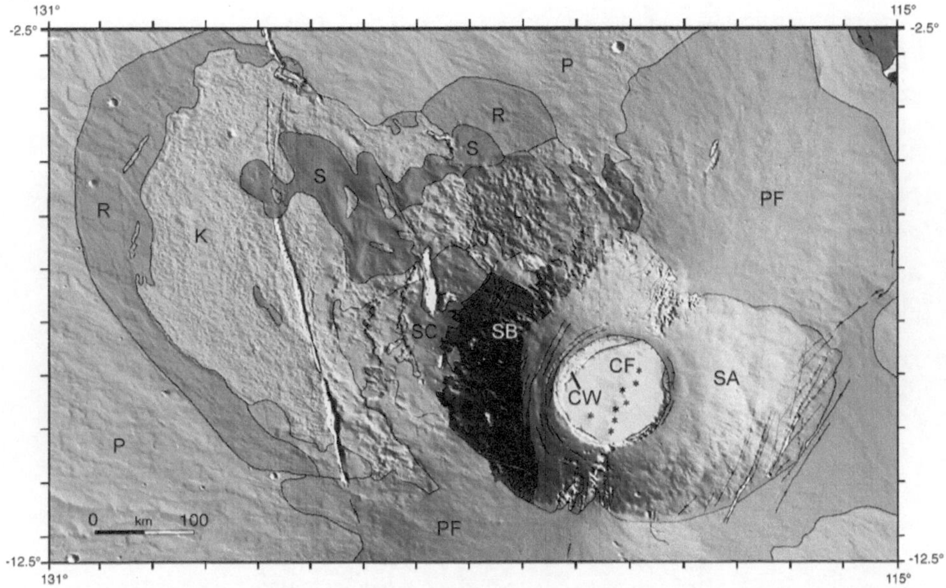

(e)

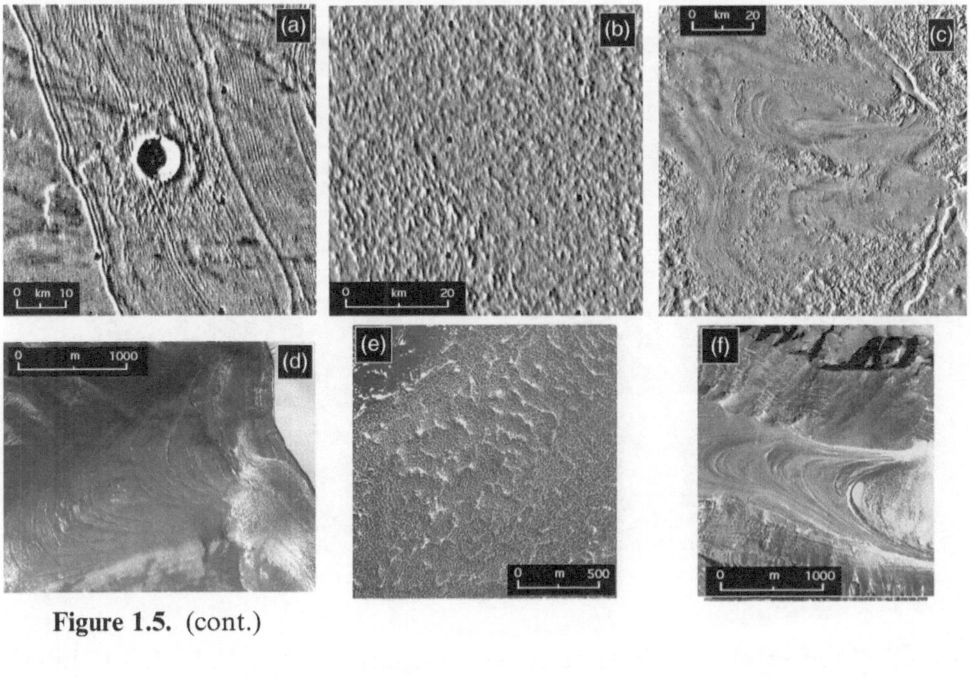

Figure 1.5. (cont.)

recognized on Mars, including possible composite cones, small shield volcanoes and fields of small cones with summit craters. Many of these cratered cones are analogous to structures called pseudocraters in Iceland, which result from local phreatic eruptions as lavas flow over water-saturated

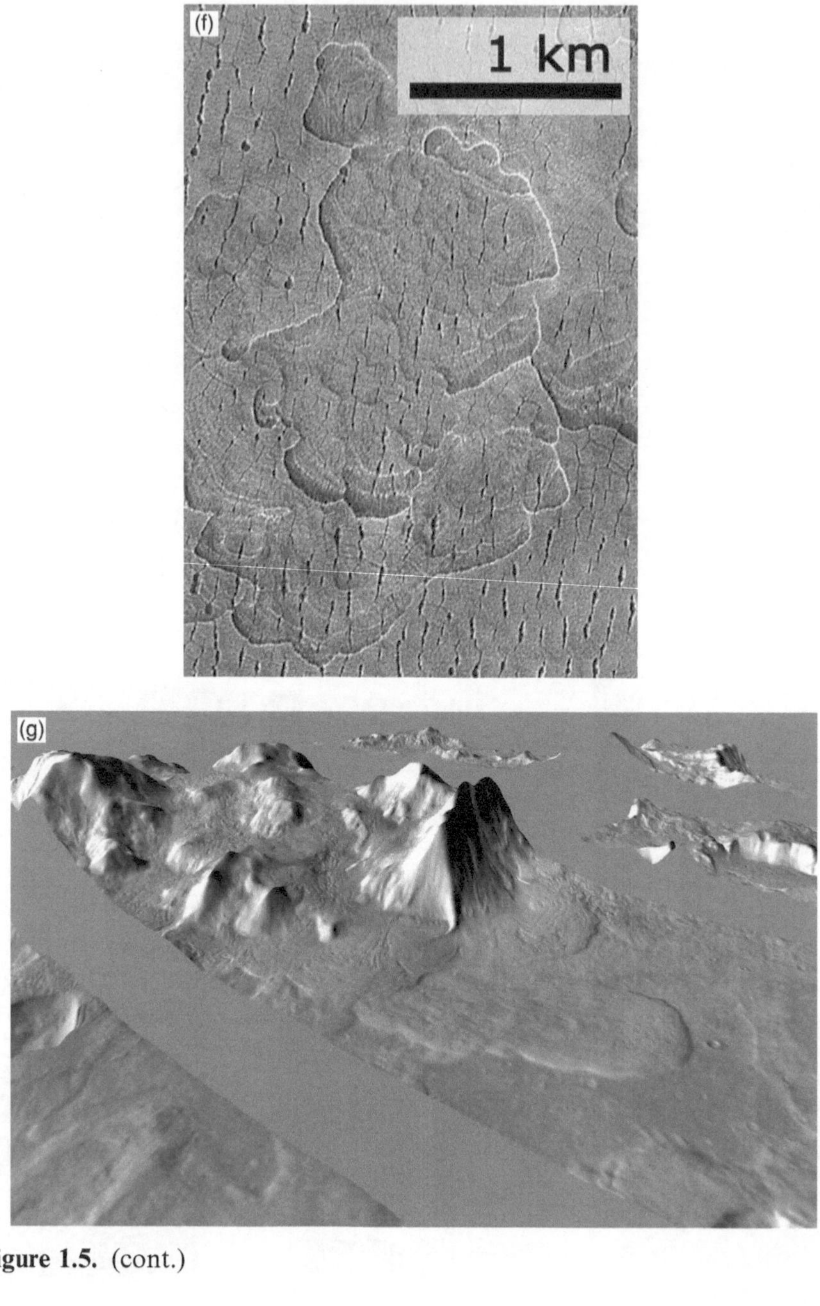

Figure 1.5. (cont.)

ground (Frey *et al.*, 1979; Greeley and Fagents, 2001). Indeed, the interaction of magma and groundwater and ice may have been critical in the evolution of Mars and produced a host of landforms for which terrestrial analogs are becoming increasingly useful (e.g., Chapman and Tanaka, 2001, 2002; Head and Wilson, 2002; Wilson and Head, 2002;

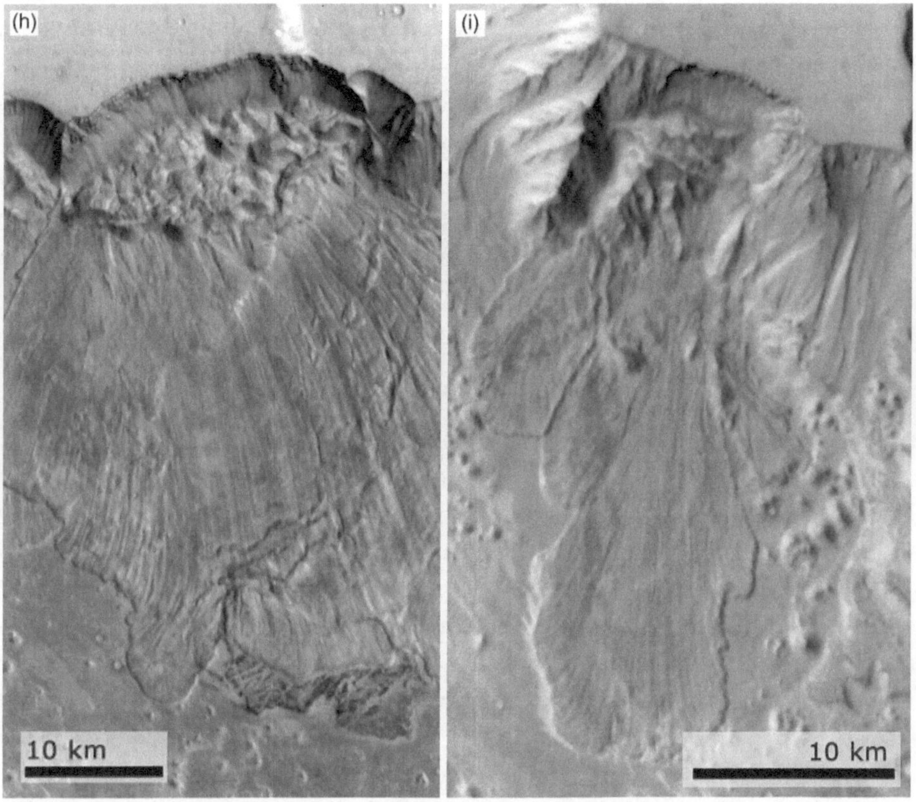

Figure 1.5. (cont.)

Chapman *et al.*, 2003; Ghatan *et al.*, 2003; Ivanov and Head, 2003; Wilson and Mouginis-Mark, 2003a, b).

In summary, volcanic processes have operated throughout the history of Mars, being particularly important globally in the Noachian and Hesperian, and important regionally in the Amazonian (Tharsis and Elysium). Evidence has been cited for geologically recent circumpolar volcanism (Garvin *et al.*, 2000) and volcanism appears to have occurred in the last few millions of years (Hartmann and Berman, 2000; Berman and Hartman, 2002; Burr *et al.*, 2002; Werner *et al.*, 2003). Very smooth terrain revealed by MOLA (Kreslavsky and Head, 2000) in the Elysium and Amazonis areas corresponds to very fresh platy flood basalt deposits and aqueous floods. Apparently, radial dikes in Elysium have cracked the cryosphere and released ground water and lava in the very recent history of Mars (Head *et al.*, 2003). An understanding of the nature of these unusual deposits has been possible only through the investigation of terrestrial analogs, such as the Laki fissure eruptions in Iceland (Kestzthelyi *et al.*, 2000) and many others (e.g., see in this volume, Mouginis-Mark and Rowland; Crumpler *et al.*; Keszthelyi and McEwen; Fagents and Thordarson; Chapman and Smellie; Gregg;), illustrating the importance of comparative field studies on Earth. The advent of high-resolution images and topographic data have also permitted more sophisticated assessment of lava flow rheologies (e.g., Glaze *et al.*, 2003) and comparison to terrestrial analogs (e.g., Peitersen and Crown, 2000).

1.2.3 Tectonic landforms and processes

Ample evidence for tectonic processes is seen in the morphology of the Martian surface (e.g., Carr, 1981; Banerdt *et al.*, 1992) (Figure 1.3). A variety of structural features, both extensional (Figure 1.3a) (simple graben, complex graben, rifts, tension cracks, troughs, and polygonal troughs) and contractional (Figure 1.3b) (wrinkle ridges, lobate scarps, fold belts) are testimony to brittle failure of the crust and the lithosphere. The relative ages of features may be dated by structural mapping and crater counts, but additional information on topography and gravity is required to model loads and to derive stresses in the lithosphere. The global dichotomy, formed early in Martian history, is the most fundamental physiographic feature on the planet and one or several mega-impacts (Wilhelms and Squyres, 1984; Frey and Schultz, 1988) have been invoked to account for it, but recent MOLA-based investigations (Smith *et al.*, 1999) did not find any single or several large circular topographic depressions to confirm this hypothesis (except for the Utopia basin, which had been speculated to be an impact basin even in the pre-MGS era: McGill, 1989). Endogenic processes offer alternatives, and several convection or subduction mechanisms have been proposed, including a plate-tectonic scenario (Sleep, 1994; but see Tanaka, 1995). An ancient phase of plate tectonics has also been proposed to explain magnetic anomalies (Acuna *et al.*, 2001) in the cratered highlands (Connerney *et al.*, 1999), although many alternate hypotheses are being considered to explain these features. Detailed structural mapping of key locations (e.g., the dichotomy boundary) required to further test the hypothesis is underway with new MGS and Mars Odyssey data.

Graben structures on Mars are generally narrow (a few kilometers) and long (ten to several hundred kilometers) and bounded by inward dipping normal faults and related features (Banerdt *et al.*, 1992; Mege *et al.*, 2003; Wilkins and Schultz, 2003; Schultz *et al.*, Chapter 15 this volume) (Figure 1.3a). Traditionally interpreted as purely tectonic features

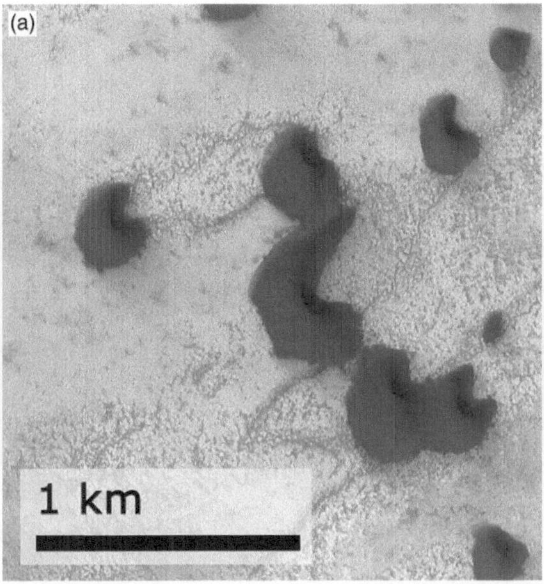

Figure 1.6. Eolian landforms and processes. (a) The north polar cap of Mars is surrounded by fields of dark sand dunes. This MOC image shows several dunes (dark) in the north polar region. The winds responsible for them blow from the lower left toward the upper right. The picture is located near 78.6° N, 243.9° W. (Illumination is from the lower left; MOC image M0201403; NASA/JPL/MSSS.) (b) A rugged surface southwest of Huygens Basin (near 20.1° S, 307.3° W) covered by large windblown ripples or small dunes. Their orientations indicate that the responsible winds came from either the northwest (upper left) or southeast (lower right), or both. The more complex ripple patterns within the two large craters result from local topographic influences on the wind. (Illumination is from the upper left; MOC image R0501809: NASA/JPL/MSSS.)

(e.g., Cailleau *et al.*, 2003), recent studies have also emphasized the role of radial dike emplacement in the formation of many of these features (e.g., Wilson and Head, 2002; Head *et al.*, 2003). More analogous to rifts on the Earth are the wider (up to 100 km) and deeper structures (many kilometers) that rupture the entire lithosphere (e.g., Tempe Terra, Valles Marineris, and Thaumasia) (e.g., Lucchitta *et al.*, 1992; Hauber and Kronberg, 2001). Wrinkle ridges, linear to arcuate asymmetric topographic highs (Figure 1.3b), are the most common contractional structures and form patterns of distributed deformation (e.g., Chicarro *et al.*, 1985; Plescia and Golombek, 1986; Watters and Maxwell, 1986; Watters, 1988, 1993). MOLA topographic data of the northern plains show a previously unreported system of ridges generally concentric to Tharsis (Smith *et al.*, 2001; Head *et al.*, 2002), and a

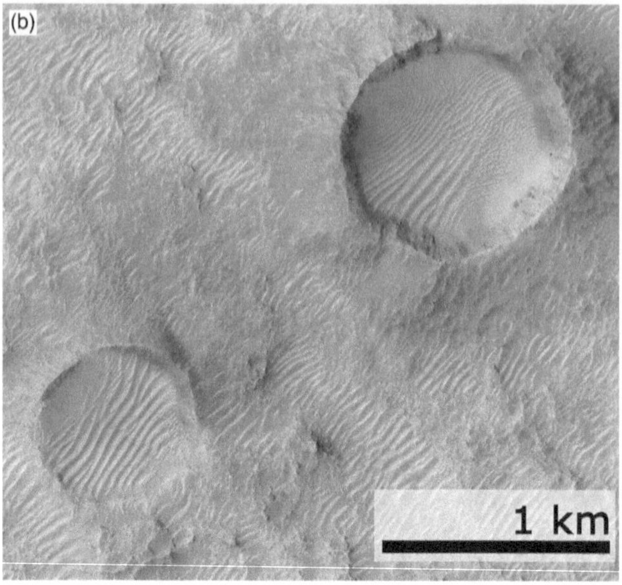

Figure 1.6. (cont.)

system of large contractional ridges and buckles to the south and southwest of Tharsis have been described (Schultz and Tanaka, 1994).

Most observed tectonic features can be explained by regional-scale deformation and by far the dominant element in Martian tectonics is the Tharsis bulge (e.g., Carr, 1981; Banerdt *et al.*, 1982, 1992; Anguita *et al.*, 2001; Lowry and Zhong, 2003). An abundance of extensional structures (simple and complex graben, rifts, and troughs) radiate outwards from Tharsis which is also the center of a concentric pattern of contractional structures (wrinkle ridges) (Maxwell, 1982; Chicarro *et al.*, 1985). Among the processes proposed to explain the formation of the huge topographic bulge are domal uplifting (e.g., Phillips *et al.*, 1973; Carr, 1973; Hartmann, 1973), magmatic intrusion (Sleep and Phillips, 1979, 1985; Willemann and Turcotte, 1982), and volcanic loading (Solomon and Head, 1982) (see Mège and Masson, 1996 for a comprehensive review). The concept of Tharsis as the center of volcano-tectonic activity throughout most of its history is complicated by the fact that there are several local and regional sub-centers and such centers seem to have changed in space and time. Statistical analyses of recent hemispheric-scale structural mapping indicate five successive stages of volcano-tectonic activity in Tharsis (Anderson *et al.*, 2001).

MGS gravity and topography-based lithospheric deformation models appear to have simplified the stress states required to explain most of the tectonic features around Tharsis. Flexural loading stresses based on present day gravity and topography appear to explain the type, location, orientation, and strain of most tectonic features around Tharsis. These models require the load to be of the scale of Tharsis and geologic mapping constrains the load to have been in place by the Middle to Late Noachian (>3.8–4.3 Ga). Thin layers within Valles Marineris interpreted as Late Noachian lava flows (McEwen *et al.*, 1999) are likely part of the load that caused the flexure around Tharsis and an antipodal dome that explains the first-order topography and gravity of the planet. The dichotomy boundary also shows much evidence for tectonic modification and complex degradation.

1.2.4 Fluvial landforms and processes

Giant channels and smaller branching valley networks on Mars suggest that unlike today (Hecht, 2002), the climate in the past may have supported warmer and wetter conditions and precipitation and surface runoff (see, for example, Masursky *et al.*, 1977; Lucchitta and Anderson, 1980; Pieri, 1980; Carr, 1981; Baker *et al.*, 1992; Carr, 1996). This evidence (Figure 1.4) suggests that liquid water was stable on the surface in the past (see, for example, Sharp and Malin, 1975; Carr, 1981; Mars Channel Working Group, 1983; Baker *et al.*, 1992). There are many differences in size, small-scale morphology, inner channel structure, and source regions compared to terrestrial analogs on Earth, indicating that fluvial erosion on Mars has a distinctive genesis and evolution. Martian valleys have been categorized as outflow channels, fretted channels, runoff channels, and quasi-dendritic networks (see, for example, Sharp and Malin, 1975; Carr, 1981). Researchers distinguish between a fluvial channel (the conduit through which a river flows), a valley (a linear depression), and a fluvial valley (generally contains many channels) (e.g., Mars Channel Working Group, 1983; Carr, 1996) although the terms outflow channels and valley networks are more commonly used to characterize linear erosion features on Mars (Masson *et al.*, 2001).

Outflow channels (Figure 1.4b) are up to tens of kilometers across and reach lengths of a few hundreds to thousands of kilometers, with gradients of channel floors ranging from 0 to 2.5 m/km (Baker *et al.*, 1992). Tributaries are rare, but branching downstream is common, resulting in an anastomosing pattern of channels (Carr, 1996). The channels tend to be deeper at their source than downstream and in general have high width to depth ratios and low sinuosities (Baker *et al.*, 1992; Carr, 1996). The distribution of outflow channels is restricted to four main areas: the vicinity of the Chryse-Acidalia basin (e.g., Rotto and Tanaka, 1995; Ivanov and

Head, 2001; Williams and Malin, 2004), west of the Elysium volcano complex in Elysium Planitia (e.g., Berman and Hartmann, 2002; Burr *et al.*, 2002; Fuller and Head, 2002; Plescia, 2003), the eastern part of Hellas basin (Mest and Crown, 2001), and along the western and southern border of Amazonis Planitia. Although all these channels are denoted as outflow channels, their occurrence, source regions, and geological context differ. Most around Chryse emanate fully developed from circular to elliptical depressions termed chaotic terrain, which has formed by collapse rather than by removal of material from above, indicating the involvement of groundwater in the channel formation process. In Elysium, large outflow channels originate at fractures oriented radial to the flanks of the volcanic complex and some have a lahar-like nature (Russell and Head, 2003).

Catastrophic release of groundwater (Baker *et al.*, 1992) is the origin most frequently called on and outflow channel dimensions indicate discharges of about 10^7–10^9 m^3s^{-1} (Carr, 1979; Komar, 1979; Baker, 1982; Robinson and Tanaka, 1990; Dohm *et al.*, 2001), about two orders of magnitude larger than the largest known flood events on Earth (the Channeled Scabland or the Chuja Basin flood in Siberia; Baker, 1973; Baker *et al.*, 1992; see Chapters 11 and 12 this volume). Candidate processes include groundwater under high artesian pressure confined below a permafrost zone which may break out, triggered by events which disrupt the permafrost seal (e.g., such as impacts, faults, or dikes) either by breaking the surface (e.g., Head *et al.*, 2003) or sending a large pressure pulse through the aquifer (Carr, 1979, 1995, 1996, 2002). A water release process through melting of ground ice by volcanic heat may also operate (Baker *et al.*, 1991).

Valley networks (Figure 1.4a), on the other hand, are common in the southern highlands and are open branching valleys in which tributaries merge downstream. Although they resemble terrestrial stream valleys they are far less complex than their analogs on Earth (Carr, 1995). The number of branches is low with large undissected areas between individual branches and the networks themselves are spaced apart leaving large areas of undissected highland between them (Pieri, 1980; Stepinski *et al.*, 2004) indicating that there has been little or no competition between adjacent drainage basins (Carr and Clow, 1981). Their U-shaped form with flat floors and steep walls is the main characteristic of individual branches. They are distinguished from channels by the absence of bedforms (Mars Channel Working Group, 1983). Some valley systems, such as Nirgal Vallis, are several hundreds of kilometers in length and a few tens of kilometers wide with short accordant tributaries. The majority of the valley systems, however, are typically no longer than

200 km and only a few kilometers wide. Valley networks are mainly located in the cratered uplands at elevations ranging from 2 km to over 5 km. Some drain into craters, but most start at local highs and drain as winding streams with relatively large branch junction angles into low areas between large impact craters where they terminate against smooth plains (Carr, 1995). Another set of channel networks occurs on volcanoes such as Hecates, Ceraunius, and Alba (Wilson and Mouginis-Mark, 1987; Mouginis-Mark *et al.*, 1988; Gulick and Baker, 1990).

Valley networks are much smaller than outflow channels indicating smaller corresponding discharges. The pattern of dissection, the drainage development, and the general valley morphologies are similar to those of many terrestrial rivers and a similar origin by running water has been implied (e.g., Craddock and Howard, 2002) with discussion centering on the relative importance of runoff versus sapping (Sharp and Malin, 1975; Pieri, 1976, 1980; Carr, 1981; Baker, 1982, 1990; Laity and Malin, 1985; Goldspiel and Squyres, 2000; Williams and Phillips, 2001; Grant and Parker, 2002; Irwin and Howard, 2002). Consequently these features have implications for the evolution of the Martian climate (Carr, 1983; Squyres, 1989; Baker *et al.*, 1991; Squyres and Kasting, 1994; for a summary see Baker *et al.*, 1992 and Carr, 1996) and a simple explanation is that the valleys are old themselves and the climatic requirements for valley formation were met early in the history of Mars and rapidly declined during the subsequent evolution. Based on the evaluation of high-resolution Mars Observer Camera (MOC) images Malin and Carr (1999) concluded that the valleys were formed by fluid erosion, and in most cases the source was groundwater. Alternatively, some heat may have been provided by high global (Carr and Head, 2003) or local (Travis *et al.*, 2003) heat flux and melting of surface snowpack or ground ice. Relatively young small-scale alcove-like gullies (Figure 1.4c) combined with small channels and aprons in the walls of some impact craters indicate that groundwater seepage or short-term surface runoff may occur under recent climatic conditions (Malin and Edgett, 2000; Gaidos, 2001; Mellon and Phillips, 2001; Andersen *et al.*, 2002; Mangold *et al.*, 2003).

1.2.5 Lake and ocean-related landforms and processes

Abundant evidence for flowing water leads to the question of its fate (Clifford, 1993; Clifford and Parker, 2001). Numerous upland craters show evidence of channels entering the crater and flat floors due to the apparent deposition of sediment. These regions are interpreted to be the sites of standing bodies of water or lakes (e.g., Carr, 1996; Cabrol and Grin, 2001; Newsom *et al.*, 2003), sometimes with deltas (e.g., Ori *et al.*, 2000) and recent MOC data show evidence for layered deposits in many impact craters (e.g., Malin and Edgett, 2000). Large outflow channels emptied into the northern lowlands (and Hellas; see Moore and Wilhelms, 2001) and their floods must have left extended deposits, large lakes, and possibly oceans. The northern plains cover an older rougher terrain that survived as hills and knobs commonly outlining old pre-plains impact craters. The plains are complex deposits probably formed by many processes, such as sedimentation from outflow channels, volcanism, and mass wasting from the adjacent highlands modified by impact craters. Recent analyses using MOLA altimetry data suggest that the present surface deposits (mostly the Hesperian-aged Vastitas Borealis Formation) are underlain by Noachian-aged cratered surfaces (Frey *et al.* 2002). Early Hesperian ridged plains of volcanic origin, and the overlying Late Hesperian Vastitas Borealis Formation which is at least 100 m thick (Head *et al.*, 2002; Kreslavsky and Head, 2002). Many of the detailed surface features seen in MOC images seem to be recent and formed by action of ground ice and debris mantles (see Kreslavsky and Head, 2000). Some features surrounding the northern plains form contacts interpreted as ancient shorelines (Parker *et al.*, 1989, 1993) marking the boundaries of former lakes or a northern ocean. Analysis of the elevations of the contacts and surface roughness using MOLA data shows that some of these data are consistent with the presence of a large standing body of water, while others are not (Head *et al.*, 1999; Carr and Head, 2003; Leverington and Ghent, 2004). Terrestrial analogs for shoreline processes may provide the critical observations that will help inform the interpretations of the range of features seen on Mars, and terrestrial analogs have also continued to be useful in studies of aqueous flooding (e.g., Komatsu *et al.*, Chapter 13 this volume) and lacustrine studies (e.g., Komatsu *et al.*, Chapter 13 this volume; Cabrol, Chapter 14 this volume).

1.2.6 Polar, circumpolar, periglacial, glacial, and mass wasting landforms and processes

Study of the poles of Mars and related glacial and periglacial processes have undergone a renaissance in recent years due to new data (Figures 1.5). More sophisticated questions can now be posed (e.g., Clifford *et al.*, 2000), and the detailed geology of polar and circumpolar deposits can be assessed (e.g., Vasavada *et al.*, 1999; Fishbaugh and Head, 2000, 2001; Schenk and Moore, 2000; Head, 2001; Head and Pratt, 2001; Kolb and Tanaka, 2001;

Murray *et al.*, 2001; Tanaka and Kolb, 2001; Byrne and Murray, 2002; Fishbaugh and Head, 2002; Ghatan and Head, 2002; Koutnik *et al.*, 2002; Milkovich *et al.*, 2002; Ghatan *et al.*, 2003). The stratigraphy of the layered terrain (Figure 1.5a), which holds the key to the recent climate history of Mars, is becoming better known (e.g., Fenton and Herkenhoff, 2000), and the structure and morphology of polar and circumpolar deposits is beginning to reveal important characteristics of the south polar residual ice cap (e.g., Thomas *et al.*, 2000) and layered terrain (Head, 2001; Murray *et al.*, 2001). The north polar cap structure is now much more well known (e.g., Fishbaugh and Head, 2000, 2001; Kolb and Tanaka, 2001; Tanaka and Kolb, 2001; Byrne and Murray, 2002) and a major distinctive basal unit has been discovered and documented (Figure 1.5b). Indeed, detailed analysis (Bland and Smith, 2000) suggests that the poles of Mars may offer future explorers the possibility of collecting meteorites typical of those in circum-Martian space, just as polar explorers do in the Antarctic regions of the Earth.

Permanently frozen ground (permafrost) and freezing and thawing of the ground are the characteristics of periglacial surfaces on Earth (Tricart, 1968, 1969). On Mars, a number of surface features (Figure 1.5f) have been attributed to these processes (e.g., Carr and Schaber, 1977; Squyres, 1978; Lucchitta, 1981; Rossbacher and Judson, 1981; Lucchitta, 1985; Squyres and Carr, 1986; Squyres *et al.*, 1992; Carr, 1996) (Figures 1.6 and 1.7). Most of these landforms are the result of freeze–thaw cycles in the active layer above the permanently frozen ground which is not common under present Mars climatic conditions (e.g., Carr, 1996). A wide range of terrain-softening features are seen in a latitudinal belt between 30° and 60° (Squyres, 1979; Squyres and Carr, 1986; Carr, 1996).

Lobate debris aprons with well-defined flow fronts and convex-upward surfaces are seen at the upland/lowland boundary and extend from the highlands and isolated mesas into the low-lying plains (Sharp, 1973) (Figure 1.5g). In valleys, opposing walls confine material flows and the flow fronts often converge to form lineated valley fill which resembles terrestrial median moraines on glaciers (Carr, 2001). It is not clear, however, to what level transverse versus longitudinal flow contributed to lineation generation (Carr and Schaber, 1977). In some places, terrain softening appears also in craters where material has moved down the inner crater wall and probably formed concentric crater fill (Squyres, 1989; Jankowski and Squyres, 1992, 1993). Lobate debris aprons have been interpreted to result from gelifluction, frost creep (Carr and Schaber, 1977) or movement of ice-lubricated erosional debris (Squyres, 1978; Pierce and Crown, 2003;

Mangold *et al.*, 2003). Viscous flow features on Mars also have been compared to rock glaciers on Earth (e.g., Squyres, 1989; Colaprete and Jakosky, 1998; Milliken *et al.*, 2003), that are lobate accumulations of angular rock debris that move downslope due to deformation of internal ice or frozen sediments (e.g., Wahrhaftig and Cox, 1959; Barsch, 1988; Benn and Evans, 1998; Degenhardt and Giardino, 2003; Whalley and Azizi, 2003). Recent high-resolution MOC images (Carr, 2001) (Figure 1.5g) have shown the importance of high-resolution data in the assessment of the floors of channels and valleys and THEMIS and HRSC data show new evidence for extensive viscous flow and suggest a much more important role for glaciation than thought in the past (Figure 1.5).

Rampart crater ejecta is marked by distinct lobes and provides strong evidence for subsurface ice or water (Carr, 1977, 1996; Carr *et al.*, 1977; Wohletz and Sheridan, 1983; Costard, 1989), although the entrainment of atmospheric gases is a possible alternative mechanism (Schultz and Gault, 1979). Costard and Kargel (1999) found a correlation between craters with high mobility ejecta and their distribution at low altitude particularly in areas of fractured ground and thermokarst, further supporting the idea that the formation of those craters is connected to subsurface water or ice.

Fractured plains composed of giant polygons 30 km across (McGill, 1986; Pechmann, 1980; Lane and Christensen, 2000) are located preferentially in Acidalia, Elysium, and Utopia Planitiae (Pechmann, 1980). MOLA and MOC data established the average width and depth of troughs bordering giant polygons in Utopia Planitia, leading Hiesinger and Head (2000) to propose a tectonic origin consistent with Pechmann (1980), who argued that thermal contraction on Mars should be confined to the uppermost 10 m of the surface and thus not to produce deep cracks. Much smaller polygons with diameters of 10–100 m are abundant and are interpreted to be the result of thermal contraction in ice-rich soils based on the similar scales as compared to terrestrial ice-wedge polygons (Mutch *et al.*, 1977; Evans and Rossbacher, 1980).

Thaw lakes and alases in Alaska and Yakutia have provided a basis for comparison to some features on Mars (Sharp, 1973; Gatto and Anderson, 1975; Carr and Schaber, 1977; Theilig and Greeley, 1979; Lucchitta, 1981). Earlier interpretations of Martian pits and depressions as thermokarst features have recently received support from Costard and Kargel (1995) and thermokarst processes have also influenced the surfaces of lobate debris aprons and lineated valley fill (Mangold *et al.*, 2000a, b). Pingo-like features on Mars have been reported by several authors (e.g., Lucchitta, 1981; Rossbacher and Judson, 1981). Cabrol *et al.* (2000), Lucchitta (1982, 2001), and Tricart (1988) found that some characteristics of the outflow channels might be explained by ice sculpture. Springs generating icings, which began to flow as glaciers, could have been the source, as well as rivers, floods, or mudflows (Williams and Malin, 2004) that built ice dams and plugs moving through the channel. Further examination of periglacial terrestrial analogs is critical in analyzing and understanding these features.

Glaciation as a process to account for specific landforms on Mars has been discussed by a number of authors (e.g., Allen, 1979; Clifford, 1980; Lucchitta, 1981; Tanaka and Scott, 1987; Baker *et al.*, 1991; Head and Pratt, 2001). Kargel and Strom (1992) argued that only glaciation can account in a simple and unifying way for many of the features covering wide areas in Argyre, Hellas, and the south polar region. Some of the most convincing morphological evidence for ancient glaciation on Mars are long sinuous ridges analogous to terrestrial eskers (e.g., Kargel and Strom, 1992). Eskers are the result of infillings of ice-walled river channels by subglacial, englacial, or supraglacial drainage networks (Benn and Evans, 1998; Brennand, 2000). Recent MOLA measurements of the heights and widths of the largest Martian features seem to support the esker hypothesis. Head and Pratt (2001) interpreted the Hesperian-aged Dorsa Argentea Formation to be the remnant of a south circumpolar ice sheet that has since retreated by sublimation and basal volcano-related melting and drainage. Channels emerging from the vicinity of the margins of the Dorsa Argentea Formation flow downslope into the Argyre Basin, apparently draining the meltwater (Head and Pratt, 2001; Milkovich *et al.*, 2002; Ghatan *et al.*, 2003).

Moraines have also been tentatively identified by Lucchitta (1981) at the base of the giant Martian shield volcanoes in the Tharsis region. Chapman (1994) supported the earlier reports and found evidence for moberg ridges, table mountains, and jokulhlaup deposits, proposing the existence of a paleo-ice sheet northwest of the Elysium volcanoes in Utopia Planitia.

Head and Marchant (2003), following the earlier work of Lucchitta (1981), interpreted the fan-shaped deposits on the northwest flanks of the Tharsis Montes and Olympus Mons as the remnants of cold-based glaciers (Figure 1.5d,e). The three facies commonly seen in association with these deposits seem plausibly interpreted as drop moraines, sublimation tills, and rock glaciers (Head and Marchant, 2003) (Figure 1.5d,e). The presence of these tropical mountain glaciers seems contradictory to our understanding of the current distribution of water and ice on Mars (Mellon *et al.*, 2004), but their location appears to be plausible in the context of the behavior of water in the atmosphere during periods of high obliquity (Richardson and Wilson, 2002; Haberle *et al.*, 2003).

Recent MOLA and MOC data have revealed the presence of a latitude-dependent surface mantling layer (Kreslavsky and Head, 2000, 2002) that appears to be highly correlated with distinctive morphologic features at certain latitudes (Mustard *et al.*, 2001). This deposit has been interpreted to be the residual deposit of geologically recent ice ages, when the higher obliquity of Mars caused mobilization of polar water ice and redeposition toward equatorial regions (Head *et al.*, 2003). The presence of this deposit, and the tropical mountain glaciers, suggest that the currently observed distribution of ice stability on Mars (Feldman *et al.*, 2002; Mellon *et al.*, 2004) may be considerably different from what it has commonly been in the past. In addition to an increasing awareness of the presence of water ice deposits and features in the geological evolution of Mars, the role of carbon dioxide and its relations with water in the volatile evolution of Mars has received new attention (Hoffman, 2000).

Landslides and mass wasting are a major process in the degradation of landforms on Mars (Figure 1.5h,i). Some of the most prominent examples of large-scale mass wasting and landslides occur around the margins of Olympus Mons (e.g., Francis and Wadge, 1983; Tanaka, 1985) and on the floor of Valles Marineris. Landslides on the floor of Valles Marineris appear to span a wide range of ages and the new altimetery data permit assessment of rheological constraints (e.g., Harrison and Grimm, 2003). At the other end of the spectrum, dry talus cones and dust avalanches occur over wide areas of Mars (e.g., Sullivan *et al.*, 2001; Gerstell *et al.*, 2004). In all of these areas, terrestrial analogs have been important in the interpretation and understanding of the landforms on Mars (e.g., see Costard *et al.*, Chapter 10 in this volume).

1.2.7 Eolian landforms and processes

In the absence of currently known or abundant active volcanism, tectonism, and running water on the surface, wind processes dominate the present environment on Mars (Figure 1.6). Winds are capable of creating dust storms, at times obscuring the surface from view despite the tenuous carbon dioxide atmosphere (average surface pressure is 6.5 mbar). Earth-based and orbital observations show that dust storms occur predominantly in the southern hemisphere summer but data obtained from orbiters and landers show that eolian activity can also occur at other seasons.

Greeley *et al.* (1992) documented a wide variety of wind-related features including wind depositional features, such as dunes (Figure 1.6a), and wind erosional features, such as yardangs. Various albedo patterns, such as wind streaks that change with time (e.g., Thomas *et al.*, 2003), are the most common eolian feature observed from orbit, while typical features seen on the surface include rock-related features, called wind tails; both are considered to represent the prevailing winds at the time of their formation and can be mapped to infer wind directions. Many bright streaks are thought to form by the deposition of dust in the waning stages of dust storms, while dark streaks appear to result from erosion of windblown particles, exposing darker substrate, or leaving a lag deposit of lower albedo material on the surface. General circulation models of the atmosphere show that bright streaks correlate with predicted regional wind directions (Greeley *et al.*, 1993), while dark streaks are influenced more by local topography. Images from MGS show that mantles of windblown materials, inferred to be dust deposits settled from suspension, can form both bright and dark surfaces (Edgett and Malin, 2000). Further, dunes and duneforms also occur as both bright and dark features. eolian processes can both cover and resurface by deposition of sediments (Arvidson *et al.*, 2003), and uncover and expose by deflation and exhumation (e.g., Greeley *et al.*, 2001; Fenton *et al.*, 2003), processes that can alter surface ages based on impact crater statistics. Eolian features and deposits (duneforms, windstreaks, eroded craters, ventifacts) provide important clues to past climates on Earth and Mars (e.g., Kuzmin *et al.*, 2001; Bradley *et al.*, 2002) and analyses of present and past eolian activity can give new insight into the evolution of the climate and atmosphere of Mars, and the history of the polar and circumpolar regions (Howard, 2000; Tanaka, 2000; Bridges and Herkenhoff, 2002). Detailed surface observations (e.g., Greeley *et al.*, 2000b; Golombek *et al.*, 1999) and the high-resolution data provided by the MOC (Figure 1.6b) have been instrumental in the advancement of our knowledge (e.g., Edgett and Malin, 2000; Reiss *et al.*, 2004) and to help to improve our ability to compare to terrestrial analogs (Greeley *et al.*, 2002). Terrestrial analog studies have, in turn, significantly informed us about interpreting the features on Mars (e.g., see Chapter 9 in this volume).

1.3 Summary

A "new Mars" is beginning to emerge from recent spacecraft observations; increased spatial and spectral resolution, greater areal coverage and synoptic views, detailed topographic data, knowledge of the distribution and state of water, more sophisticated atmospheric general circulation models, and long-duration surface exploration (Golombek *et al.*, 2003) all have been helpful in establishing links between orbital remote sensing data and terrain at the outcrop scale. These results have set the stage for the much more sophisticated utilization of terrestrial analogs in deconvolving the complex signal of the geological record and history of the planet, as outlined in this volume for the full range of geomorphic landforms and geologic processes.

Acknowledgments

The author thanks James Dickson for help in preparation of figures and references and for general support during the preparation of the manuscript.

References

Acuna, M. H., Connerney, J. E. P., Wasilewski, P. *et al.* (2001). Magnetic field of Mars: summary of results from the aerobraking and mapping orbits. *Journal of Geophysical Research (Planets)*, **106** (E10), 23403–17.

Allen, C. C. (1979). Volcano-ice interactions on Mars. *Journal of Geophysical Research*, **84**, 8048–59.

Anderson, R. C., Dohm, J. M., Golombek, M. P. *et al.* (2001). Primary centers and secondary concentrations of tectonic activity through time in the western hemisphere of Mars. *Journal of Geophysical Research (Planets)*, **106** (E9), 20563–85.

Anguita, F., Farelo, A., Lopez, V. *et al.* (2001). Tharsis dome, Mars: new evidence for Noachian-Hesperian thick-skin and Amazonian thin-skin tectonics. *Journal of Geophysical Research (Planets)*, **106** (E4), 7577–89.

Arvidson, R. E., Seelos, F. P. IV, Deal, K. S. *et al.* (2003). Mantled and exhumed terrains in Terra Meridiani, Mars. *Journal of Geophysical Research (Planets)*, **108** (E12), doi: 10.1029/2002JE001982.

Baker, V. R. (1973). *Paleohydrology and Sedimentology of Lake Missoula Flooding of Eastern Washington.* Geological Society of America SP-144. Boulder: Geological Society of America.

Baker, V. R. (1982). *The channels of Mars.* Austin: University of Texas Press.

Baker, V. R. (1990). *Spring Sapping and Valley Network Development.* Geological Society of America SP-252, 235–265. Boulder: Geological Society of America.

Baker, V. R., Carr, M. H., Gulick, V. C., Williams, C. R., and Marley, M. S. (1992). Channels and valley networks. In *Mars*, ed. H. Kieffer, B. Jakosky, C. Snyder, and M. Matthews. Tucson: University of Arizona Press, pp. 493–522.

Baker, V. R., Strom, R. G., Gulick, V. C. *et al.* (1991). Ancient oceans, ice sheets and hydrological cycle on Mars. *Nature*, **352**, 589–94.

Banerdt, W. B., Phillips, R. J., Sleep, N. H., and Saunders, R. S. (1982). Thick shell tectonics on one-plate planets: application to Mars. *Journal of Geophysical Research*, **87**, 9723–33.

Banerdt, W. B., Golombek, M. P., and Tanaka, K. L. (1992). Stress and tectonics on Mars. In *Mars*, ed. H. H. Kieffer, B. M. Jakosky, C. W. Snyder, and M. S. Matthews. Tucson: University of Arizona Press, pp. 249–97.

Barlow, N. G. and Perez, C. B. (2003). Martian impact crater ejecta morphologies as indicators of the distribution of subsurface volatiles. *Journal of Geophysical Research (Planets)*, **108** (E8), doi: 10.1029/2002JE002036.

Barlow, N. G., Boyce, J. M., Costard, F. M. *et al.* (2000). Standardizing the nomenclature of Martian impact crater ejecta morphologies. *Journal of Geophysical Research (Planets)*, **105** (E11), 26733–8.

Barsch, D. (1988). Rock glaciers. In *Advances in Periglacial Geomorphology*, ed. M. J. Clark. Chichester: Wiley, pp. 69–90.

Benn, D. I. and Evans, D. J. A. (1998). *Glaciers and Glaciation*. London: Arnold.

Berman, D. C. and Hartman, W. K. (2002). Recent fluvial, volcanic, and tectonic activity on the Cerberus Plains of Mars. *Icarus*, **159** (1), 1–17.

Bland, P. A. and Smith, T. B. (2000). Meteorite accumulations on Mars. *Icarus*, **144** (1), 21–6.

Bradley, B. A., Sakimoto, S. E., Frey, H., and Zimbelman, J. R. (2002). Medusae Fossae Formation: new perspectives from Mars Global Surveyor. *Journal of Geophysical Research (Planets)*, **107** (E08), doi: 10.1029/2001JE001537.

Brennand, T. A. (2000). Deglacial meltwater drainage and glaciodynamics; inferences from Laurentide eskers, Canada. *Geomorphology*, **32** (3–4), 263–96.

Bridges, N. T. and Herkenhoff, K. E. (2002). Topography and geologic characteristics of aeolian grooves in the south polar layered deposits of Mars. *Icarus*, **156** (2), 387–98.

Burr, D. M., Grier, J. A., McEwen, A. S., and Keszthelyi, L. P. (2002). Repeated aqueous flooding from the Cerberus Fossae: evidence for very recently extant, deep groundwater on Mars. *Icarus*, **159** (1), 53–73.

Byrne, S. and Murray, B. (2002). North polar stratigraphy and the paleo-erg of Mars. *Journal of Geophysical Research (Planets)*, **107** (E06), doi: 10.1029/2001JE001615.

Cabrol, N. A. and Grin, E. A. (2001). The evolution of Lacustrine environments on Mars: is Mars only hydrologically dormant? *Icarus*, **149** (2), 291–328.

Cabrol, N. A., Grin, E. A., and Pollard, W. H. (2000). Possible frost mounds in an ancient Martian lake bed. *Icarus*, **145**, 91–107.

Cailleau, B., Walter, T. R., Janle, P., and Hauber, E. (2003). Modeling volcanic deformation in a regional stress field: implications for the formation of graben structures on Alba Patera, Mars. *Journal of Geophysical Research (Planets)*, **108** (E12), doi: 10.1029/2003JE002135.

Carr, M. H. (1973). Volcanism on Mars. *Journal of Geophysical Research*, **78**, 4049–62.

Carr, M. H. (1977). Distribution and emplacement of ejecta around Martian impact craters. In *Impact and Explosion Cratering*, ed. D. J. Roddy, R. O. Pepin, and R. B. Merrill. New York: Pergamon Press, pp. 593–602.

Carr, M. H. (1979). Formation of Martian flood features by release of water from confined aquifers. *Journal of Geophysical Research*, **84**, 2995–3007.

Carr, M. H. (1981). *The Surface of Mars*. New Haven: Yale University Press.

Carr, M. H. (1983). The stability of streams and lakes on Mars. *Icarus*, **56**, 476–95.

Carr, M. H. (1995). The Martian drainage system and the origin of valley networks and fretted channels. *Journal of Geophysical Research*, **100**, 7479–507.

Carr, M. H. (1996). *Water on Mars*. New York: Oxford University Press.

Carr, M. H. (2001). Mars Global Surveyor observations of Martian fretted terrain. *Journal of Geophysical Research (Planets)*, **106** (E10), 23571–94.

Carr, M. H. (2002). Elevations of water-worn features on Mars: implications for circulation of groundwater. *Journal of Geophysical Research (Planets)*, **107** (E12), doi: 10.1029/2002JE001845.

Carr, M. H. and Clow, G. D. (1981). Martian channels and valleys: their characteristics, distribution and age. *Icarus*, **48**, 91–117.

Carr, M. H. and Head, J. W. (2003). Oceans on Mars: an assessment of the observational evidence and possible fate. *Journal of Geophysical Research (Planets)*, **108** (E5), doi: 10.1029/2002JE001963.

Carr, M. H. and Schaber, G. G. (1977). Martian permafrost features. *Journal of Geophysical Research*, **82**, 4039–54.

Carr, M. H., Crumpler, L. S., Cutts, J. A. *et al.* (1977). Martian impact craters and emplacement of ejecta by surface flow. *Journal of Geophysical Research*, **82**, 4055–65.

Cattermole, P. (1987). Sequence, rheological properties, and effusion rates of volcanic flows at Alba Patera, Mars. *Journal of Geophysical Research*, **92**, E553–60.

Chapman, M. G. (1994). Evidence, age, and thickness of a frozen paleolake in Utopia Planitia, Mars. *Icarus*, **109**, 393–406.

Chapman, M. G. and Tanaka, K. L. (2001). Interior trough deposits on Mars: subice volcanoes? *Journal of Geophysical Research (Planets)*, **106** (E5), 10087–100.

Chapman, M. G. and Tanaka, K. L. (2002). Related magma-ice interactions: possible origins of chasmata, chaos, and surface materials in Xanthe, Margaritifer, and Meridiani Terrae, Mars. *Icarus*, **155** (2), 324–39.

Chapman, M. G., Gudmundsson, M. T., Russell, A. J., and Hare, T. M. (2003). Possible Juventae Chasma subice volcanic eruptions and Maja Valles ice outburst floods on Mars: implications of Mars Global Surveyor crater densities, geomorphology, and topography. *Journal of Geophysical Research (Planets)*, **108** (E10), doi: 10.1029/2002JE002009.

Chicarro, A. F., Schultz, P. H., and Masson, P. (1985). Global and regional ridge patterns on Mars. *Icarus*, **63**, 153–74.

Christensen, P. R., Bandfield, J. L., Smith, M. D., Hamilton, V. E., and Clark, R. N. (2000a). Identification of a basaltic component on the Martian surface from Thermal Emission Spectrometer data. *Journal of Geophysical Research (Planets)*, **105** (E4), 9609–22.

Christensen, P. R., Bandfield, J. L., Clark, R. N. *et al.* (2000b). Detection of crystalline hematite mineralization on Mars by the Thermal Emission Spectrometer: evidence for near-surface water. *Journal of Geophysical Research (Planets)*, **105** (E4), 9623.

Christiansen, E. H. (1989). Lahars in the Elysium region of Mars, *Geology*, **17**, 203–6.

Clifford, S. M. (1980). A model for the removal and subsurface storage of a primitive Martian ice sheet. In *Reports of Planetary Geology and Geophysics Program-1990*, NASA-TM 82385, Washington, DC, pp. 405–7.

Clifford, S. M. (1993). A model for the hydrologic and climatic behaviour of water on Mars. *Journal of Geophysical Research*, **98**, 10973–1016.

Clifford, S. M. and Parker, T. J. (2001). The evolution of the Martian hydrosphere: implications for the fate of a primordial ocean and the current state of the northern plains. *Icarus*, **154** (1), 40–79.

Clifford, S. M., Crisp, D., Fisher, D. A. *et al.* (2000). The state and future of Mars polar science and exploration. *Icarus*, **144** (2), 210–42.

Colaprete, A. and Jakosky, B. M. (1998). Ice flow and rock glaciers on Mars. *Journal of Geophysical Research*, **103**, 5897–909.

Connerney, J. E. P., Acuna, M. H., Wasilewski, P. J. *et al.* (1999). Magnetic lineations in the ancient crust of Mars. *Science*, **284**, 794–8.

Costard, F. M. (1989). The spatial distribution of volatiles in the Martian hydrolithosphere. *Earth, Moon, and Planets*, **45**, 265–90.

Costard, F. M. and Kargel, J. S. (1995). Outwash plains and thermokarst on Mars. *Icarus*, **114**, 93–112.

Costard, F. M. and Kargel, J. S. (1999). New evidences for ice rich sediments in the northern plains from MGS data. *Fifth Int. Conference on Mars*, **972** (CD-ROM), #6088.

Craddock, R. A. and Howard, A. D. (2002). The case for rainfall on a warm wet early Mars. *Journal of Geophysical Research (Planets)*, **107** (E11), doi: 10.1029/2001JE001505.

Crown, D. A. and Greeley, R. (1993). Volcanic geology of Hadriaca Patera and the eastern Hellas region of Mars. *Journal of Geophysical Research (Planets)*, **98** (E2), 3431–51.

Degenhardt, J. J. and Giardino, J. R. (2003). Subsurface investigation of a rock glacier using ground-penetrating radar: implications for locating stored water on Mars. *Journal of Geophysical Research (Planets)*, **108** (E4), doi: 1029/2002JE001888.

Dohm, J. M., Ferris, J. C., Baker, V. R. *et al.* (2001). Ancient drainage basin of the Tharsis region, Mars: potential source for outflow channel systems and putative oceans or paleolakes. *Journal of Geophysical Research (Planets*), **106** (E12), 32943–58.

Edgett, K. S. and Malin, M. C. (2000). New views of Mars eolian activity, materials, and surface properties: three vignettes from the Mars Global Surveyor Mars Orbiter Camera. *Journal of Geophysical Research (Planets)*, **105** (E1), 1623–50.

Evans, N. and Rossbacher, L. A. (1980). The last picture show: small-scale patterned ground in Lunae Planum. In *Reports of Planetary Geology Program-1980*, NASA-TM 82385, NASA, Washington, DC, pp. 376–8.

Feldman, W. C., Boynton, W. V., Tokar, R. L. *et al.* (2002). Global distribution of neutrons from Mars: results from Mars Odyssey. *Science*, **297** (5578), 75–8.

Fenton, L. K. and Herkenhof, K. E. (2000). Topography and stratigraphy of the northern Martian polar layered deposits using photoclinometry, stereogrammetry, and MOLA altimetry. *Icarus*, **147** (2), 433–43.

Fenton, L. K., Bandfield, J. L., and Ward, A. W. (2003). Aeolian processes in Proctor Crater on Mars: sedimentary history as analyzed from multiple data sets. *Journal of Geophysical Research (Planets)*, **108** (E12), doi: 10.1029/2002JE002015.

Fishbaugh, K. E. and Head, J. W. (2000). North polar region of Mars: topography of circumpolar deposits from Mars Orbiter Laser Altimeter (MOLA) data and evidence for asymmetric retreat of the polar cap. *Journal of Geophysical Research (Planets)*, **105** (E9), 22455–86.

Fishbaugh, K. E. and Head, J. W. (2001). Comparison of the north and south polar caps of Mars: new observations from MOLA data and discussion of some outstanding questions. *Icarus*, **154** (1), 145–61.

Fishbaugh, K. E. and Head, J. W. (2002). Chasma Boreale, Mars: topographic characterization from Mars Orbiter Laser Altimeter data and implications for mechanisms of formation. *Journal of Geophysical Research (Planets)*, **107** (E03), doi: 10.1029/2000JE001351.

Forsberg-Taylor, N. K., Howard, A. D., and Craddock, R. A. (2004). Crater degradation in the Martian highlands: morphometric analysis of the Sinus Sabaeus region and simulation modeling suggest fluvial processes. *Journal of Geophysical Research (Planets)*, **109** (E5), doi: 10.1029/2004JE002242.

Francis, P. W. and Wadge, G. (1983). The Olympus-Mons aureole: formation by gravitational spreading. *Journal of Geophysical Research*, **88**, 8333–44.

Frey, H. and Schultz, R. A. (1988). Large impact basins and the mega-impact origin for the crustal dichotomy on Mars. *Geophysical Research Letters*, **15**, 229–32.

Frey, H., Lowry, B. L., and Chase, S. A. (1979). Pseudocraters on Mars. *Journal of Geophysical Research*, **84**, 8075–68.

Frey, H., Roark, V. J. H., Shockey, K. M., Frey, E. L., and Sakimoto, S. E. H. (2002). Ancient lowlands on Mars. *Geophysical Research Letters*, **29** (10), doi: 10.1029/2001GL013832.

Fuller, E. R. and Head, J. W. (2002). Amazonis Planitia: the role of geologically recent volcanism and sedimentation in the formation of the smoothest plains on Mars. *Journal of Geophysical Research (Planets)*, **107** (E10), doi: 10.1029/2002JE001842.

Gaidos, E. J. (2001). Cryovolcanism and the recent flow of liquid water on Mars. *Icarus*, **153** (1), 218–23.

Garvin, J. B., Sakimoto, S. E. H., Frawley, J. J., Schnetzler, C. C., and Wright, H. M. (2000). Topographic evidence for geologically recent near-polar volcanism on Mars. *Icarus*, **145** (2), 648–52.

Gatto, L. W. and Anderson, D. M. (1975). Alaskan thermokarst terrain and possible Martian analogs. *Science*, **188**, 255–7.

Gerstell, M. F., Aharonson, O., and Schorghofer, N. (2004). A distinct class of avalanche scars on Mars. *Icarus*, **168** (1), 122–30.

Ghatan, G. J. and Head, J. W. (2002). Candidate subglacial volcanoes in the south polar region of Mars: morphology, morphometry, and eruption conditions. *Journal of Geophysical Research (Planets)*, **107** (E07), doi: 10.1029/2001JE001519.

Ghatan, G. J., Head, J. W., and Pratt, S. (2003). Cavi Angusti, Mars: characterization and assessment of possible formation mechanisms. *Journal of Geophysical Research (Planets)*, **108** (E5), doi: 10.1029/2002JE001972.

Glaze, L. S., Baloga, S. M., and Stofan, E. R. (2003). A methodology for constraining lava flow rheologies with MOLA. *Icarus*, **165** (1), 26–33.

Goldspiel, J. M. and Squyres, S. W. (2000). Groundwater sapping and valley formation on Mars. *Icarus*, **148** (1), 176–92.

Golombek, M. P., Anderson, R. C., Barnes, J. R. *et al.* (1999). Overview of the Mars Pathfinder Mission: launch through landing, surface operations, data sets, and science results. *Journal of Geophysical Research*, **104** (E4), 8523–54.

Golombek, M. P., Grant, J. A., Parker, T. J. *et al.* (2003). Selection of the Mars Exploraiton Rover landing sites. *Journal of Geophysical Research (Planets)*, **108** (E12), doi: 10.1029/2003JE002074.

Grant, J. A. and Parker, T. J. (2002). Drainage evolution in the Margaritifer Sinus region, Mars. *Journal of Geophysical Research (Planets)*, **107** (E9), doi: 10.1029/2001JE001678.

Greeley, R. (1973). Mariner 9 photographs of small volcanic structures on Mars. *Geology*, **1**, 175–80.

Greeley, R. and Guest, J. E. (1987). Geologic map of the eastern equatorial region of Mars, scale 1:15,000,000. *US Geological Survey Miscellaneous Investigation Series Map*. I-1802-B.

Greeley, R. and Fagents, S. A. (2001). Icelandic pseudocraters as analogs to some volcanic cones on Mars. *Journal of Geophysical Research (Planets)*, **106** (E9), 20527–46.

Greeley, R. and Spudis, P. D. (1978). Volcanism in the cratered terrain hemisphere of Mars. *Journal of Geophysical Research*, **5**, 453–5.

Greeley, R., Lancaster, N., Lee, S., and Thomas, P. (1992). Martian aeolian processes, sediments, and features. In *Mars*, ed. H. H. Kieffer, B. M. Jakosky, C. W. Snyder, and M. S. Mathews. Tucson: University of Arizona Press, pp. 730–66.

Greeley, R., Skypeck, A., and Pollack, J. B. (1993). Martian aeolian features and deposits: comparisons with general circulation model results. *Journal of Geophysical Research*, **98**, 3183–93.

Greeley, R., Fagents, S. A., Bridges, N. *et al.* (2000a). Volcanism on the Red Planet: Mars. In *Environmental Effects on Volcanic Eruptions: From Deep Oceans to Deep Space*, ed. J. R. Zimbelman and T. K. P. Gregg. New York: Kluwer Academic/Plenum, pp. 75–112.

Greeley, R., Kraft, M. D., Kuzmin, R. O., and Bridges, N. T. (2000b). Mars Pathfinder landing site: evidence for a change in wind regime from lander and orbiter data. *Journal of Geophysical Research (Planets)*, **105** (E1), 1829–40.

Greeley, R., Kuzmin, R. O., and Haberle, R. M. (2001). Aeolian processes and their effects on understanding the chronology of Mars. In *Chronology and Evolution of Mars*, ed. R. Kallenbach, J. Giess, and W. K. Hartmann. Dordrecht: Kluwer Academic, pp. 393–404.

Greeley, R., Bridges, N. T., Kuzmin, R. O., and Laity, J. E. (2002). Terrestrial analogs to wind-related features at the Viking and Pathfinder landing sties on Mars. *Journal of Geophysical Research (Planets)*, **107** (E01), doi: 10.1029/2000JE001481.

Gulick, V. C. and Baker, V. R. (1990). Origin and evolution of valleys on Martian volcanoes. *Journal of Geophysical Research*, **95**, 14325–44.

Haberle, R. M., Murphy, J. R., and Schaeffer, J. (2003). Orbital change experiments with a Mars general circulation model. *Icarus*, **161** (1), 66–89.

Harrison, K. P. and Grimm, R. E. (2003). Rheological constraints on Martian landslides. *Icarus*, **163** (2), 347–62.

Hartmann, W. K. (1973) Martian surface and crust: review and synthesis. *Icarus*, **19**, 550–75.

Hartmann, W. K. and Berman, D. C. (2000). Elysium Planitia lava flows: crater count chronology and geological implications. *Journal of Geophysical Research (Planets)*, **105** (E6), 15011–25.

Hartmann, W. K. and Neukum, G. (2001). Cratering chronology and the evolution of Mars. *Space Science Reviews*, **96**, 165–94.

Hauber, E. and Kronberg, P. (2001). Tempe Fossae, Mars: a planetary analog to a terrestrial continental rift? *Journal of Geophysical Research (Planets)*, **106** (E9), 20587–602.

Head, J. W. (2001). Mars: evidence for geologically recent advance of the south polar cap. *Journal of Geophysical Research (Planets)*, **106** (E5), 10075.

Head, J. W. and Marchant, D. R. (2003). Cold-based mountain glaciers on Mars: western Arsia Mons. *Geology*, **31** (7), 641–4.

Head, J. W. and Pratt, S. (2001). Extensive Hesperian-aged south polar ice sheet on Mars: evidence for massive melting and retreat, and lateral flow and ponding of meltwater. *Journal of Geophysical Research (Planets)*, **106** (E6), 12275.

Head, J. W. and Wilson, L. (2002). *Mars: A Review and Synthesis of General Environments and Geological Settings of Magma–H_20 Interactions.* Geological Society Special Publication 202, pp. 27–57.

Head, J. W. III, Hiesinger, H., Ivanov, M. A. *et al.* (1999). Possible ancient oceans on Mars: evidence from Mars Orbiter Laser Altimeter data. *Science*, **286**, 2134–7.

Head, J. W., Kreslavsky, M. A., and Pratt, S. (2002). Northern lowlands of Mars: evidence for widespread volcanic flooding and tectonic deformation in the Hesperian Period. *Journal of Geophysical Research (Planets)*, **107** (E01), doi: 10.1029/2000JE001445.

Head, J. W., Wilson, L., and Mitchell, K. L. (2003). Generation of recent massive water floods at Cerberus Fossae, Mars by dike emplacement, cryosphere cracking, and confined aquifer groundwater release. *Geophysical Research Letters*, **30** (11), 31.

Hecht, M. H. (2002). Metastability of liquid water on Mars. *Icarus*, **156** (2), 373–86.

Hiesinger, H. and Head, J. W. (2000). Characteristics and origin of polygonal terrain in southern Utopia Planitia, Mars: results from Mars Orbiter Laser Altimeter and Mars Orbiter Camera data. *Journal of Geophysical Research (Planets)*, **105** (E5), 11999.

Hiesinger, H. and Head, J. W. (2004). The Syrtis Major volcanic province, Mars: synthesis from Mars Global Surveyor data. *Journal of Geophysical Research (Planets)*, **109** (E1), doi: 10.1029/2003JE002143.

Hoffman, N. (2000). White Mars: a new model for Mars' surface and atmosphere based on CO_2. *Icarus*, **146** (2), 326–42.

Howard, A. D. (2000). The role of eolian processes in forming surface features of the Martian polar layered deposits. *Icarus*, **144** (2), 267–88.

Hynek, B. M., Phillips, R. J., and Arvidson, R. E. (2003). Explosive volcanism in the Tharsis region: global evidence in the Martian geologic record. *Journal of Geophysical Research (Planets)*, **108** (E9), doi: 10.1029/2003JE002062.

Irwin, R. P. and Howard, A. D. (2002). Drainage basin evolution in Noachian Terra Cimmeria, Mars. *Journal of Geophysical Research (Planets)*, **107** (E7), doi: 10.1029/2001JE001818.

Ivanov, M. A. and Head, J. W. (2001). Chryse Planitia, Mars: topographic configuration, outflow channel continuity and sequence, and tests for hypothesized ancient bodies of water using Mars Orbiter Laser Altimeter (MOLA) data. *Journal of Geophysical Research (Planets)*, **106** (E2), 3275.

Ivanov, M. A. and Head, J. W. (2003). Syrtis Major and Isidis Basin contact: morphological and topographic characteristics of Syrtis Major lava flows and material of the Vastitas Borealis Formation. *Journal of Geophysical Research (Planets)*, **108** (E6), doi: 10.1029/2002JE001994.

Jankowski, D. G. and Squyres, S. W. (1992). The topography of impact craters in "softened" terrain on Mars. *Icarus*, **100**, 26–39.

Jankowski, D. G. and Squyres, S. W. (1993). "Softened" impact craters on Mars: implications for ground ice and the structure of the Martian megaregolith. *Icarus*, **106**, 365–79.

Kargel, J. S. and Strom, R. G. (1992). Ancient glaciation on Mars. *Geology*, **20**, 3–7.

Keszthelyi, L. P., McEwen, A. S., and Thordarson, T. (2000). Terrestrial analogs and thermal models for Martian flood lavas. *Journal of Geophysical Research (Planets)*, **105** (E6), 15027–50.

Kolb, E. J. and Tanaka, K. L. (2001). Geologic history of the polar regions of Mars based on Mars Global Surveyor data: II. Amazonian period. *Icarus*, **154** (1), 22–39.

Komar, P. D. (1979). Comparison of the hydraulics of water flows in Martian outflow channels with flows of similar scale on Earth. *Icarus*, **37**, 156–81.

Koutnik, M., Byrne, S., and Murray, B. (2002). South polar layered deposits of Mars: the cratering record. *Journal of Geophysical Research (Planets)*, **107** (E11), doi: 10.1029/2001JE001805.

Kreslavsky, M. A. and Head, J. W. (2000). Kilometer-scale roughness on Mars: results from MOLA data analysis. *Journal of Geophysical Research (Planets)*, **105** (E11), 26695–712.

Kreslavsky, M. A. and Head, J. W. (2002). Fate of outflow channel effluents in the northern lowlands of Mars: the Vastitas Borealis Formation as a sublimation residue from frozen ponded bodies of water. *Journal of Geophysical Research (Planets)*, **107** (E12), doi: 10.1029/2001JE001831.

Kuzmin, R. O., Greeley, R., Rafkin, S. C. R., and Haberle, R. (2001). Wind-related modification of some small impact craters on Mars. *Icarus*, **153** (1), 61–70.

Laity, J. E. and Malin, M. C. (1985). Sapping processes and the development of theater-headed valley networks on the Colorado Plateau. *Geological Society of American Bulletin*, **96**, 203–17.

Lane, M. D. and Christensen, P. R. (2000). Convection in a catastrophic flood deposit as the mechanism for the giant polygons on Mars. *Journal of Geophysical Research (Planets)*, **105** (E7), 17617–28.

Leverington, D. W. and Ghent, R. R. (2004). Differential subsidence and rebound in response to changes in water loading on Mars: possible effects on the geometry of ancient shorelines. *Journal of Geophysical Research (Planets)*, **109** (E1), doi: 10.1029/2003JE002141.

Lowry, A. R. and Zhong, S. (2003). Surface versus internal loading of the Tharsis rise, Mars. *Journal of Geophysical Research (Planets)*, **108** (E9), doi: 10.1029/2003JE002111.

Lucchitta, B. K. (1981). Mars and Earth: comparison of cold-climate features. *Icarus*, **45**, 264–303.

Lucchitta, B. K. (2001). Antarctic ice streams and outflow channels on Mars. *Geophysical Research Letters*, **28** (3), 403–6.

Lucchitta, B. K. (1982). Ice sculpture in the Martian outflow channels. *Journal of Geophysical Research (Planets)*, **87**, 9951–73.
Lucchitta, B. K. (1985). Geomorphologic evidence for ground ice on Mars. In *Ices in the Solar System*, ed. J. Klinger, D. Benest, A. Dollfus, and R. Smoluchowski. Dorcrecht: D. Reidel, pp. 583–604.
Lucchitta, B. K. and Anderson, D. M. (1980). Martian outflow channels sculptured by glaciers. In *Reports of Planetary Geology Program*, NASA TM-**81776**, pp. 271–3.
Lucchitta, B. K., Clow, G. D., Geissler, P. E. *et al.* (1992). The canyon system on Mars. In *Mars*, ed. H. H. Kieffer, B. M. Jakosky, C. W. Snyder, and M. S. Matthews. Tucson: University of Arizona Press, pp. 453–92.
Malin, M. C. and Carr, M. H. (1999). Groundwater formation of Martian valleys. *Nature*, **397**, 589–91.
Malin, M. C. and Edgett, K. S. (2000). Evidence for recent groundwater seepage and surface runoff on Mars. *Science*, **288**, 2330–5.
Malin, M. C. and Edgett, K. S. (2001). Mars Global Surveyor Mars Orbiter Camera: interplanetary cruise through primary mission. *Journal of Geophysical Research (Planets)*, **106** (E10), 23429–570.
Malin, M. C. *et al.* (1998). Early views of the Martian surface from the Mars Orbiter Camera of Mars Global Surveyor. *Science*, **279**, 1681–5.
Mangold, N., Costard, F., and Peulvast, J. -P., (2000a). Thermokarstic degradation of lobate debris aprons and fretted channels. 2nd Mars Polar Science Conference, Abstract 4032.
Mangold, N., Costard, F., and Peulvast, J.-P. (2000b). Thermokarstic degradation of the Martian surface. 2nd Mars Polar Science Conference, Abstract 4052.
Mangold, N., Costard, F., and Forget, F. (2003). Debris flows over sand dunes on Mars: evidence for liquid water. *Journal of Geophysical Research (Planets)*, **108** (E4), doi:10.1029/2002JE001958.
Mars Channel Working Group (1983). Channels and valleys on Mars. *Geological Society of American Bulletin*, **94**, 1035–54.
Masson, P., Carr, M. H., Costard, F. *et al.* (2001). Geomorphologic evidence for liquid water. In *Chronology and Evolution of Mars*, ed. R. Kallenbach, J. Giess, and W. K. Hartmann. Dordrecht: Kluwer Academic, pp. 333–64.
Masursky, H., Boyce, J. V., Dial, A. L. Jr., Schaber, G. G., and Strobell, M. E. (1977). Classification and time of formation of Martian channels based on Viking data. *Journal of Geophysical Research*, **82**, 4016–37.
Maxwell, T. A. (1982). Orientation and origin of ridges in the Lunae Palus-Coprates region of Mars. *Journal of Geophysical Research*, **87**, A97–108.
McEwen, A. S., Malin, M. C., Carr, M. H., and Hartmann, W. K. (1999). Voluminous volcanism on early Mars revealed in Valles Marineris. *Nature*, **397**, 584–6.
McGill, G. E. (1986a). The giant polygons of Utopia, Northern Martian Plains. *Geophysical Research Letters*, **13**, 705–8.
McGill, G. E. (1986b). The giant polygons of Utopia, Northern Martian Plains. *Journal of Geophysical Research (Planets)*, **108** (E5), doi: 10.1029/2002JE001852.
McGill, G. E. (1989). Buried topography of Utopia, Mars: persistence of a giant impact depression. *Journal of Geophysical Research*, **94**, 2753–9.
McSween, H. Y. Jr., Murchie, S. L. III, and Britt, D. T. (1999). Chemical, multipectral, and textural constrains on the composition and origin of rocks at the Mars Pathfinder landing site. *Journal of Geophysical Research*, **104**, 8679–716.

Mège, D. and Masson, P. (1996). Stress models for Tharsis formation, Mars. *Planetary Space Science*, **44**, 1471–97.

Mege, D., Cook, A. C., Garel, E., Lagabrielle, Y., and Cormier, M.-H. (2003). Volcanic rifting at Martian grabens. *Journal of Geophysical Research (planets)*, **108** (ES), 5044.

Mest, S. C. and Crown, D. A. (2001). Geology of the Reull Vallis region, Mars. *Icarus*, **153** (1), 89–110.

Mellon, M. T. and Phillips, R. J. (2001). Recent gullies on Mars and the source of liquid water. *Journal of Geophysical Research (Planets)*, **106** (E10), 23165–80.

Mellon, M. T., Feldman, W. C., and Prettyman, T. H. (2004). The presence and stability of ground ice in the southern hemisphere of Mars. *Icarus*, **169** (2), 324–40.

Milkovich, S. M., Head, J. W., and Pratt, S. (2002). Meltback of Hesperian-aged ice-rich deposits near the south pole of Mars: evidence for drainage channels and lakes. *Journal of Geophysical Research (Planets)*, **107** (E06), doi: 10.1029/2001JE001802.

Milliken, R. E., Mustard, J. F., and Goldsby, D. L. (2003). Viscous flow features on the surface of Mars: observations from high-resolution Mars Orbiter Camera (MOC) images. *Journal of Geophysical Research (Planets)*, **108** (E6), doi: 10.1029/2002JE002005.

Moore, J. M. and Wilhelms, D. E. (2001). Hellas as a possible site of ancient ice-covered lakes. Abstracts of papers submitted to the 32nd Lunar and Planetary Science Conference, CD #32, abstract 1446.

Mouginis-Mark, P. J., Wilson, L., and Head, J. W. (1982). Explosive volcanism on Hecates Tholus, Mars: investigation of eruption conditions. *Journal of Geophysical Research*, **87**, 9890–904.

Mouginis-Mark, P. J., Wilson, L., and Zuber, M. T. (1992). The physical volcanology of Mars. In *Mars*, ed. H. H. Kieffer, B. M. Jakosky, C. W. Snyder, and M. S. Matthews. Tucson: University of Arizona Press, pp. 424–52.

Mouginis-Mark, P. J., Wilson, L., and Zimbelman, R. J. (1988). Polygenic eruptions on Alba Patera, Mars: evidence of channel erosion on pyroclastic flows. *Bulletin of Volcanology*, **50**, 361–79.

Murray, B., Koutnik, M., Byrne, S. *et al.* (2001). Preliminary geological assessment of the northern edge of Ultimi Lobe, Mars south polar layered deposits. *Icarus*, **154** (1), 80–97.

Mustard, J. F., Cooper, C. D., and Rifkin, M. K. (2001). Evidence for recent climate change on Mars from the identification of youthful near-surface ground ice. *Nature*, **412**, 411–14.

Mutch, T. A., Arvidson, R. E., Head, J. W., Jones, K. L., and Saunders, R. S. (1976). *Geology of Mars*. Princeton: Princeton University Press.

Mutch, T. A., Arvidson, R. E., Binder, A. B., Guiness, E. A., and Morris, E. C. (1977). The geology of the Viking Lander 2 site. *Journal of Geophysical Research*, **82**, 4452–67.

Newsom, H. E., Barber, C. A., Hare, T. M. *et al.* (2003). Paleolakes and impact basins in southern Arabia Terra, including Meridiani Planum: implications for the formation of hematite deposits on Mars. *Journal of Geophysical Research (Planets)*, **108** (E12), doi: 10.1029/2002JE001993.

Ori, Marinangeli, G. G. L., and Baliva, A. (2000). Terraces and Gilbert-type deltas in crater lakes in Ismenius Lacus and Memnonia (Mars). *Journal of Geophysical Research (Planets)*, **105** (E7), 17629–42.

Parker, T. J., Saunders, R. S., and Schneeberger, D. M. (1989). Transitional morphology in the west Deuteronilus Mensae region of Mars: implications for modification of the lowland/upland boundary. *Icarus*, **82**, 111–45.

Parker, T. J., Gorsline, D. S., Saunders, R. S., Pieri, D. C., and Schneeberger, D. M. (1993). Coastal geomorphology of the Martian northern plains. *Journal of Geophysical Research*, **98**, 11061–78.

Pechmann, J. C. (1980). The origin of polygonal troughs in the Northern Plains of Mars. *Icarus*, **42**, 185–210.

Peitersen, M. N. and Crown, D. A. (2000). Correlations between topography and intraflow width behavior in Martian and terrestrial lava flows. *Journal of Geophysical Research (Planets)*, **105** (E2), 4123–34.

Pelkey, S. M. and Jakosky, B. M. (2002). Surficial geologic surveys of Gale Crater and Melas Chasma, Mars: integration of remote-sensing data. *Icarus*, **160** (2), 228–57.

Pelkey, S. M., Jakosky, B. M., and Christensen, P. R. (2003). Surficial properties in Melas Chasma, Mars, from Mars Odyssey THEMIS data. *Icarus*, **165** (1), 68–89.

Phillips, R. J., Saunders, R. S., and Conel, J. E. (1973). Mars: crustal structure inferred from Bouguer gravity anomalies. *Journal of Geophysical Research*, **78**, 4815–20.

Pierce, T. L. and Crown, D. A. (2003). Morphologic and topographic analyses of debris aprons in the eastern Hellas region, Mars. *Icarus*, **163** (1), 46–65.

Pieri, D. C. (1976). Martian channels: distribution of small channels on the Martian surface. *Icarus*, **27**, 25–50.

Pieri, D. C. (1980). Martian valleys: morphology, distribution, age and origin. *Science*, **210**, 895–7.

Plescia, J. B. (2000). Geology of the Uranius group volcanic constructs: Uranius Patera, Ceraunius Tholus, and Uranius Tholus. *Icarus*, **143** (2), 376–96.

Plescia, J. B. (2003). Cerberus Fossae, Elysium, Mars: a source for lava and water. *Icarus*, **164** (1), 79–95.

Plescia, J. B. and Golombek, M. P. (1986). Origin of planetary wrinkle ridges based on the study of terrestrial analogs. *Geological Society of American Bulletin*, **97**, 1289–99.

Rathbun, J. A. and Squyres, S. W. (2002). Hydrothermal systems associated with Martian impact craters. *Icarus*, **157** (2), 362–72.

Reiss, D., van Gasselt, S., Neukum, G., and Jaumann, R. (2004). Absolute dune ages and implications for the time of formation of gullies in Nirgal Vallis, Mars. *Journal of Geophysical Research (Planets)*, **109** (E6), doi: 10.1029/2004JE002251.

Richardson, M. I. and Wilson, R. J. (2002). Investigation of the nature and stability of the Martian seasonal water cycle with a general circulation model. *Journal of Geophysical Research (Planets)*, **107** (E5), doi: 10.1029/2001JE001536.

Robinson, M. S. and Tanaka, K. L. (1990). Magnitude of a catastrophic flood event at Kasei Valles, Mars. *Geology*, **18**, 902–5.

Rossbacher, L. A. and Judson, S. (1981). Ground ice on Mars: inventory, distribution, and resulting landforms. *Icarus*, **45**, 39–59.

Rotto, S. and Tanaka, K. L. (1995). Geologic/geomorphic map of the Chryse Planitia Region of Mars. *US Geological Survey Miscellaneous Investigation Series Map*. I-2441-A.

Russell, P. S. and Head, J. W. (2003). Elysium-Utopia flows as mega-lahars: a model of dike intrusion, cryosphere cracking, and water-sediment release. *Journal of Geophysical Research (Planets)*, **108** (E6), doi: 10.1029/2002JE001995.

Schenk, P. M. and Moore, J. M. (2000). Stereo topography of the south polar region of Mars: volatile inventory and Mars Polar Lander site. *Journal of Geophysical Research (Planets)*, **105** (E10), 24529–44.

Schultz, P. H. and Gault, D. E. (1979). Atmospheric effects on Martian ejecta emplacement. *Journal of Geophysical Research*, **84**, 7669–87.

Schultz, P. H. and Mustard, J. F. (2004). Impact melts and glasses on Mars. *Journal of Geophysical Research (Planets)*, **109** (E1), doi: 10.1029/2002JE002025.

Schultz, R. A. and Tanaka, K. L. (1994). Lithospheric-scale buckling and thrust structures on Mars: the Coprates rise and south Tharsis ridge belt. *Journal of Geophysical Research*, **99**, 8371–85.

Scott, D. H. and Tanaka, K. L. (1986). Geologic map of the western hemisphere of Mars; scale 1:15M. *US Geological Survey Miscellaneous Investigation Series Map.* I-1802-A.

Scott, E. D. and Wilson, L. (2002). Plinian eruptions and passive collapse events as mechanisms of formation for Martian pit chain craters. *Journal of Geophysical Research (Planets)*, **107** (E04), doi: 10.1029/2000JE001432.

Sharp, R. P. (1973). Mars: fretted and chaotic terrain. *Journal of Geophysical Research*, **78**, 4073–83.

Sharp, R. P. and Malin, M. C. (1975). Channels on Mars. *Geological Society of American Bulletin*, **86**, 593–609.

Sleep, N. H. (1994). Martian plate tectonics. *Journal of Geophysical Research*, **99**, 5639–55.

Sleep, N. H. and Phillips, R. J. (1979). An isostatic model for the Tharsis Province, Mars. *Geophysical Research Letters*, **6**, 803–6.

Sleep, N. H. and Phillips, R. J. (1985). Gravity and lithospheric stress on the terrestrial planets with reference to the Tharsis Region of Mars. *Journal of Geophysical Research*, **90**, 4469–89.

Smith, D. E., Zuber, M. T., Frey, H. V. *et al.* (1998). Topography of the northern hemisphere of Mars from the Mars Orbiter Laser Altimeter. *Science*, **279**, 1686–92.

Smith, D. E., Zuber, M. T., Solomon, S. C. *et al.* (1999). The global topography of Mars and implication for surface evolution. *Science*, **284**, 1495–503.

Smith, D. E., Zuber, M. T., Frey, H. V. *et al.* (2001). Mars Orbiter Laser Altimeter: experiment summary after the first year of global mapping of Mars. *Journal of Geophysical Research (Planets)*, **106** (E10), 23689–722.

Solomon, S. C. and Head, J. W. (1982). Evolution of the Tharsis Province of Mars: the importance of heterogeneous lithospheric thickness and volcanic construction. *Journal of Geophysical Research*, **87**, 9755–74.

Squyres, S. W. (1978). Martian fretted terrain: flow of erosional debris. *Icarus*, **34**, 600–13.

Squyres, S. W. (1979). The distribution of lobate debris aprons and similar flows on Mars. *Journal of Geophysical Research*, **84**, 8087–96.

Squyres, S. W. (1989). Water on Mars. *Icarus*, **79**, 229–88.

Squyres, S. W. and Carr, M. H. (1986). Geomorphic evidence for the distribution of ground ice on Mars. *Science*, **231**, 249–52.

Squyres, S. W. and Kasting, K. F. (1994). Early Mars: how warm and how wet? *Science*, **265**, 744–8.

Squyres, S. W., Clifford, S. M., Kuzmin, R. O., Zimbelman, J. R., and Costard, F. M. (1992). Ice in the Martian regolith. In *Mars*, ed. H. H. Kieffer, B. M. Jakosky, C. W. Snyder, and M. S. Matthews. Tucson: University of Arizona Press, pp. 523–54.

Stepinski, T. J., Collier, M. L., McGovern, P. J., and Clifford, S. M. (2004). Martian geomorphology from fractal analysis of drainage networks. *Journal of Geophysical Research (Planets)*, **109** (E2), doi: 10.1029/2003JE002193.

Stewart, E. M. and Head, J. W. (2001). Ancient Martian volcanoes in the Aeolis region: new evidence from MOLA data. *Journal of Geophysical Research (Planets)*, **106** (E8), 17505–14.

Sullivan, R., Thomas, P., Veverka, J., Malin, M., and Edgett, K. S. (2001). Mass movement slope streaks imaged by the Mars Orbiter Camera. *Journal of Geophysical Research (Planets)*, **106** (E10), 23607–34.

Tanaka, K. L. (1985). Ice-lubricated gravity spreading of the Olympus Mons aureole deposits. *Icarus*, **62**, 191–206.

Tanaka, K. L. (1986). The stratigraphy of Mars. *Lunar and Planetary Science XVII* (CD-ROM), #E139–58.

Tanaka, K. L. (1995). Did Mars have plate tectonics? *Mercury*, **24**, 11.

Tanaka, K. L. (2000). Dust and ice deposition in the Martian geologic record. *Icarus*, **144** (2), 254–66.

Tanaka, K. L. and Kolb, E. J. (2001). Geologic history of the polar regions of Mars based on Mars Global Surveyor data: I. Noachian and Hesperian Periods. *Icarus*, **154** (1), 3–21.

Tanaka, K. L. and Scott, D. H. (1987). Geologic map of the polar region of Mars, scale 1:15,000,000. *US Geological Survey Miscellaneous Investigation Series Map*. I-1802-C.

Tanaka, K. L., Isbell, N. K., Scott, D. H., Greeley, R., and Guest, J. E. (1987). The resurfacing history of Mars: a synthesis of digitized Viking-based geology. *Lunar and Planetary Science XVIII* (CD-ROM), 665–78.

Tanaka, K. L., Scott, D. H., and Greeley, R. (1992). Global stratigraphy. In *Mars*, H. H. Kieffer, B. M. Jakosky, C. W. Snyder, and M. S. Matthews. Tucson: University of Arizona Press, pp. 354–82.

Theilig, E. and Greeley, R. (1979). Plains and channels in the Lunae Planum-Chryse Planitia region of Mars. *Journal of Geophysical Research*, **84**, 7994–8010.

Thomas, P. C., Malin, M. C., and Edgett, K. S. (2000). North–south geological differences between the residual polar caps on Mars. *Nature*, **404**, 161–4.

Thomas, P. C., Gierasch, P., Sullivan, R. *et al.* (2003). Mesoscale linear streaks on Mars: environments of dust entrainment. *Icarus*, **162** (2), 242–58.

Travis, B. J., Rosenberg, N. D., and Cuzzi, J. N. (2003). On the role of widespread subsurface convection in bringing liquid water close to Mars' surface. *Journal of Geophysical Research (Planets)*, **108** (E4), doi: 10.1029/2002JE001877.

Tricart, J. (1968). Periglacial landscapes. In *Encyclopedia of Geomorphology*, ed. R. W. Fairbridge. New York, NY: Reinhold, pp. 829–33.

Tricart, J. (1969). *Geomorphology of Cold Environments*. London: Macmillan.

Tricart, J. L. F. (1988). Environmental change of planet Mars demonstrated by landforms. *Zeitschrift fur Geomorphologie*, **32**, 385–407.

Vasavada, A. R., Paige, D. A., and Wood, S. E. (1999). Near-surface temperatures on Mercury and the Moon and the stability of polar ice deposits. *Icarus*, **141**, 179–93.

Wahrhaftig, C. and Cox, A. (1959). Rock glaciers in the Alaska Range. *Bulletin of the Geology Society of America*, **70**, 383–436.

Warner, N. H. and Gregg, T. K. P. (2003). Evolved lavas on Mars? Observations from southwest Arsia Mons and Sabancaya volcano, Peru. *Journal of Geophysical Research (Planets)*, **108** (E10), doi: 10.1029/2002JE001969.

Watters, T. R. (1988). Wrinkle ridge assemblages on the terrestrial planets. *Journal of Geophysical Research*, **93**, 15599–616.

Watters, T. R. (1993). Compressional tectonism on Mars. *Journal of Geophysical Research*, **98** (E9), 17049.

Watters, T. R. and Maxwell, T. A. (1986). Orientation, relative age, and extent of the Tharsis plateau ridge system. *Journal of Geophysical Research*, **91**, 8113–25.

Werner, S. C., van Gasselt, S., and Neukum, G. (2003). Continual geological activity in Athabasca Valles, Mars. *Journal of Geophysical Research (Planets)*, **108** (E12), doi: 10.1029/2002JE002020.

Whalley, W. B. and Azizi, F. (2003). Rock glaciers and protalus landforms: analogous forms and ice sources on Earth and Mars. *Journal of Geophysical Research (Planets)*, **108** (E4), doi: 10.1029/2002JE001864.

Wilhelms, D. E. and Squyres, S. W. (1984.) The Martian hemispheric dichotomy may be due to a giant impact. *Nature*, **309**, 138–40.

Wilkins, S. J. and Schultz, R. A. (2003). Cross faults in extensional settings: stress triggering, displacement localization, and implications for the origin of blunt troughs at Valles Marineris, Mars. *Journal of Geophysical Research (Planets)*, **108** (E6), doi: 10.1029/2002JE001968.

Willeman, R. J. and Turcotte, D. L. (1982). The role of lithospheric stress in the support of the Tharsis Rise. *Journal of Geophysical Research*, **82**, 9793–801.

Williams, R. M. E. and Malin, M. C. (2004). Evidence for late stage fluvial activity in Kasei Valles, Mars. *Journal of Geophysical Research (Planets)*, **109** (E6), doi: 1029/2003JE002178.

Williams, R. M. E. and Phillips, R. J. (2001). Morphometric measurements of Martian valley networks from Mars Orbiter Laser Altimeter (MOLA) data. *Journal of Geophysical Research (Planets)*, **106** (E10), 23737–52.

Wilson, L. and Head, J. W. (1994). Mars: review and analysis of volcanic eruption theory and relationships to observed landforms. *Reviews of Geophysics*, **32**, 221–63.

Wilson, L. and Head, J. W. (2002). Tharsis-radial graben systems as the surface manifestation of plume-related dike intrusion complexes: models and implications. *Journal of Geophysical Research (Planets)*, **107** (E08), doi: 10.1029/2001JE001593.

Wilson, L. and Mouginis-Mark, P. J. (1987). Volcanic input to the atmosphere from Alba Patera on Mars. *Nature*, **330**, 354–7.

Wilson, L. and Mouginis-Mark, P. J. (2001). Estimation of volcanic eruption conditions for a large flank event on Elysium Mons, Mars. *Journal of Geophysical Research (Planets)*, **106** (E9), 20621–28.

Wilson, L. and Mouginis-Mark, P. J. (2003a). Phreatomagmatic dike–cryosphere interactions as the origin of small ridges north of Olympus Mons, Mars. *Icarus*, **165** (2), 242–52.

Wilson, L. and Mouginis-Mark, P. J. (2003b). Phreatomagmatic explosive origin of Hrad Vallis, Mars. *Journal of Geophysical Research (Planets)*, **108** (E8), doi: 10.1029/2002JE001927.

Wilson, L. Scott E. D., and Head, J. W. III (2001). Evidence for episodicity in the magma supply to the large Tharsis volcanoes. *Journal of Geophysical Research (Planets)*, **106** (E1), 1423–34.

Wohletz, K. H. and Sheridan, M. F. (1983). Martian rampart crater ejecta: experiments and analysis of melt water interaction. *Icarus*, **56**, 15–37.

Zimbelman, J. R. and Edgett, K. S. (1992). The Tharsis Montes, Mars; comparison of volcanic and modified landforms. *Proceedings of the Lunar and Planetary Science Conference*, **22**, 31–44.

2

Impact structures on Earth and Mars

Nadine G. Barlow
Dept. Physics and Astronomy, Northern Arizona University

Virgil Sharpton
Geophysical Institute, University of Alaska

and

Ruslan O. Kuzmin
Vernadsky Institute, Russian Academy of Sciences

2.1 Introduction

Every solid-surfaced body in the Solar System except Io shows evidence of the impact cratering process, and Comet Shoemaker-Levy 9 showed that impacts can even temporarily leave their mark on gas planets. Earth's active geologic environment has erased much of its cratering record, particularly from the early episode of high impact rates known as the late heavy bombardment period (>3.8 Gyr ago). In comparison, ~60% of the Martian surface preserves the late heavy bombardment record. Mars retains the most complete record of impact cratering in the entire Solar System (Barlow, 1988) and these craters display a range of morphologic features seldom seen on other solid-surface bodies. Comparison of terrestrial and Martian craters provides a more thorough understanding of impact structures: Mars preserves the pristine morphologic features which erosion has largely destroyed for terrestrial craters, but terrestrial studies allow us to understand subsurface structures and materials resulting from impact for which we currently have no information on Mars. Presence of an atmosphere and subsurface volatiles suggests that crater formation may be more similar on these two bodies than between Earth and Moon.

Understanding how impact craters form results from laboratory experiments, computer simulations, nuclear and chemical explosions, and terrestrial crater studies. Laboratory experiments were instrumental in realizing that high-velocity impacts create approximately circular craters except at low impact angles (Gault and Wedekind, 1979). Nuclear and large chemical explosions provided the first opportunity to study the physics of crater

The Geology of Mars: Evidence from Earth-based Analogs, ed. Mary Chapman. Published by Cambridge University Press.

formation (Oberbeck, 1977). Field studies of terrestrial impact craters revealed that high-pressure mineral phases are diagnostic features of impacts (Shoemaker, 1963). Computer simulations allow us to study the stages of crater formation in various environments (Melosh, 1989). The combination of these techniques continues to advance our understanding of crater formation on Earth and other bodies.

2.2 Characteristics of impact craters

2.2.1 General characteristics of terrestrial and Martian craters

The process of impact crater formation consists of three stages: (1) shock compression of target material, (2) excavation and formation of a crater cavity, and (3) post-event crater modification (Gault *et al.*, 1968; Melosh, 1989). Crater formation is a very rapid process, reaching completion in 60–100 s for 40–80 km diameter craters (Melosh, 1989; Ivanov, 1999). Large energies and pressures associated with impact crater formation produce diagnostic features in target rocks which distinguish these structures from depressions formed by other geologic processes.

Impact craters display a range of morphologic features as their size increases. The smallest craters, called simple craters, display a bowl-shaped appearance with little additional modification. The slightly larger complex craters display more complicated structures, including central peaks/central pits, wall terraces, floor deposits, and shallow depths. As crater size increases, central peaks are replaced by an inner ring of mountains, called a peak ring. The largest craters (basins) display multiring structures, which result from excavation through a substantial portion of the body's lithosphere (Melosh and McKinnon, 1978). The transition diameter between these different morphologies is largely dependent on the body's gravity, although target characteristics also contribute. On Earth, the transition between simple craters and complex craters occurs near a diameter of 2.5 km in sedimentary material and about 4.5 km in crystalline targets. Peak ring morphology begins to appear in craters around 40 km in diameter. No multiring basins have been conclusively identified on Earth. The simple-to-complex transition diameter for Mars is estimated at 7 km (Garvin *et al.*, 2002) while peak-ring basins begin to appear in 50 km diameter craters and multiring basins appear at diameters larger than 300 km (Strom *et al.*, 1992).

Pristine impact craters are generally circular depressions with a raised rim and surrounded by excavated debris (the ejecta blanket). The ratio of terrestrial crater depth to diameter decreases from about 1/5 for small simple craters to 1/30–1/40 for the larger complex craters. The surrounding ejecta blanket produced during crater excavation is derived mostly from materials originally in the upper 1/3 of the crater (Melosh, 1989).

Morphometric properties of Martian impact craters are statistically different from analogous features in lunar impact craters (Garvin *et al.*, 2002; Melosh, 1989). The depths of fresh Martian impact craters have been determined using Mars Orbiter Laser Altimeter (MOLA) near-centerline topographic cross-sections and gridded digital elevation models. Depths (d) of simple Martian craters (in km) are related to the crater diameter (D) (in km) by

$$d_s = 0.26\, D^{0.67}$$

while the depth–diameter relationship for complex craters between 7 and 110 km is given by

$$d_c = 0.36\, D^{0.49}$$

These depths are shallower than the depths of lunar craters ($d_s = 0.196\, D^{1.01}$; $d_c = 1.044\, D^{0.301}$). Similarly the heights of central peaks in Martian complex craters ($h_{cp} = 0.04\, D^{0.51}$) tend to be greater than those in lunar craters ($h_{cp} = 0.0006\, D^{1.97}$). These differences cannot be explained entirely by differences in gravitational acceleration, and thus target material properties, particularly the presence of subsurface volatiles, are likely responsible.

Earth's active erosional environment rapidly degrades ejecta blankets surrounding terrestrial craters, destroying primary morphologic features and leading to considerable controversy as to whether ballistic or flow processes are responsible for ejecta emplacement (cf., Hörz *et al.*, 1983; Newsom *et al.*, 1986). As noted above, ejecta material is primarily derived from the top 1/3 of the crater depth and consists of a mixture of target and meteoritic material, ranging in size from dust to large boulders. The thickest and roughest part of the ejecta blanket lies within about 1 to 1.5 crater radii from the rim and is called the continuous ejecta deposit. Thin, radially extensive but patchy ejecta material and secondary craters may extend beyond the continuous ejecta blanket to form the discontinuous ejecta deposits (Melosh, 1989). A few terrestrial impacts contain tektites as part of their ejecta deposits. Tektites are silica-rich (68–82% SiO_2) black or green translucent glass created from terrestrial materials during large impact events. They are melted and ejected by the impact, solidifying into glass during passage through the Earth's atmosphere. Four tektite-strewn fields have been identified on Earth and source craters for three of the four have been identified: Ivory Coast (Bosumtwi, Ghana), Central Europe (moldavites) (Ries, Germany), North America (bediasites) (Chesapeake Bay, USA), and Australasia (crater unknown) (French, 1998).

Ejecta blankets surrounding fresh Martian impact craters display a wider range of morphologies than those seen on other planets. Most fresh impact craters between ~3 and 50 km diameter display a unique type of layered ejecta morphology, resembling a mud slurry (Barlow *et al.*, 2000). Smaller and larger fresh craters are typically surrounded by a radial pattern, similar to ejecta blankets seen around lunar craters. Two mechanisms have been proposed to explain the formation of layered ejecta morphologies on Mars: impact into and melting–vaporization of subsurface ice (Carr *et al.*, 1977; Greeley *et al.*, 1980; Stewart *et al.*, 2001), and ejecta entrainment in the thin Martian atmosphere (Schultz and Gault, 1979; Barnouin-Jha *et al.*, 1999a, b). The diameter at which a particular morphology begins to appear (the "onset diameter") and the lateral extent of ejecta (a measure of its mobility at the time of formation) show correlations with latitude and depth of excavation (Kuzmin *et al.*, 1988; Costard, 1989; Barlow and Bradley, 1990), which strongly support their formation through ground ice excavation. In many cases, a flow-like ejecta pattern skirts the relief of pre-impact terrain rather than being superposed as a mantle, which suggests ejecta emplacement primarily as surface flow instead of a ballistically deposited debris blanket.

Erosional processes gradually rework ejecta material, breaking down larger debris and reducing vertical relief. Terrestrial and Martian impact craters display a wide range of preservation states, from pristine to almost completely destroyed. Pristine craters display sharp rims, some type of ejecta blanket, crisp interior features such as central peaks and wall terraces, and little to no evidence of floor deposits. As crater age increases and degradation occurs, the ejecta blanket and interior morphologies are destroyed, the crater floor becomes infilled with sedimentary deposits, and the rim is stripped away. Eolian, fluvial, glacial, volcanic, tectonic, and impact processes have operated episodically throughout terrestrial and Martian histories to produce this range of crater degradation (Grant and Schultz, 1993a). Craddock and Howard (2002) have developed models which simulate crater degradation due to different geologic processes and compared the resulting topography with MOLA profiles of Martian impact craters. Their results suggest that precipitation could have played a major role in the degradation of craters during early Martian history.

Gravity and magnetic anomalies are often associated with impact craters. These approximately circular anomalies result from intense fracturing and refilling of the crater cavity with lower density breccias and impact melts.

2.2.2 Diagnostic features

Studies of terrestrial crater materials reveal that mechanical, thermal, and chemical transformation of the target rocks during crater formation produces a series of diagnostic features which can be used to confirm an impact origin. These features include shatter cones, shock metamorphism, planar deformation features, and high-pressure mineral phase transformations.

Shatter cones arise where shock pressures reach 20–100 kbar (Dietz, 1959; Roddy and Davis, 1977). Shatter cones are curved, striated fractures forming partial to complete cones in target rocks and they form when shock waves meet the rock boundaries (Figure 2.1). Cone angles range from 15° to 120°, being wider in massive and coarse-grained consolidated rocks and narrower in fine-grained and soft rocks. The vertex of the cone in a fractured rock is oriented towards the source of the advancing shock front.

Shock pressures of 50–150 kbar in quartz and feldspars create microscopic fissures, called planar cracks. Higher shock pressures (up to 300 kbar) produce planar deformation features by deforming the atomic structure, which affects the optical properties of the mineral. Very high shock pressures can convert

Figure 2.1. Shatter cone from the Kara crater, Siberia (69° N 63° E). Image courtesy of Teemu Ohman.

the entire crystal into glass (Stöffler, 1972). These high pressures affect the crystalline structure of minerals, converting quartz (SiO_2) into the high-pressure phases of coesite and stishovite and carbon into diamond and lonsdaleite. Because stishovite and lonsdaleite cannot form in the pressure-temperature environments of the terrestrial crust and mantle, their detection is diagnostic of shock metamorphism resulting from impact.

Extreme energies and pressures associated with impact processes create shock-metamorphosed rocks collectively known as impactites. Impact melts are glassy fragments which were melted by the impact's high temperatures and pressures. Crushing and fragmenting rocks during the impact process create breccias, which are rocks composed of angular fragments. Breccias can be authigenic (derived in place from local target rocks) or allogenic (derived from other locations). Breccias are characterized by the number of fragment lithologies (monomict versus polymict), the character of the fragments, and characteristics of the matrix (French, 1998).

2.2.3 Terrestrial crater studies and their implications to Mars

Because of their great size – and the energy levels required to generate them – planetary-scale craters cannot be replicated in the laboratory. Current technology cannot produce experimental projectile speeds faster than $\sim$6 km s^{-1}, well below the expected range of most planetary impacts. Additionally, there are substantial uncertainties in comparing centimeter-scale craters produced in experiments to scales appropriate to planetary cratering. Even with the high-resolution imagers at Mars today, the smallest resolvable craters are two orders of magnitude larger than those that can be generated in the laboratory. Furthermore, laboratory-scale experiments cannot replicate the complexities in target conditions that planetary craters experience. Consequently, extracting the implications of the morphological, lithological, and structural variabilities shown in the Martian cratering record requires access to detailed analysis that only terrestrial craters provide.

Barringer crater, Arizona, USA

Barringer, or Meteor Crater, Arizona (35°1' N, 111°1' W, Figure 2.2), is one of the best-preserved impact craters on Earth (Shoemaker, 1963). The 1.2 km diameter simple crater formed $\sim$50 000 years ago (Sutton, 1985) when an $\sim$50 m iron meteoroid struck the Colorado Plateau. The target material consists of 1400 m of nearly flat-lying Paleozoic and Mesozoic sedimentary rocks, underlain by Precambrian crystalline basement. Rocks exposed at Barringer crater consist of Permian Coconino sandstone, Permian Toroweap

Figure 2.2. Oblique aerial photograph of Meteor or Barringer Crater taken by David Roddy.

formation (interbedded sandstone and dolomite), Permian Kaibab Formation (limestone, sandstone, and fossiliferous dolomite), and Triassic Moenkopi sandstone and siltstone. All strata exposed in the crater walls are uplifted and thickened; this deformation increases dramatically from the crater floor to the rim.

The strata along the crater walls are broken by a number of small, almost vertical faults with scissors-type displacement. Regional jointing appears to have controlled the shape of the crater, since the diagonals of the crater's square shape coincide with the trend of the two major joint sets.

The rim rises between 30 and 65 m above the surrounding terrain. One of the more remarkable contributions that Barringer crater studies have provided is the presence of an inverted or overturned stratigraphic succession comprising the crater rim (Figure 2.3) (Shoemaker, 1963). Particularly evident in the north and east crater walls, the Moenkopi Formation has been folded back upon itself, with the upper limb of the fold consisting of a flap that has been rotated as much as 180° away from the crater.

Vestiges of the ejecta blanket consist of disturbed Moenkopi and Kaibab strata, overlain by a sequence of Quaternary debris and alluvium. Target stratigraphy is discernable in the radial distribution of rock types in the ejecta. Near the crater rim, ejecta are enriched in fragments of the Coconino and Toroweap layers. Kaibab-derived debris dominates an outer 100–300 m wide annulus. Beyond this zone, and particularly evident to the south at

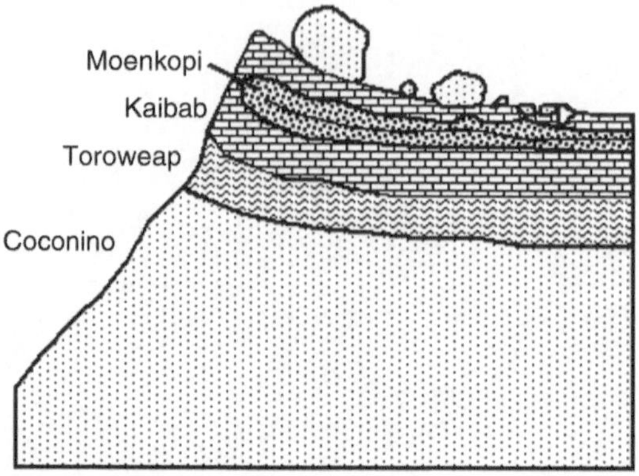

Figure 2.3. Schematic cross section through the wall of Meteor Crater showing inverted stratigraphy characteristic of the crater's rim zone. Modified from Melosh (1989).

Figure 2.4. Aerial panoramic oblique aerial view of the Ries Crater (from the south) shows the outline of the crater rim emphasized by clouds. The town of Nördlingen is visible in the southwestern portion of the crater. This historical image was provided courtesy of Rieskrater Museum, Nördlingen, Germany.

distances of up to 1 km, lies a discontinuous series of deposits composed primarily of Moenkopi fragments. Grant and Schultz (1993b) suggest that the hummocky, lobate morphology exhibited by this ejecta deposit is primary, leading to speculation that Barringer crater might be similar to Martian layered ejecta craters.

Ries, Bavaria, Germany

Ries crater is located in Bavaria, southern Germany (48°53' N, 10°37' E) (Figure 2.4). The crater is 24 km in diameter and has a 12 km diameter irregular inner ring composed of a few hills rising about 50 m above the surrounding surface. The crater rim has a maximum elevation of ~200 m above the crater floor. Ries formed ~14.9 million years ago (Chao, 1977; von Englehardt, 1990) from the hypervelocity impact of a stony iron meteorite ~1 km in diameter. Although its size would suggest that Ries should be a central peak crater, geophysical studies indicate that it is morphologically consistent with a peak ring crater (Pohl *et al.*, 1977).

The target rocks at Ries are a series of Triassic, Jurassic, and Miocene sediments ~0.5 km thick. Excavation penetrated this sequence and uplifted and excavated material from an underlying crystalline basement of granites, gneisses, and amphibolites.

Ejecta deposits at Ries have retained close compositional affiliations with underlying target stratigraphy (Table 2.1). Moldavite tektites formed during the early stages of the impact event from unconsolidated sediments located at the surface of the target sequence. The majority of what would be the continuous ejecta deposit in planetary images is made up of "bunte breccia" consisting of weakly shocked sedimentary rock fragments with a small proportion of highly shocked crystalline basement fragments. The bunte breccia shows clear signs of having been emplaced in a manner that incorporated large amounts of the local target in a process of "ballistic sedimentation" (Oberbeck, 1975). Brecciated but virtually unshocked megablocks of sediments and crystalline basement form a discontinuous distribution within and around the crater.

Suevite, a clay-rich polymict breccia consisting principally of highly shocked fragments derived from the basement sequence and impact melt inclusions,

Table 2.1. *Characteristics of Ries crater ejecta units*

Ries crater ejecta unit	Shock level	Source level	Ejecta characteristics
Moldavites	Very high (completely melted)	Very shallow	Discontinuous; range: 500 km[a]
Bunte breccia	Weak	Shallow	Continuous blanket; Range:[a] to 40 km[b] Thickness: ≤100 m[b]
Megablocks (monomict)	Weak	Shallow and deep	Discontinuous; Range:[a] 6 to 100 km Size: to 2 km
Suevite	High (melt inclusions and abundant solid state changes)	Deep	Discontinuous(?); Range: to ~25 km[b] Thickness: ≤80 m[b]

[a] Measured from center of crater; [b] presently.
Source: von Englehardt (1990).

Figure 2.5. Ejecta layering at the Aumuhle quarry near the northern rim of the crater. Dark lower unit is bunte breccia, light upper unit is suevite.

is common around the Ries. In several outcrops, suevite is preserved directly in contact with the underlying bunte breccia (Figure 2.5). The smooth contact between bunte breccia and suevite indicates that the suevites landed on the ground as a high-density turbulent suspension of solid particles in vapor (von Engelhardt *et al.*, 1995). The suevite contains chimney-like degassing pipes which have been proposed to result from devolatilization of the basement inclusions (Newsom *et al.*, 1986).

Applications to Mars

Martian impact craters appear morphologically similar to terrestrial craters at the resolutions currently available. Detailed analysis of the mineralogy and structure of terrestrial impact craters provides clues of what features we should look for with future missions to Mars. Mars Global Surveyor imagery reveals widespread sedimentary deposits (Malin and Edgett, 2000) and thermal infrared analysis suggests that the major compositional units are "basalt" and "andesite" (Bandfield *et al.*, 2000). Diagnostic features useful on Earth such as shocked quartz and high-pressure phases such as coesite and stishovite may not be common on Mars due to compositional variations. However, other features such as shatter cones and planar deformation features might be present and distinguishable by rover instruments. Materials similar to suevite and bunte breccia might be identifiable in Martian crater ejecta deposits by rover missions and human explorers. Complete analysis of the mineralogic and lithologic properties of Martian crater materials will have to await the return of samples to Earth. Subsurface sounding, such as that which will be conducted by the Mars Express MARSIS and Mars Reconnaisance Orbiter SHARAD instruments, may help constrain the geophysical characteristics of Martian craters. Similarities and differences between terrestrial and Martian craters will provide important new insights into how environment affects crater formation.

2.3 Effects of volatiles on crater features

2.3.1 Martian craters

Martian craters are better analogs for craters on Earth than are lunar craters because they form in the presence of an atmosphere and probable subsurface volatile reservoirs. Atmospheres can be eroded by large impacts (Melosh and Vickery, 1989) or enhanced by release of subsurface volatiles during impact (Carr, 1996). Winds and atmospheric circulation help to disperse impact debris and may play a role in the formation of certain ejecta features (Barnouin-Jha *et al.*, 1999a, b). Atmospheres extend the geologic processes operating on a planetary surface to include eolian, fluvial, and glacial processes, which all operate to degrade craters and their associated ejecta blankets. Impact into targets containing water and/or ice creates unique features not seen on dry bodies such as the Moon.

Certain features associated with Martian impact craters are among the arguments inferring the presence of subsurface volatiles (Carr, 1996). Global mapping and statistical analysis of all Martian fresh craters in the 1–80 km diameter size range have shown that onset diameters for layered ejecta craters decrease from the equator towards the poles, indicating that the top of an ice-rich layer lies at shallower depths at higher latitudes (Figure 2.6) (Kuzmin *et al.*, 1988). Qualitative information about subsurface ice distribution in Martian permafrost as a function of depth and latitude has been determined using the ejecta mobility (EM) ratio. EM is the ratio of layered ejecta diameter to crater diameter and is believed to provide information about the relative concentration of ice in the excavated regolith at the time of crater formation (Mouginis-Mark, 1979, 1987; Costard, 1989). A strong correlation of EM with latitude has been found: the ratio increases in a regular manner from equatorial to polar latitudes in both hemispheres of the planet, being notably higher in the northern hemisphere than in the southern.

Fresh Martian impact craters display three major types of layered ejecta morphologies (Figure 2.7): a single layer of ejecta material (single-layer

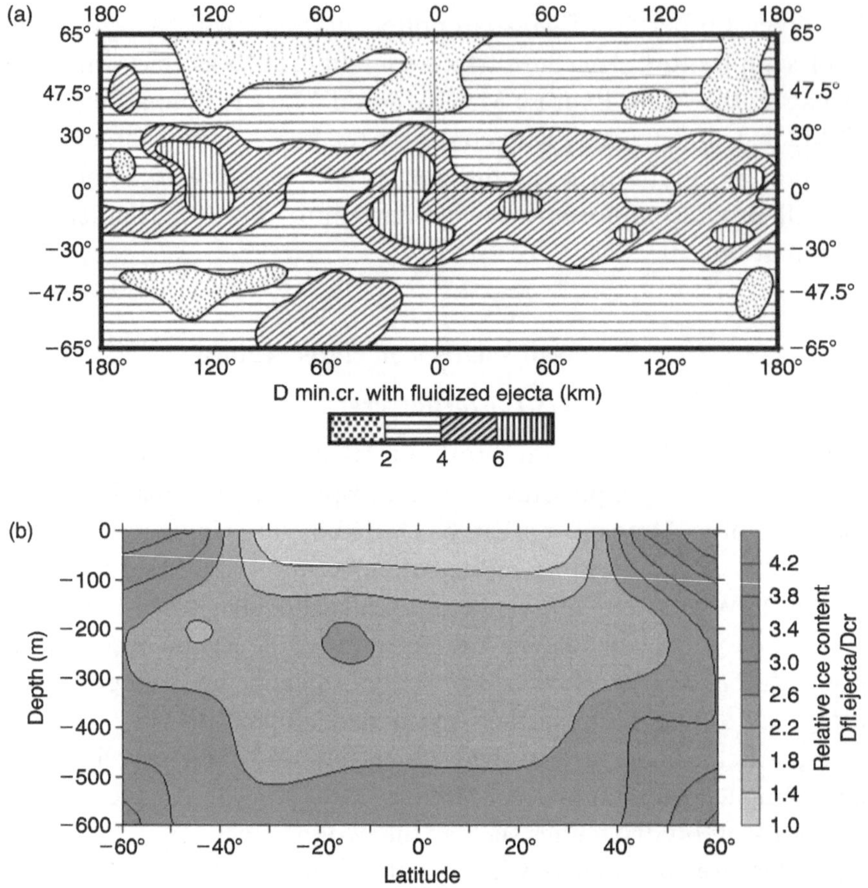

Figure 2.6. Proposed distribution of subsurface ice derived from Martian crater analysis. (a) Distribution of layered ejecta morphology onset diameters. (b) Relative ice content of Martian substrate as a function of latitude and depth, as determined from the onset diameters of layered ejecta craters with excavation depths up to 600 m.

ejecta, or SLE), two ejecta layers (double-layer ejecta, or DLE), and three or more complete or partial layers of ejecta (multiple-layer ejecta, or MLE) (Barlow *et al.*, 2000). The subsurface volatile theory argues that the SLE morphology results from impact into subsurface ice while the DLE morphology results from impact into layered target materials with varying concentrations of ice (Mouginis-Mark, 1979, 1987; Costard, 1989; Barlow and Bradley, 1990). Processes responsible for the MLE morphology are more controversial. Barlow and Bradley (1990) argue that excavation depths and regional variations seen in the distribution of MLE craters indicate this morphology results from excavation into liquid water reservoirs.

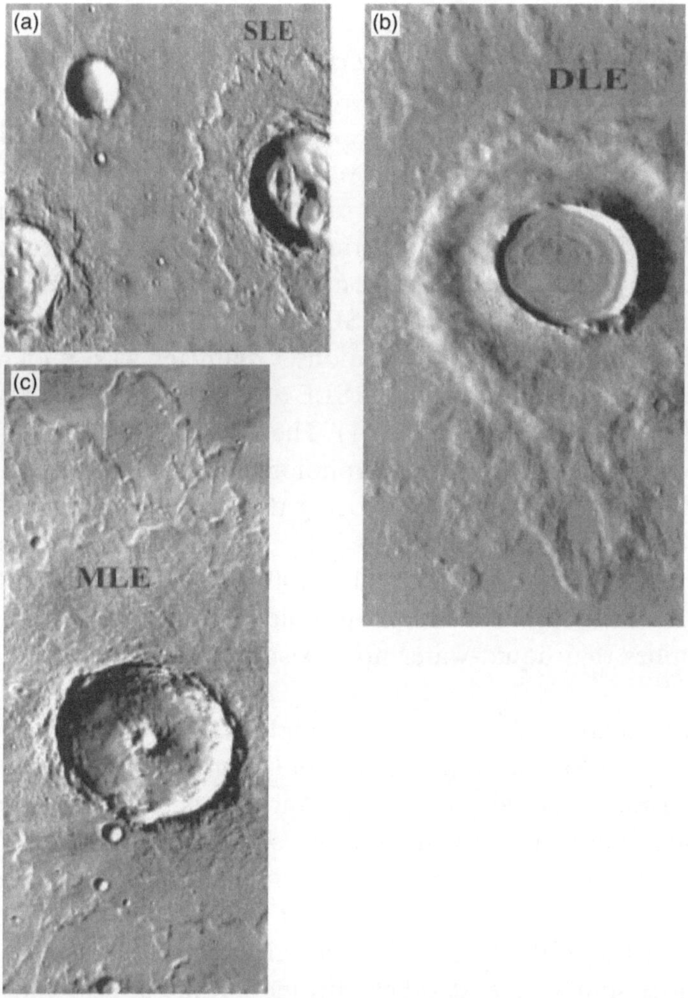

Figure 2.7. (a) Example of a SLE morphology crater. Crater is 12.7 km in diameter and located at 24° N 102° E. (THEMIS image I02493005) (b) DLE crater, located near 43° N 212° E. Outer ejecta layer is prominent near bottom of the image. (THEMIS image I03051002). (c) This MLE crater is 22 km in diameter and located near 6° N 70° E. Note the central pit in this crater, which may be another indicator of subsurface volatiles. (THEMIS image I03218002)

Alternately, recent hydrocode models suggest the MLE morphology results from impact into ice and subsequent interaction of a vapor plume with the atmosphere (Stewart *et al.*, manuscript submitted to *Nature*). The distribution of DLE and MLE morphologies generally correlates with surface locations interpreted to be H_2O-rich, based on Mars Odyssey Gamma Ray Spectrometer (GRS) results (Feldman *et al.*, 2002; Mitrofanov *et al.*, 2002), suggesting possible interactions between proposed deep volatile reservoirs excavated by craters and the uppermost meter of Mars (Barlow and Perez, 2003).

Layered ejecta morphologies are seen for craters in very specific diameter ranges (Mouginis-Mark, 1979; Barlow and Bradley, 1990). SLE and DLE morphologies are typically seen for craters 3–25 km in diameter while the MLE morphology is associated with craters 25–50 km in diameter. There are, however, strong latitudinal and regional variations. Viking image analyses suggested that the smallest equatorial SLE craters were between 3 and 5 km in diameter while closer to the poles the onset diameter was <1 km (Kuzmin *et al.*, 1988). Regional variations in the SLE onset diameter have been reported (Boyce *et al.*, 2000; Barlow *et al.*, 2001). The DLE onset diameter appears to be <3 km while that for the MLE morphology in equatorial regions is around 20 km (Barlow and Bradley, 1990). Using these onset diameters, the depths to buried ice-rich reservoirs implied by SLE and DLE morphologies is <250 m in equatorial regions and <100 m near the poles. If the MLE morphology is due to excavation into liquid-water reservoirs, its onset diameter implies that liquid-water may exist at depths of <1.5 km near the equator.

Subsurface volatiles also have been implicated in the preponderance of central pits within Martian impact craters (Figure 2.7) (Wood *et al.*, 1978). Barlow and Bradley (1990) noted that central pits are found preferentially in craters superposed on the outer rings of multiring basins. These areas would have undergone substantial fracturing during basin formation and may have become repositories for enhanced volatile concentrations.

Craters at mid-latitudes (± 30–$55°$) typically exhibit more subdued features than craters of similar age at other latitudes. Rims appear more rounded, floor deposits are common, and the overall crater topographic profile is more subdued. This is likely the result of quasi-viscous relaxation of topography due to creep of ice-rich regolith (Jankowski and Squyres, 1992). This creep process also may be responsible for the variations in crater morphometric properties seen between the equatorial and polar regions (Garvin *et al.*, 2000).

2.3.2 Terrestrial craters in volatile-rich environments

Knowledge about the structure and composition of terrestrial impact craters is necessary to understand the evolution of Martian impact craters and those conditions under which they formed. We highlight three complex terrestrial impact structures which have formed at various times and in different volatile-rich environments to demonstrate the variety of structures which can be produced.

Haughton crater, northern Canada

Haughton (75°22' N; 89°41' W) is a 24 km diameter impact structure located on Devon Island in the Canadian Arctic (Figure 2.8) (Robertson and Grieve, 1978; Robertson and Sweeney, 1983). It is the northernmost known terrestrial impact crater located in a permafrost environment. The crater is 23 Myr old (Jessberger, 1988) and formed in target rocks consisting of Lower Paleozoic sedimentary rocks overlying the Precambrian metamorphic basement comprising the Canadian shield (Grieve, 1988).

Geologic mapping reveals that the nearly flat Paleozoic sequence excavated by the crater is ~1.8 km thick (Sharpton *et al.*, 1998). These deposits consist of limestone, dolomites, shale, carbonates, and gypsum. The sequence is underlain by undifferentiated Lower Ordovician-Cambrian shale, sandstone, dolomite, and conglomerates. The central part of the crater consists of a 3.6–5.5 km basin filled by allogenic breccias and surrounded by an interior ring of uplifted rocks. Some of the rocks in the interior ring were uplifted from ~600 m depth. The middle crater ring (14–15 km in diameter) morphologically consists of a chain of hills, elevated 140–175 m above the central basin. The exterior ring of the Haughton structure represents the rim of an impact crater, which is elevated above the central basin floor by about 200 m.

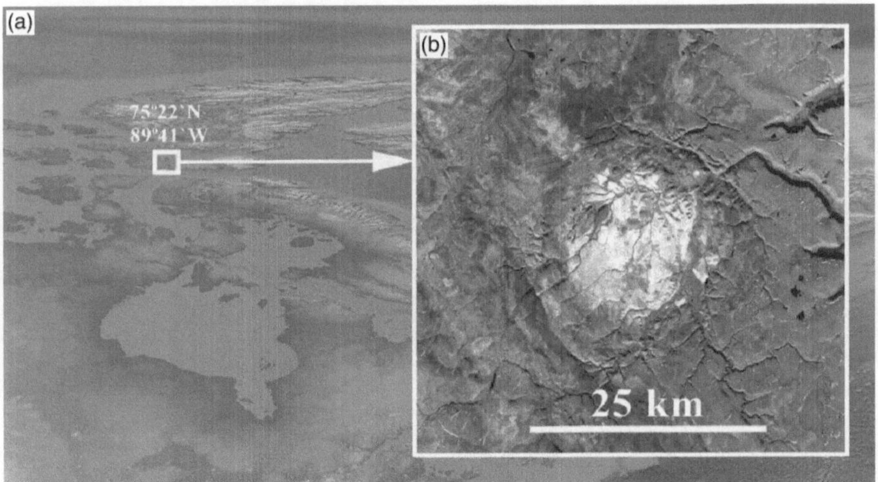

Figure 2.8. (a) The geographic position of Haughton Crater on Devon Island in northern Canada. (b) Landsat view of the crater.

Recent field studies within Haughton crater (Lee *et al.*, 1998; Sharpton *et al.*, 1998) show several characteristics which make the crater a particularly suitable terrestrial analog for similarly sized Martian craters in terms of formation and subsequent modification. These characteristics include its location in the Arctic permafrost zone with an arid, cool climate (similar to modern climate conditions on Mars), impact-induced hydrothermal activity (Osinski *et al.*, 2001), and distinctive surface features associated with periglacial processes. The general terrain morphology in the area of the crater is reminiscent of terrain softening seen at mid-latitudes on Mars, suggesting viscous creep of ice-rich regolith. Moreover, studies of the drainage system developed within the crater are very useful for understanding fluvial processes that have operated on Mars and which produced modifications seen in many Martian craters (Figure 2.9). Paleo-lacustrian records found in Haughton offer a possible analog to paleolake sediments which formed in many older Martian craters (Lee *et al.*, 1998). The high level of scientific interest in Haughton crater is related to its wide suite of periglacial, fluvial,

Figure 2.9. Gullies seen along rim of Martian crater near 39° S 159° E. Similar features in Haughton may provide important insights on the mechanism(s) producing such gullies. (MOC Release image MOC2-388)

mass-wasting, and paleo-lacustrian features and processes, which provide an important analog to the variety of processes and features which have affected Martian impact craters.

Popigai crater, Siberia, Russia

Popigai Crater is located on the northern part of the Middle Siberian plateau (71°39' N; 111°11' E) in the basin of the Popigai River, a tributary of the Khatanga River (Figure 2.10). Field investigations and analysis of existing geologic data convincingly show that the ~100 km diameter Popigai depression is one of Earth's largest impact craters (Masaitis *et al.*, 1975).

Popigai formed ~35.7 Myr ago (Bottomley *et al.*, 1997) and is now one of the best-preserved complex craters on Earth. Target rocks consist of two layers: the lower layer is dense Achaean gneiss with a thickness of about 15 km, while the upper layer consists of less dense Proterozoic, Paleozoic, and

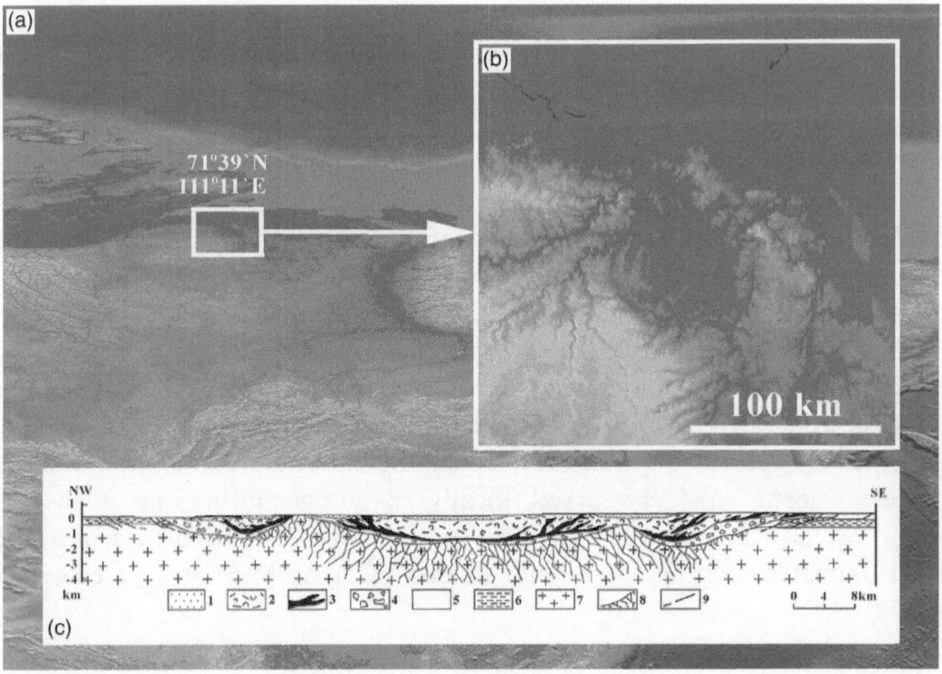

Figure 2.10. (a) The geographic position of Popigai Crater in the Middle Siberian plateau. (b) Shaded topographic map of Popigai. (c) Cross-section of crater (from Masaitis *et al.*, 1980). Legend: 1, coptoclastite; 2, suevite; 3, tagamite; 4, allogenic breccia; 5, Paleozoic and Mesozoic sedimentary, volcanic-sedimentary, and igneous rocks; 6, Upper Proterozoic sedimentary rocks; 7, Archean crystalline rocks; 8, authigenic breccias; 9, ruptured disturbances (thrusts).

Mesozoic sedimentary rocks about 1.2 km thick (Masaitis *et al.*, 1975). The crater has a circular-shaped negative gravity anomaly which stands out from a general background of linear positive gravity anomalies associated with crystalline foundation rocks. A negative magnetic field within the Popigai depression has an isometric shape, which contrasts with the NW-trending positive and negative anomalies caused by the composition and structure of surrounding crystalline foundation rocks.

Primary structural features of Popigai are a 75 km diameter circular central depression with a distinct 45 km diameter peak ring surrounded by a 60 km diameter annular trough (Figure 2.10). The central depression cuts into rocks of the crystalline foundation and is partially filled by a complicated 2 km thick layer of impact melt and allogenic breccias. The floor of the depression is 300–400 m lower than the surrounding plateau. Geophysical data suggest a low central uplift (10–15 km diameter) in the main depression (Masaitis *et al.*, 1980). This central uplift consists of authigenic breccias covered by a thick (< 2 km) layer of suevite derived from Pliocene-Quaternary sediments filling the crater.

Material covering the crater floor consists of two layers. The bottom layer (< 1.5 km thick) contains blocky allogenic breccias which are cemented by impact melt in places. These breccias are overlain by a stratum of impactites < 600 m thick. Impact diamonds, produced from graphite in the target gneiss, are common in the impact-melt rocks (Masaitis, 1998). The crater center is covered by suevite up to 1 km thick. Masaitis *et al.* (1975) estimate that the rocks were subjected to a pressure of more than 600 kbar and a temperature ~2000 °C. The total volume of preserved impact melted products is about 1750 km^3 (Masaitis *et al.*, 1980).

The entire crater is surrounded by a 100 km diameter external moat, similar to moats seen in MOLA profiles of Martian impact craters. The moat occurs in intensively deformed sedimentary rocks. Beyond the moat is the ~150 km diameter ejecta zone, preserved locally as allogenic breccias (Masaitis *et al.*, 1980; Masaitis, 1999). The ratio of the diameter of visible ejecta to the crater diameter is about 1.6, similar to that of many Martian layered ejecta craters.

Puchezh-Katunk crater, Russia

Puchezh-Katunk is one of the few terrestrial impact structures whose ejecta may be a very close analog for layered ejecta craters on Mars. Puchezh-Katunk (56°58' N; 43°43' E) has a rim diameter of about 40 km and its ejecta blanket extends to 80 km. The crater is ~175 Myr old (Masaitis *et al.*, 1996). Its interior structure indicates that Puchezh-Katunk is a typical

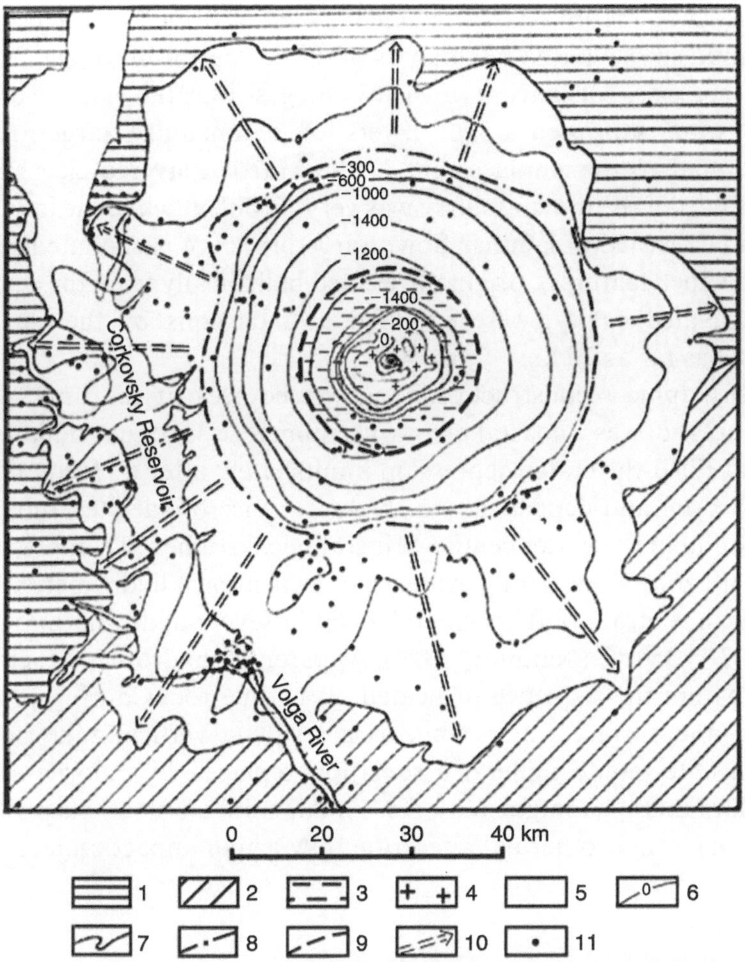

Figure 2.11. Geologic-structural sketch map of Puchezh-Katunki impact crater (from Mashchak and Naumov, 1999). The external lobate-like edge of the allogenic breccia is easily noted. Legend: 1, Lower Triassic rocks; 2, Upper Permian rocks; 3, Vendian rocks; 4, Archean rocks; 5, undivided unit of Permian, Carbonic, Devonian, Vendian, and Archean rocks; 6, surface equal area lines; 7, external boundary of the allogenic breccia distribution; 8, external crater moat; 9, axis of the circular trench; 10, axes of the radial trenches; 11, location of drilling holes.

complex crater (Figure 2.11). The central uplift (12–14 km basal diameter and 2–2.5 km high) of crystalline basement is surrounded by a circular trench and has a distinctive pit on the top, similar to the summit pits seen on central peaks in some Martian craters (Masaitis, 1999; Mashchak and Naumov, 1999). The crater's ejecta blanket is overlain by a sheet of variegated allogenic breccias (up to 200 m thick). The structural and textural characteristics of the breccia sheet suggest that it formed through the radial flow of a mud slurry. Based on analysis of the structural-lithologic features of the ejecta breccias, Masaitis *et al.* (1993) suggest that the ejecta was derived from the water-saturated upper layers of sedimentary target materials, perhaps similar to the emplacement of the Martian layered ejecta deposits. The upper part of the moving slurry was very turbulent while the lower part of the column consisted of laminar flow. Large blocks of carbonate rocks (with significant shock features), originally ejected ballistically from the central part of the excavated cavity, were engulfed as inclusions by the slurries and transported as far as 25 km.

Paleogeographic reconstruction suggests the area at the time of impact (Middle Jurassic) was a coastal plain with numerous lagoons and lakes. Thus, waters that filled the crater depression immediately after its formation could have washed in, and deposited, the thick sequence of suevites and polymict breccias found in the crater center. Mineralogical studies of samples from the central uplift zone show strong evidence of post-impact hydrothermal activity during which central uplift rocks underwent a severe hydrothermal transformation at $T \leq 350$ °C (Naumov, 1999). Apparently, hydrothermal circulation occurred within the sequence of heated crystalline rocks during percolation of ground and surface waters and, as a result, a vertical zone now exists with a smectite-zeolite mineral association replaced by chlorite-anhydrite. Future Mars missions might look for similar mineral assemblages to determine the role of hydrothermal alteration in Martian impact craters.

2.4 Discussion

Studies of terrestrial impact structures provide important insights into the fundamental mechanisms affecting formation and modification of Martian impact craters. Topographic and high-resolution multispectral data are providing new details about the morphologies and morphometries of Martian impact craters. These new data, combined with continuing studies of terrestrial impact structures and advanced computer models, are enhancing our understanding of the environments in which these craters formed. This multidisciplinary approach to studying Martian impact craters promises to resolve many of the remaining questions about these features, such as the mechanism responsible for formation of layered ejecta morphologies, what morphometric variations imply about target characteristics, and the concentrations and physical states of subsurface volatiles which affect observed crater features.

Acknowledgments

The authors thank John McHone and Horton Newsom for their many constructive comments which improved the manuscript.

References

Bandfield, J. L., Hamilton, V. E., and Christensen, P. R. (2000). A global view of Martian surface compositions from MGS-TES. *Science*, **287**, 1626–30.

Barlow, N. G. (1988). Crater size-frequency distributions and a revised Martian relative chronology. *Icarus*, **75**, 285–305.

Barlow, N. G. and Bradley, T. L. (1990). Martian impact craters: correlations of ejecta and interior morphologies with diameter, latitude, and terrain. *Icarus*, **87**, 156–79.

Barlow, N. G. and Perez, C. B. (2003). Martian impact crater ejecta morphologies as indicators of the distribution of subsurface volatiles. *Journal of Geophysical Research*, **108**, 5085, doi: 10.1029/2002JE002036.

Barlow, N. G., Koroshetz, J., and Dohm, J. M. (2001). Variations in the onset diameter for Martian layered ejecta morphologies and their implications for subsurface volatile reservoirs. *Geophysical Research Letters*, **28**, 3095–8.

Barlow, N. G., Boyce, J. M., Costard, F. M., Craddock, R. A., Garvin, J. B., Sakimoto, S. E. H. *et al.* (2000). Standardizing the nomenclature of Martian impact crater ejecta morphologies. *Journal of Geophysical Research*, **105**, 26733–8.

Barnouin-Jha, O. S., Schultz, P. H., and Lever, J. H. (1999a). Investigating the interactions between an atmosphere and an ejecta curtain. 1. Wind tunnel tests. *Journal of Geophysical Research*, **104**, 27105–15.

Barnouin-Jha, O. S., Schultz, P. H., and Lever, J. H. (1999b). Investigating the interactions between an atmosphere and an ejecta curtain. 2. Numerical experiments. *Journal of Geophysical Research*, **104**, 27117–31.

Bottomley, R., Grieve, R., York, D., and Masaitis, V. (1997). The age of the Popigai impact event and its relation to events in the stratigraphic column. *Nature*, **388**, 365–8.

Boyce, J. M., Roddy, D. J., Soderblom, L. A., and Hare, T. (2000). Global distribution of on-set diameters of rampart ejecta craters on Mars: their implications to the history of Martian water. *Abstracts of Papers Submitted to the 31st Lunar and Planetary Science Conference*. Houston: Lunar and Planetary Institute, CD 31, Abstract 1167.

Carr, M. H. (1996). *Water on Mars*. New York: Oxford University Press.

Carr, M. H., Crumpler, L. S., Cutts, J. A., Greeley, R., Guest, J. E., and Masursky, H. (1977). Martian impact craters and emplacement of ejecta by surface flow. *Journal of Geophysical Research*, **82**, 4055–65.

Chao, E. C. (1977). The Ries Crater of southern Germany, a model for large basins on planetary surface. *Geologisches Jahrbuch*, **44**, 3–81.

Costard, F. M. (1989). The spatial distribution of volatiles in the Martian hydrolithosphere. *Earth Moon and Planets*, **45**, 265–90.

Craddock, R. A. and Howard, A. D. (2002). The case for rainfall on a warm, wet early Mars. *Journal of Geophysical Research*, **107**, 10.1029/2001JE001505.

Dietz, R. S. (1959). Shatter cones in cryptoexplosion structures (meteorite impact?). *Journal of Geology*, **67**, 496–505.

Feldman, W. C., Boynton, W. V., Tokar, R. L. *et al.* (2002). Global distribution of neutrons from Mars: results from Mars Odyssey. *Science*, **297**, 75–8.
French, B. M. (1998). *Traces of Catastrophe: A Handbook of Shock-Metamorphic Effects in Terrestrial Meteorite Impact Structures*, Lunar and Planetary Science Contribution 954. Houston: Lunar and Planetary Institute.
Garvin, J. B., Sakimoto, S. E. H., Frawley, J. J., and Schnetzler, C. (2000). North polar region craterforms on Mars: geometric characteristics from the Mars Orbiter Laser Altimeter. *Icarus*, **144**, 329–52.
Garvin, J. B., Sakimoto, S. E. H., Frawley, J. J., and Schnetzler, C. (2002). Global geometric properties of Martian impact craters. *Abstracts of Papers Submitted to the 33rd Lunar and Planetary Science Conference*. Houston: Lunar and Planetary Institute, CD 33, Abstract 1255.
Gault, D. E. and Wedekind, J. A. (1979). Experimental studies of oblique impact. *Proceedings of the 9th Lunar and Planetary Science Conference*. Houston: Lunar and Planetary Institute, pp. 3843–75.
Gault, D. E., Quaide, W. L., and Oberbeck, V. R. (1968). Impact cratering mechanics and structures. In *Shock Metamorphism of Natural Materials*, ed. B. M. French and N. M. Short. Baltimore, MD: Mono Book, pp. 87–99.
Grant, J. A. and Schultz, P. H. (1993a). Degradation of selected terrestrial and Martian impact craters. *Journal of Geophysical Research*, **98**, 11025–42.
Grant, J. A. and Schultz, P. H. (1993b). Erosion of ejecta at Meteor Crater, Arizona. *Journal of Geophysical Research*, **98**, 15033–47.
Greeley, R., Fink, J., Gault, D. E. *et al.* (1980). Impact cratering in viscous targets: laboratory experiments. *Proceedings of the 11th Lunar and Planetary Science Conference*. Houston: Lunar and Planetary Institute, pp. 2075–97.
Grieve, R. A. F. (1988). The Haughton impact structure: summary and synthesis of the results of the HISS project. *Meteoritics*, **23**, 249–54.
Hörz, F., Ostertag, R., and Rainey, D. A. (1983). Bunte Breccia of the Ries: continuous deposits of large impact craters. *Reviews of Geophysics and Space Physics*, **21**, 1167–725.
Ivanov, B. A. (1999). Numerical modeling of cratering (in Russian). In *Deep Drilling in the Puchezh-Katunki Impact Structure*, ed. V. L. Masaitis and L. A. Pevzner. Special Publication of the VSEGEI Press, pp. 276–86.
Jankowski, D. G. and Squyres, S. W. (1992). The topography of impact craters in "softened" terrain on Mars. *Icarus*, **100**, 26–39.
Jessberger, E. K. (1988). ^{40}Ar-^{39}Ar dating of the Haughton impact structure. *Meteoritics*, **23**, 233–4.
Kuzmin, R. O., Bobina, N. N., Zabalueva, E. V., and Shashkina, V. P. (1988). Structural inhomogeneities of the Martian cryosphere. *Solar System Research*, **22**, 121–33.
Lee, P., Bunch, T. E., Cabrol, N. *et al.* (1998). Haughton-Mars 97 – I: overview of observations at Haughton Impact Crater, a unique Mars analog site in the Canadian High Arctic. *Abstracts of Papers Submitted to the 29th Lunar and Planetary Science Conference*. Houston: Lunar and Planetary Institute, CD 29, Abstract 1973.
Malin, M. C. and Edgett, K. S. (2000). Sedimentary rocks of early Mars. *Science*, **290**, 1927–37.

Masaitis, V. L. (1998). Popigai crater: origin and distribution of diamond-bearing impactites. *Meteoritics & Planetary Science*, **33**, 349–59.

Masaitis, V. L. (1999). Impact structures of northeastern Eurasia: the territories of Russia and adjacent countries. *Meteoritics & Planetary Science*, **34**, 691–711.

Masaitis, V. L., Mikhailov, M. V., and Selivanovskaya, T. V. (1975). *The Popigai Meteorite Crater* (in Russian). Moscow: Nauka Press.

Masaitis, V. L., Naumov, M. V., and Ivanov, B. A. (1993). Fluidized ejecta of Puchezh-Katunki impact crater: possible implications to rampart craters of Mars. *Vernadsky Institute-Brown University Microsymposium*, **18**, 45–56.

Masaitis, V. L., Mashchak, M. S., and Naumov, M. V. (1996). Puchezh-Katunki astrobleme: structural model of the gigantic impact crater (in Russian). *Solar System Research*, **30**, 5–13.

Masaitis, V. L., Danilin, V. N., Mashchak, M. S. *et al.* (1980). *The Geology of Astroblemes* (in Russian). Leningrad: Nedra Press.

Mashchak, M. C. and Naumov, M. B. (1999). Morphology and interior structure of the Puchezh-Katunki impact structure (in Russian). In *Deep Drilling in the Puchezh-Katunki Impact Structure*, ed. V. L. Masaitis and L. A. Pevzner. Special Publication of the VSEGEI Press, pp. 187–99.

Melosh, H. J. (1989). *Impact Cratering: A Geologic Process*. New York: Oxford University Press.

Melosh, H. J. and McKinnon, W. B. (1978). The mechanics of ringed basin formation. *Geophysical Research Letters*, **5**, 985–8.

Melosh, H. J. and Vickery, A. M. (1989). Impact erosion of the primordial atmosphere of Mars. *Nature*, **338**, 487–9.

Mitrofanov, I. G., Anfimov, D., Kozyrev, A. *et al.* (2002). Maps of subsurface hydrogen from the High Energy Neutron Detector, Mars. *Science*, **297**, 78–81.

Mouginis-Mark, P. (1979). Martian fluidized crater morphology: variations with crater size, latitude, altitude, and target material. *Journal of Geophysical Research*, **84**, 8011–22.

Mouginis-Mark, P. J. (1987). Water or ice in the Martian regolith? Clues from rampart craters seen at very high resolution. *Icarus*, **71**, 268–86.

Naumov, M. V. (1999). Hydrothermal-metasomatic mineral formation (in Russian). In *Deep Drilling in the Puchezh-Kutunki Impact Structure*, ed. V. L. Masaitis and L. A. Pevzner. Special Publication of the VSEGEI Press, pp. 276–86.

Newsom, H. E., Graup, G., Sewards, T., and Keil, K. (1986). Fluidization and hydrothermal alteration of the suevite deposit at the Ries Crater, West Germany, and implications for Mars. *Journal of Geophysical Research*, **91**, E239–51.

Oberbeck, V. R. (1975). The role of ballistic erosion and sedimentation in lunar stratigraphy. *Reviews of Geophysics & Space Physics*, **13**, 337–62.

Oberbeck, V. R. (1977). Applications of high explosion cratering data to planetary problems. In *Impact and Explosion Cratering*, ed. D. J. Roddy, R. O. Pepin, and R. B. Merrill. New York: Pergamon Press, pp. 45–65.

Osinski, G. R., Spray, J. G., and Lee, P. (2001). Impact-induced hydrothermal activity within the Haughton impact structure, arctic Canada: generation of a transient, warm, wet oasis. *Meteoritics & Planetary Science*, **36**, 731–45.

Pohl, J., Stöffler, D., Gall, H., and Ernstson, K. (1977). The Ries impact crater. In *Impact and Explosion Cratering*, ed. D. J. Roddy, R. O. Pepin, and R. B. Merrill. New York: Pergamon Press, pp. 343–404.

Robertson, P. B. and Grieve, R. A. F. (1978). The Haughton impact structure. *Meteoritics*, **13**, 615–19.

Robertson, P. B. and Sweeney, J. F. (1983). Haughton impact structure: structural and morphological aspects. *Canadian Journal of Earth Science*, **20**, 1134–51.

Roddy, D. J. and Davis, L. K. (1977). Shatter cones formed in large-scale experimental explosion craters. In *Impact and Explosion Cratering*, ed. D. J. Roddy, R. O. Pepin, and R. B. Merrill. New York: Pergamon Press, pp. 715–50.

Schultz, P. H. and Gault, D. E. (1979). Atmospheric effects on Martian ejecta emplacement. *Journal of Geophysical Research*, **84**, 7669–87.

Sharpton, V. L., Dressler, B. O., and Sharpton, T. J. (1998). Mapping the Haughton Impact Crater, Devon Island, NWT: implications for the shape and size of the excavation cavity. *Abstracts of Papers Submitted to the 29th Lunar and Planetary Science Conference*. Houston: Lunar and Planetary Institute, CD 29, Abstract 1867.

Shoemaker, E. M. (1963). Impact mechanics at Meteor Crater, Arizona. In *The Moon, Meteorites, and Comets*, ed. B. M. Middlehurst and G. P. Kuiper, Chicago: University of Chicago Press, pp. 301–36.

Stewart, S. T., O'Keefe, J. D., and Ahrens, T. J. (2001). The relationship between rampart crater morphologies and the amount of subsurface ice. *Abstracts of Papers Submitted to the 32nd Lunar and Planetary Science Conference*. Houston: Lunar and Planetary Institute, CD 32, Abstract 2092.

Stöffler, D. (1972). Deformation and transformation of rock-forming minerals by natural and experimental shock processes: I. Behavior of minerals under shock compression. *Fortschritte der Minerae*, **49**, 50–113.

Strom, R. G., Croft, S. K., and Barlow, N. G. (1992). The Martian impact cratering record. In *Mars*, ed. H. H. Kieffer, B. M. Jakosky, C. W. Snyder, and M. S. Matthews. Tucson: University of Arizona Press, pp. 383–423.

Sutton, S. R. (1985). Thermoluminescence measurements on shock-metamorphosed sandstone and dolomite from Meteor Crater, Arizona: 2. Thermoluminescence age of Meteor Crater. *Journal of Geophysical Research*, **90**, 3690–700.

von Engelhardt, W. (1990). Distribution, petrography and shock metamorphism of the ejecta of the Ries crater in Germany – a review. *Tectonophysics*, **171**, 259–73.

von Engelhardt, W., Arndt, J., Fecker, B., and Pankau, H. G. (1995). Suevite breccia from the Ries crater, Germany: origin, cooling history and devitrification of impact glasses: *Meteoritics & Planetary Science*, **30**, 279–93.

Wood, C. A., Head, J. W., and Cintala, M. J. (1978). Interior morphology of fresh Martian craters: the effects of target characteristics. *Proceedings of the 9th Lunar and Planetary Science Conference*. Houston: Lunar and Planetary Institute, pp. 3691–709.

3

Terrestrial analogs to the calderas of the Tharsis volcanoes on Mars

Peter J. Mouginis-Mark, Andrew J. L. Harris and Scott K. Rowland
Hawaii Institute of Geophysics and Planetology
University of Hawaii at Manoa

3.1 Introduction

The structure and morphology of Martian calderas have been well studied through analysis of the Viking Orbiter images (e.g., Mouginis-Mark, 1981; Wood, 1984; Mouginis-Mark and Robinson, 1992; Crumpler *et al.*, 1996), and provide important information on the evolution and eruptive styles of the parent volcanoes. Using Viking data it has been possible, for numerous calderas, to define the sequence of collapse events, identify locations of intra-caldera activity, and recognize post-eruption deformation for several calderas. Inferences about the geometry and depth of the magma chamber and intrusions beneath the summit of the volcano can also be made from image data (Zuber and Mouginis-Mark, 1992; Scott and Wilson, 1999). In at least one case, Olympus Mons, analysis of compressional and extensional features indicates that, when active, the magma chamber was located within the edifice (i.e., at an elevation above the surrounding terrain). The summit areas of Olympus and Ascraeus Montes provide evidence of a dynamic history, with deep calderas showing signs of having been full at one time to the point that lava flows spilled over the caldera rim (Mouginis-Mark, 1981). Similarly, shallow calderas contain evidence that they were once deeper (e.g., the western caldera of Alba Patera; Crumpler *et al.*, 1996). Some of the best evidence for circumferential vents on Mars can be found on Pavonis Mons, where several sinuous rilles can be identified that must have originated from vents close to the rim (Zimbelman and Edgett, 1992).

Mouginis-Mark and Rowland (2001) reviewed the geomorphic information for Martian calderas that can be determined from the Viking Orbiter data.

The Geology of Mars: Evidence from Earth-based Analogs, ed. Mary Chapman. Published by Cambridge University Press.

They made certain predictions about what might be learnt from analysis of higher spatial resolution data provided by the Mars Orbiter Camera (MOC) and the visible camera that is part of the Thermal Emission Imaging System (THEMIS), as well as the analysis of digital topographic data from the Mars Orbiter Laser Altimeter (MOLA). MOC images have a spatial resolution of ~1.5 to 6.0 m/pixel, and THEMIS visible wavelength images have a resolution of 19 m/pixel; in both instances, this coverage is much better than the 40–200 m/pixel data that were typically obtained by the Viking Orbiters. As this chapter will show, these data do indeed provide a wealth of new information on the structure and evolution of calderas on Mars, and importantly this new information makes direct comparison between calderas on Earth and Mars more instructive. Nowhere is this opportunity for utilizing terrestrial analogs greater than with the four giant shield volcanoes in the Tharsis region of Mars: Olympus, Arsia, Pavonis, and Ascraeus Montes (Figure 3.1). This chapter concentrates on these four volcanoes and identifies clear differences in the subsidence and/or infilling history of each. For each key caldera feature we have identified a terrestrial analog. Coupled with field observations of these analogs, MOC, MOLA, and THEMIS observations provide new insights into the subsurface structure and magma supply rate for each Martian volcano.

This comparative study also underscores the importance of continued study of equivalent caldera features and processes on Earth. Although only one large-scale caldera collapse event at a basaltic shield has occurred in the historic record (the 1968 collapse of Fernandina caldera, Galapagos; Simkin and Howard, 1970), fieldwork on terrestrial calderas continues to provide insight into these dynamic features that can be applied to comparable features on Mars. Few of the terrestrial examples have been studied with the goal of furthering an understanding of their planetary analogs, thereby opening the possibility of future productive analysis of calderas on basaltic volcanoes on Earth.

In each instance in the following discussion, attention is drawn to how particular terrestrial analogs are instructive for understanding the following attributes of Martian calderas:

(1) The occurrence and distribution of intra-caldera slumps and avalanches;
(2) The variability in the location and size of multiple collapse centers;
(3) The occurrence of intra-caldera benches;
(4) Broad-scale sagging of the summit;
(5) The magnitude and variability in the tilting of caldera floors;
(6) The location and magnitude of intra-caldera eruptions.

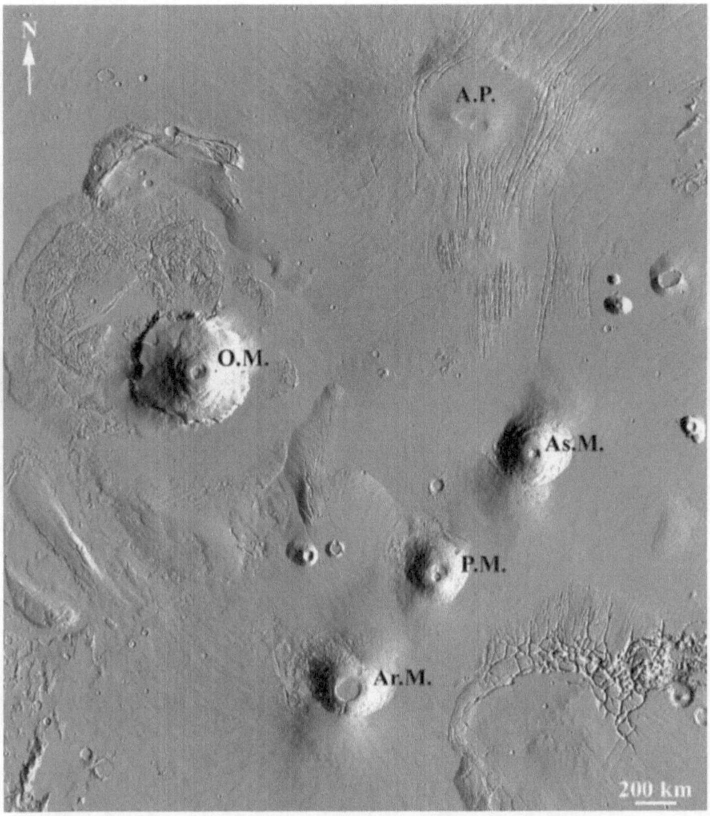

Figure 3.1. Location map for the Tharsis Montes. Image is a shaded relief version of the 128th degree MOLA digital elevation model. Field of view extends from 20° S to 50° N, and 210 to 270° E. The main volcanoes are indicated ("O.M." is Olympus Mons, "Ar.M." is Arsia Mons, "P.M." is Pavonis Mons, "As.M." is Ascraeus Mons, and "A.P." is Alba Patera).

This chapter concludes with a more detailed description of the caldera at Masaya volcano, Nicaragua. Here, multiple collapse events and ease of access allow for detailed study of numerous collapse and infilling events and the ways in which they interact. Many features at Masaya assist in understanding the dynamics of Martian caldera floors.

3.2 Observations of deformation and infilling on the Tharsis shields

3.2.1 Arsia Mons

The interior of the Arsia Mons caldera (Figure 3.2a) is nearly horizontal, varying in elevation by only ~300 m across the 110 km diameter caldera. First recognized in low-Sun angle Viking images (Carr *et al.*, 1977; Crumpler

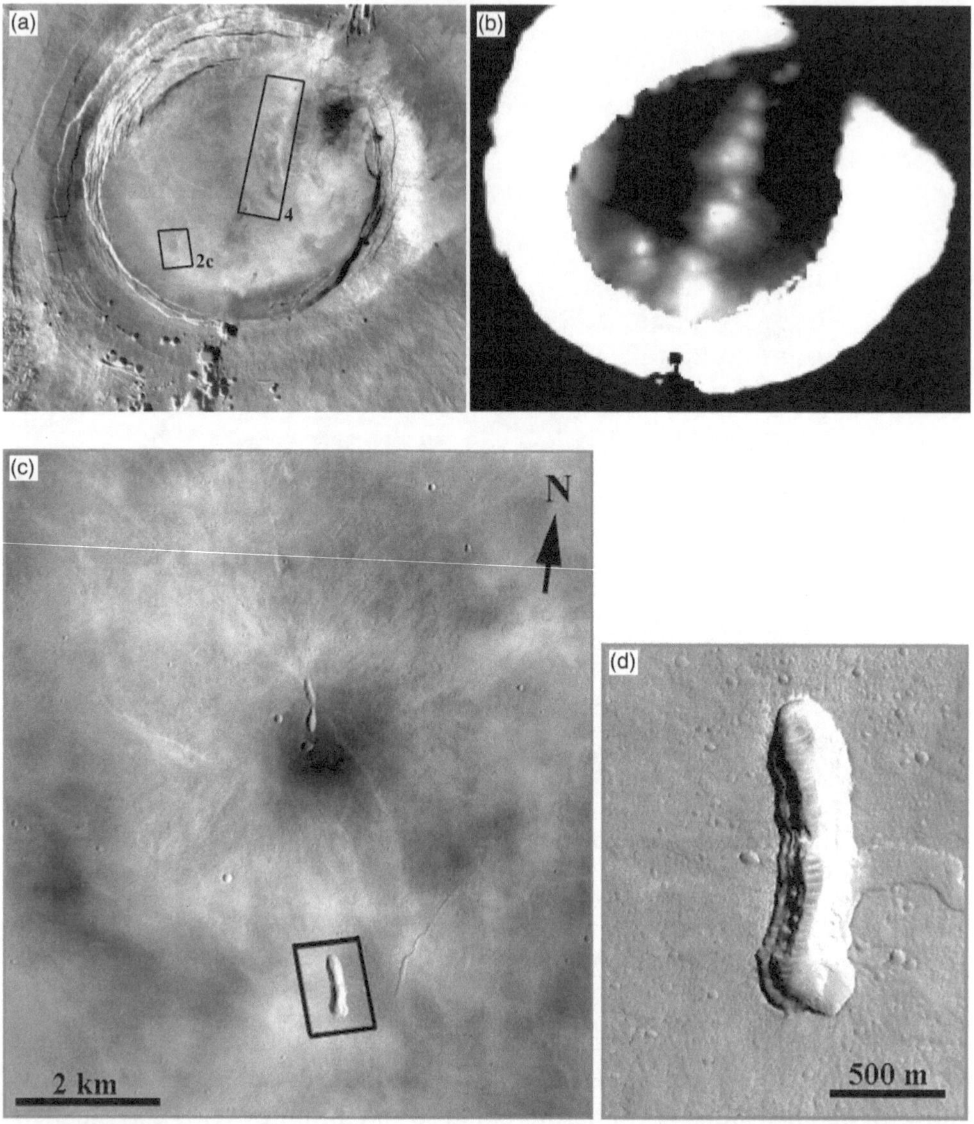

Figure 3.2. (a) Viking Orbiter image of Arsia Mons caldera. Boxes show the locations of Figure 3.2c and 3.4. (b) Image of the floor of Arsia Mons, derived by stretching the MOLA elevation data such that black is at 16 250 m and white is at 16 650 m. This rendition enhances a series of low hills extending roughly north–south across the floor. (c) THEMIS VIS image of a cone on the floor of Arsia Mons (see (a) for location). Box indicates the area of coverage of the MOC image in (d). THEMIS frame no. V01453002 (d) MOC image of an elongate fissure at the summit of a lava shield on the floor of Arsia Mons, see (c) for location. MOC frame number E1003391, image resolution is 6.6 m/pixel.

and Aubele, 1978), there is a series of low hills aligned in a general NNE–SSW orientation across the caldera floor (Figure 3.2b). Similar intra-caldera hills have not been found at any of the other large Martian volcanoes with the possible exception of Ulysses Patera (Plescia, 1994). MOLA data show that these hills are shield-like in morphology, with heights of ~200 m and basal diameters of ~20 km, yielding average flank slopes of well less than a tenth of a degree. THEMIS visible images confirm the interpretation of Carr *et al.* (1977) that these hills are small lava shields. Each has an elongate crater at the summit and a subtle radial texture consisting of lava flows, most of which are < 250 m wide (Figure 3.2c). MOC data reveal layers within the summit craters (Figure 3.2d). Using Viking Orbiter images, Mouginis-Mark (1981) identified a few individual lava flows on the NW inner wall and floor of the caldera. Subsequent analysis of MOC and THEMIS data show that flow margins are only visible on the floor immediately surrounding the lava shields, no other flows are visible either in the limited coverage provided by MOC, or the more extensive THEMIS coverage.

One of the best terrestrial analogs to this line of cones occurs on the floor of Mokuaweoweo caldera, Mauna Loa volcano, Hawaii. Here, a line of fractures runs across the floor of the caldera in an extension of the SW and NE rift zones. At the SW end of this intra-caldera fracture system are two cones that formed during the 1940 eruption (Macdonald 1954; Figure 3.3). These cones have low-angle slopes near their bases, steepen upward, and are considerably smaller than the intra-caldera shields of Arsia Mons. Flooding of the floor by 1984 lava (Lockwood *et al.*, 1987) has buried much of the lower,

Figure 3.3. Low lava shields capped by spatter cones on the floor of Mokuaweoweo caldera, Mauna Loa volcano, Hawaii. View is taken in 1985 from the eastern caldera rim, looking southwest towards the cone of 1940, with the western rim of the caldera in the background. Notice that the lower flanks of the cone have formed a low lava shield similar to the examples found on the floor of Arsia Mons.

gentler slopes of these cones, accentuating their overall steepness and making them less similar in appearance to the Arsia Mons cones. However, it is evident that the base of each 1940 cone is actually a low lava shield and the upper (steeper) portion is a cap of spatter. Unlike the Arsia Mons cones, the 1940 cones are fairly symmetric, but like Arsia Mons the 1940 cones lie along a fundamental structural lineament of the volcano. THEMIS data of the Arsia Mons caldera (Figures 3.2a and 3.4) show that the cones lie along a structural trend that includes a breach in the SW caldera wall through which voluminous lavas have issued and produced a lava fan on the SW flanks (Crumpler *et al.*, 1996) and a set of deep pits/fractures on the NE flank that are associated with a similar lava fan. Additionally, this generally SW–NE trend is the same as that along which Arsia, Pavonis, and Ascraeus Montes are distributed. It is interesting to note that although the intra-caldera Arsia Mons lava shields are located along this fundamental trend, the elongate summit craters are in an *en echelon* orientation, showing vergence to the east.

3.2.2 Pavonis Mons

The summit of Pavonis Mons contains an obvious, deep, roughly circular, 45–50 km diameter caldera, termed the "small caldera" by Crumpler *et al.* (1996). Inspection of MOLA topography shows that the northern wall of the small caldera is ~2700 m high, and the southern wall is ~4600 m high. The floor of the small caldera is remarkably featureless and horizontal, with a change in elevation across the floor of only ~40–50 m. Crumpler *et al.* (1996) interpreted the lack of features on the small caldera floor to indicate either a young age or mantling by dust. The walls of the small caldera are bimodal in the sense that the western wall is characterized by numerous slump blocks stepping from the western rim to the floor whereas the eastern wall is essentially a single drop. Crumpler *et al.* (1996) interpreted the western slumps as terraces but MOLA topography indicates that their top surfaces are not horizontal. These slumps appear to be the result of post-caldera mass-wasting and are very similar to those found on the northern wall of the Fernandina caldera (Galapagos; Rowland and Munro, 1992).

Additionally, a larger but shallower collapse feature, the large caldera, can be identified in Viking images, lying immediately to the north of the small caldera (Figure 3.5). The large caldera is also roughly circular with a diameter of 80–100 km. Unlike the deeper, small caldera it is much more structurally complex. The west, north, and most of the east boundaries of this larger collapse structure have a ridge-like morphology, down-dropped not only

Figure 3.4. THEMIS image of six lava shields (arrowed) aligned across the floor of Arsia Mons caldera. See Figure 3.2a for location. Note that the summit of each shield has an elongate crater, but that this elongation direction is not the same as the principal axis of the line of cones. THEMIS image V04399002.

inward towards the center but outward and downslope as well. To the southwest the margin is truncated by the small caldera. To the south the obvious ridge-like margin becomes indistinct and Crumpler *et al.* (1996) interpreted this to be due to burial by post-large-depression lava flows.

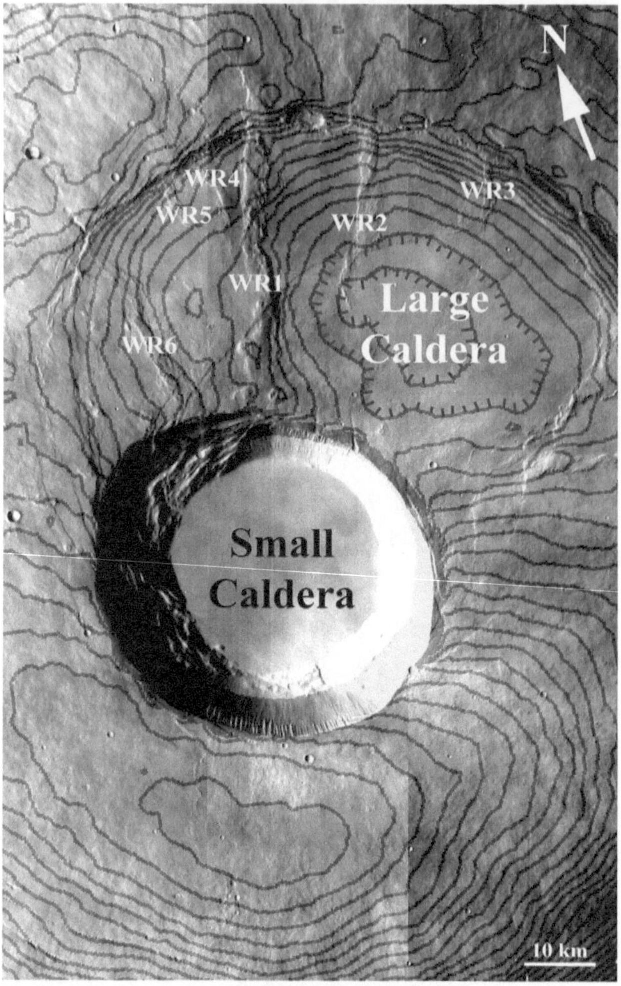

Figure 3.5. THEMIS mosaic of the Pavonis Mons caldera, showing the small and large calderas. Contours are derived from MOLA data, and are at 100 m intervals. The lowest point of the small caldera is 9230 m above the MOLA datum, and the high point to the south of the caldera rim is 14 036 m. THEMIS frame numbers I01639007, I02001008, and I02725005. WR1 to WR3 are parallel wrinkle ridges, WR4 to WR6 are almost radial to a point within the small caldera.

Numerous slightly arcuate faults occur in the south portion of the large caldera. They are roughly parallel to the western margin and down-dropped toward the center of the volcano. Most of these faults are truncated by the small caldera. At least four eruptive vents are located in the vicinity of these arcuate faults (Wood, 1979). The longest of these faults can be traced NE almost all the way across the large depression. Within the large caldera are three roughly parallel, NW-trending, wrinkle ridges. The longest arcuate fault cuts across the westernmost of three roughly parallel wrinkle ridges (see below) and appears to truncate the other two although it does not extend obviously to the easternmost one.

At least six wrinkle ridges can be identified within and just outside the large caldera (Figure 3.5). These were interpreted as radial structures by Crumpler *et al.* (1996); however, as noted above, three of them are parallel to one another and therefore cannot be radial. A couple of the other wrinkle ridges do have an orientation radial to a point in the NE part of the small caldera but if all of them are considered together they are not in a radial orientation. Instead, the three parallel wrinkle ridges lie along the same fundamental NW–SE structural trend seen at Arsia Mons and along which Arsia, Pavonis, and Ascraeus Montes are located. Two more of the wrinkle ridges are roughly parallel to this trend, leaving only one of the six that is not. That these wrinkle ridges are cut by intra-caldera faults and possibly also buried by lava flows indicates that they pre-date formation of the large caldera. Additionally, at least one and perhaps a second of the wrinkle ridges extends beyond the NE margin of the large caldera, making it unlikely that it is associated with intra-caldera compression. The small caldera truncates the large caldera's outer margin, the arcuate faults, and the wrinkle ridges, clearly making its formation the most recent large-scale tectonic event.

Neither the large nor the small caldera rim is at a constant elevation. Instead the highest elevation of the rim occurs SW of the small caldera. MOLA topography indicates that this southern part of the summit is the highest part of the volcano (~14 050 m), supporting the idea that this is an area where recent construction occurred. Such an offset was also noted at Olympus Mons (Mouginis-Mark and Robinson, 1992), and Walker (1988) observed that both Mauna Loa and Kilauea show a similar relationship. In both of these Hawaiian examples the calderas are offset to the SE from the topographic highest points on their respective volcanoes. This offset relationship may develop for any of three possible reasons. First it could indicate that construction by lava flows and pyroclastics was not favored directly above the magma chamber, and instead non-vertical eruptive dikes were most common. Second it might indicate migration of the magma chamber from an originally more central location. Finally, consider a spherical magma chamber centered within a right circular cone. The closest distance from the magma chamber wall to the surface (and hence the thinnest carapace of solid rock) will not be directly upwards but instead at an angle towards the side. It may be that when collapse stresses are accumulating around an evacuated or partially evacuated magma chamber it is these locations that are most likely to collapse.

The large caldera at Pavonis Mons shows clear evidence of centripetal sagging because its surface slopes inward. In fact, the center of the shallow caldera is almost a kilometer (~840 m) lower than the outer edge of the floor. This sagging has been noted in the exhumed Koolau caldera (Hawaii) by Walker (1988) and attributed to the gravitational sinking of a cumulate pile that develops within the core of the volcano.

3.2.3 Ascraeus Mons

The summit caldera complex of Ascraeus Mons consists of four large, coalesced collapse structures (Figure 3.6a), the deepest and most central of which is the most recent (Mouginis-Mark, 1981; Crumpler *et al.*, 1996).

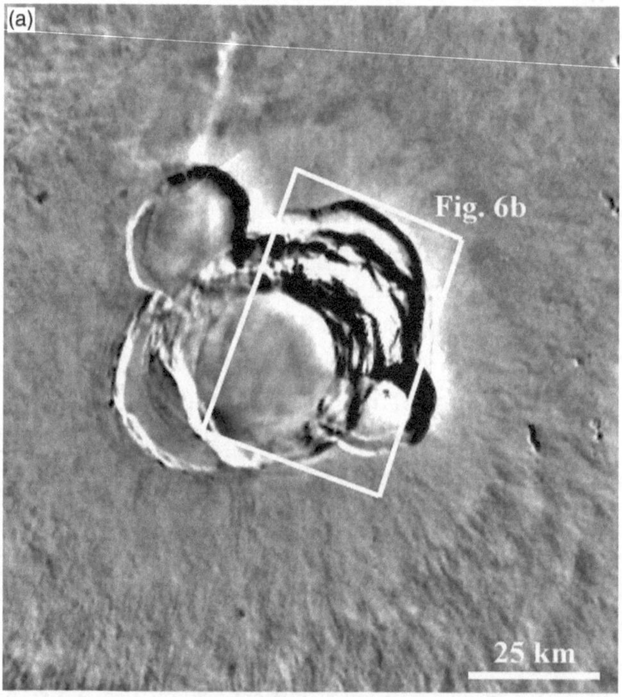

Figure 3.6. (a) View of the summit area of Ascraeus Mons, showing the four prominent collape pits that comprise the summit caldera. The location of (b) is also shown. Image part of the Viking digital image model compiled by the US Geological Survey. (b) Oblique view looking northward across the caldera of Ascraeus Mons. Image was derived by merging two THEMIS VIS images (frame numbers V01464013 and V01826008) with the 128th degree MOLA DEM. The main segment of the caldera is ~40 km in diameter, and the height difference of the main wall at left is ~3.4 km. Two different styles of collapse are evident in this image: (1) collapse as

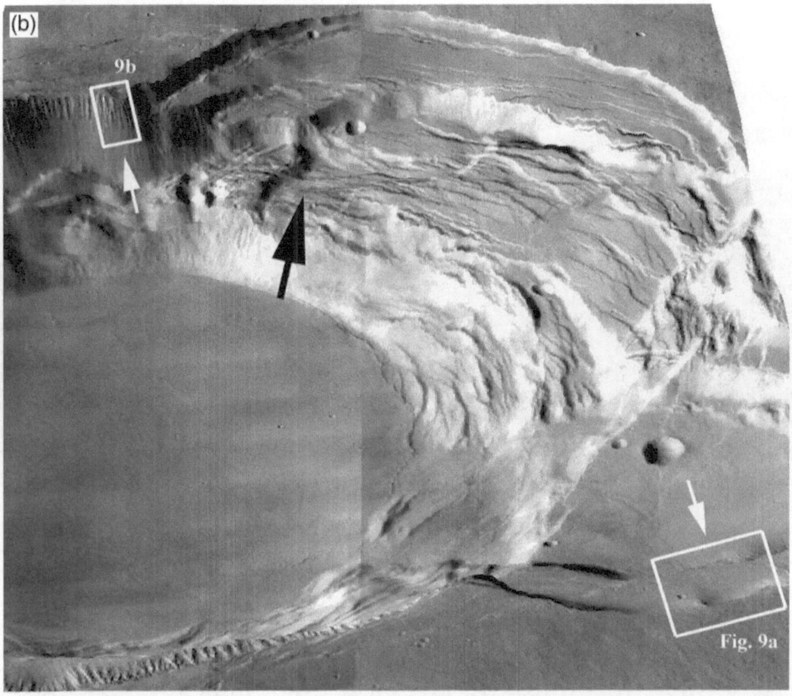

Figure 3.6. (cont.) individual pits and (2) progressive slumping as a series of narrow slump blocks. The large arrow indicates an area where both processes have occurred, with the slumping modifying the original pit. The locations of the MOC data shown in Figure 3.9a and 3.9b are indicated, with small arrows indicating the viewing directions. Vertical exaggeration is 8×.

The four older calderas are located roughly to the NE, SE, SW, and NW of the deepest central caldera. MOLA topography indicates that the floors of all four of these older calderas are at different elevations, indicating either that they were never continuous or that if they were, their order of most recent collapse is in an order from currently shallowest to currently deepest (NW, NE, SW, SE). The caldera of Karthala (Grand Comoros Islands) is similarly complex and Mouginis-Mark and Rowland (2001) presented an order of collapse feature formation based on a similar comparison of floor depth.

As noted above the central caldera is the deepest and most recent, based on the fact that it truncates the four more outer calderas. This central caldera is ~24 km across and ~3.4 km deep, and presents evidence of significant postformation mass wasting, on both large and small scales. Most notably, more than half of the NE caldera has slumped into the central caldera, producing a series of downward-stepping blocks separated by normal

Figure 3.7. Air view of block faulting of the northern wall of Kilauea volcano, Hawaii, beneath Volcano House. View is towards the NNW. The height of the wall at this part of the caldera is ~120 m.

faults and deep fractures (Figure 3.6b). A very similar set of down-dropped blocks occurs in the NE portion of Kilauea caldera (Figure 3.7). In the Kilauea case, age relationships are clear and indicate that the current caldera floor has ponded against and partially buried the slumped blocks. Clearly this means that the caldera was at one time deeper and this correlates with the earliest written descriptions of the Kilauea caldera, which can be interpreted to suggest a caldera at least twice as deep as at present (Macdonald and Abbot, 1970, pp. 69–77).

At the base of the lowermost of these NE slump blocks at Ascraeus Mons, a narrow ridge has formed in the material filling the central caldera. This has the form of a compressional feature and it may indicate that some movement of the slump blocks occurred after the most recent central-caldera resurfacing event. In contrast, the remaining walls of the innermost caldera, both where they truncate the NW, SW, and SE calderas, as well as where they truncate the entire edifice, consist of much steeper and more continuous drops. Slump blocks do occur at these locations but their sizes relative to the caldera walls are small. These smaller slump blocks are also similar to those comprising much of the northern wall of the Fernandina caldera (Rowland and Munro, 1992).

These different styles of post-central-caldera mass wasting are probably due to different mechanical strengths of the wall material as well as to the geometry of the youngest caldera wall relative to the older calderas that it truncates. The central caldera only truncates a small portion of the NW caldera, for example, and if the NW caldera walls angle inward

Figure 3.8. Multiple benches within the caldera of Fernandina volcano, Galapagos, can be seen in this view from the caldera rim, taken in 1989. The height of the wall is ~900 m.

(which is likely) they will act against the tendency for the NW caldera fill material to slump en masse into the central caldera. A similar argument can be made for the SE caldera although the youngest caldera truncates a slightly larger percentage of it. A terrestrial example of this situation occurs at the SE end of the Fernandina caldera (Figure 3.8). Here a large, partially infilled, pre-caldera pit crater was truncated by caldera formation. That the pit crater existed prior to caldera formation and is not a down-dropped block can be seen by the lack of correlation between layers exposed in the pit-crater fill and in its back wall (Rowland and Munro, 1992). In this case caldera formation truncated slightly more than half of the pit crater and apparently sufficient support is provided by the remaining pit crater walls so that the non-truncated fill material has not failed and collapsed into the caldera.

On the other hand, the youngest caldera truncates more than half of both the NE and SW calderas. Their remaining caldera walls probably dip in the same direction normal to the unbuttressed material exposed by collapse of the youngest caldera. They are therefore much more likely to collapse into the youngest caldera. It is not immediately clear why this has occurred

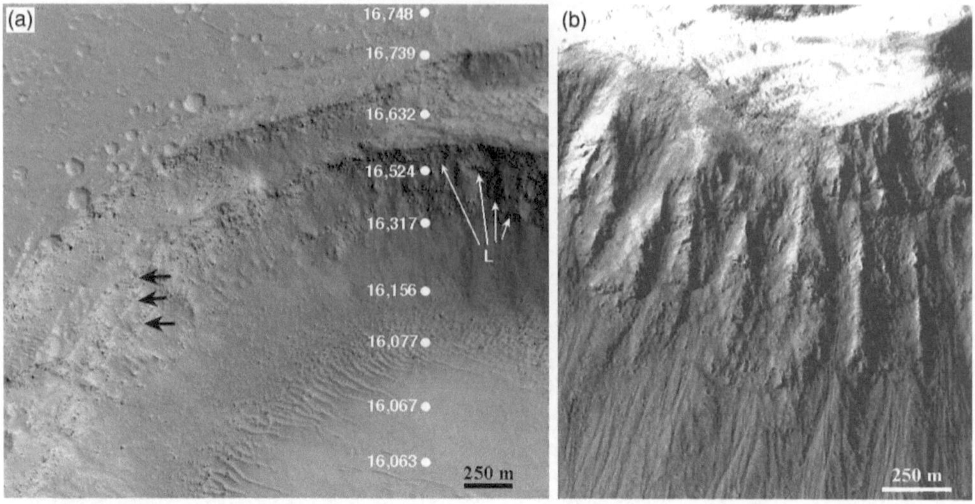

Figure 3.9. At left: (a) Portion of the SE caldera wall of Ascraeus Mons, see Figure 3.6b for location. Notice also that the layering ("L") in the lower wall continues around to the left of this image (arrows). The location of selected individual MOLA elevation measurements is shown (elevations in meters). MOC image number M080521, 4.40 m/pixel. At right: (b) Portion of the northern rim of Ascraeus Mons caldera. The field of view includes approximately the top kilometer of the section. See Figure 3.6b for location. Part of MOC image FHA-00952, 2.95 m/pixel.

at the NE caldera and not the NW. A MOC image of the southwestern caldera rim shows multiple layers within the wall (Figure 3.9a). Four prominent layers can be identified within a 315 m high section of the crater wall, which has a total height of ~660 m. The average thickness is therefore ~80 m. These layers can be seen to continue in the eastern wall of the crater, indicating that the units are contiguous on a horizontal distance of several kilometers; our preferred interpretation is that they are ponded lava flows. In contrast, the northern wall (Figure 3.9b) plunges almost directly to the caldera floor, and has a height of ~3375 m. This wall shows no similar laterally extensive layering, and is instead characterized by spur-and-gully morphology to a level ~1 km below the rim, followed by talus and then larger slump blocks at the floor.

3.2.4 Olympus Mons

Mouginis-Mark (1981) showed that there are six major collapse structures comprising the 60 × 80 km diameter caldera complex of Olympus Mons (Figure 3.10). Viking Orbiter images first indicated that the central floor

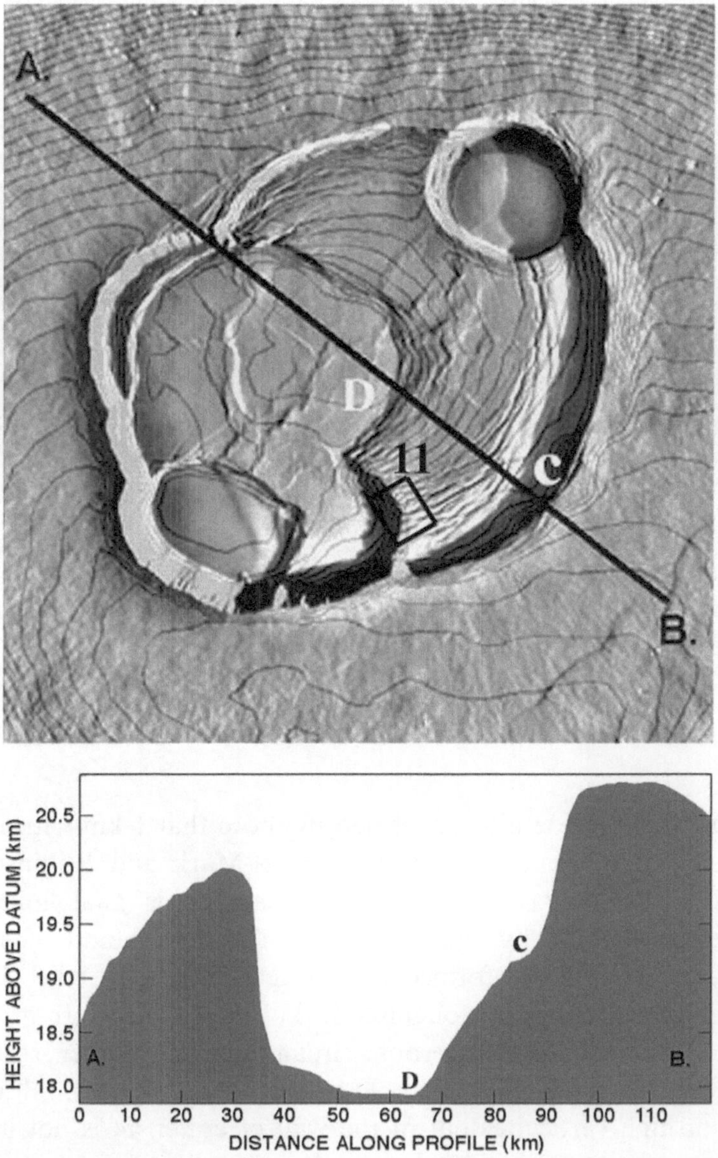

Figure 3.10. Image and contour map of Olympus Mons caldera. Topographic profile oriented approximately NW–SE across Olympus Mons caldera, derived from gridded MOLA data. Contour interval is 100 m. Location of the profile is between points A and B. Vertical exaggeration is 20×. The prominent down-sag on the floor is evident here, with the same unit extending from foot of the caldera at point C to the center of the caldera at point D. Small box indicates the location of Figure 3.11.

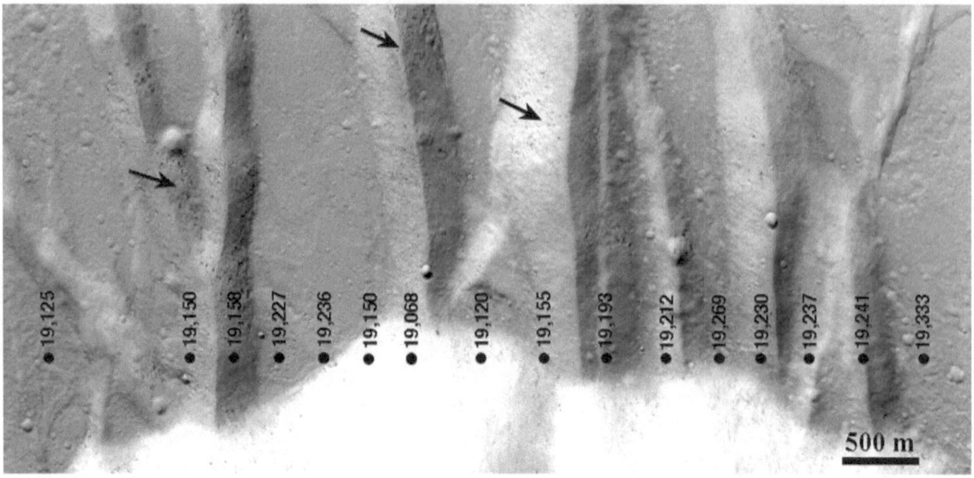

Figure 3.11. Section of the floor of Olympus Mons caldera. See Figure 3.10 for location. Illumination direction is from the bottom of the image. The sawtooth texture is a series of graben cut by a subsequent collapse event, which now forms the bright wall at the bottom of this image. These graben are up to ~90 m deep, and some contain boulders on their floor (arrowed). Also shown is location of selected individual MOLA elevation measurements is shown (elevations in meters). MOC image number M0402248, 5.94 m/pixel.

of Olympus Mons' caldera has subsided by more that 1 km with respect to the perimeter of the caldera floor (Mouginis-Mark and Robinson, 1992; Figure 3.10). As at Pavonis Mons, this sagging is analogous to that documented at Koolau by Walker (1988). The largest and oldest of the segments appears to have formed as a single large lava lake (Crater #1 using the nomenclature of Mouginis-Mark, 1981). This part of the floor is heavily fractured by numerous circumferential graben. A MOLA profile reveals these graben to be ~90 m deep (Figure 3.11). Although the entire 670 m high vertical extent of the wall of crater #4 is not imaged by MOC, at least the topmost 170 m section of the wall shows no signs of layering (Figure 3.11).

Calderas on active Hawaiian shield volcanoes are very dynamic features (Decker, 1987) and represent the interplay between infilling by erupted materials, collapse and sagging due to magma withdraw and the weight of deep cumulate material, and inflation and deflation of active magma chambers. Although large-scale sagging of caldera floors has not been documented at active terrestrial volcanoes, the decrease in outer flank slopes towards the rims of the Kilauea and Mauna Loa calderas suggests broad-scale

Figure 3.12. Air photo (looking east) of the main caldera of Masaya. San Pedro pit crater is in the foreground, then Nindiri, with Santiago crater the most distant of the three connected pits. The isolated pit crater at top of image is the Masaya pit. The two-lane road (R) on the rim gives sense of scale. Dots and arrows give camera positions and viewing directions, respectively, for Figures 3.13 and 3.14.

downward sagging (Rowland and Garbeil, 2000). The inward sagging recorded in the lavas of Koolau volcano (Walker, 1988) also record this gentle downward motion. That most active calderas do not show this behavior in their caldera floors is almost certainly due to the regular resurfacing that occurs.

Fortuitously, subsidence and the ability to observe the crater floor in cross-section at Masaya volcano (Nicaragua) presents an opportunity to study the same type of subsidence in the field. Masaya is a persistently active basaltic volcano that comprises four main pit craters, which are named (from east to west) Masaya, Santiago, Nindiri, and San Pedro (Figure 3.12). Rymer *et al.* (1998) reviewed the geologic history of Masaya volcano, and showed that lava lakes have been common features in historic times at this volcano. Two lava flows have also been erupted (in 1670 and 1772 from Nindiri and Masaya pits respectively) onto the northern flanks of the volcano. Despite the large differences in scale, because of the striking similarity in the evolution of Masaya and Olympus Mons, it is possible to investigate the features of the Nindiri lava lakes to further elucidate the evolution of calderas on Mars.

3.3 Field investigation of terrestrial analogs: Masaya volcano, Nicaragua

There are numerous strong similarities between Masaya and the Olympus Mons caldera. The most obvious comparison can be seen within Nindiri crater (Figures 3.12 and 3.13), which is ~550 m in diameter. Nindiri crater is partially filled by frozen lava lakes that formed between 1570 and 1670. These comprise three main ponded units, units A, B, and C in Figure 3.13. Assuming that the scale of deformation is valid, these three units subsequently sagged downward elastically before failure in a brittle fashion along circular bounding faults (Figure 3.12), producing fractures that are morphologically similar to the circumferential fractures seen around the perimeter of the floor of Olympus Mons caldera (Mouginis-Mark and Robinson, 1992; Zuber and Mouginis-Mark, 1992). The faults extend down to the former (pre-1570) crater floor of Nindiri, where they widen into

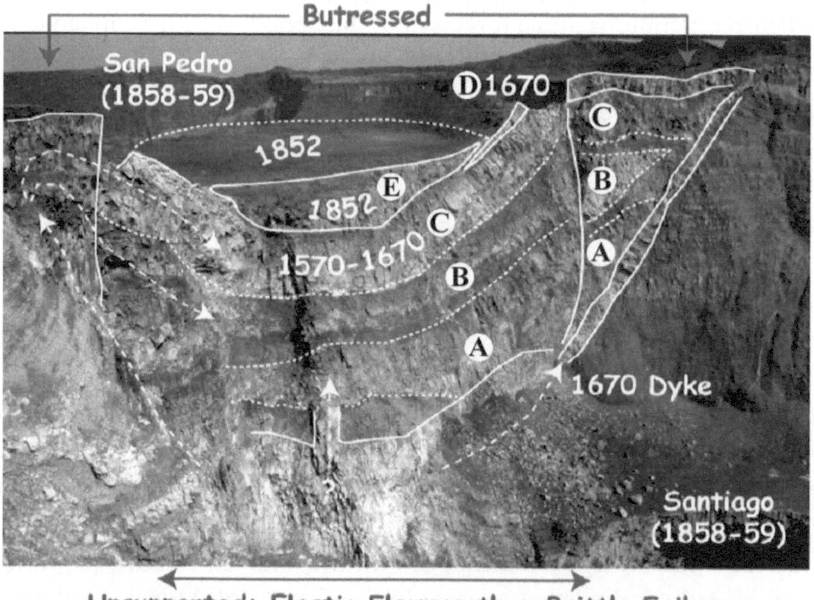

Figure 3.13. Field photo of the eastern wall of Nindiri pit crater, taken in February 2002, looking northwest towards San Pedro crater. The timing of events run from (A) the eruption of the pre-1670 lavas, to (E) the eruption of the 1852 flow that is now ponded within the subsided sequence of lavas. The section height is about 300 m, but better field measurements are needed of individual flow thickness in order to calculate the volume for each eruption. See Figure 3.12 for viewing geometry.

broad fracture zones on the north side and join with a few main faults on the south side. The final unit in this sequence erupted into a nearly full crater, such that lava spilled from a low point on the northern crater rim to feed lava onto the outer flank. A more recent lava lake was subsequently erupted in 1852 onto the crater floor where it ponded within the sag-structure.

San Pedro and Santiago pit craters formed subsequently, cutting the lavas ponded in Nindiri. Fortuitously, these allow vertical sections in excess of 300 m to be studied at the eastern and western edges of the pile. The uppermost ponded unit, the 1852 unit, is ~20 m thick and has a diameter of 268 m. This yields a bulk volume of 1.1×10^6 m^3. Ponded units A, B, and C have a total thickness of ~100 m which, with a diameter of ~550 m, gives a combined bulk volume of 7.6×10^6 m^3. Units A, B, and C, as well as the 1670 overflow unit, all appear to have sagged together. For such elastic sagging to have occurred, it is necessary that all three units must still have been molten at the time of deformation. This means that they must have been erupted in close succession. At the 1963 Alae lava lake (Kilauea, Hawaii) the 980 °C isotherm, the boundary between the brittle and molten layers in the lake, increased in depth (d) depending on $d = 0.00132\sqrt{t} - 0.18$ (Peck, 1978). Given a ~30 m thickness for the lowermost Masaya unit, and assuming that the magma effusion rates of Masaya and Kilauea are comparable, this places a limit of ~16 years on the maximum age difference between the oldest, lowermost, unit (A) and the youngest, uppermost, unit (C). It is, however, likely that heating and insulation by B and C prolonged the ductile lifetime of A.

The following sequence of events can be inferred. The three thick lava pond units erupted over a < 16 year time span. The final eruption in this sequence (possibly intra-caldera activity) poured into a full crater such that lava overflowed the northern rim. Withdrawal of shallow conduit magma, contraction due to degassing, thermal contraction, or compaction of the underlying talus at the end of this eruption sequence caused these units to then sag and fail together. Finally, after a hiatus of 82 years the 1852 eruption ponded in the sag-structure. Following 1852, the San Pedro and Santiago pit craters formed (Rymer *et al.*, 1998). These craters are 260 and 280 m deep and 358 and 494 m wide, respectively, giving volumes of 3.6 and 5.4×10^7 m^3. These volumes are far in excess of the Nindiri lava lake volumes. Thus drainage of these voids must also have involved flank eruptions and/or drain-back into the deeper system.

3.4 Implications for Mars

The terrestrial features and insights presented here can be used as guides and analogs when examining Olympus Mons, where the Martian lakes and pit craters may have formed in a similar way. There are, however, some drawbacks. If Nindiri crater is an analog to the summit of Olympus Mons, extensional features should be found around the perched outer perimeter of the crater and compressional features towards the center. Much of the surface of each of the Nindiri lakes is buried by more recent units or tephra, making it difficult to construct a geologic map of the 1570–1670 units and the 1852 ponded lava flow within Nindiri. This creates problems when trying to identify any extensional and compressional features that may have formed due to continuing subsidence of the underlying surface. However, some surfaces are accessible from the southern or northern rims. Here parts of the lake surface exposed at these locations have numerous surface features that warrant close inspection as they may have Martian analogs (Figure 3.14).

Mouginis-Mark and Robinson (1992) identified a transition between an extensional regime around the outer parts of the Olympus Mons caldera floor that was replaced by a compressional environment closer to the center of the caldera. This distribution of tectonic features was then exploited by Zuber

Figure 3.14. View of the 1852 ponded lava flow in Nindiri crater, as seen from the south in February 2002. The diameter of Nindiri crater is ~1 km. Note the circumferential graben (arrowed) on the far side, and various small-scale topographic features on the floor. See Figure 3.12 for viewing geometry.

and Mouginis-Mark (1992), who used a finite-element model to calculate the probable depth to the roof of the magma chamber at the time of subsidence. Zuber and Mouginis-Mark (1992) calculated that the magma chamber roof was ~16 km beneath the floor of the solidified lava lake, placing the magma chamber within the 20 km high volcanic edifice. The cooling histories of terrestrial and Martian lava lakes may be different, due at least to the absence of percolating rainwater on Mars. Even if the deep interior of lava ponds on both planets is liquid for the same time, some of the exterior units of terrestrial examples should be much colder for a given time and, therefore, should have failed more easily in a brittle fashion. Brittle deformation may thus take longer to initiate on Mars. Such detailed comparisons await future modeling efforts.

It is also important to understand the relative timing between different collapse events on Mars, because they are probably indicators of the magma production rate, and hence the total age of the volcano (Wilson *et al.*, 2001). Given that we have evidence for elastic failure of the Olympus Mons floor, we can suggest that the crater floor deformed while the lake was still mostly molten, as at Nindiri. Although it is highly likely that the cooling conditions of lava lakes on Mars were different from those on Earth, it is possible that the subsidence took place only a few years after the lava lake was formed and was thus at a high enough temperature to allow elastic deformation. Brittle deformation, however, is more likely to imply a longer time period between lava lake emplacement and collapse. The extreme case, as discussed by Wilson *et al.* (2001) is that the magma chamber freezes between events so that a new cycle of magma chamber formation, lava eruption, and pit crater collapse has to take place.

Some volumetric and thermal constraints can also be placed on the sequence of events. As at Masaya, by calculating the time it takes for a lake of a given thickness to completely cool below the brittle–ductile transition, the maximum time can be calculated during which plastic deformation can occur (i.e., plastic deformation could only occur while the lake core was still partially molten). This time estimate is relevant to other Martian lava lakes that underwent plastic deformation, which should be recognized as bowl-shaped surfaces in the MOLA elevation measurements. For instance, the caldera floor of Ascraeus Mons reveals several thinner ponded lava flows (Figure 3.9) compared to Olympus Mons (Figure 3.11). Five prominent layers can be identified within a 315 m high section of the wall of the Ascraeus Mons crater (Figure 3.9). This implies an average thickness of the ponded lavas of ~63 m. Given the Alae relationship of Peck (1978) this thickness would take ~70 years to freeze. Reconstructing an original diameter of ~20 km for this crater (number 3 of Mouginis-Mark, 1981) implies an average volume for each ponded layer of ~19.8 km^3. Using the same hypothetical dimensions for an equivalent lava flow on the flanks (i.e., 50 m thick and 10 km wide) as Olympus Mons would suggest that a single intra-caldera eruption of Ascraeus Mons might have produced a flow ~40 km long, which is typical of lava flows observed on the flanks (Zimbelman, 1985).

3.5 Conclusions

The data for the Tharsis Montes provided by MOC, THEMIS, and MOLA present unparalleled new opportunities for the identification of different stages in the evolution of Martian calderas. Fieldwork on terrestrial volcanoes also points the direction for further studies. A key component of the terrestrial studies is that they provide a temporal view of the way that calderas evolve, which no doubt has been an important factor in the development of sag features at both Olympus and Pavonis Montes. By virtue of the frequent eruptions at Kilauea, Fernandina, Masaya, and Mauna Loa, it is possible to better understand which landforms can form rapidly (such as a caldera collapse) or the longer-term subsidence of the caldera floor. This current survey using THEMIS and MOC images is inevitably limited to areas where image data have been obtained, and where the lighting geometry allows the identification of structures. Furthermore, because image data are still being acquired through at least the spring of 2004, additional scenes not included in this study will become available. An important aspect of any future study of planetary calderas would be an attempt to identify the three-dimensional extent of some of the units identified here, particularly those revealed in the walls of the Olympus Mons and Ascraeus Mons calderas.

Acknowledgments

The NASA Mars Data Analysis Program supported this work under grant NAG5-9576. We thank Laszlo Keszthelyi and Jim Zimbelman for helpful reviews.

References

Carr, M. H., Greeley, R., Blasius, K. R., Guest, J. E., and Murray, J. B. (1977). Some Martian volcanic features as viewed from the Viking Orbiters. *Journal of Geophysical Research*, **82**, 3985–4015.

Crumpler, L. S. and Aubele, J. C. (1978). Structural evolution of Arsia Mons, Pavonis Mons, and Ascraeus Mons: Tharsis region of Mars. *Icarus*, **34**, 496–511.

Crumpler, L. S., Head, J. W., and Aubele, J. C. (1996). Calderas on Mars: characteristics, structure, and associated flank deformation. In *Volcano Instability on Earth & Other Planets*, ed. W. J. McGuire, A. P. Jones, and J. Neuberg, Geological Society Special Publication, 110, Geological Society of London, pp. 307–48.

Decker, R. W. (1987). Dynamics of Hawaiian volcanoes: an overview. In *Volcanism in Hawaii*, ed. R. W. Decker, T. L. Wright, and P. H. Stauffer, US Geological Survey Professional Paper 1350, pp. 997–1018.

Lockwood, J. P., Dvorak, J. J., English, T. T. *et al.* (1987). Mauna Loa 1974–1984: a decade of intrusive and extrusive activity. In *Volcanism in Hawaii*, ed. R. W. Decker, T. L. Wright, and P. H. Stauffer, US Geological Survey Professional Paper 1350, pp. 537–70.

Macdonald, G. A. (1954). Activity of Hawaiian volcanoes during the years 1940–50. *Bulletin of Volcanology*, series 2, **15**, 120–79.

Macdonald, G. A. and Abbott, A. T. (1970). *Volcanoes in the Sea*. Honolulu: University of Hawaii Press.

Mouginis-Mark, P. J. (1981). Late-stage summit activity of Martian shield volcanoes. *Proceedings of the 12th Lunar and Planetary Science Conference*. Houston: Lunar and Planetary Institute, 12B, pp. 1431–47.

Mouginis-Mark, P. J. and Robinson, M. S. (1992). Evolution of the Olympus Mons caldera, Mars. *Bulletin of Volcanology*, **54**, 347–60.

Mouginis-Mark, P. J. and Rowland, S. K. (2001). The geomorphology of planetary calderas. *Geomorphology*, **37**, 201–23.

Peck, D. L. (1978). Cooling and vesiculation of Alae lava lake, Hawaii. *US Geological Survey Professional Paper* 935-B, p. 59.

Plescia, J. B. (1994). Geology of the small Tharsis volcanoes: Jovis Tholus, Ulysses Patera, Biblis Patera. *Icarus*, **111**, 246–69.

Rowland, S. K. and Garbeil, H. (2000). Slopes of oceanic basalt volcanoes. In *Remote Sensing of Active Volcanoes*, ed. P. J. Mouginis-Mark, J. A. Crisp, and J. H. Fink, American Geophysical Union Monograph 116, pp. 223–47.

Rowland, S. K. and Munro, D. C. (1992). The caldera of Volcan Fernandina: a remote sensing study of its structure and recent activity. *Bulletin Volcanology*, **55**, 97–109.

Rymer, H., van Wyk de Vries, B., Stix, J., and Williams-Jones, G. (1998). Pit crater structure and processes governing persistent activity at Masaya volcano, Nicaragua. *Bulletin Volcanology*, **59**, 345–55.

Scott, E. D. and Wilson, L. (1999). Evidence for a sill emplacement event on the upper flanks of the Ascraeus Mons shield volcano, Mars. *Journal of Geophysical Research*, **104**, 27079–89.

Simkin, T. and Howard, K. A. (1970). Caldera collapse in the Galapagos Islands, 1968. *Science*, **169**, 429–37.

Walker, G. P. L. (1988). Three Hawaiian calderas: an origin through loading by shallow intrusions? *Journal of Geophysical Research*, **93**, 14773–84.

Wilson, L., Scott, E. D., and Head, J. W. (2001). Evidence for episodicity in the magma supply to the large Tharsis volcanoes. *Journal of Geophysical Research*, **106**, 1423–33.

Wood, C. A. (1979). Monogenetic volcanoes of the terrestrial planets. *Proceedings of the 10th Lunar & Planetary Science Conference*, **10**, 2815–40.

Wood, C. A. (1984). Calderas: a planetary perspective. *Journal of Geophysical Research*, **89**, 8391–406.

Zimbelman, J. R. (1985). Estimates of rheologic properties for flows on the Martian volcano Ascraeus Mons. *Journal of Geophysical Research*, **90**, D157–62.

Zimbelman, J. R. and Edgett, K. S. (1992). The Tharsis Montes, Mars: comparison of volcanic and modified landforms. *Proceedings of the 22nd Lunar and Planetary Science Conference*, vol. 22. Houston: Lunar and Planetary Institute, pp. 31–44.

Zuber, M. T. and Mouginis-Mark, P. J. (1992). Caldera subsidence and magma depth of the Olympus Mons volcano, Mars. *Journal of Geophysical Research*, **97**, 18295–307.

4

Volcanic features of New Mexico analogous to volcanic features on Mars

Larry S. Crumpler, Jayne C. Aubele
New Mexico Museum of Natural History and Science, Albuquerque
and
James R. Zimbelman
Center for Earth and Planetary Studies, National Air and Space Museum, Smithsonian Institution, Washington

4.1 Introduction

The arid climate, extensional rift setting, range in type and age of volcanic eruptions, and generally widespread and geologically youthful volcanism in New Mexico contribute to an environment rich in geologic processes and landforms analogous to many of those on Mars. Young (<5 Ma) volcanoes and associated volcanic rocks are more widely distributed throughout the state than in many other volcanic localities on the North American continent. All of the principal volcanic landforms occur including long lava flows, viscous domes, calderas, composite volcanoes, monogenetic scoria cones, small shield volcanoes, and numerous hydromagmatic vents. The morphologies, volcanic emplacement processes, and dissected structures, and the arid environment, result in many volcanic landforms analogous to those on Mars. These features provide some clues to the details of geologic processes responsible for their Martian counterparts that are uncommon in areas where volcanism is less abundant and where the environments are less arid.

The largest young caldera (Valles Caldera), largest young lava flows (McCartys and Carrizozo), abundance of Quaternary volcanic fields, volatile-rich magmatism, including non-juvenile (maars) and juvenile types (Shiprock-Narbona Pass), spring deposits, and one of the great modern rift valleys on Earth (Rio Grande rift) occur in an arid setting where annual precipitation is between 8 and 15 inches (20–40 cm) per year. Combined with arid dissection and eolian in-fill, these contribute to a landscape that mimics the appearance of many volcanic terrains on Mars. In addition to surficial geologic processes, the interaction between subsurface ground water

The Geology of Mars: Evidence from Earth-based Analogs, ed. Mary Chapman. Published by Cambridge University Press.

flow and emergence with the arid climate results in spring deposits that may be analogs for Martian groundwater circulation.

4.2 Distribution and characteristics of volcanism in New Mexico

New Mexico is divided into several broad geologic provinces: Rio Grande rift, Colorado Plateau, Eastern High Plains, Rocky Mountains, and Basin and Range. In New Mexico, volcanic landforms occur in all of these provinces (Figure 4.1). The most recent volcanism is associated with the Rio Grande rift and Colorado Plateau margins, whereas slightly older mid-Tertiary volcanism occurs within the Basin and Range, Rocky Mountains, and Colorado Plateau interior. Several dominantly basaltic and recent volcanic fields occur on the margins of the Rio Grande rift or within the rift interior: the Taos Plateau

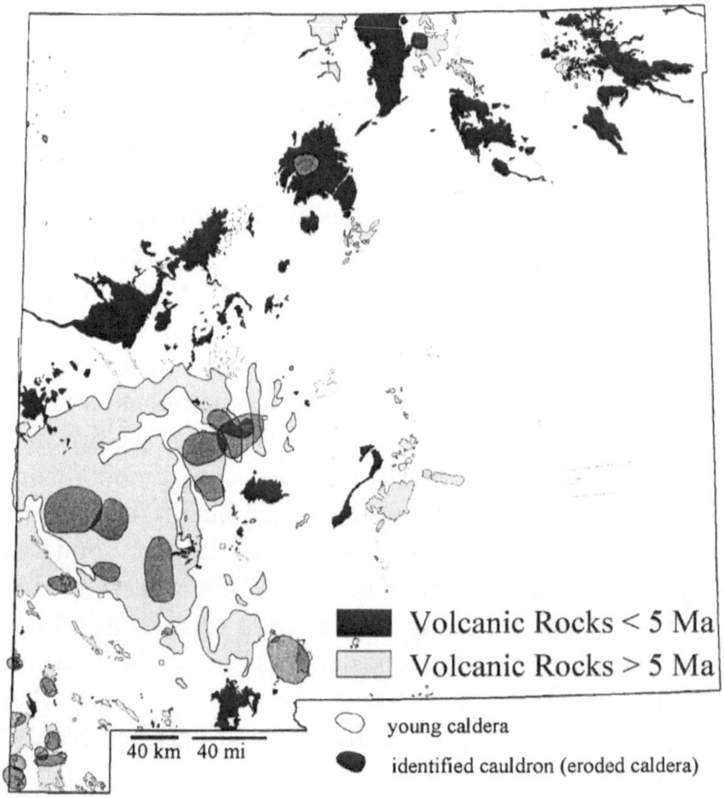

Figure 4.1. Distribution of late and middle Cenozoic volcanic rocks in New Mexico. The presence of a diverse range of volcano morphologies, the arid climate, and one of the few great continental rift valleys on Earth results in many volcanic landforms and processes that are analogous to those on Mars.

field, the Cerros del Rio field, the Jemez field, volcanoes within the Albuquerque basin part of the rift, and scattered fields and lava flows within the southern rift such as the Jornado del Muerto field, Carrizozo lava flow, and Engle (Elephant Butte) field. Several other volcanic fields, such as the Potrillo field and Palomas field, and isolated cones and flows, occur where the Rio Grande rift merges with the Basin and Range province. Eruptions within all of the late Cenozoic fields are widely dispersed in late Cenozoic time; ages of individual eruption within many of these volcanic fields cluster near 2 Ma, but range from 14 Ma to 0.03 Ma. Older, mid-Tertiary volcanism, associated with widespread silicic volcanism common throughout the Basin and Range province of the southwest, resulted in formation of many calderas that are now eroded, although their extensive sheets of ash flow tuffs are widespread and well exposed. Effusive and pyroclastic landforms within all of these volcanic areas have some bearing on understanding volcanic processes on Mars.

The younger volcanic fields around the margins of the Colorado Plateau in New Mexico are dominantly basaltic, but include large volcanic centers or isolated domes of more silicic magma, similar to other fields on the southern (Condit *et al.*, 1999) and southwestern (Best and Brimhall, 1974) margins of the Colorado Plateau. Most of the young New Mexico fields occur along a broad alignment known as the "Jemez zone" or "Jemez lineament" (Smith and Luedke, 1984). The Jemez zone includes the Red Hill, Zuni-Bandera (El Malpais), Mount Taylor, Jemez, Cerros del Rio, Ocaté, and Raton-Clayton fields. The Ocaté and Raton-Clayton fields are isolated from the Colorado Plateau margin and occur within the high plains of northeastern New Mexico. Although the Jemez zone has been described as a hot spot trace, and is roughly parallel to the Yellowstone hot spot track, ages throughout the Jemez zone are randomly distributed. The alignment of fields may instead represent a linear tectonically defined boundary between fundamentally different ancient lithosphere sections (McMillan *et al.*, 2000) with differing melting and magma emplacement properties. The Jemez field itself is the site of one of the largest young morphologically clear examples of an ash-flow caldera. It is an area where volcanism has occurred for over 13 Ma and has erupted basaltic, intermediate, and rhyolitic magma compositions. The development of large silicic calderas such as the Valles Caldera requires a significant, long-term melting anomaly in order to generate large magma chamber volumes associated with ash flow volcanism. The precise origin of the melting anomaly in the case of the Valles Caldera is unclear, but the location of the Jemez volcanic field at a point on the west margin of the Rio Grande rift where the rift steps east, suggests that the interaction between structures and the high regional geothermal gradients associated with the rift combined to generate a significant melt source and pathways for the melts into the upper crust (Self *et al.*, 1986).

Widespread volcanism in the Southwestern USA throughout the Cenozoic (Christiansen and Yeats, 1992) is attributed to high regional geothermal gradients and upwelling of mantle across the western USA following subduction of the East Pacific rise. Together with lithospheric extension and corresponding enhancement of magma ascent, the mantle melts generated by upwelling and heating resulted in volcanism both within the areas affected by lithospheric extension and within the interior of the Colorado Plateau. Volcanism in New Mexico may be categorized as "continental magmatism" and is distinct from that associated with plate margins. Although mid-Tertiary volcanism was voluminous and characterized by calc-alkalic affinities that include a variety of more silicic magma types, late Cenozoic magmas were largely alkali basaltic (Christiansen and McKee, 1978).

The regionally clustered large volcanoes of Mars have more in common with hot spot distributions controlled by broad upwelling of plumes (Crumpler and Revenaugh, 1997; Greeley *et al.*, 2000), such as the mechanism producing major volcanic provinces within the Southwestern USA, than they do to plate margin volcanism.

4.3 Mega morphology and outcrop scale morphology: scale-dependent preservation of volcanic morphology

The scale-dependent state of preservation of volcanic morphology on planetary surfaces in general is an important, but underappreciated aspect of volcanic analog studies. Many analog studies make use of relatively young and uneroded volcanic surfaces on Earth for comparison with Martian volcanic terrains. However, the great age and wide distribution of mantling deposits on Mars, combined with mechanical erosion by impact and environmental effects on rocky surfaces on Mars, likely result in significant degradation of primary volcanic landforms. Older volcanic terrains, such as those occurring in the arid environments in New Mexico, are better analogs for purposes of predicting the types of characteristics that occur in well-preserved but slightly degraded Martian volcanic terrains.

Primary volcanic features in New Mexico are preserved at several length scales and the degree of apparent preservaton of primary volcanic characteristics is often scale dependent. In other words, preserved morphologies and structures frequently appear youthful at scales exceeding 1 m (air photo resolutions), but may be soil-covered, rocky, and almost unrecognizable as volcanic landforms at outcrop (i.e., less than 1 m) scale (Figure 4.2). A similar state of preservation may characterize volcanic features on Mars. Volcanic features in excess of several hundred million to a billion years old on Mars may be usefully compared with primary morphologies of volcanic surfaces in New Mexico that are of the order of two million years old. In the following, we examine a few examples.

The general characteristics of young and unmodified lava flows are relatively well documented (e.g., Kilburn, 2000) from many studies. The particular morphologic characteristics present may vary from flow to flow due to differences in their emplacement conditions and composition and fundamental lava type (pahaoehoe or aa), but in general all recently erupted flows (less than 1000 to 10 000 years old) preserve millimeter- and centimeter-scale features of lava surfaces superimposed on larger-scale features such as tumuli, collapse depressions, channels, and individual flow margins (Elston *et al.*, 1976a). A survey of the surfaces of lava flows in New Mexico with reference to their individual ages identifies a progression with age in which progressively larger-scale features are degraded (Figure 4.3). Surface glass and centimeter-scale roughness characteristic of recently erupted flows are preserved up to tens of thousands of years. Meter-scale features, such as tumuli and larger structures only, are present after one to two hundred thousand years. At an age of one million years, few surface topographic features remain, although the individual flow margins are generally well preserved and are easily mapped in air photos or traced by ground traverses. At ages greater than about 5 million years, lava surfaces are generally flat and featureless in which even flow margins may be difficult to distinguish and the flow is frequently eroded and occurs as the caprock of isolated mesas.

The correlation between age and surface characteristics is not precise due to differences in the initial surface features and topographic roughness of individual flows, differences in the local environment, such as proximity to regional drainages, differences in local eolian deposition of silt, and differences in altitude and differences in global climate over the range of ages and corresponding variations in rainfall and vegetation at local and regional scales. Nonetheless, in New Mexico it is possible to infer an approximate age, to better than the closest order of magnitude, of many lava flows based on their surface preservation characteristics for a region as a whole.

Processes responsible for the degradation in the arid climate of New Mexico are largely mechanical weathering of the surfaces and airfall of silt derived from surrounding loose fines that compose many of the regional Mesozoic

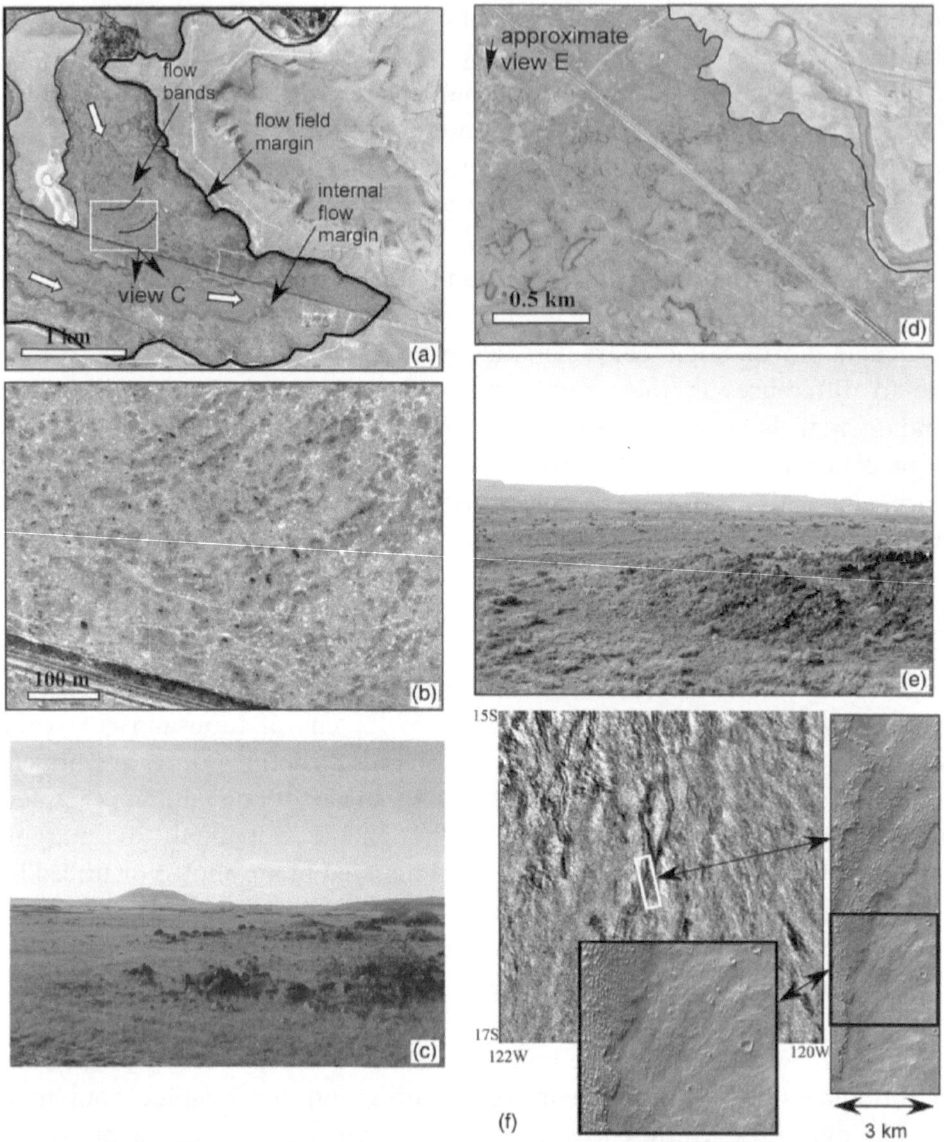

Figure 4.2. Examples of scale-dependent state of preservation in lava flows. (a–c) Sequence of progressively higher resolution images of a flow dated at <1.0 Ma. Flow margins and surface waves or bands are prominent in low-resolution images. At the surface outcrops are sparse and their relation to the flow surface structure is unclear. (d, e) Aerial photo of a flow dated at ~100 ka. Collapse depressions and tumuli are prominent, yet the surface appears as an undulating rubble sheet with low areas filled by fines; arrow in (d) indicates approximate direction of view shown in (e). (f) MOC image (FHA 01222) and corresponding Viking context of flows south of Arsia Mons. High-resolution perspective (inset) of these fresh-appearing flows shows that areas between relief features are smooth and appear mantled.

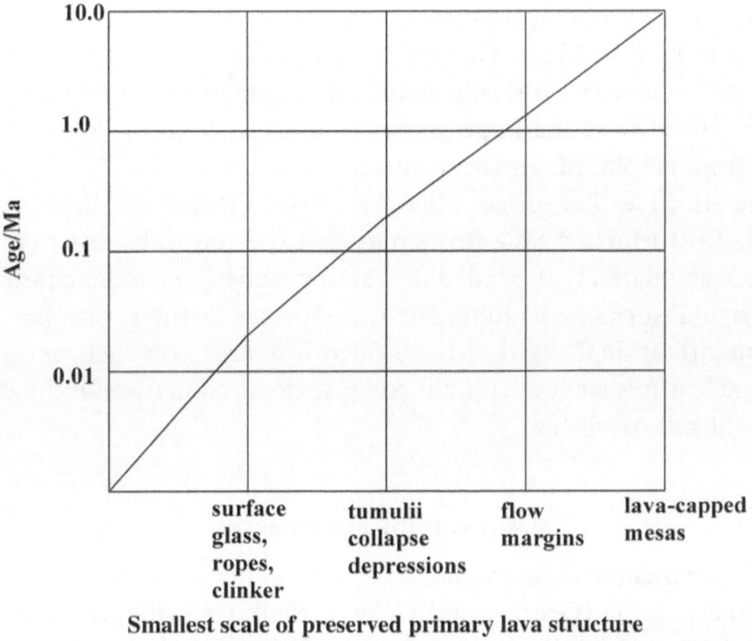

Figure 4.3. Proposed sequence of degradation on lava flow surfaces for New Mexico (after Elston *et al.*, 1976a). With increasing age, similar progressions of erosion on Mars will result in diminishing preservation of short-wavelength landforms. Lava flow margins and patterns of flow will be preserved much longer than detailed morphologic features characteristic of recent lava flows. Expectations of volcanological variety based on orbital images can thus be misleading.

sedimentary rocks. Low areas within a flow are gradually filled by spallation of surface glass (tachylite) and dislodged centimeter-scale fragments of surface clinker and fragmented glassy surfaces, and organic debris from sparse but widespread vegetation growing in cracks. Further infill occurs from airfall of silt. Relief features are in this way gradually lowered, the surfaces of low areas are raised, and the overall topographic roughness is smoothed. In the absence of significant rainfall and runoff that would carry accumulated silt away, alluvium accumulates in low areas of the flow surfaces to depths capable of burying many smaller-scale surface features. Despite overall weathering and degradation, a substantial component of the small-scale features may be preserved in low areas by these burial processes such that many small-scale features may actually be preserved even at great ages (>1 Ma) but are simply covered. Older flows are therefore not necessarily devoid of primary features, and detailed examination of outcrops and dissected exposures may still reveal characteristics of the primary surfaces.

Many of these or similar processes may have contributed to degradation of lava flow surfaces on Mars. Despite the relatively lower rates of weathering on Mars as compared with Earth, significant degradation of flows is predicted to have occurred over the large geologic times since most volcanic surfaces on Mars were emplaced. In the youngest lava flows, mantling by dust may be a more effective than direct ablative erosion. Based on the New Mexico analogies, most Martian lava flows may be expected to be less than pristine and to appear relatively degraded at outcrop scales, yet some characteristics of their original surfaces, including delicate surface textures, may be preserved beneath mantling deposits that have filled low areas on their surfaces; and larger-scale features such as tumuli, collapse depressions, and individual flow margins will still be visible.

4.4 Ash flows and calderas

Several large volcanic edifices and associated calderas on Mars, particularly the highland paterae (Greeley and Crown, 1990; Crown and Greeley, 1993) and several Tharsis (Gulick and Baker, 1990) and Elysium (Mouginis-Mark *et al.*, 1982) volcanoes have been compared with more explosive terrestrial volcanic centers. Radial deposits common to these volcanoes appear distinct from obvious lava flows associated with many shield volcanoes on Mars in that V-shaped valleys and flat inter-valley surfaces are consistent with erosion of more sheet-like and less digitate materials than lava flows (Figure 4.4).

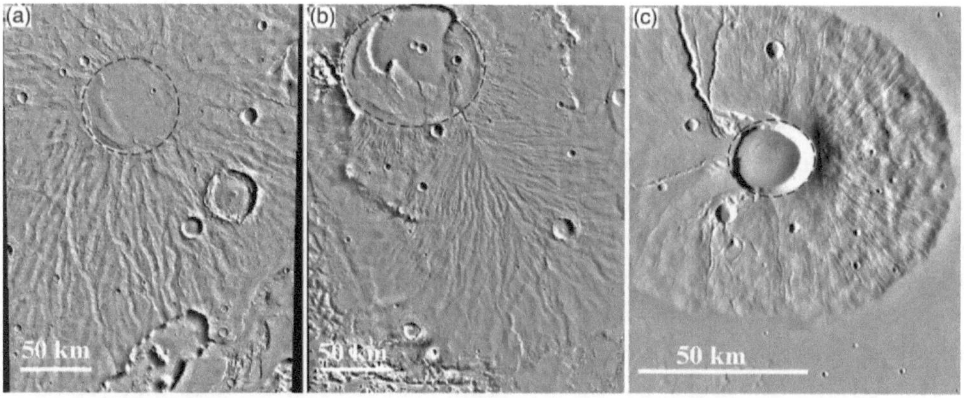

Figure 4.4. Examples of possible ash-flow type morphologies of Martian volcano flanks. (a) Hadriaca Patera. Shallow caldera and V-shaped valleys are analogous to the characteristics of many terrestrial ash-flow calderas and flows. The large valley to the south is a candidate for hydrothermal discharge. (b) Apollinaris Patera. (c) Ceraunius Tholus.

This characteristic has been compared with the morphology of radiating sheets of ash flow tuffs formed from pyroclastic flow mechanisms associated with many terrestrial explosive (plinian) volcanic centers (Reimers and Komar, 1979; Greeley and Crown, 1990; Gulick and Baker, 1990). Although terrestrial pyroclastic flows resulting in ash-flow tuffs are almost exclusively the result of silicic eruptions, on Mars the tendency for greater magma fragmentation (Wilson and Head, 1994) and the presence of geologic settings capable of delivering large volumes of volatiles to a high volume mafic eruption (Crown and Greeley, 1993) makes more tenable the suggestion that some deposits with ash-flow-like morphology may be mafic pyroclastic flows. The Valles Caldera, an almost circular, 20 km diameter caldera (Figure 4.5), bears many similarities to patera-type volcanoes of Mars in this respect. In diameter alone it is also comparable to the smaller Martian calderas (Crumpler *et al.*, 1996; Mouginis-Mark and Rowland, Chapter 3, this volume).

The Valles Caldera is the type example of resurgent ash-flow calderas (Smith and Bailey, 1968). Eruptions of the Valles Caldera culminated with the emplacement of at least two stratigraphic members of a broad apron of ash flows, the Bandelier ash-flow tuff, during collapse of the caldera at about 1.1 Ma ago. The Bandelier ash-flow tuffs radiate outward from the caldera forming a broad apron of thick ash-flow tuff sheets. The ash-flow tuffs have been dissected subsequent to their emplacement resulting in a distinct radially patterned morphology of alternating V-shaped canyons and flat-topped interfluves (Figure 4.5d) analogous to the pattern of valleys surrounding highland paterae on Mars. Resurgent doming of the center of the Valles Caldera over the next several hundred thousand years following caldera formation resulted in uplift of the floor in a rounded central peak that exceeds the elevation of the caldera rim. The resurgent doming was followed by a series of extrusions of several rhyolite domes on the caldera floor along the collapse ring fracture and encircling the resurgent dome. The latest rhyolite dome eruptions occurred approximately 0.04 Ma. Resurgent doming in the Valles Caldera and subsequent extrusive dome formation is unlike the relatively flat-lying interiors of Martian paterae. This difference reflects the differing dynamics and shallow location of the associated Valles Caldera magma chamber and its rhyolitic composition as opposed to the likely mafic nature of Martian magmatism. However, the explosive characteristics of terrestrial rhyolite ash eruptions and gas-charged Martian mafic magmas may be comparable (Crown and Greeley, 1993).

At geologically long time scales the radial ash sheets of the Valles Caldera appear easily eroded and less competent than lava flows because of the

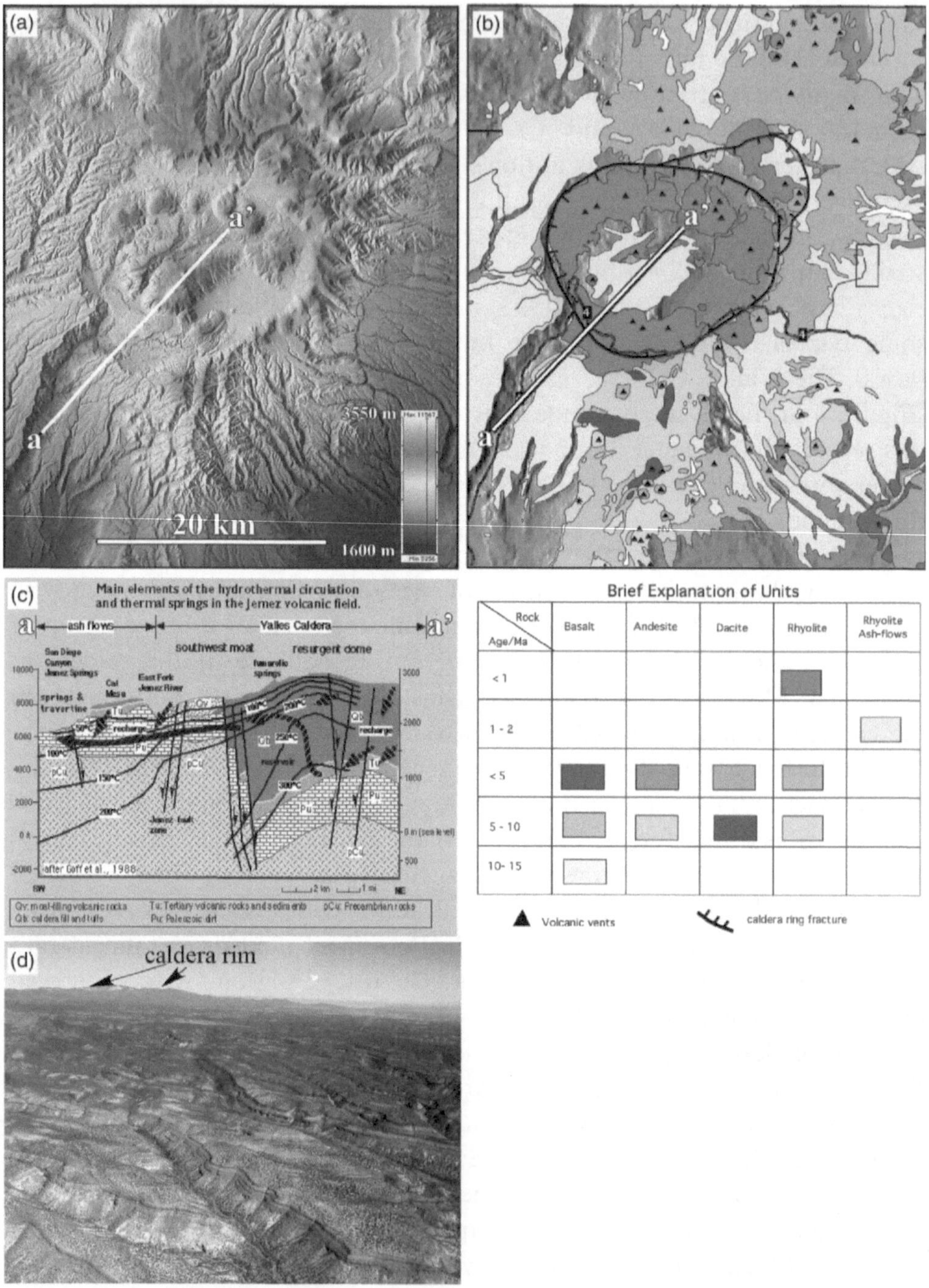

Figure 4.5. (a) Digital shaded relief map of the Valles Calderas, near Los Alamos, New Mexico. (b) Generalized geologic map of the Valles Caldera. After Luedke and Smith (1978) and Smith *et al.* (1970). (c) Schematic section (after Goff *et al.*, 1989) from the caldera center through

development of deep canyons with steep walls. However, the Bandelier ash-flow tuffs are in fact relatively competent rock formed from compaction and welding of ash at the time of emplacement. The combination of relatively low permeability and an underlying substrate and basal layers of loosely consolidated ash have contributed to fluvial runoff along general gradients directed outward from the caldera margins and corresponding rapid incision and valley wall retreat. Because of the impermeability, aqueous flow from the caldera interior tends to be directed along the base of the ash flow sheets further enhancing formation of springs along the base of the ash flows and enhanced erosion along the base of valley walls.

Many such springs around the margins of the ash sheets are anomalously warm due to deep penetration of water at its source in an area of high residual magmatic heat. These observations have significance in the context of current Mars exploration goals. If compaction and welding of mafic ash flows around explosive volcanic centers on Mars contributed to deposits with similar hydrologic properties, thermal springs may have been common in the valleys and distal reaches of the tuffs. Early in the history of caldera formation, any groundwater that emerged from the base of ash flow sheets around Martian paterae may have similarly inherited heat from the adjacent large volcanic centers, thus leading to numerous hot or thermal springs where the aquifer emerged in the valleys and edges of the ash flow sheets. These are the types of spring environments that are targets for investigation as potential sites for early pre-biotic and biotic processes (McKay and Stoker, 1989; Boston *et al.*, 1992). Thus, based on our understanding of an analogous caldera on Earth, the margins of radiating ash sheets around Martian paterae may be useful sites for concentrated investigations that seek sites of former thermal springs.

4.5 Large radial dikes

Radial fractures are a prominent characteristic of the Tharsis region of Mars (Figure 4.6a). Possible interpretations of the Tharsis pattern include

Caption for Figure 4.5. (cont.) the south margin and down the prominent valley on the southwest flank illustrating the principal characteristics of hydrothermal circulation and relationship to down-gradient springs and spring deposits. Similar circulation patterns may have existed shortly after formation of the highland paterae type calderas on Mars (see, for example, Figure 5.4a). (d) Oblique air photo of valleys cut into the extensive layered sheets of Bandelier ash flow tuffs that were erupted during caldera formation; view west from vicinity of the Rio Grande; note distance from the caldera rim. (For a color version of this figure, please refer to color plate section.)

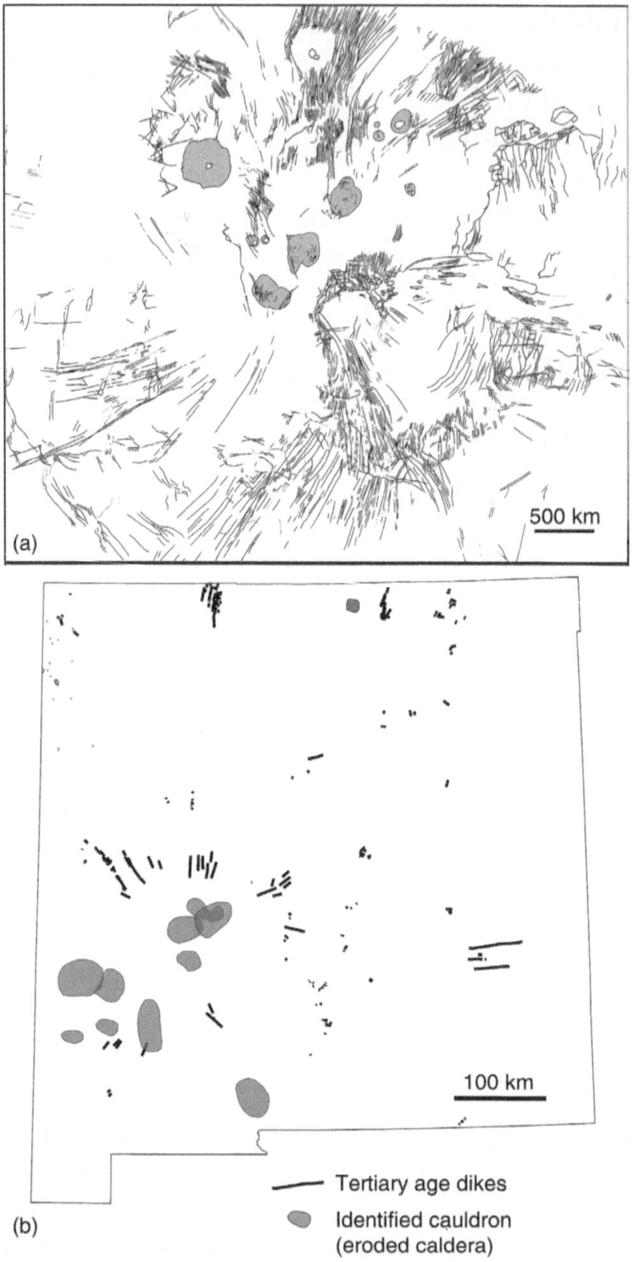

Figure 4.6. (a) Distribution of fractures radial to the Tharsis region of Mars (after Hamblin and Christiansen, 1990). (b) Distribution of mid-Tertiary mafic dikes in New Mexico. The pattern of dike orientations is generally radial to the central complex of mid-Tertiary ash flow calderas (the Mogollon volcanic province). Dike locations from Elston *et al.* (1976b) and New Mexico Bureau of Geology and Mineral Resources (2003).

strain associated with uplift (Banerdt *et al.*, 1992) and radial dikes associated with volcanism (Wilson and Head, 2002). Radiating fracture patterns interpreted as the surface expression of radial dikes occur on Venus, Mars, and Earth (Ernst *et al.*, 2001; Wilson and Head, 2002). Terrestrial examples may be up to several hundred to a thousand kilometers in overall diameter. Most examples are from older volcanic regions where erosion has been great enough to expose dikes and intrusions that were formerly at shallow depths within the crust. Some are associated with individual volcanoes, but the more intriguing examples radiate from large igneous provinces on Earth where magmatism was protracted and voluminous.

A potential example of the latter type occurs in association with the largest mid-Tertiary Mogollon magmatic province in New Mexico. A map of mid-Tertiary dike distribution in New Mexico (Figure 4.6; Elston *et al.*, 1976b) is similar to radial dike systems described from many igneous provinces on Earth. If all mid-Tertiary dikes in New Mexico are mapped, the distribution appears to radiate from a center located within the Datil-Mogollon ash-flow caldera volcanic complex (Elston *et al.*, 1976b), a cluster of silicic and intermediate composition ash-flow caldera of mid-Tertiary age.

Many stratigraphic, petrologic, chemical, and isotopic studies of the associated volcanic rocks of the Mogollon volcanic region suggest that the thermal source for production of many of the voluminous non-basaltic melts may have been extensive basaltic magmas injected into the lower crust at or slightly before the period of great ash-flow volcanism that occurred throughout the southwest in mid-Tertiary time. Many of the earliest eruptions were basalts and basaltic andesite, followed by development of many intrusions of magmas with mixed or intermediate compositions. Upward progression and mixing of renewed basaltic intrusions into previously melted crustal materials, and previously mixed magma volumes in the upper crust resulted in hybrid magmas across a broad range of compositions. Ultimately, these magmas developed as extensive plutons that fed overlying dominantly silicic centers and corresponding ash-flows and calderas. The volume of magma necessary for the development of such an extensive field of plutons was presumably large and resided in the lower crust as an extensive series of magma chambers of dominantly mafic composition. In many ways this is analogous to the situation envisoned for magmas within the Martian crust during emplacement of the Tharsis province (Wilson and Head, 2000).

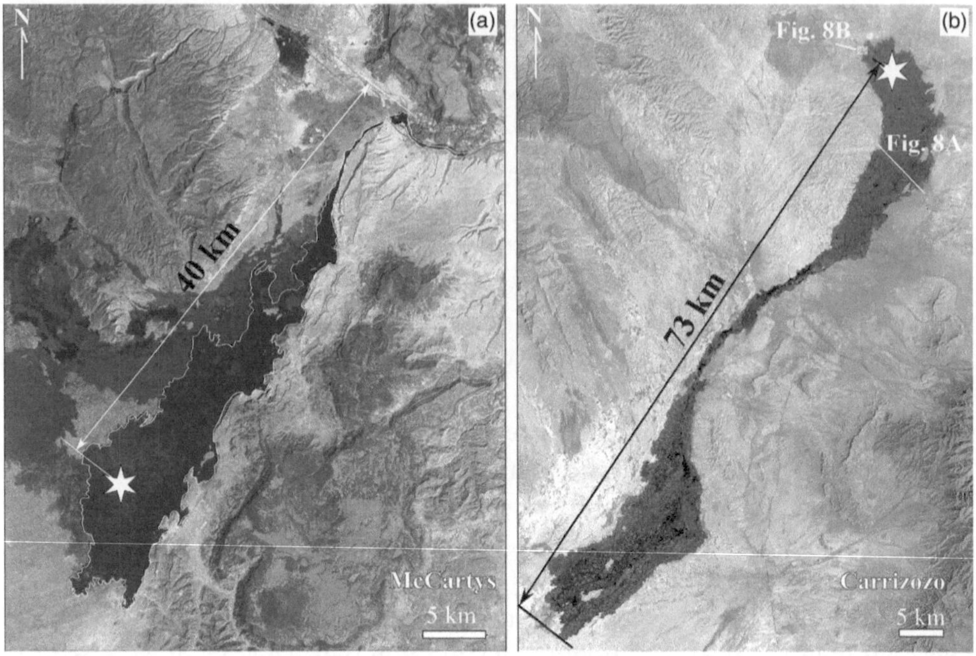

Figure 4.7. Two of the longest young lava flows in North America, the McCartys and Carrizozo flows, provide insights into possible emplacement mechanisms for large lava flows common on Mars. (a) McCartys lava flow (white outline), Grants, New Mexico. Older flows of differing age and surface brightness lie to the left (west). (b) Carrizozo lava flow. Location of two precison topographic traverse are shown as lines near the north source region and north central portions of the flow.

4.6 Large lava flows and flow fields

The McCartys and Carrizozo lava flows (Figure 4.7) are two of the largest young lava flows in North America. The flows are unrelated and occur in widely separated areas of the state; however, both are compound, tube-fed pahoehoe type flows. Because of their great volume and youth they are exceptional analogs for large lava flows common on Mars and the other terrestrial planets (Zimbelman, 1998; Zimbelman and Johnston, 2001). For example, the overall planimetric shapes appear as though the flows are the result of a single fluid emplacement of lava, but in fact each developed from multiple overlapping segments and through branching tube arrangements. By understanding these and other details of the actual emplacement process, the actual factors controlling emplacement of large lava flows common on Mars may be better constrained.

The McCartys flow is the latest eruption in the extensive field of lava flows of the Zuni-Bandera field and is 48 km long from its small source vent to its terminus in the San Jose Valley near the present day route of interstate I-40 (Nichols, 1946; Maxwell, 1982, 1986; Thelig, 1990). The entire flow covers ~189 km^2, as outlined using satellite image data aided by published geologic mapping (Maxwell, 1986), a value smaller than the one reported by Nichols (1946). By assuming an average thickness of 23 m, Nichols (1946) estimated a total volume for the McCartys flow of 7.9 km^3. By comparison the somewhat less well-preserved 1783 flow of the Laki basaltic fissure eruption in Iceland (Thordarson and Self, 1993) is estimated to have a volume of 13 km^3. Basaltic glass (tachylite) is common throughout the surface of the McCartys flow which preserves a myriad of small-scale features resulting from plastic deformation of a fluid, hot silicate melt (Nichols, 1946). The flow surface is largely pahoehoe, although in many places the pahoehoe is broken up into jumbled plates that transition to a true a'a texture. The McCartys lava consists of tholeiitic basalt, initially considered to be undifferentiated (Renault, 1970). Subsequent extensive sampling and analysis revealed that the McCartys lavas are quartz-normative tholeiite with plagioclase phenocrysts within 4 km of the vent and olivine-normative tholeiite containing olivine phenocrysts at distances greater than 4 km (Carden and Laughlin, 1974). The detailed sample study also revealed considerable longitudinal variation in both major- and trace-element chemistry along the flow (Carden and Laughlin, 1974, table 1), but vertical chemical variations were found not to be significant. Likewise, a study of the vertical density and vesiculation along the length of the McCartys flow (Aubele *et al.*, 1988) indicated no significant variation with distance from the vent in the vesicularity. Cosmogenic and radiocarbon dating methods both give a young age of ~3000 yr for the McCartys flow, providing agreement within the analytical uncertainties of both methods (Laughlin *et al.*, 1994).

The Carrizozo flow is 75 km long from the vent area to the distal margin in the Tularosa Valley (Keszthelyi and Pieri, 1993; Zimbelman and Johnston, 2003). The entire flow covers ~330 km^2 to an estimated thickness of 10 to 15 m, for a total erupted volume of ~4.3 km^3 (Allen, 1951). The lava is intermediate in composition between alkalic and tholeiitic basalt and is consistent with the regional volcanism associated with the Rio Grande rift (Renault, 1970; Faris, 1980; Anthony *et al.*, 1998). Various researchers have distinguished between upper and lower Carrizozo flow units, separated by a narrow "neck" in the medial reach (e.g., Anthony *et al.*, 1998; Dunbar, 1999). Chemical analyses to date have measured small differences between the upper and lower lavas, although it is unclear whether the measurements are statistically significant. Cosmogenic (isotopic changes induced by exposure to high-energy particles) studies indicate exposure ages of 4800 yr (Anthony *et al.*, 1998) to 5200 yr (Dunbar, 1999) for the Carrizozo flow, well within the 1700 and 700 yr error estimates, respectively. These results make the Carrizozo flow the second youngest volcanism in New Mexico (Anthony *et al.*, 1998), after only the McCartys flow.

Recent Differential Global Positioning System (DGPS) data of both flows (Zimbelman and Johnston, 2003) have revealed new details about the relief and emplacement of these compound basaltic flows. A topographic transect across the Carrizozo flow along Highway 380, ~10 km down flow from the vent, supports the interpretation that multiple flow elements banked against earlier episodes to build the field from east to west (Figure 4.8a). The topographic data, with horizontal and vertical precision ~2 to 4 cm, show distinctive terracing along the flow margins in the proximal and distal portions of both flows (Figure. 4.8b), interpreted to be indicative of multiple episodes or scales of the local flow emplacement. The origin of these terraces is unclear but they may be associated with changes in local gradient that resulted in the fluid flow interiors exceeding the crustal strength during inflation of the flow and corresponding breakouts. Topographic measurements of the McCartys flow are consistent with results obtained for the Carrizozo flow, and help to constrain how the eruptions may have produced both flows. The dimensions of a very narrow neck on the McCartys flow, ~40 km down flow from the vent, provide strong constraints on the lava tube that must have fed the distal portions the flow, resulting in a maximum likely effusion rate of ~500 m^3/s. At Carrizozo, a single medial ridge along the narrow central portion of the flow can be interpreted as a lava tube similar to the McCartys flow neck, which suggests a maximum likely effusion rate of ~800 m^3/s. Both flows could have been emplaced at these rates within a period of a few months, or of the order of 20 years at sustained lower rates (Keszthelyi and Pieri, 1993).

4.7 Hydromagmatic volcanism

Evidence for fluvial and ice-related processes on Mars (Fanale *et al.*, 1986; Squyres and Carr, 1986; Squyres *et al.*, 1992; Clifford, 1993; Carr, 1996), together with widespread evidence for volcanoes at many scales, raises the possibility that the interaction between water and magma, or hydromagmatic eruptions, may have occurred on Mars (Frey and Jarosevich, 1982; Mouginis-Mark *et al.*, 1982; Mouginis-Mark, 1985; Squyres *et al.*, 1987; Mouginis-Mark *et al.*, 1992; Crown and Greeley, 1993; Fagents and

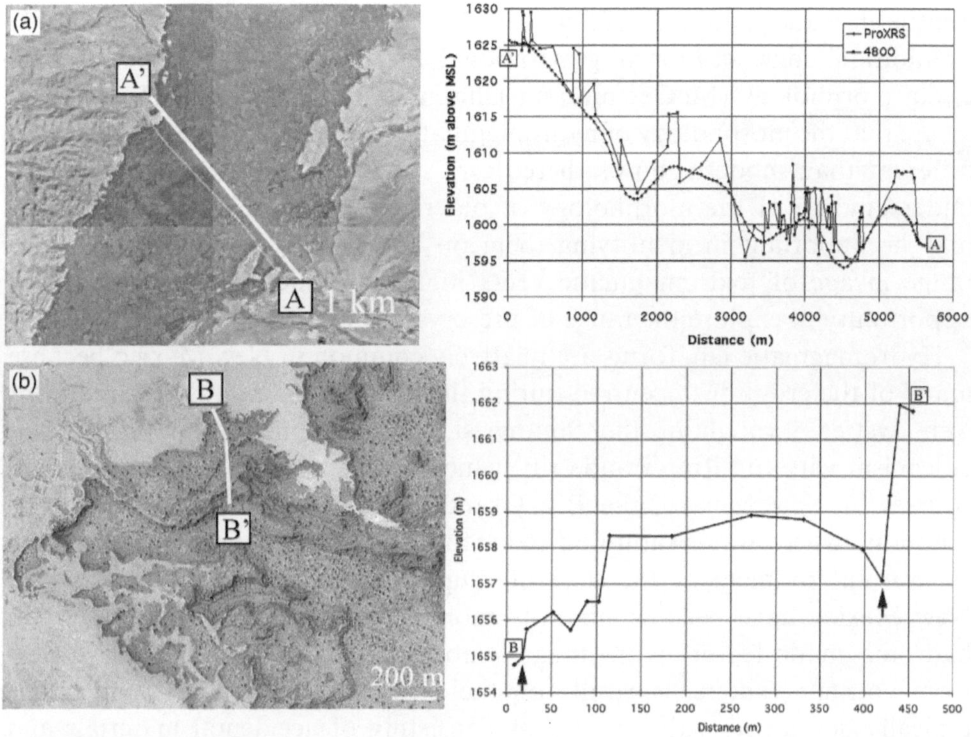

Figure 4.8. Precision topographic transects across portions of the Carrizozo lava flow; locations indicated on Figure 4.7(b). (a) DGPS data along US Highway 380, across the Carrizozo flow ~10 km down flow from the vent at Little Black Peak. A pronounced slope from the right (east) to left (west) is interpreted as the result of multiple flows banked against one another. (b) Terraced margin, NW corner of Carrizozo flow. Line shows location of DGPS profile. Left: Portion of USGS Digital Orthophoto data of Black Peak Quadrangle, New Mexico. Right: DGPS data along indicated traverse.

Thordarson, Chapter 6; Chapman and Smellie, Chapter 7, this volume). Hydromagmatic eruptions are a type of plinian to vulcanian eruption in which near-surface water is vaporized when magma ascends into the saturated layer. The rapid flashing of the water to vapor drives powerful explosive bursts (Sheridan and Wohletz, 1983; Wohletz, 1986) capable of fragmentation and deposition of significant quantities of both conduit wall rock and the magma itself. Landforms associated with magma–water interaction are common and distinctive on Earth, and if present on Mars should be detectable based on the predictable characteristics of deposits and structures associated with hydromagmatic eruptions. Relief on any hydromagmatic vents on Mars is likely to have been subdued over geologic time through mechanical weathering and gradual infilling of the crater floors with airfall dust. In addition, consideration of gravitational and environmental effects on rim ejecta morphology (McGetchin and Ullrich, 1973; Wilson and Head, 1994) insure that the morphology of hydromagmatic vents are likely to be somewhat different than modern, unweathered terrestrial counterparts. Therefore an understanding of the morphology of older and eroded examples on Earth may be important in identifying them on Mars. The abundance and wide range in age of hydromagmatic vents in New Mexico offers an excellent opportunity to explore the range of preservation characteristics.

Hydromagmatic landforms are relatively common in New Mexico because many of the eruptions occurred during the late Pleistocene when conditions were wetter throughout the Southwest. In addition, the association of volcanism with the Rio Grande rift, which confines one of the major water courses in New Mexico, the Rio Grande, has also resulted in numerous eruptions where water-saturated sediments were characteristic of the near-surface environment at the time of eruption. Because many eruptions in New Mexico have been dominantly monogenetic and basaltic, a common hydromagmatic feature is the maar (Aubele *et al.*, 1976). Maars are crater-form volcanic centers, generally on the order of a kilometer in diameter, typically developed with raised rims consisting of accidental materials and ash ejected during multiple violent steam-blast explosions (Lorenz, 1973). Subsequent to eruption most maars are occupied at least temporarily by ponded water and are thus sites of pluvial deposition, partial infilling, and back-wasting of the initially steep crater walls.

On Mars, because of the prevalence of small impact craters, the ability to distinguish between small craters of hydromagmatic origin and small craters of impact origin will be critical if maars are to be identified on Mars at all. Several studies of maar craters in New Mexico have helped to define the physical characteristics of these volcanic craters. Relatively youthful and morphologically typical maars occur within the Zuni-Bandera volcanic region, the Potrillo volcanic field, and within the Mount Taylor volcanic field.

The Zuni Salt Lake maar (Figure 4.9a; Cummings, 1968) is an excellent example. In size and age (1.5 to 2 km and 0.024 Ma) (Bradbury, 1966) and in its arid Colorado Plateau setting, it is comparable to Meteor Crater (1.3 km and 0.08 Ma), making it an exceptional model for comparison and discrimination of the morphology of both types of crater. Zuni Salt Lake also illustrates the type of characteristics that are most useful in determining the volcanic origin of maars as compared with impact craters. The most prominent difference is the relatively shallow relief of the maar floor,

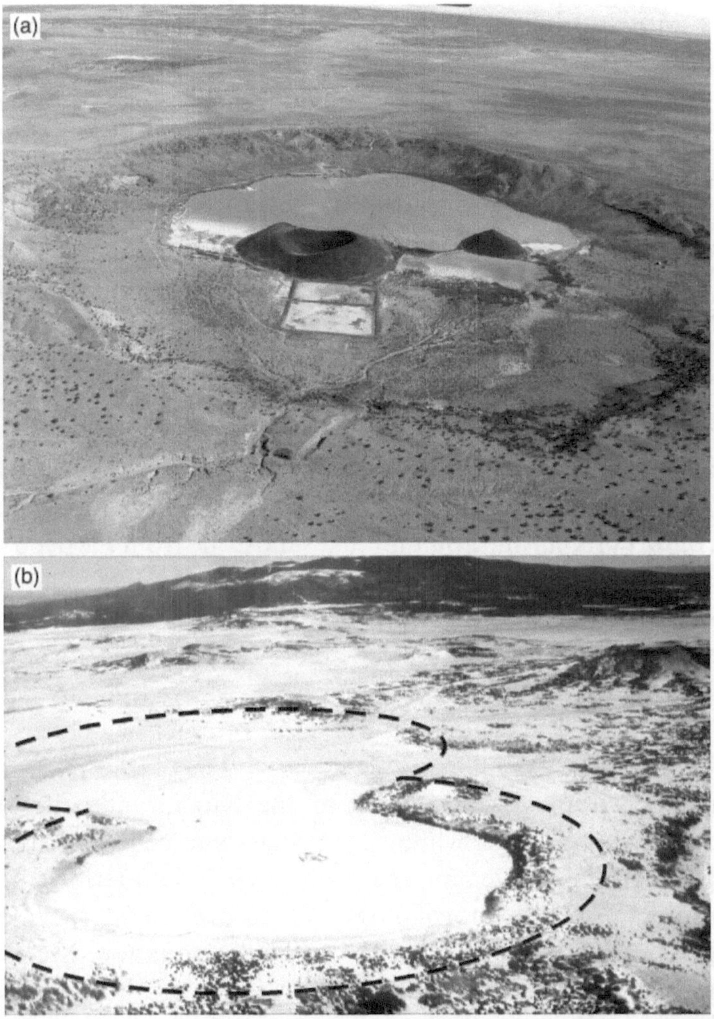

Figure 4.9. Although dry now, many volcanic vents in New Mexico were emplaced during the extended pluvial periods of the late Cenozoic in New Mexico. Centers of former hydromagmatic eruptions are common in New Mexico. Similarly, volcanism during the early wetter geologic past of Mars may have resulted in analogous landforms. (a) Zuni Salt Lake maar, Quemado, New Mexico. (b) Laguna de Alejandro maars, Mount Taylor field, San Mateo, New Mexico.

approximately 60 m from rim to floor, compared with Meteor Crater (200 m). The original depth is estimated as only slightly deeper considering that the base of the interior scoria cones appear not to have been buried. Erosion of the loosely consolidated rim ejecta and the long-term presence of an interior saline lake have resulted in flattening of the floor and retreat of the crater walls. More distinctive is the presence of two small scoria cones on the floor. Clear evidence of association with interior volcanic features together with the layered and easily eroded rim deposits such as this is a salient characteristic that would distinguish Martian hydromagmatic craters from those of impact origin.

Maars within the Mount Taylor field (Crumpler, 1980a, b) are of interest because of their great age and location in an area where dissection has been minimal. The eruptions that formed the maars of the Mount Taylor field occurred approximately 2 million years ago within a volcanic field that is dominated by scoria cones and extensive alkali basalt flows. Several are elongated or overlap along distinct fissure lines (Figure 4.9b). Almost all are a least 1 km in diameter and were occupied for extensive time by interior lakes. Because of the great thickness of basaltic rocks in the field, the rims consist largely of accidental blocks of underlying basalt and trachytic rocks in a matrix of tuff breccia and comminuted sandstone from the deep substrate. These maars are now situated on an extensive volcanic plateau that is essentially a 2 million year old surface isolated 300 m above the surrounding deserts. Because of their subdued, yet otherwise preserved form, they may be somewhat better examples of the state of preservation to be expected for small hydromagmatic centers on Mars where relief has been "softened" yet dissection has been minimal.

A few maars within volcanic fields of the Rio Grande rift have been dissected by canyon erosion, while preserving some characteristics of their surface expression, thus revealing the geometry of the interior deposits and its relationship to surface morphology. These are useful sites for understanding the complex internal dynamics of maar formation. Montoso maar in the Cerros del Rio field (Aubele, 1978) is an example in the middle Rio Grande rift and lies on the margins of a deep side canyon cut into the walls of a narrow gorge (White Rock Canyon) cut by the Rio Grande. Here a side canyon has completely cut into the relatively soft interior of a maar, exposing layers of inward-dipping ash and tuff breccias, several subsidence faults, and interior basaltic intrusions, yet rim ejecta typical of maars, consisting of accidental blocks of basalt, sandstone, and tuffs mixed with juvenile ash and palagonite, is preserved on the top of the canyon walls.

Maars in the Potrillo field (Hoffer, 1976) include the irregular-shaped Potrillo maar and Hunt's Hole, a deep maar with numerous ultramafic blocks in the rim ejecta. Portrillo maar illustrates the fact that not all maars are circular but can have complexly scalloped margins resulting from transient explosions appearing at different locations across the interior of the active vent.

4.8 Tertiary Colorado Plateau volcanism

Because of the relatively great crustal thickness (Bills and Ferrari, 1978) and potentially more iron-rich mafic and ultramafic mantle melt compositions of Mars (McGetchin and Smyth, 1978; Treiman, 1986; Bertka and Holloway, 1989; Longhi, 1990), erupted melts and eruption processes are potentially different from those on Earth. One consequence of this is the possibility of the eruption of high-fluidity melts and the inclusion of mantle volatiles leading to explosive eruptions with pelean and vulcanian characteristics (Wilson and Head, 1994). Such eruptions would be more explosive and compositionally distinct from common mafic eruptions on Earth. One problem with identifying this type of volcanic vent on Mars is the near absence of information about the morphology of the vents that they would produce. For this reason, volcanic styles on Earth that appear to have resulted from unusual volatile rich mafic eruptions are of potential importance in understanding some eruption processes on Mars.

Volcanism in the interior of the Colorado Plateau, represented by the mid-Tertiary Navajo volcanic field (Williams, 1936; Semken, 2001), is an example of a type of volcanism which has not been observed in the historic record. The Navajo volcanic field consists of approximately 80 eroded volcanic centers of Oligocene to Miocene age distributed throughout the Four Corners region, now preserved as volcanic necks or eroded depressions. They are characterized by unusual compositions, evidence for vulcanian eruption, and association with large sub-circular depressions. These eruption centers are also unusual in that they have occurred in a region of stable and thick lithosphere in the near absence of strong extensional stresses leading to the conclusion that magma was deep-seated and ascended by volatile-aided processes uncommon in most modern volcanic regions.

Many of these eruption centers consist of unusually mafic to ultramafic compositions. The exposures of the conduits from which they erupted are composed almost entirely of tuff breccias of highly comminuted mafic rocks suggesting extremely explosive emplacement styles and deep mantle sources (Delaney and Pollard, 1981). Inferred surface characteristics based on the overwhelming abundance of tuff breccia imply a vent that was perhaps similar to maars although possibly driven by juvenile volatiles more than near-surface hydromagmatic processes. Because most of these centers are deeply eroded, their value from a morphological standpoint is limited. However, at least one example of the near-surface characteristics of these volcanic vents remains in the form of an unusual crater-shaped depression (Narbona Pass volcano; Figure 4.10) on the summit of the Chuska Mountains

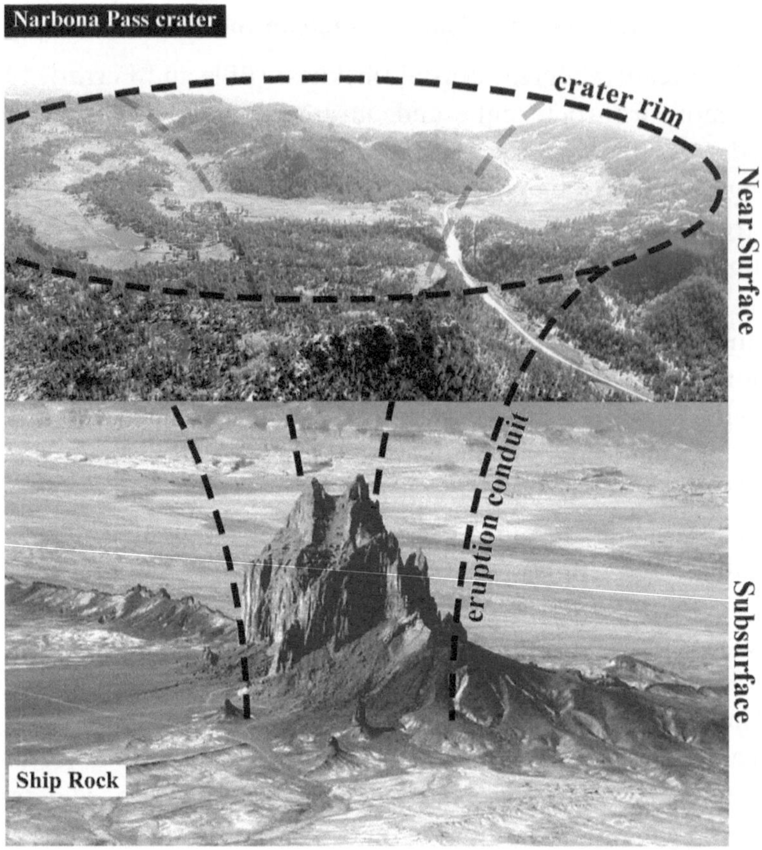

Figure 4.10. Comparison of Ship Rock and Narbona Pass volcano, Chuska Mountains, between Gallup and Shiprock, New Mexico. Ship Rock is interpreted (Semken, 2001) to be the near-surface conduit fill for a former hydromagmatic volcano. The surface expression of the former Ship Rock volcanic vent may have appeared similar to the Narbona Pass crater. The Narbona Pass crater is preserved because it is perched on the summit of an elevated plateau, the Chuska Mountains, which is an erosional remnant of a formerly thicker section of sediments within the central Colorado Plateau.

of western New Mexico (Ehrenberg, 1978; Semken, 2001). Narbona Pass volcano is a nearly circular depression 3 km in diameter. The absence of significant volcanic materials in the areas outside the depression leads to the conclusion that at least some vertical erosion of what is essentially a Tertiary surface has occurred since its emplacement. Within the depression there are several intrusions of basaltic composition as well as horizontal trachyte flows surrounded by tuff breccias analogous to that in other Navajo field volcanic necks. The relatively large size compared with modern hydromagmatic craters (maars) attest to the fact that the original vent was substantial.

Because of the association with a relatively uncommon magma type and unusual conduit characteristics, the Narbona volcano is likely to be one of the more unusual volcanoes in the terrestrial record and relates to a type potentially present on Mars, dating from early Noachian time when explosive volcanism may have been more common. In this respect, not all maar-like features on Mars may have formed from interaction with near-surface volatiles. Juvenile volatiles from a primitive volatile-rich Martian mantle also could have played an important part in possible explosive volcanism on Mars. Centers such as the Narbona Pass volcano, and other eroded centers of the Colorado Plateau, may be among the few terrestrial examples of this type of volcanism.

4.9 Spring deposit cones

Deposits associated with natural springs are another form of endogenic geologic process that can be similar, in both their mechanisms of formation and their resultant landforms, to volcanic vents and volcanic terrains. In New Mexico, these deposits create a variety of morphologies including large sheets and ledges and conical edifices surmounted by summit "craters" and fissures (Figure 4.11) that strongly mimic volcanic vents. Some of the New Mexico spring deposits are associated with hydrothermal circulation. Others are the result of natural ground water flow in an arid environment. Both are conditions that may have counterparts on Mars.

In New Mexico, spring deposits occur where a significant vertical discontinuity interrupts the groundwater flow and causes the water to emerge at the surface (Figure 4.12). (See Figure 4.4 for another example of thermal spring deposits associated with the Valles Caldera.) Spring deposits are particularly common in association with high-angle faults, such as the marginal fault zones of the Rio Grande rift, which act as high hydraulic conductivity conduits for aquifers confined by overlying aquitards and aquicludes. Faults on the margins of basins and other discontinuous structures, such as anticlines, where aquifers are brought in contact with less permeable rocks or where an aquifer is abruptly terminated within existing topographic slopes by recent faulting are also common sites of spring deposits. Spring deposits are common in New Mexico, therefore, because they are associated with tectonically active or formerly active areas and both evaporation rates and mineral content of spring water are high. These controls on spring formation mean that spring deposits mimic many of the characteristics of volcanic vents: most importantly, they occur in association with linear structures and faults, and they form volcano-like deposits.

Figure 4.11. Constructional edifices at spring emergence sites may emulate volcanoes, including conical forms, summit craters, and fissured mounds. (a) Conical spring mound; and (b) Summit "crater" on edifice shown in (a). Spring deposits are developed along the axis of a prominent anticline. San Ysidro, New Mexico. (c) Fissured mound. Note crater near far end (arrow 1) and graben along apex (arrow 2). San Ysidro, New Mexico.

The upper few kilometers of the Martian crust are likely to be highly fractured (Fanale, 1976; Carr, 1979) and thus permeable to fluid flow and aquifer development (Clifford, 1993). If water is present in the subsurface as a fluid, it will accumulate within the permeable zone and, under the force of gravity, water will flow in the substrate from topographically high regions towards regions of discharge in topographically low regions. At that point

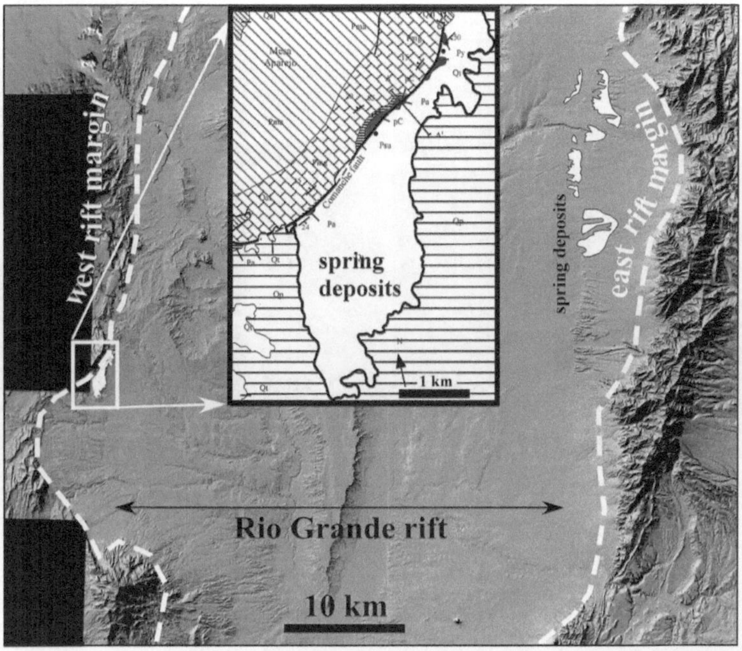

Figure 4.12. Springs are commonly located along the trace of significant regional faults. In this case, springs have developed along the western boundary fault of the Rio Grande rift near Belen, New Mexico. Continuous discharge, high mineral content of the spring waters, and the arid environment have resulted in the accumulation of significant deposits.

it is either discharged or accumulated in the subsurface. The gradient on the upper surface of the water-saturated region defines the top of the water table (Hubbert, 1953; Freeze and Cherry, 1979). The flow is controlled by the relief on the water table, or the potentiometric surface, which is generally a subdued reflection of the surface topography. At the largest scales, we may expect that water in a Martian aquifer of regional extent will flow from high elevations, such as the highlands to lowlands or local basins. There are many physical features over the surface of Mars that are similar to the morphologies typical of terrestrial spring mounds. For example, large areas of pitted mounds, generally interpreted as hydromagmatic (e.g., "pseudocraters"; Frey and Jarosevich, 1982) or ice-related phenomena (pingos), bear many of the characteristics of summit-pitted spring mounds (Figure 4.13a). Other anomalous mounds in apparently non-volcanic terrain on Mars occur in young basins or around basin margins (Figure 4.13b).

The influence of environment of formation on spring deposition, such as ambient pressure and temperature, is unknown, as are the chemical

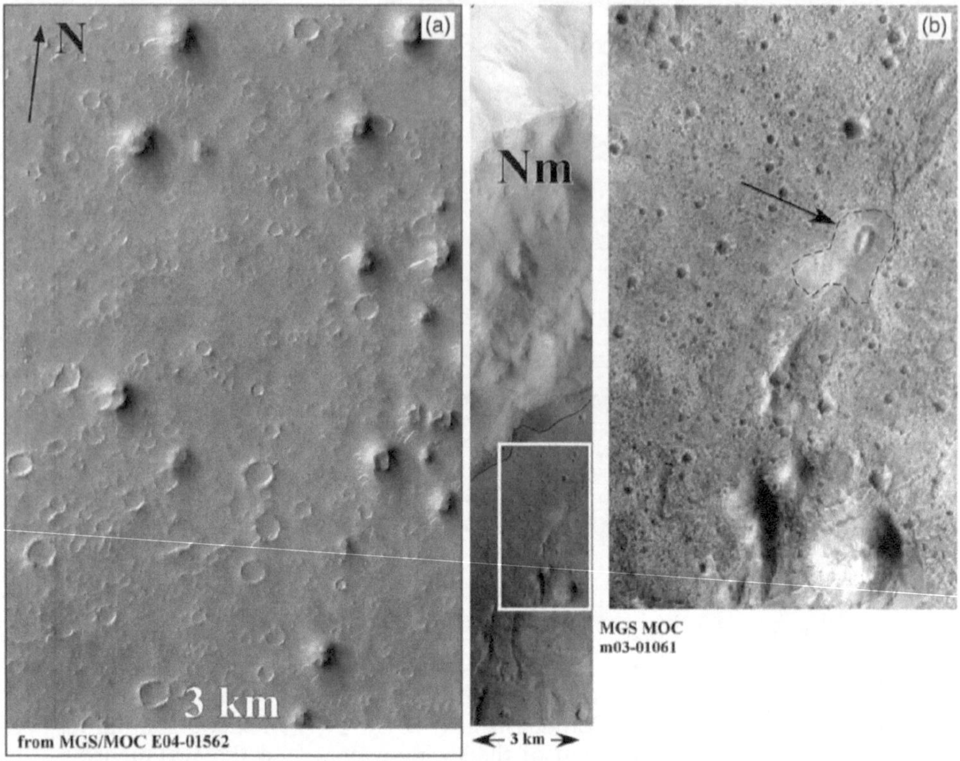

Figure 4.13. (a) Pitted mounds within the Isidis basin. (b) Isolated cratered mound within the highland–lowland transition zone. Down-gradient from an area of abundant valley networks.

compositions of materials that might comprise spring deposition environments on Mars. The corresponding deposits may reflect differences in their detailed environment and composition; however, they must form by a similar mechanism and with similar associations to major tectonic features. Spring deposits formed from carbonates appear relatively unlikely on Mars because there is as yet little evidence for extensive carbonate in the surface. Nonetheless many minerals are soluble in water and will respond similarly to dissolution, transportation, and deposition during desiccation. The abundance of dissolvable compounds in the Martian crust must be great. Given the deeply brecciated nature likely for the highlands and the widespread distribution of atmospherically transported volatiles and dust, including volcanic and impact-generated aerosols, the crust is likely to be liberally mixed with compounds that are unstable in water. Sulfur, sulfides, and related iron-rich materials may be important water-soluble materials that could constitute spring deposits on Mars.

Acknowledgments

The results in this discussion were supported in part by National Aeronautics and Space Administration Grants NAG5 9785 and NAG5 12850 (LSC) from the Mars Data Analysis Program and Mars Fundamental Research Program, and NAG5-4164 and NAG5-10375 (JRZ) from the Planetary Geology and Geophysics Program.

References

Allen, J. E. (1951). The Carrizozo malpais. *Roswell Geological Society 8th Field Conference Guidebook, Carrizozo-Capitan-Chupaddera Mesa Region*, pp. 9–11.

Anthony, E. Y., Hoffer, J., Williams, W. J., Poth, J., and Penn, B. (1998). Geochemistry and geochronology of Quaternary mafic volcanic rocks in the vicinity of Carrizozo, New Mexico. *New Mexico Geological Society 49th Field Conference Guidebook, Las Cruces Country II*, pp. 117–22.

Aubele, J. C. (1978). Geology of Cerros del Rio volcanic field. In *Guidebook to Rio Grande Rift in New Mexico and Colorado*, ed. J.W. Hawley. New Mexico Bureau of Mines and Mineral Resources Circular 163, pp. 198–201.

Aubele, J. C., Crumpler, L. S., Loeber, K. N., and Kudo, A. M. (1976). Maars and tuff rings of New Mexico. *New Mexico Geological Society Special Publication*, **6**, 109–14.

Aubele, J. C., Crumpler, L. S., and Elston, W. E. (1988). Vesicle zonation and vertical structure of basalt flows. *Journal Volcanology & Geothermal Research*, **35**, 349–74.

Banerdt, W. B., Golombek, M. P., and Tanaka, K. L. (1992). Stress and tectonics on Mars. In *Mars*, ed. H. H. Kieffer, B. M. Jakosky, C. W. Snyder, and M. S. Matthews. Tucson; University of Arizona Press, pp. 249–97.

Bertka, C. M. and Holloway, J. R. (1989). Matian mantle primary melts: an experimental study of melt density and viscosity at 23 kb. *Abstracts of Papers Submitted to the 20th Lunar and Planetary Science Conference*. Houston: Lunar and Planetary Institute, pp. 69–70.

Best, M. G. and Brimhall, W. H. (1974). Late Cenozoic alkali basaltic magmas in the western Colorado Plateaus and the Basin and Range transition zone, USA, and their bearing on mantle dynamics. *Geological Society of America Bulletin*, **85**, 1677–90.

Bills, B. G. and Ferrari, A. J. (1978). Mars topography and geophysical implications. *Journal of Geophysical Research*, **83**, 497–508.

Boston, P. J., Ivanov, M. V., and McKay, C. P. (1992). On the possibility of chemosynthetic ecosystems in subsurface habitats on Mars. *Icarus*, **95**, 300–8.

Bradbury, J. P. (1966). Pleistocene-Recent geologic history of Zuni Salt lake, New Mexico. *New Mexico Geological Society 17th Field Conference Guidebook, Taos-Raton-Spanish Peaks Country, New Mexico and Colorado*, p. 119.

Carden, J. R. and Laughlin, A. W. (1974). Petrochemical variations within the McCartys lava flow, Valencia County, New Mexico. *Geological Society of America Bulletin*, **85**, 1479–84.

Carr, M. H. (1996). *Water on Mars*. New York: Oxford University Press.

Carr, M. H. (1979). Formation of Martian flood features by release of water from confined aquifers. *Journal of Geophysical Research*, **84**, 2995–3007.

Christiansen, R. L. and McKee, E. H. (1978). Late Cenozoic volcanic and tectonic evolution of the Great Basin and Columbia Intermontane regions. In *Cenozoic Tectonics and Regional Geophysics of the Western Cordillera*, ed. R. B. Smith, and G. P. Eeaton, Geological Society of America Memoir 152, pp. 283–311.

Christiansen, R. L. and Yeats, R. S. (1992). Post-Laramide geology of the US Cordilleran region. In *The Geology of North America, The Cordilleran Orogen; Conterminous U.S.* Geological Society of America, G-3, pp. 261–406.

Clifford, S. M. (1993). A model for the hydrologic and climatic behavior of water on Mars. *Journal of Geophysical Research*, **98**, 10973–1016.

Condit, C. D., Crumpler, L. S., and Aubele, J. C. (1999). Lithologic, geochemical, and paleomagnetic maps of the Springerville volcanic field, Arizona. *US Geological Survey Miscellaneous Investigations map.* I-2431, scale 1:50,000.

Crown, D. A. and Greeley, R. (1993). Volcanic geology of Hadriaca Patera and the eastern Hellas region of Mars. *Journal of Geophysical Research*, **98**, 3441–51.

Crumpler, L. S. (1980a). Alkali basalt through trachyte suite and volcanism, Mesa Chivato, Mount Taylor volcanic field, New Mexico, part I. *Geological Society of America Bulletin*, **91**, 253–5.

Crumpler, L. S. (1980b). Alkali basalt through trachyte suite and volcanism, Mesa Chivato, Mount Taylor volcanic field, New Mexico, part II. *Geological Society of America Bulletin*, **91**, 1293–313.

Crumpler, L. S., Head, J. W., and Aubele, J. C. (1996). Calderas on Mars: characteristics, structural evolution, and associated flank structures. In *Volcanic Instability on Earth and Other Planets*. Geological Society of London Special Publication 110, 307–48.

Crumpler, L. S. and Revenaugh, J. (1997). Hot spots on Earth, Venus, and Mars: spherical harmonic spectra. *Abstracts of Papers Submitted to the 27th Lunar and Planetary Science Conference*, pp. 275–6.

Cummings, D. (1968). Geologic map of Zuni Salt Lake volcanic crater, Catron County, New Mexico. *US Geological Survey Miscellaneous Investigations map. I-544*, scale 1:6,000.

Delaney, P. T. and Pollard, D. D. (1981). Deformation of host rocks and flow of magma during growth of minette dikes and breccia-bearing intrusions near Ship Rock, New Mexico. US Geological Survey Professional Paper, 1202, 61 pp.

Dunbar, N. W. (1999). Cosmogenic ^{36}Cl-determined age of the Carrizozo lava flows, south-central New Mexico. *New Mexico Geology*, **21**, 25–8.

Ehrenberg, S. N. (1978). Petrology of potassic volcanic rocks and ultramafic xenoliths from the Navajo volcanic field, New Mexico and Arizona. Ph.D. dissertation, University of California, Los Angeles.

Elston, W. E., Aubele, J. C., Crumpler, L. S., and Eppler, D. B. (1976a). Influence of tectonic setting, composition, and erosion on basaltic landforms, New Mexico and Mars. *Reports of Planetary Geology Programs, NASA Technical Memo*, TMX-3364, pp. 129–32.

Elston, W. E., Rhodes, R. C., Coney, P. J., and Deal E. G. (1976b). Progress report on the Mogollon Plateau volcanic field southwestern New Mexico: 3. Surface expression of a pluton. *New Mexico Geological Society Special Publication*, **5**, 3–28, Fig. 18.

Ernst, R. E., Grosfils, E., and Mége, D. (2001). Giant dyke swarms: Earth, Venus, Mars. *Annual Reviews of Earth & Planetary Science*, **29**, 489–534.

Fanale, F. P. (1976). Martian volatiles: their degassing history and geochemical fate. *Icarus*, **28**, 179–202.

Fanale, F. P., Savail, J. R., Zent, A. P., and Postawko, S. E. (1986). Global distribution and migration of subsurface ice on Mars. *Icarus*, **67**, 1–18.

Faris, K. B. (1980). A geochemical model for the magmatic history of the Carrizozo basalt field, south-central New Mexico. M.S. thesis, New Mexico Institute of Mining and Technology, Socorro.

Freeze, R. A. and Cherry, J. A. (1979). *Groundwater*. Englewood Cliffs, NJ: Prentice-Hall.

Frey, H. and Jarosevich, M. (1982). Sub-kilometer Martian landscapes: properties and possible terrestrial analogs. *Journal of Geophysical Research*, **87**, 9867–79.

Goff, F., Gardner, J. N., Baldridge, W. S. *et al.* (1989). Excursion 17B: volcanic and hydrothermal evolution of the Valles Caldera and the Jemez volcanic field. In *Field Excursions to Volcanic Terranes in the Western United States, Volume I, Southern Rocky Mountains*. New Mexico Bureau of Mines and Mineral Resources Memoir 46, pp. 381–434.

Greeley, R. and Crown, D. A. (1990). Volcanic geology of Tyrrhena Patera, Mars. *Journal of Geophysical Research*, **95**, 7133–49.

Greeley, R., Fagents, S. A., Bridges, N. *et al.* (2000). Volcanism on Mars. In *Environmental Effects on Volcanic Eruptions: From the Deep Ocean to Deep Space*, ed. T. Gregg and J. Zimbelman. Kluwer Academic/Plenum Publishers, pp. 75–112.

Gulick, V. C. and Baker, V. R. (1990). Origin and evolution of valleys on Martian volcanoes. *Journal of Geophysical Research*, **95**, 14325–44.

Hamblin, W. K. and Christiansen, E. H. (1990). *Exploring the Planets*. New York: Macmillan.

Hoffer, J. M. (1976). Geology of the Potrillo basalt field. *New Mexico Bureau of Mines & Minerals Research Circular*, **149**, 30.

Hubbert, M. K. (1953). Entrapment of petroleum under hydrodynamic conditions. *American Association of Petroleum Geology Bulletin*, **37**, 1954–2026.

Keszthelyi, L. P. and Pieri, D. C. (1993). Emplacement of the 75-km-long Carrizozo lava flow field, south-central New Mexico. *Journal Volcanology & Geothermal Research*, **59**, 59–75.

Kilburn, C. R. J. (2000). Lava flows and flow fields. In *Encyclopedia of Volcanology*, ed. H. Sigurdsson *et al.* Academic Press, pp. 291–305.

Laughlin, A. W., Poths, J., Healey, H. A., Reneau, S., and WoldeGabriel, G. (1994). Dating of Quaternary basalts using the cosmognic 3He and 14C methods with implications for excess 40Ar. *Geology*, **22**, 135–8.

Longhi, J. (1990). Magmatic processes on Mars: insights from SNC meteorites. *Abstracts of Papers Submitted to the 21st Lunar and Planetary Science Conference*. Houston: Lunar and Planetary Institute, pp. 716–17.

Lorenz, V. (1973). On the formation of maars. *Bulletin Volcanology*, **37**, 183–204.

Luedke, R. G. and Smith, R. L. (1978). Map showing distribution, composition, and age of late-Cenozoic volcanic centers in Arizona and New Mexico. *US Geological Survey, Miscellaneous Geological Investigations Map*. I-1091, 2 sheets.

Maxwell, C. H. (1982). El Malpais. *New Mexico Geological Society, 33rd Field Conference Guidebook, Albuquerque Country II*, pp. 299–301.

Maxwell, C. H. (1986). Geologic map of El Malpais lava field and surrounding areas, Cibola County, New Mexico. *US Geological Survey Miscellaneous Geological Investigations Map*. I-1595, 1:62,500 scale.

McGetchin, T. and Smyth, T. R. (1978). The mantle of Mars: some possible geologic implications of its high density. *Icarus*, **34**, 512–36.

McGetchin, T. and Ullrich, G. W. (1973). Xenoliths in maars and diatremes, with inferences for the moon, Mars, and Venus. *Journal of Geophysical Research*, **78**, 1833–53.

McKay, C. P. and Stoker, C. R. (1989). The early environment and its evolution on Mars: implications for life. *Reviews of Geophysics*, **27**, 189–214.

McMillan, N. J., Dickin, A. P., and David Haag, D. (2000). Evolution of magma source regions in the Rio Grande rift, southern New Mexico. *Geological Society of America Bulletin*, **112**, 1582–93.

Mouginis-Mark, P. J. (1985). Volcano/ground ice interactions in Elysium Planitia, Mars. *Icarus*, **64**, 265–84.

Mouginis-Mark, P. J., Wilson, L., and Head, J. W. (1982). Explosive volcanism at Hecates Tholus, Mars: investigation of eruptiuon conditions. *Journal of Geophysical Research*, **87**, 9890–904.

Mouginis-Mark, P. J., Wilson, L., and Zuber, M. (1992). The physical volcanology of Mars. In *Mars*, ed. H. Kieffer, B. Jalosky, C. Snyder, and M. Matthews. Tucson: University of Arizona Press, pp. 424–52, 523–54.

New Mexico Bureau of Geology and Mineral Resources (2003). Geologic map of New Mexico, 1:500,000 scale. New Mexico Bureau of Geology and Mineral Resources, 2 sheets.

Nichols, R. L. (1946). McCartys basalt flow, Valencia County, New Mexico. *Geological Society of America Bulletin*, **57**, 1049–86.

Reimers, C. E. and Komar, P. D. (1979). Evidence for explosive volcanic density currents on certain Martian volcanoes. *Icarus*, **39**, 88–110.

Renault, J. (1970). Major-element variations in the Potrillo, Carrizozo, and McCarty's basalt fields, New Mexico. New Mexico Bureau Mines Mineral Resources, Circular 113, 22 pp.

Self, S., Goff, F., Gargner, J. N., Wright, J. V., and Kite, W. M. (1986). Explosive rhyolitic volcanism in the Jemez Mountains: vent locations, caldera development, and relation to regional structure. *Journal of Geophysical Research*, **91**, 1779–98.

Semken, S. C. (2001). The Navajo volcanic field. In *Volcanology of New Mexico*, ed. L. S. Crumpler and S. G. Lucas, New Mexico Museum of Natural History and Science Bulletin, 18, pp. 79–83.

Sheridan, M. F. and Wohletz, K. H. (1983). Hydrovolcanism: basic considerations and review. *Journal Volcanology & Geothermal Research*, **17**, 1–29.

Smith, R. L. and Bailey, R. A. (1968). Resurgent cauldrons. In *Studies in Volcanology – A Memoir in Honor of Howel Williams*. Geological Society of America Memoir 116, pp. 613–62.

Smith, R. L. and Leudke, R. G. (1984). Potentially active volcanic lineaments and loci in western conterminous United States. In *Explosive Volcanism; Inception, Evolution, and Hazards*. Washington, DC: National Academy of Sciences, pp. 47–66.

Smith, R. L., Bailey, R. A., and Ross, C. S. (1970). Geologic map of the Jemez Mountains, New Mexico. *US Geological Survey Miscellaneous Geologic Investigations Map*. I-571, 1:125,000 scale.

Squyres, S. W. and Carr, M. H. (1986). Geomorphic evidence for the distribution of ground ice on Mars. *Science*, **231**, 249–52.
Squyres, S. W., Wilhelms, D. E., and Moosman, A. C. (1987). Large-scale volcano-ground ice interaction on Mars. *Icarus*, **70**, 385–408.
Squyres, S. W., Clifford, S. M., Kuzmin, R. O., Zimbelman, J. R., and Costard, F. M. (1992). Ice in the Martian regolith. In *Mars*, ed. H. Kieffer, B. Jalosky, C. Snyder, and M. Matthews. Tucson: University of Arizona Press, pp. 523–54.
Thelig, E. (1990). The Zuni-Bandera field. In *Volcanoes of North America*, ed. C. Wood and J. Kienle. New York, NY: Cambridge University Press, pp. 303–5.
Thordarson, T. and Self, S. (1993). The Laki (Skaftár Fires) and Grímsvötn eruptions in 1783–1785. *Bulletin of Volcanology*, **55**, 233–63.
Treiman, A. H. (1986). The parental magma of the Nakhla achondrite: ultrabasic volcanism on the shergottite parental body. *Geochimica Cosmochimica Acta*, **50**, 1061–70.
Williams, H. (1936). Pliocene volcanoes of the Navajo-Hopi country. *Geological Society of America Bulletin*, **47**, 111–72.
Wilson, L. and Head J. W. (1994). Mars: review and analysis of volcanic eruption theory and relationships to observed landforms. *Reviews of Geophysics*, **32**, 221–63.
Wilson, L. and Head, J. W. (2002). Tharsis-radial graben systems as the surface manifestation of plume-related dike intrusion complexes: models and implications. *Journal of Geophysical Research*, **107** (E8), 101029/2001JE001593.
Wohletz, K. H. (1986). Explosive magma–water interactions: thermodynamics, explosion mechanisms, and field studies. *Bulletin of Volcanology*, **48**, 245–64.
Zimbelman, J. R. (1998). Emplacement of long lava flows on planetary surfaces. *Journal of Geophysical Research*, **10**, 27503–16.
Zimbelman, J. R. and Johnston, A. K. (2001). Improved topography of the Carrizozo lava flow: implications for emplacement conditions. In *Volcanology in New Mexico*, ed. L. Crumpler and S. Lucas. New Mexico Museum of Natural History and Science Bulletin 18, 131–6.
Zimbelman, J. R. and Johnston, A. K. (2003). New precision topographic measurements of the Carrizozo and McCartys basalt flows, New Mexico. *New Mexico Geological Society 53rd Field Conference, Geology of White Sands*, pp. 121–7.

5

Comparison of flood lavas on Earth and Mars

Laszlo Keszthelyi
US Geological Survey, Flagstaff

and

Alfred McEwen
University of Arizona

5.1 Introduction

Flood lavas, by definition, cover vast areas in great sheets of lava, without the construction of major edifices (e.g., Geikie, 1880; Washington, 1922; Tyrrell, 1937; Self *et al.*, 1997). The flat terrain that flood lavas produce has led to the term "plateau volcanism" to be used as a synonym for flood volcanism. In addition, the classic erosion pattern of flood lavas leaves a series of topographic steps. Thus many flood basalt provinces are known as "traps" from the Scandinavian word for steps. Plateau volcanism transitions to "plains" volcanism when low shields become common (Greeley and King, 1977). It is not surprising that these large-volume eruptions are usually composed of the most common of volcanic rocks: basalt. Thus, the term "flood basalt" is often used interchangeably with "flood volcanism." However, there can be interesting and significant compositional variability within flood "basalt" provinces. The most general term to describe all large-volume volcanism is "Large Igneous Province" (LIP) (e.g., Coffin and Eldholm, 1994).

LIPs represent a major geologic event with significant repercussions on the interior of a planetary body. The extraction of such large volumes of magma can alter the thermal state of the mantle, indicate major changes in the convection patterns within the mantle, and lead to geochemical evolution of the mantle on a regional scale (e.g., Coffin and Eldholm, 1994 and references therein). Flood lavas also alter the face of a planet for geologically significant time. On Earth, it takes of the order of a hundred million years to erode a flood basalt province, and the lunar mare have survived for billions of years. These massive eruptions can also have very significant impacts on

The Geology of Mars: Evidence from Earth-based Analogs, ed. Mary Chapman. Published by Cambridge University Press.

a planet's atmosphere and climate (e.g., Plescia, 1993; Thordarson and Self, 1996). Thus flood volcanism affects a wide range of geologic processes and cannot be ignored by anyone who wants to understand a large rocky planetary body.

In this chapter we provide a brief overview of flood volcanism on Mars and the Earth and then delve into some specific similarities and differences in the lavas on the two planets. We end with a discussion of the major unanswered questions regarding these immense geologic features.

5.2 General observations of flood lavas on Mars

Mars is famous for having the Solar System's largest shield volcanoes, but it also boasts immense flood lavas. Mapping suggests that much of the Martian plains are covered by flood lavas (e.g., Scott and Carr, 1978; Scott and Tanaka, 1986; Greeley and Guest, 1987; Tanaka and Scott, 1987). A very conservative estimate by Greeley and Schneid (1991) states that at least 46% of Mars is covered by volcanics and 82.4% of this area is composed of volcanic plains and plateaus. Thus, a minimum of 38% of Mars' surface is covered by flood lavas. A look at the units that have been mapped as likely to contain fresh or modified flood lavas (Figure 5.1) indicates that nearly half of Mars was covered by flood lavas.

Recent results from Mars Global Surveyor indicate an even greater role for flood volcanism than indicated by Figure 5.1. Most of the Northern Plains are covered by tectonic ridges (Withers and Neumann, 2001), which have characteristics consistent with buried wrinkle ridges, and are laterally continuous with Early Hesperian-aged ridged plains mapped as flood lavas (Head *et al.*, 2002b). However, Tanaka *et al.* (2003) suggest that these units could be sedimentary. Head *et al.* estimate that an additional 10% of Mars' surface was buried by lavas to an average thickness of 800–1000 m. In addition, several workers (e.g., Malin and Edgett, 2001; Bradley *et al.*, 2002) have noted that the Medusae Fossae Formation (MFF) overlies lava in places, and Keszthelyi *et al.* (2000) suggested that the MFF consists of tephra produced by the Cerberus plains flood lavas. The MFF covers 2.1×10^6 km^2, but how much of this terrain is underlain by lava is poorly known.

In addition, much of the southern highlands, although mapped as undivided Noachian units, have the compositional signature of basalt (Christensen *et al.*, 2001). These regions may consist of Noachian flood basalts (and other volcanics) that have been heavily modified by impact and fluvial processes. Indeed, a significant result from the Thermal Emission

Figure 5.1. Distribution of possible flood lavas on the surface of Mars. Simplified from global maps (Scott and Carr, 1978; Scott and Tanaka, 1986; Tanaka and Scott, 1987; Greeley and Guest, 1987) and placed over MOLA global shaded relief topographic map. Areas in red have been either mapped as flood lavas or as likely to contain modified flood lavas. Yellow shows other primarily volcanic units. All remaining materials are shown in light blue. The red areas cover 47% of the surface of Mars, which is marginally more than the very conservative estimate of 38% by Greeley and Schneid (1991). (For a color version of this figure, please refer to color plate section.)

Spectrometer (TES) is that only volcanic compositions have been detected, aside from dust and volatiles. Even the relatively bright and finely layered deposits like "White Rock" lack spectral features of carbonates or sulfates and appear similar to Martian dust (Ruff *et al.*, 2001).

Flood lavas may also make up a significant volume fraction of Mars' crust. McEwen *et al.* (1999) reported >8 km of layers in the outer slopes of Valles Marineris (Figure 5.2); they favored the interpretation that this is primarily a stack of flood lavas. Malin and Edgett (2001) disputed that interpretation because there are few large boulders (or chunks of detached bedrock >5 m diameter) on the floor of Valles Marineris, except on landslides. We think this distribution of large boulders is compatible with a flood lava interpretation. The lavas will tend to break up into sizes dictated by the spacing of cooling joints, typically 1–2 m or smaller. Basaltic lava rocks that tumble many kilometers to the floor of Valles Marineris, or are carried for kilometers in outflow channel floods, will break into boulders 1–2 m in diameter or smaller. Large boulders resolvable by the Mars Orbital Camera (MOC) have either slid downhill only a small distance from their source, or were carried in landslides that largely preserved the

Figure 5.2. Layered deposits in the walls of Valles Marineris. Portion of MOC image E11-03502 showing layers in the outer slopes of Coprates Chasma near 14.4° S, 63° W. Scene is 3 km wide, and image was acquired at 3.6 m/pixel; north is down.

original stratigraphy. Beyer and McEwen (2004) reported that blocks can be seen within a few hundred meters downslope from strong (cliff-forming) layers within Valles Marineris, but not near the thicker stacks of weaker layers.

Several recent studies support the flood lava interpretation for the crustal layers exposed by Valles Marineris. Schultz (2002) showed that wallrock strengths are consistent with layered igneous rocks. However, Beyer and McEwen (2004) suggested that much of the overall wall strength in Coprates Chasma may be controlled by a few especially strong layers, so the majority of the layers could be much weaker, perhaps rich in tephra. The crustal layers in Valles Marineris are rather coarse and thick bedded (Figure 5.2), distinct from the finer layers seen inside the canyons and elsewhere (Malin and Edgett, 2001). Ori and Karna (2003) have mapped out the global distribution of thick-bedded units; they are concentrated around Tharsis (including Valles Marineris) and Elysium, the major volcanic regions of Mars.

Mapping shows that flood volcanism has been important throughout the geologic history of Mars. Not surprising is the indication that volcanism was more vigorous earlier in Mars' history and that it seems to have been gradually decaying as the planet cools (Tanaka, 1986; Greeley and Schneid, 1991). Some workers have suggested the fact that surface units mapped as lavas are dominantly Hesperian indicates that there was a peak in volcanism in the early Hesperian (Greeley and Schneid, 1991; Head *et al.*, 2002b). However, these surface lavas may cap thick sequences that extend back into the Noachian, as suggested by the crustal layers exposed in Valles Marineris (McEwen *et al.*, 1999). This, and the possibility that most Noachian terrains are modified flood lavas, would eliminate the putative Hesperian peak.

Perhaps more intriguing is the indication of very young flood lavas (Plescia, 1990). A region in the Cerberus and Amazonis plains, approximately the size of Canada, is covered with relatively fresh-looking flood lavas (Keszthelyi *et al.*, 2000) (Figure 5.3). The sparsity of craters on some of these lavas has been modeled to indicate surface exposure ages of less than 10 million years (Hartmann *et al.*, 1999; Hartmann and Berman, 2000). These models may be misleading if small craters on Mars are dominated by secondaries (McEwen, 2003); the few craters that are found on these young lavas could all be secondary craters, most from the 10 km diameter impact crater Zunil (McEwen *et al.*, 2003). Another source of age uncertainty is the possible role of burial and exhumation by a mobile mantling deposit. However, based on the fact that no crater larger than 500 m in diameter has been found superimposed on the freshest lavas, they must have been erupted within the last ~100 Ma (P. Lanagan and A. S. McEwen, submitted).

Further evidence for extensive flood volcanism in the recent past comes from the Martian meteorites. There are ~16 meteorites in 4 groups found on the Earth that are thought to have come from Mars (e.g., Treiman *et al.*, 2000; Nyquist *et al.*, 2001). Table 5.1 shows the ages of these. Impacts that launch material from Mars to Earth may provide a highly non-random

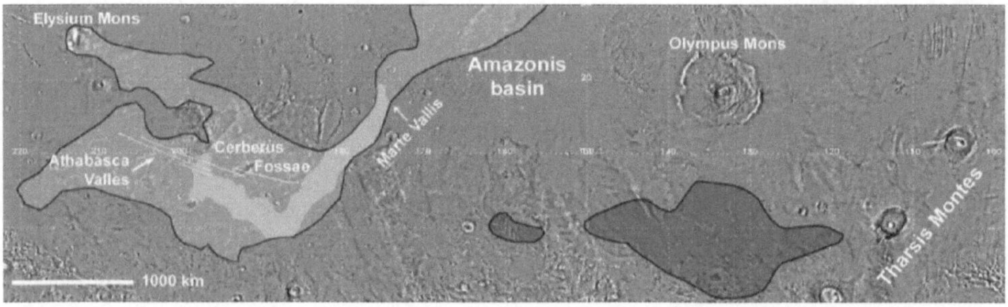

Figure 5.3. Location map of the Cerberus plains and associated features. Light tone shows radar-rough area that appears to correspond to the recent flood lava flows. Lightest tone is the approximate outline of what may be a single, fresh, flood lava flow. The dark shading shows the area with very low radar return that may be related to pyroclastic deposits from the flood lava eruptions. After Keszthelyi *et al.* (2000).

Table 5.1. *Crystallization and exposure ages of Martian meteorites*

Meteorite group	Crystallization age (Ga)	Exposure (~ejection) age (Ma)
ALH84001	~4.5	15
Nakhlites	~1.3	11
Chassignites	~1.3	11
Shergottites	~0.2–0.5	0.7–20

sample, favoring young lavas where a strong surface layer enhances spallation (Head *et al.*, 2002a). Nevertheless, the dominance of Martian meteorites with ages less than 2 Ga is strong evidence that significant parts of Mars are covered by relatively young igneous rocks. In several cases, petrographic investigations of these meteorites favor thick lava flows as the source rocks (e.g., Nyquist *et al.*, 2001). This suggests that we already have samples of geologically young Martian flood lavas on the Earth. These samples indicate that the Martian lavas are basaltic, but with interesting geochemical differences from common terrestrial basalts.

The young age of many of the Martian flood lavas provides a unique opportunity to observe the surface morphology of flood lavas. The best-preserved flood lava flows are in the Elysium Planitia, and are fed from the Cerberus Fossae (Plescia 1990; Keszthelyi *et al.*, 2000; P. Lanagan and A. S. McEwen, submitted), located near the equator and southeast of Elysium Mons (Figure 5.3). These flows have a distinctive "platy-ridged" surface morphology (Figure 5.4), that is uncommon on terrestrial lava flows (Keszthelyi *et al.*, 2000). We hypothesized that the plates form in a two-stage process: first the lava advances relatively slowly and a thick, coherent, stable crust is able to form; later a surge in the lava flux disrupts this crust and transports large pieces as rafts on a wave of molten lava. The ridges would form when rafts crash against each other and push up a pile of broken and crushed lava. Modeling of this process suggests that eruption rates of $10^3 - 10^6$ m^3/s are needed during the surges and that the lava rheology must be broadly similar to terrestrial basalts. Individual eruptions may have lasted of the order of a decade to output the voluminous (~10^4 km^3) lava flows (Keszthelyi *et al.*, 2000).

One consequence of these massive eruptions should have been dramatic (short-term) changes to the Martian atmosphere and climate. Plescia (1993) estimated that the eruption of the group of flood lavas that covered the Cerberus plains could have produced ~10^{16} kg of water and a similar amount of CO_2. This is enough CO_2 to double the atmospheric pressure

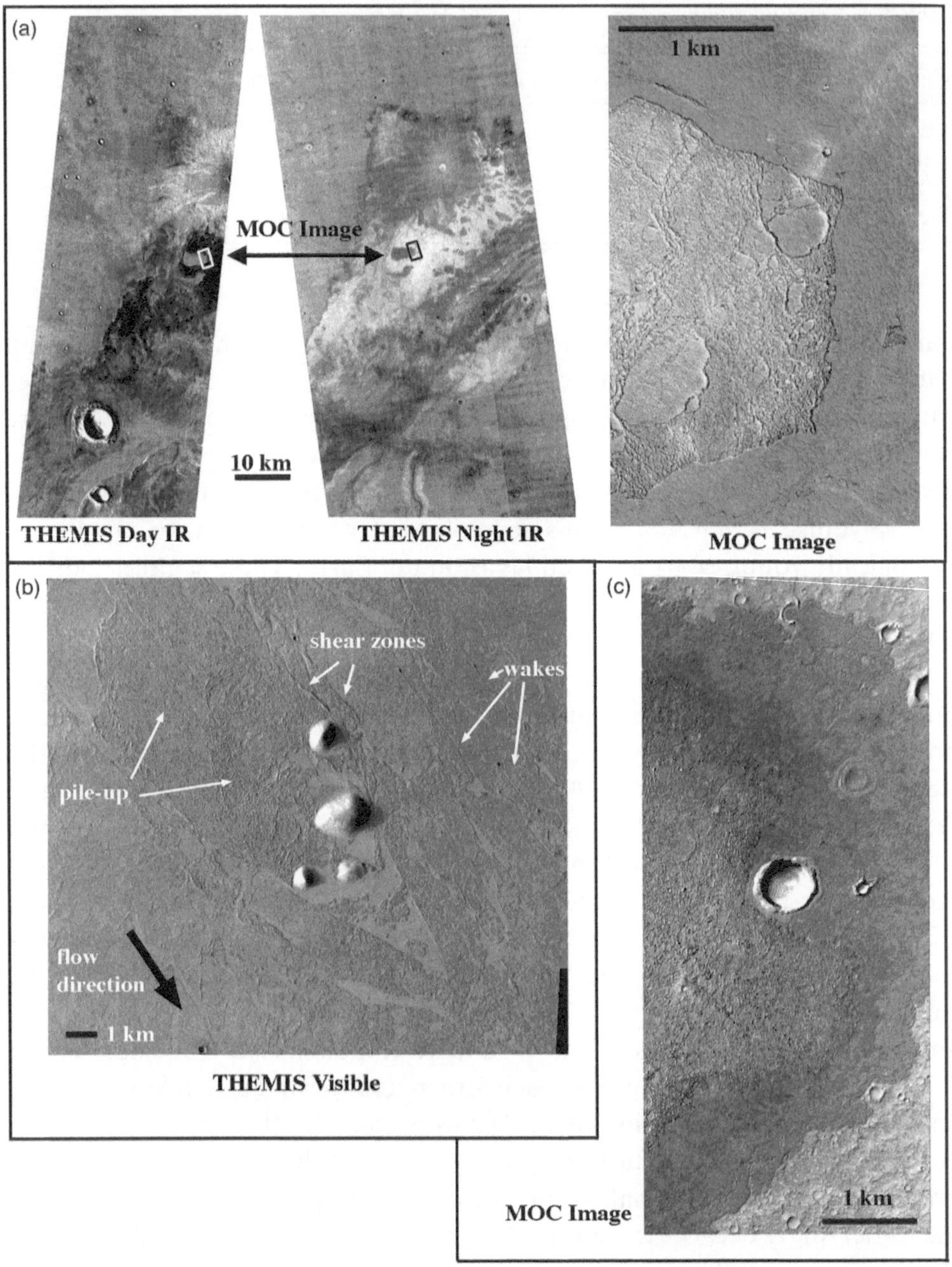

Figure 5.4. Examples of Martian platy-ridged flood lava flows. (a) Portions of THEMIS infrared observations I01118002 (day), I00875002 (night) and I01237007 (night) and MOC image E11-03799 from near 9° N, 205° W, Athabasca Valles, Elysium Planitia, Mars. THEMIS thermal infrared images show surface temperatures at ~4am and 4pm local time. Note that much of the flat lava is relatively cool in the day and warm at night, indicating a high

and enough water to blanket all of Mars in a layer of ice ~10 cm deep. However, this activity is likely to have been distributed as discrete events across tens or hundreds of millions of years. While smaller, the impact of individual eruptions could still have been global (Keszthelyi and McEwen, 2001; P. Lanagan and A. S. McEwen, submitted). The same system of fissures that fed the most recent flood lavas also seem to have fed the most recent water floods on Mars (Burr *et al.*, 2002). It is likely that the dike intrusions that fed the lava eruptions also provided the pathway for subsurface water to erupt to the surface (Mitchell *et al.*, 2003). The giant system of radial graben surrounding Tharsis has also been interpreted as the surface expression of dikes (Mege and Masson, 1996; Wilson and Head, 2002), which may have fed much of the Noachian and Hesperian-age flood lavas.

How common are andesitic flood lavas on Mars? Evidence for andesitic lavas has been reported from the Mars Pathfinder APXS experiment, in spite of uncertainties in calibration and weathering (Wanke *et al.*, 2001). In addition, TES has shown two major rock spectral types, interpreted as more basaltic and more andesitic lava compositions (Christensen *et al.*, 2001). Large volumes of andesite on Earth are associated with subduction zones, and are not expected on Mars. However, an alternative explanation for type 2 ("andesitic") spectra is alteration increasing the silica content of surface coatings on basalt (Wyatt and McSween, 2002). The type 2 regions consist of apparent sediments on top of ridged plains in the northern plains (Wyatt and Tanaka, 2003), so these may be largely transported and weathered materials, not indicative of the composition of the flood lavas in the northern plains (Head *et al.*, 2002b). Preliminary results from

Caption for Figure 5.4. (cont.) thermal inertia and significant dense rock in the upper few centimeters. However, the ridged section, in this case, has relatively low thermal inertia. This could be due to trapping of large amounts of dust, which would also explain why the area seems bright in the MOC image taken in visible wavelengths. North is to the top in the THEMIS images. THEMIS images are 100 m/pixel; MOC image is 3.63 m/pixel. (b) Portion of THEMIS visible image V06012001, near latitude 2.5, longitude 195.7° W, central Cerberus plains. Lava shows shear structures and wakes in the lava as well as the pile-up of debris on the flow top due to topographic obstacles. The flow direction and some properties of the translating upper crust can be determined from this observation. Image 18 m/pixel, north is to the top. Solar incidence angle of nearly 80° from the west (left) accentuates the topography. (c) Portion of MOC image M01-00111, at 5.8° N, 215° W, western Cerberus plains. Image shows lava flow transition from pahoehoe along margins to platy-ridged interior, similar to flows in Iceland. Flow margin is no more than a few tens of meters thick. Image 5.9 m/pixel, north is approximately up.

the Gamma Ray Spectrometer (GRS) on Mars Odyssey suggests very low silica abundances, lower than the Mars meteorites (Boynton *et al.*, 2003), whereas the TES data indicate compositions richer in silica than the Mars meteorites. Both of the MER landers have detected olivine in the soils, suggesting mafic compositions (Klingerhaufer *et al.*, 2004). McSween *et al.* (2003) examine the recent data in a global geochemical context and find that lava compositions that are dominantly basalt-like provide a tidy, self-consistent model. In summary, we clearly do not yet understand the lava compositions of Mars, but suggest that the Mars meteorites provide the most reliable clues.

5.3 General observations of flood lavas on Earth

Large igneous provinces can be found scattered across the Earth, but make up only a few percent of its surface area (Figure 5.5). The less than a dozen well-preserved and well-exposed continental flood basalt provinces (Table 5.2) are perhaps the best known of these LIPs, but there are also traces of many older flood basalts have been largely erased by the vigorous geologic activity on Earth. Some of the largest and best-preserved flood

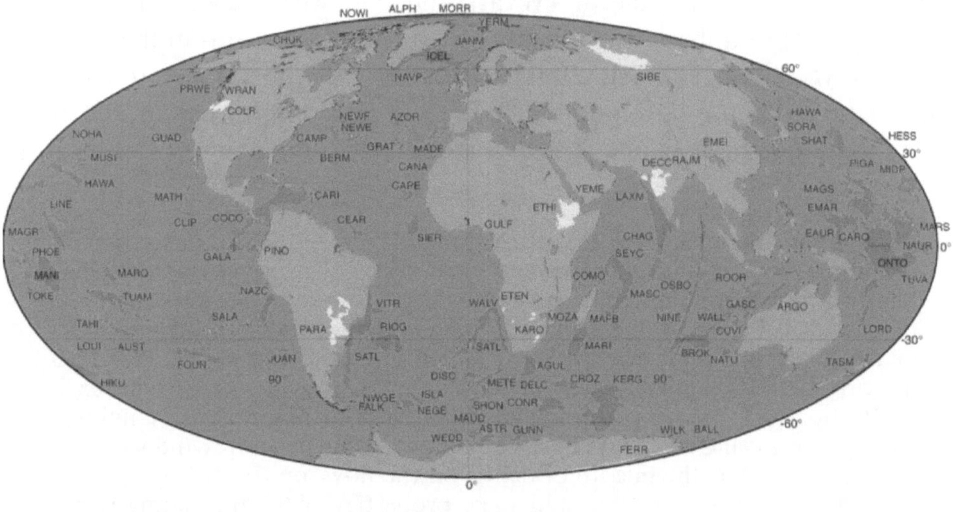

Figure 5.5. Map of Recent (≤250 Ma) Large Igneous Provinces (LIPs) on Earth. Well-preserved continental flood basalt provinces listed in Table 5.2 are depicted in yellow, other LIPs in red. While recent LIPs only cover a few percent of Earth's surface, it is important to note that this only represents a few percent of Earth's geologic history. Map after Coffin and Eldholm (1994). Abbreviated names listed in full in Coffin and Eldholm (1994). (For a color version of this figure, please refer to color plate section.)

Table 5.2. *Well-preserved continental flood basalt provinces*

Name and Figure 5.4 abbreviation	Age of main activity (Ma)	Area (10^6 km^2)	Volume (10^6 km^3)
Columbia River (COLR)	16	0.164	0.175
Ethiopian (ETHI)	30	0.6	~1
Deccan (DECC)	66	≥1.5	>8
Etendeka-Parana (ETEN & PARA)	132	~3	>1.2
Karoo-Ferrar (KARO & FERR)	183	>3	~2.5
Siberian (SIBE)	249	>0.34	>2

Data from Macdougall 1988; Rampino and Stothers, 1988; Tolan *et al.*, 1989; Coffin and Eldholm, 1994; Haggerty, 1996; Mahoney and Coffin, 1997; Duncan *et al.*, 1997. In all cases, the majority of the lava volume was erupted in under 1 million years.

basalt provinces (e.g., the Ontong-Java and Kerguelen Plateaus) are found on the ocean floor, limiting access to them. There are also many other LIPs, some associated with mantle hotspots and others associated with rifting of continents (e.g., Coffin and Eldholm, 1994).

While activity often persists at a low level for ~10 million years in an individual province, it appears that the bulk (~90%) of the volume is erupted in less than 1 million years. The most detailed age information comes from the most recent and best-studied flood basalt province, the Columbia River Basalts. There, the lava extrusion rate built up over about a million years, then flux remained high for several hundred thousand years, and finally tapered off in a roughly exponential fashion over the next several million years (e.g., Tolan *et al.*, 1989). This pattern of lava production has been attributed to a new mantle plume reaching the base of the crust. The head of the plume is very broad and is capable of generating hundreds of thousands of km^3 of magma. Once the plume head has dissipated, the narrow tail of the plume can continue to feed volcanism at a lower rate for many tens of millions of years (e.g., Coffin and Eldholm, 1994; White and McKenzie, 1995).

Magma rising from the mantle is almost invariably mafic in composition. However, in the case of continental flood basalts, the magmas are often significantly contaminated by assimilation of crustal materials during ascent (e.g., Macdougall, 1988; Mahoney and Coffin, 1997). In fact, the bulk of the Columbia River Basalt Group consists of basaltic andesite, not basalt (e.g., Hooper, 1997). In other cases, the primary mafic magmas pond in the lower crust and provide heat for melting the continental rocks. The result is that there can be considerable diversity in the types of lavas erupted in a continental flood basalt province. For example, the Paraná-Etendeka

Figure 5.6. Photos from various flood basalt provinces. (a) Deccan Traps, near Mahabaleshwar, India. (b) Deccan Traps, Ajunta Caves, India. (c) Etendeka Basalts, near Huab, Namibia. (d) Columbia River Basalts,

province has large volumes of lava with compositions ranging from 48 to 72 wt.% SiO_2. The lavas include numerous examples of basalts, basaltic-andesites, dacites, trachydacites, and rhyolites (Marsh *et al.*, 2001). The Siberian Traps similarly contain a mix of lavas dominated by basalts and basaltic-andesites, but also including significant volumes of picrites, trachybasalts, trachyandesites, basanites, nephelinites, and maimechites (e.g., Sharma, 1997).

While the terrestrial flood lavas generally do not have their surfaces exposed, it is possible to infer their morphology from cross-sectional views afforded by erosion (Figure 5.6). These views show that the majority of the flood lavas are inflated pahoehoe flows. They have the smooth, piece-wise continuous surfaces that define pahoehoe. They also have a three-part internal vesicle distribution that is diagnostic of the inflation process. Thus the lavas were transported hundreds of kilometers from their vents underneath an insulating crust. The emplacement of a typical $\sim$1000 km^3 flood lava flow is estimated to have taken about a decade (Self *et al.*, 1997; Thordarson and Self, 1998). There are also flows that have brecciated flow tops and other morphologic features suggesting more rapid emplacement (Keszthelyi *et al.*, 2001). The techniques used to infer the emplacement style from lava morphology is discussed in more detail later in this chapter.

One intriguing aspect of flood basalt eruptions is their apparent temporal coincidence with mass extinctions (Rampino and Stothers, 1988; Haggerty, 1996; Wignall, 2001). All major extinction events in the past 300 Ma have flood basalt eruptions occurring at the same time. However, the correlation is not seen in the preserved geologic record older than 300 Ma. Furthermore, the more recent and smaller flood basalt eruptions were not concurrent with major mass extinctions (Wignall, 2001). The precise role of flood lava eruptions on mass extinctions is still highly uncertain. It is unambiguous that these massive eruptions had significant and deleterious effect on global climate (Thordarson and Self, 1996). While this should stress the ecosystem and kill many individuals of many different species, this may not be sufficient to cause a mass extinction. We speculate

Caption for Figure 5.6. (cont.) near Dry Falls, Washington, USA. Note the step-like erosion of the lava flows that gave flood basalt provinces the name "traps." Flood basalt flows are typically a few to several tens of meters in thickness and are hundreds of kilometers long. Individual lobes can be traced for many tens of kilometers. Outcrops available for study are usually controlled by erosion. The temples carved into the Deccan Traps provide a unique opportunity to see the lava structures in three dimensions.

that if any additional stress is applied to the ecosystem (e.g., a bolide impact) at the same time as a flood basalt province is being emplaced, the result will be catastrophic for life on Earth. Calling upon two processes to create mass extinctions may help explain the complex spatial and temporal pattern of extinction seen at both the Permo-Triassic and Cretaceous-Tertiary boundaries (e.g., Erwin, 1994; Sutherland, 1994; Johnson *et al.*, 2002).

5.4 Interpreting lava morphologies on Earth and Mars

Terrestrial flood lavas are predominantly inflated pahoehoe but the most recent Martian flood lavas are dominated by platy-ridges. Why? How can these flow morphologies be used to estimate eruption parameters such as effusion rates?

The formation of inflated pahoehoe flows is now quite well understood, thanks to observations of active eruptions in Hawaii (Hon *et al.*, 1994). However, some misconceptions based on earlier Hawaii studies continue to confuse the issue of how terrestrial flood basalt flows were emplaced (e.g., Anderson *et al.*, 1999; Self *et al.*, 2000). Classic pahoehoe forms when relatively fluid lavas are emplaced under relatively low strain rates (Peterson and Tilling, 1980). In Hawaii, these conditions have generally been only met for eruptions with effusion rates $<10-15$ m^3/s (Rowland and Walker, 1990). However, flood basalt lava flows >100 km long could not form at such low effusion rates because the lava would freeze before it traveled 100 km (Keszthelyi and Self, 1998; Anderson *et al.*, 1999). Minimum effusion rates to achieve lengths >100 km are of the order of $100-1000$ m^3/s, depending on the shape of the insulated conduit through which the lava flows (Keszthelyi, 1995; Keszthelyi and Self, 1998; Sakimoto and Zuber, 1998).

At first glance, this seems to imply that flood basalt lava flows cannot be dominantly pahoehoe. This apparent discrepancy between observation and theory is easy to explain. In Hawaii, flows are typically advancing across a $\leq$100 m wide front. Terrestrial flood basalt flows appear to have been 10–100 km wide. Thus, if the lava is subjected to the same strain rate (i.e., rate of advance) in both cases, the volume flux in the flood basalt case should be 100–1000 times that of a Hawaiian flow. To state it differently, a flood basalt eruption with an effusion rate of 1000 m^3/s should generate the same lava morphology as a Hawaiian eruption with an effusion rate of 1–10 m^3/s. Thus the formation of pahoehoe at eruption rates of 2000–4000 m^3/s for flood basalt lava flows (e.g., Self *et al.*, 1997; Thordarson and Self, 1998) is consistent with the result that effusion rates $>15\ m^3/s$ tend to form aa in Hawaii (Rowland and Walker, 1990). Future studies of the pahoehoe to aa transition and the rheology of flood basalt lavas may allow more quantitative constraints to be placed on the way the terrestrial pahoehoe flood lavas advanced.

For now, the best quantitative data come from the inflation process. Pahoehoe lobes and sheet flows 20–50 cm thick have been directly observed to grow into lobes 5–10 m thick. This growth takes place by having new liquid lava enter the interior of the lobe and push the solidified crust upward. The result is a distinctive and diagnostic set of surface features (Figure 5.7). If the inflation is spatially localized, a hummocky surface is created, with numerous tumuli (synonymous with "lava-rise"). When the inflation takes place over a broader region, "lava-rise-plateaus" form and "lava-rise-pits"

Figure 5.7. Photos of features diagnostic of inflation. (a) Aerial photo from the Laki Flow Field, Iceland. (b) Tumulus on Kilauea Volcano, Hawaii. (c) Edge of an inflation pit, Grants Flow Field, New Mexico, USA. Inflation generates a suite of diagnostic 1–100 m scale morphologic features. When large areas are uniformly uplifted, an inflation (or "lava-rise") plateau forms. Local areas that are not uplifted are converted into inflation pits. It is common for most inflation plateaus on a flow to reach the same height, indicating that an interconnected sheet of liquid lava (under a solid crust) provides a means of keeping hydrostatic pressure relatively uniform across a wide portion of many lava flows. Where such efficient interconnectivity is lacking, the lava will inflate to form a scattering of smaller tumuli, as in the Kilauea example. Person is 1.5 m tall in both (b) and (c).

are created over the localized areas that do not inflate. Often inflation is concentrated in areas that were depressions in the underlying topography and inflation ceases most rapidly where the flow is unusually thin over high points in the pre-existing topography (Self *et al.*, 1998). Such inverted topography is seen in the inflated pahoehoe margins of the flood lavas in the western Cerberus plains of Mars (P. Lanagan and A. S. McEwen, submitted).

It should be noted that inflation is not restricted to pahoehoe flows. Inflated aa and slab pahoehoe flows can be found in many locations on the Earth. However, nearly every pahoehoe flow has undergone some degree of inflation.

The inflation process tends to form three distinct layers within a flow: an upper vesicular crust, a dense core, and a lower vesicular crust (Self *et al.*, 1997; Thordarson and Self, 1998). While this three-part structure was recognized much earlier (e.g., Aubele *et al.*, 1988), its full implications were not clear until the inflation process was understood. The upper vesicular crust forms as long as fresh, bubble-laden, lava is transported into the lobe. Once the flow stagnates, bubbles rise to the top of the liquid layer within a matter of days. Thus the transition from the upper vesicular crust to the dense core marks the time lava stagnated, to within a few days. The time it took for this upper crust to form is known to be well approximated by the empirical relationship

$$t = 164.8\, H_c^2$$

where t is time in hours and H_c is the thickness of the upper crust (Hon *et al.*, 1994). When compared to complex numerical cooling models, this empirical relationship was found to have an error of $<25\%$ (Self *et al.*, 1998). Thus, it is possible to accurately estimate how long an inflated lava flow was active by measuring the thickness of the upper vesicular crust (Self *et al.*, 1997; Thordarson and Self, 1998). From a careful compilation of the flow duration at many different outcrops and good volume estimates, it is possible to place the best constraints on flood basalt eruption rates. The 1300 km^3 Roza flow must have taken >10 years to emplace and thus the average effusion rate could be no more than 4000 m^3/s (Self *et al.*, 1997; Thordarson and Self, 1998).

While we know how to extract quantitative information (with good estimates of uncertainties) from outcrops of inflated pahoehoe flows, the same cannot be said for platy-ridged flows on Mars. Instead, previous work has relied primarily on theoretical modeling. Keszthelyi *et al.* (2000) assumed that the platy-ridged lavas were emplaced in a two-stage process: initial flow under a stationary crust and secondary flow with a disrupted crust during surges in flow rate. It was assumed that the better-insulated flow under a stationary crust was the dominant mode of transporting lava up to 2000 km from the vents. A simple heat balance model for an insulating sheet flow (Keszthelyi and Self, 1998) was used to calculate the flow thickness and flow rate needed to transport lava under a stationary insulting crust 2000 km without freezing. It was found that the lava rheology needed to be similar to that of terrestrial basalts (of the order of 100 Pa s) if the sheet was the observed thickness of a few tens of meters. Flow rates needed to be 1–5 m/s and for ~10 km wide flows, this translates to an effusion rate of $1-5 \times 10^4$ m^3/s.

Flow during the surges was more difficult to model, because the parameters describing the behavior of the disrupted, mobile, crust are largely unconstrained. However, the thermal model for aa flows produced by Crisp and Baloga (1994) was modified to examine platy-ridged lavas during surges. Many of the plates on Martian flows can be pieced back together, like a jigsaw puzzle, indicating transport distances of a few tens of kilometers. The single most important unknown was the time scale over which the crust is stirred. A wide range of parameter space was examined, and it was determined that lava would be able to advance 10–100 km without freezing during the surges only if the time scale for mixing was >1000 s and flow rates were 10^5-10^6 m^3/s.

While these theoretical calculations produce many quantitative results, they are all dependent on our conceptual model for how platy-ridged lava actually forms. These ideas need to be tested against observations of terrestrial platy-ridged lavas. Keszthelyi *et al.* (2000) found that the 1783–4 Laki Flow Field was an excellent terrestrial analog. In particular, plates, ridges, shear zones, and wakes of similar scale to the Martian features are seen in aerial photos from Laki. Furthermore, the platy-ridged portions of the Laki flow were found to transition into pahoehoe near the flow margins, as is seen on Mars. Observing that these different lava morphologies are spatially related to each other in the same way on both Mars and the Earth is very important because it makes it exceedingly difficult to create hypotheses where the terrestrial and Martian examples are formed by fundamentally different processes.

Fieldwork in Iceland in 2000 provided important additional information to previous knowledge about the Laki lava flows. Specifically, we obtained detailed observations of the crust on a platy-ridged lava flow. These observations provide key constraints on the behavior of this crust before, during, and after it was translated. In some locations, the crust is composed of broken plates but in other areas is a breccia consisting primarily of smooth-skinned clasts that were deformed while plastic (Figure 5.8a). The plates we saw are buckled sections of the top of a typical pahoehoe sheet flow. The thickness of the sheets suggests that the lava in these areas behaved like a normal inflating pahoehoe sheet flow for a few to several hours before the surface was disrupted by compressional forces. In other areas the clasts showed evidence of both plastic and brittle deformation and the pressing of hot clasts onto cooled clasts (Figure 5.8b). Pahoehoe-like lobes of lava could also be seen intruding into the breccia, presumably as fingers of liquid pushing up from below (Figure 5.8c). We interpret these observations as showing evidence for brecciation over an extended period of time. Interestingly, in no case did we find the spinose surfaces and gnarled clast shapes of classic Hawaiian aa flows. This suggests that the deformation within Laki's brecciated flow top took place at low strain rates. The term "rubbly pahoehoe" has been used to describe this autobreccia with pahoehoe clasts (Keszthelyi *et al.*, 2001; Keszthelyi, 2002).

Another observation supporting slow churning of the breccia comes from long grooves carved into the flow top (Figure 5.9). These grooves are over a kilometer long, a few tens of meters wide, and ~4 m deep. The floor of the grooves consists of smooth pahoehoe (Th. Thordarson, personal communication) but the walls and high-standing areas are all composed of breccia. These grooves apparently formed as the crust was pushed into

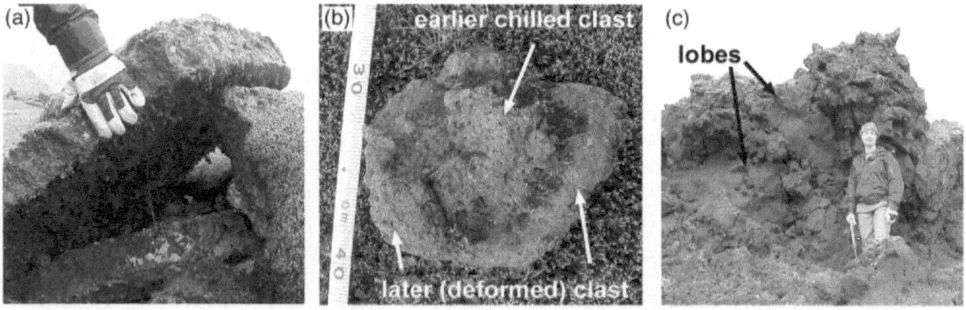

Figure 5.8. The brecciated top of the platy-ridged portion of the Laki Flow Field, Iceland. (a) Broken slabs on the top of the Laki flow. The presence and size of these slabs indicates that the flow was able to cool quietly for a few hours before the surface was disrupted by a surge in lava flux. (b) Ten centimeter clast from a more intensely brecciated flow top with a clast plastically deformed around an earlier chilled clast. This requires that the brecciation process had been sustained for some significant amount of time. (c) Cross-section through the breccia showing two large, pahoehoe lobes intruded into the breccia. These lobes require low strain rate within the breccia, at least locally.

stationary obstacles. This scraped the brecciated material off the top of the flow, forming 15 to 25 m tall mounds of breccia. Fluid lava welled-up into the breach in the crust but the adjacent brecciated crust did not slump into the gap. This means that the brecciated crust had significant strength even while flowing and that the crust was not rapidly overturning. The fact that the grooves gradually healed over a distance of $\sim$1 km suggests that the crust was translated about a kilometer before movement within the breccia exceeded $\sim$10 m. From historical observations, it is estimated that the lava with the grooves advanced about 5 km/day ($<$ 10 cm/s). If this is the case, then the brecciated crust flowed into the grooves over $\geq$5 hours. If the motion within the breccia was on of order 10 m in this time, the internal velocities should have been $<$ 1 mm/s. This very slow internal movement may help explain why classic aa textures were absent from the Laki flow top breccia.

These field observations from Laki provide us with a data point with which to test the plausibility of the theoretical model we used to estimate the eruption rates for the Martian platy-ridged flood lavas. The Laki data suggest that it took of the order of 10^4 s to overturn the brecciated crust of this platy-ridged lava flow. Thus the Laki observations indicate that (at least one) platy-ridged lava did form with a crustal mixing time scale $>$1000 s, providing some confidence that the theoretical modeling of Keszthelyi *et al.* (2000) is reasonable.

5.5 Conclusions

The similarities in flood lava eruptions on Earth and Mars are obvious – both involved the eruption of enormous volumes of mafic lava that inundated vast areas and converted them into plains and plateaus. Vent structures are small and difficult to find in both cases. Flood lava eruptions have played a major role in shaping the surfaces of both planets and the physiographic effect of massive outpourings of mafic lava are similar on the Earth and Mars.

Some differences are also readily apparent. Many Martian flood lavas have extended almost an order of magnitude further than typical terrestrial flood lavas. The dominant lava morphologies are also different, with the Martian platy-ridged lavas being emplaced at higher flow rates than the terrestrial inflated pahoehoe flows. It is likely that the higher flow rates and the longer flow lengths on Mars are related.

The way we have been able to study the lavas on Earth and Mars have also been fundamentally different. On Earth, most of the work has

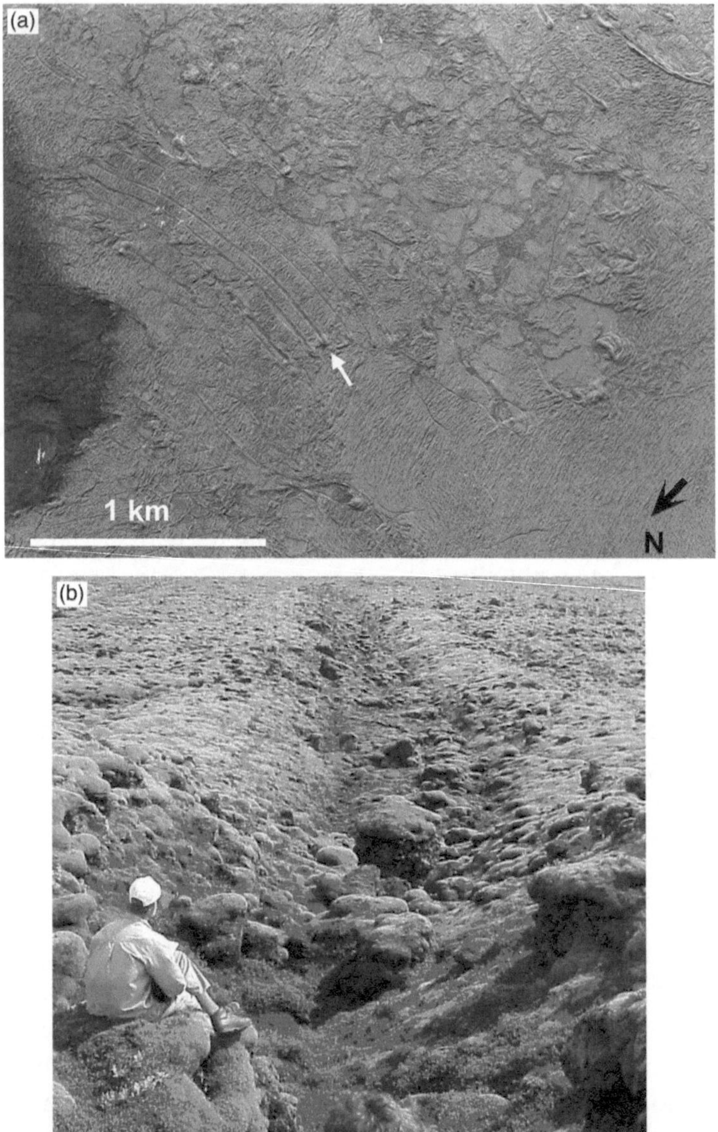

Figure 5.9. Aerial and ground photos of grooves in the flow top of the Laki Flow Field, Iceland. (a) Aerial photo of lavas emplaced in a surge of activity on June 21, 1783. Note the plates and ridges similar to the platy-ridges flood lavas on Mars. Grooves are >1 km long and follow the flow lines of the advancing lava. (b) Photo from location indicated by white arrow in (a). This groove is ~20 m wide and 4 m deep. The walls of the groove consist of breccia

been done on outcrops provided by natural and human-made erosion. The planimetric form of the lavas, and their surface morphology has to be pieced together from what is seen in isolated cross-sections. At the same time, the ready availability of samples has allowed detailed radioisotope dating and geochemical analysis of the mantle source regions. On Mars, the understanding of the flood lavas comes primarily from orbital remote sensing data and theoretical modeling. While the models are guided and constrained by experience with terrestrial analogs, the lack of the kinds of detailed observations that are possible on Earth leave significant uncertainty in the modeling results. However, there are flood lavas on Mars that appear to be much better preserved than any flood lavas on Earth. Thus the spatial distribution and relationships between different surface features on flood lavas can be better observed on Mars than on Earth.

While there have been great discoveries related to flood lavas on Earth and Mars in the past decade, major questions also remain. For Mars, the lack of precise age dates creates great difficulties in unraveling the detailed geologic history of the flood lava eruptions. Are Martian flood lava provinces emplaced in a geologically short pulse of activity, as they are on the Earth? If so, it is plausible that they are associated with discrete events in the mantle, such as a plume arriving at the base of the lithosphere. If instead the rate of Martian flood volcanism has instead only gradually varied, the dynamics of Mars' interior must be quite different from the Earth's. The production of lava versus time also has strong implications for the way in which the interior of Mars has geochemically differentiated and how the flux of volatiles to the surface has changed with time. Clearly, measuring the chemical composition of the lavas is also of great interest.

On both Earth and Mars there remain major questions about what these immense eruptions were like. Based on the largest historical basaltic eruptions, we expect a significant fraction of the erupted lava was converted to pyroclastics. On Mars, the low atmospheric pressure should have led to even more vigorous fountaining than on Earth, if volatile contents were similar. Can these pyroclastics be found? While they are likely to be largely weathered beyond recognition on the Earth, the pyroclastics from the most recent Martian flood lava eruptions should be still recognizable.

Caption for Figure 5.9. (cont.) but the floor is smooth pahoehoe. We interpret these features to have formed by scraping the flow top breccia off of the flow as it advanced past a pre-existing obstacle. The breccia was rigid enough to not slump; instead the gap was filled by fluid lava upwelling from below.

Indeed, there are already suggestions of widespread pyroclastic deposits on Mars from a number of different studies (e.g., Scott and Tanaka, 1982; Edgett *et al.*, 1997; Tanaka, 2000; Bishop *et al.*, 2002; Mouginis-Mark, 2002; Hynek *et al.*, 2003).

Finally, these massive fountains of lava must have released prodigious amounts of acidic precipitation and altered the planets' climates for decades at a time. The potential effects that these eruptions may have had on life are a topic of much speculation at this point in time. It is plausible that flood lava eruptions were catastrophic for life on Earth but a time for blooming on Mars. Clearly, much work remains to be done to understand flood lavas on Earth and Mars.

References

Anderson, S. W., Stofan, E. R., Smrekar, S. E., Guest, J. E., and Wood, B. (1999). Pulsed inflation of pahoehoe lava flows: implications for flood basalt emplacement. *Earth & Planetary Science Letters*, **168**, 7–18.

Aubele, J. C., Crumpler, L. S., and Elston, W. E. (1988). Vesicle zonation and vertical structure in basalt flows. *Journal of Volcanology & Geothermal Research*, **35**, 349–74.

Beyer, R. and McEwen, A. (2004). Stratigraphy of Eastern Coprates Chasma, Mars. *Abstracts of Papers Submitted to the 35th Lunar and Planetary Science Conference*. Houston: Lunar and Planetary Institute, CD 35, Abstract 1430.

Bishop, J. L., Murchie, S. L., Pieters, C. M., and Zent, A. P. (2002). A model for the formation of dust, soil, and rock coatings on Mars: physical and chemical processes on the Martian surface. *Journal of Geophysical Research*, **107**, JE001581.

Boynton, W. V., Taylor, G. J., Hamara, D. *et al.* (2003). Compositional diversity of the Martian crust: preliminary data from the Mars Odyssey Gamma-Ray Spectrometer. *Abstracts of Papers Submitted to the 34th Lunar and PlanetaryScience Conference*. Houston: Lunar and Planetary Institute, CD 34, Abstract 2108.

Bradley, B. A., Sakimoto, S. E. H., Frey, H., and Zimbelman, J. R. (2002). Medusae Fossae Formation: new perspectives from Mars Global Surveyor. *Journal of Geophysical Research*, **107**, paper 2 (doi: 10.1029/2001JE001537).

Burr, D. M., Grier, J. A., McEwen, A. S., and Keszthelyi, L. P. (2002). Repeated aqueous flooding from the Cerberus Fossae: evidence for very recently extant, deep groundwater on Mars. *Icarus*, **159**, 53–73.

Christensen, P. R., Bandfield, J. L., Hamilton, V. E. *et al.* (2001). Mars Global Surveyor Thermal Emission Spectrometer experiment: investigation description and surface science results. *Journal of Geophysical Research*, **106**, 23823–71.

Coffin, M. F. and Eldholm, O. (1994). Large igneous provinces: crustal structure, dimensions, and external consequences. *Reviews of Geophysics*, **32**, 1–36.

Crisp, J. and Baloga, S. M. (1994). Influence of crystallization and entrainment of cooler material on the emplacement of basaltic aa flows. *Journal of Geophysical Research*, **99**, 11819–31.

Duncan, R. A., Hooper, P. R., Rehacek, J., Marsh, J. S., and Duncan, A. R. (1997). The timing and duration of the Karoo igneous event, southern Gondwana. *Journal of Geophysical Research*, **102**, 18127–38.

Edgett, K. S., Butler, B. J., Zimbelman, J. R., and Hamilton, V. E. (1997). Geologic context for the Mars radar "Stealth" region in southwestern Tharsis, *Journal of Geophysical Research*, **102**, 21545–68.

Erwin, D. H. (1994). The Permian-Triassic extinction. *Nature*, **367**, 231–6.

Geikie, A. (1880). The lava-fields of North-western Europe. *Nature*, **23**, 3–5.

Greeley, R. and Guest J. E. (1987). Geologic map of the eastern equatorial region of Mars. US Geological Survey Miscellaneous Investigation Series Map I-1802-B.

Greeley, R. and King, J. S. (1977). *Volcanism of the Eastern Snake River Plain, Idaho: A Comparative Planetary Geology Guidebook*. Washington, DC: NASA CR-154621.

Greeley, R. and Schneid, B. D. (1991). Magma generation on Mars: amounts, rates, and comparisons with Earth, Moon, and Venus. *Science*, **254**, 996–8.

Haggerty, B. M. (1996). Episodes of flood-basalt volcanism defined by $^{40}Ar/^{39}Ar$ age distributions: correlation with mass extinctions? *Journal of Undergraduate Science*, **3**, 155–64.

Johnson, K. R., Nichols, D. J., and Hartman, J. H. (2002). Hell Creek Formation: a 2001 synthesis. In *The Hell Creek Formation and the Cretaceous-Tertiary Boundary in the Northern Great Plains: An Integrated Continental Record of the End of the Cretaceous*. ed. J. H. Hartman, K. R. Johnson, and D. J. Nichols. Geological Society of America Special Paper **361**, pp. 503–10.

Hartmann, W. K. and Berman, D. C. (2000). Elysium Planitia lava flows: crater count chronology and geological implications. *Journal of Geophysical Research*, **105**, 15011–26.

Hartmann, W. K., Malin, M. C., McEwen, A. S. *et al.* (1999). Evidence for recent volcanism on Mars from crater counts. *Nature*, **397**, 586–9.

Head, J. N., Melosh, H. J., and Ivanov, B. A. (2002a). Martian meteorite launch: high-speed ejecta from small craters. *Science*, **298**, 1752–6.

Head, J. W., Kreslavsky, M. A., and Pratt, S. (2002b). Northern lowlands of Mars: evidence for widespread volcanic flooding and tectonic deformation in the Hesperian Period. *Journal of Geophysical Research*, **107**, paper 3 (doi: 10.1029/2000JE001445).

Hon, K., Kauahikaua, J., Denlinger, R., and Mackay, K. (1994). Observations and measurements of active lava flows on Kilauea Volcano, Hawaii. *Geological Society of America Bulletin*, **106**, 351–70.

Hooper, P. R. (1997). The Columbia River Basalt Province: Current status. In *Large Igneous Provinces*, ed. J. J. Mahoney and M. Coffin. Geophysical Monograph Series 100. Washington, DC: AGU, pp. 1–27.

Hynek, B. M., Arvidson, R. E., and Phillips, R. J. (2003). Explosive volcanism in the Tharsis region: global evidence in the Martian geologic record. *Journal of Geophysical Research*, **108** (E9), 15–1, CiteID 5111, doi: 10.1029/2003JE002062.

Keszthelyi, L. (1995). A preliminary thermal budget for lava tubes on the Earth and planets. *Journal of Geophysical Research*, **100**, 20411–20.

Keszthelyi, L. (2002). Classification of mafic lava flows from ODP Leg 183, Scientific Results Volume, Ocean Drilling Program online at www.odp.tamu.edu.

Keszthelyi, L. and McEwen, A. S. (2001). Recent flood volcanism on Mars: implications for climate change, layered deposits, and lava–water interactions. *EOS Transactions, American Geophysical Union Spring Meeting*, **82** (no. 20), Abs. V42A-03, S438.

Keszthelyi, L. and Self, S. (1998). Some physical requirements for the emplacement of long basaltic lava flows. *Journal of Geophysical Research*, **103**, 27447–64.

Keszthelyi, L., McEwen, A. S., and Thordarson, Th. (2000). Terrestrial analogs and thermal models for Martian flood lavas. *Journal of Geophysical Research*, **105**, 15027–50.

Keszthelyi, L., Thordarson, Th., and Self, S. (2001). Rubbly pahoehoe: implications for flood basalt eruptions and their atmospheric effects. *EOS Transactions, American Geophysical Union Fall Meeting*, **82** (no. 47), Abs. V52A-1050, 1407.

Klingelhaufer, G., Morris, R. V., Bernhardt, B. *et al.* (2004). Mössbauer spectroscopy of soils and rocks at Gusev Crater and Meridiani Planum. *Abstracts of Papers Submitted to the 35th Lunar and Planetary Science Conference*. Houston: Lunar and Planetary Institute, CD 35, abstract 2184.

Lanagan, P. and McEwen, A. S. (submitted) Geomorphic analysis of the Cerberus plains: Constraints on the emplacement of the youngest lava flows on Mars. *Icarus*.

Macdougall, J. D. (ed.) (1988). *Continental Flood Basalts*. Dordrecht: Kluwer Academic.

Mahoney, J. J. and Coffin, M. F. (eds.) (1997). Large Igneous Provinces: continental, oceanic and, planetary flood volcanism. *American Geophysical Union Monograph* 100, 438 pp.

Malin, M. C. and Edgett, K. E. (2001). Mars Global Surveyor Mars Orbital Camera: interplanetary cruise through primary mission. *Journal of Geophysical Research*, **106**, 23429–570.

Marsh, J. S., Ewart, A., Milner, S. C., Duncan, A. R., and McG. Miller, R. (2001). The Etendeka Igneous Province: magma types and their stratigraphic distribution with implications for the evolution of the Paraná-Etendeka flood basalt province. *Bulletin of Volcanology*, **62**, 464–86.

McEwen, A. S., Malin, M. C., Carr, M. H., and Hartmann, W. K. (1999). Voluminous volcanism on early Mars revealed in Valles Marineris. *Nature*, **397**, 584–6.

McEwen, A. S. (2003). Secondary cratering on Mars. *6th International Conference on Mars*, Abstract 3268.

McEwen, A. S., Turtle, E., Burr, D. *et al.* (2003). Discovery of a large rayed crater on Mars: implications for recent volcanic and fluvial activity and the origin of Martian Meteorites. *Abstracts of Papers Submitted to the 34th Lunar and Planetary Science Conference*. Houston: Lunar and Planetary Institute, CD 34, Abstract 2040.

McSween, H. Y., Grove, T. L., and Wyatt, M. B. (2003). Constraints on the composition and petrogenesis of the Martian crust. *Journal of Geophysical Research*, doi10.1029/2003JE002175.

Mege, D. and Masson, P. (1996). A plume tectonics model for the Tharsis province, Mars. *Planetary Space Science*, **44**, 1499–546.

Mitchell, K. L., Wilson, L., and Head, J. W. (2003). Dike emplacement as a mechanism for generation of massive water floods at Cerberus Fossae, Mars. *Abstracts of Papers Submitted to the 34th Lunar and Planetary Science Conference*. Houston: Lunar and Planetary Institute, CD 34, Abstract 1332.

Mouginis-Mark, P. J. (2002). Prodigious ash deposits near the summit of Arsia Mons volcano, Mars. *Geophysical Research Letters*, **29**, GL015296.

Nyquist, L. E., Bogard, D. D., Shih, C. Y. *et al.* (2001). Ages and geologic histories of Martian meteorites. *Space Science Reviews*, **96**, 105–64.

Ori, G. G. and Karma, A. (2003). The uppermost crust of Mars and flood basalts. *Abstracts of Papers Submitted to the 34th Lunar and Planetary Science Conference*. Houston: Lunar and Planetary Institute, CD 34, Abstract 1539.

Peterson, D. W. and Tilling, R. I. (1980). Transition of basaltic lava from pahoehoe to aa, Kilauea Volcano, Hawaii: field observations and key factors. *Journal of Volcanology & Geothermal Research*, **7**, 271–93.

Plescia, J. B. (1990). Recent flood lavas in the Elysium region of Mars. *Icarus*, **88**, 465–90.

Plescia, J. B. (1993). A assessment of volatile release from recent volcanism in Elysium Mars. *Icarus*, **104**, 20–32.

Rampino, M. R. and Stothers, R. B. (1988). Flood basalt volcanism during the past 250 million years. *Science*, **241**, 663–8.

Rowland, S. K. and Walker, G. P. L. (1990). Pahoehoe and aa in Hawaii: volumetric flow rate controls the lava structure. *Bulletin of Volcanology*, **52**, 615–28.

Ruff, S. W., Christensen, P. R., Clark, R. N. *et al.* (2001). Mars; "White Rock" feature lacks evidence of an aqueous origin: results from Mars Global Surveyor. *Journal of Geophysical Research*, **106**, 23921–7.

Sakimoto, S. E. H. and Zuber, M. T. (1998). Flow and convective cooling in lava tubes, *Journal of Geophysical Research*, **103**, 27465–87.

Scott, D. H. and Carr, M. H. (1978). Geologic Maps of Mars. *US Geological Survey Miscellaneous Investigation Series Map* I-1803.

Scott, D. H. and Tanaka, K. L. (1982). Ignimbrites of Amazonis Planitia region of Mars. *Journal of Geophysical Research*, **87**, 1179–90.

Scott, D. H. and Tanaka, K. L. (1986). Geologic map of the western equatorial region of Mars. *US Geological Survey Miscellaneous Investigation Series Map* I-1802-A.

Schultz, R. A. (2002). Stability of rock slopes in Valles Marineris, Mars. *Geophysical Research Letters*, **29**, 38–41.

Self, S., Thordarson, Th., and Keszthelyi, L. (1997). Emplacement of continental flood basalt lava flows. In *Large Igneous Provinces*, ed. J. J. Mahoney and M. Coffin. Geophysical Monograph Series 100. Washington, DC: AGU, pp. 381–410.

Self, S., Keszthelyi, L., and Thordarson, Th. (1998). The importance of pahoehoe. *Annual Reviews of Earth & Planetary Science*, **26**, 81–110.

Self, S., Keszthelyi, L. P., Thordarson, T. *et al.* (2000). Pulsed inflation of pahoehoe lava flows: implications for flood basalt emplacement: discussion and reply. *Earth & Planetary Science Letters*, **179**, 421–8.

Sharma, M. (1997). Siberian Traps. In *Large Igneous Provinces*, ed. J. J. Mahoney and M. Coffin. Geophysical Monograph Series 100. Washington, DC: AGU, pp. 273–95.

Sutherland, F. L. (1994). Volcanism around the K/T boundary time: its role in an impact scenario for the K/T extinction events. *Earth Science Reviews*, **39**, 1–26.

Tanaka, K. L. (1986). The stratigraphy of Mars. *Journal of Geophysical Research*, **91**, E139–58.

Tanaka, K. L. (2000). Dust and ice deposition in the Martian geologic record. *Icarus*, **144**, 254–66.

Tanaka, K. L. and Scott, D. H. (1987). Geologic map of the polar regions of Mars. *US Geological Survey Miscellaneous Investigation Series Map* I-1802-C.

Tanaka, K. L., Skinner Jr, J. A., Hare, T. M., Joyal, T., and Wenker, A. (2003). Resurfacing history of the northern plains of Mars based on geologic mapping of Mars Global Surveyor data. *Journal of Geophysical Research*, **108**, doi: 10.1029/2002JE001908.

Thordarson, Th. and Self, S. (1996). Sulfur, chlorine, and fluorine degassing and atmospheric loading by the Roza eruption, Columbia River Basalt Group, Washington, USA. *Journal of Volcanology & Geothermal Research*, **74**, 49–73.

Thordarson, T. and Self, S. (1998). The Roza Member, Columbia River Basalt Group: a gigantic pahoehoe lava flow field formed by endogenous processes? *Journal of Geophysical Research*, **103**, 27411–45.

Tolan, T. L., Reidel, S. P., Beeson, M. H. *et al.* (1989). Revisions to the estimates of the areal extent and volume of the Columbia River Basalt Group. In *Volcanism and Tectonism in the Columbia River Flood-Basalt Province*, ed. S. P. Reidel and P. R. Hooper. Geological Society of America Special Paper 239, pp. 1–20.

Treiman, A. H., Gleason, J. D., and Bogard, D. D. (2000). The SNC meteorites are from Mars. *Planetary & Space Science*, **48**, 1213–30.

Tyrrell, G. W. (1937). Flood basalts and fissure eruption. *Bulletin of Volcanology*, **1**, 89–111.

Wanke, H., Bruckner, J., Dreibus, G., Rieder, R., and Ryabchikov, I. (2001). Chemical composition of rocks and soils at the Pathfinder site. *Space Science Reviews*, **96**, 317–30.

Washington, H. S. (1922). Deccan Traps and the other plateau basalts. *Geological Society of America Bulletin*, **33**, 765–804.

White, R. S. and McKenzie, D. (1995). Mantle plumes and flood basalts. *Journal of Geophysical Research*, **100**, 17543–85.

Wignall, P. B. (2001). Large Igneous Provinces and mass extinctions. *Earth Science Reviews*, **53**, 1–33.

Wilson, L. and Head, J. W. (2002). Tharsis-radial graben systems as the surface manifestation of plume-related dike intrusion complexes: models and implications. *Journal of Geophysical Research*, **107** (E08), doi: 10.1029/2001JE001593.

Withers, P. and Neumann, G. A. (2001). Enigmatic northern plains of Mars. *Nature*, **410**, 651.

Wyatt, M. B. and McSween, H. Y. (2002). Spectral evidence for weathered basalt as an alternative to andesite in the northern lowlands of Mars. *Nature*, **417**, 263–6.

Wyatt, M. B. and Tanaka, K. L. (2003). Origin of MGS-TES surface compositions in the northern plains and polar region of Mars. *Third International Conference on Mars Polar Science Exploration*, Abstract 8118.

6

Rootless volcanic cones in Iceland and on Mars

Sarah A. Fagents and Thorvaldur Thordarson
University of Hawaii at Manoa

6.1 Introduction

In the 1970s, the two Viking spacecraft returned images of the surface of Mars in which numerous small domes, knobs, and mounds were visible. Based on the presence of summit depressions in many of these domes, they were interpreted to be rootless volcanic cones (Frey *et al.*, 1979; Frey and Jarosewich, 1982), by analogy with similar features found in Iceland (Thoroddsen, 1894; Thorarinsson, 1951, 1953). Rootless cones (also called pseudocraters – a literal translation of the Icelandic *gervigígar*) form as a result of explosive lava–water interaction, whereby a flowing lava encounters a waterlogged substrate, causing violent vaporization of the water and expulsion of the lava from the explosion site (Thorarinsson, 1951, 1953). Repeated explosive pulses build a cone of disintegrated liquid and solid lava debris (Thordarson *et al.*, 1992). As the activity at a given site within the flow wanes, explosions may be initiated elsewhere, leading to construction of a field of tens to hundreds of cones. Although they may bear a superficial resemblance to primary volcanic cones built over a subsurface conduit, Icelandic rootless cones are quite distinct, in that they are surface phreatomagmatic structures formed at the lava–substrate interface (Thordarson, 2000).

The identification of possible rootless cone fields at mid to low latitudes on Mars incited great interest because of the implication for the presence and distribution of volatiles (i.e., water or ice) in the near-surface environment on Mars (Frey *et al.*, 1979; Frey and Jarosewich, 1982). Prior to the recognition of the cone fields, the assumption was that the near-surface regolith was essentially devoid of water ice, owing to its instability at mid

The Geology of Mars: Evidence from Earth-based Analogs, ed. Mary Chapman. Published by Cambridge University Press.

to low latitudes under present Martian climatic conditions (Clifford and Hillel, 1983; Fanale *et al.*, 1986). The possible identification of phreatomagmatic volcanic structures produced by explosive interaction of hot lava and surface or near-surface volatiles therefore presents interesting and conflicting evidence.

With the advent of the great wealth of data acquired by the Mars Orbiter Camera (MOC) aboard the Mars Global Surveyor spacecraft, as well as the more recent Thermal Emission Imaging System (THEMIS) on the Mars Odyssey mission, we now have a much more detailed view of these cone fields than was afforded by the lower resolution of the Viking imaging system. Together with a myriad of other geologic features, these cone fields continue to strengthen the case for abundant surface and near-surface water throughout Mars' history. The presence of water is of key significance in understanding the geologic history and climatic evolution of Mars. Moreover, water is a crucial requirement for biogenic development, and acquiring knowledge of its distribution forms the keystone for NASA's exploration goals.

So what exactly can rootless cones tell us about water on Mars? To answer this, it is necessary to turn to a combination of analysis of spacecraft images, numerical modeling of the dynamics of cone formation, and studies of terrestrial field analogs. Without detailed field studies, rootless cones are difficult to distinguish morphologically from primary (rooted) cones on Earth. This problem is magnified when using imagery to study extraterrestrial examples: one is simply unable to distinguish the details of cone morphology and the relationship with host lava. What is required is to compile a suite of observations on rootless cone field geomorphology, including occurrences, clustering/dispersal of cones within a field, dimensions and morphology, and relationship to the host lava and underlying terrain.

In addition to the Icelandic cone groups, other field analogs provide insight into rootless eruptions. In some respects, Icelandic rootless cone formation is similar to the growth of littoral cones such as those observed in Hawaii, where lava flows entering the ocean interact explosively with seawater (Fisher, 1968; Jurado-Chichay *et al.*, 1996; Mattox and Mangan, 1997). The numerous observations and detailed documentation of littoral explosive activity during the past 21 years of the Pu'u 'O'o eruption of Kilauea volcano provide important constraints on mechanisms of explosive magma–water mixing and the formation of rootless cones. This is enhanced by field studies of peperites, which are clastic rocks formed by disintegration of magma upon intrusion into wet sediments (Skilling *et al.*, 2002).

Peperites can therefore provide information on contact geometries and modes of mingling that precede explosive fuel–coolant interactions (Skilling *et al.*, 2002; Zimanowski and Büttner, 2002).

However, the details of lava–water interactions remain poorly understood. A considerable body of theoretical and experimental work has investigated the key factors controlling the style and intensity of magma–water interactions (e.g., Wohletz, 1983, 1986; Wohletz and McQueen, 1984; Zimanowski *et al.*, 1991; Fröhlich *et al.*, 1993; White, 1996; Büttner *et al.*, 2000). In laboratory simulations of explosive interactions, the efficiency of conversion of thermal energy to mechanical energy (utilized to fragment the melt) reaches a maximum at a water/melt mass ratio of 0.3, when finely fragmented melt is able to transfer heat efficiently to the water and produce a violently expanding gas phase (Wohletz, 1983, 1986; Wohletz and McQueen, 1984). Ratios smaller or greater than this value produce significantly less explosive events. White (1996) has argued that irregularities and impurities in natural systems will cause maximal water/magma mass ratios to differ from those in controlled laboratory conditions. However, most studies agree that the key factors controlling explosive intensity are the mass ratio of interacting water and melt, the contact geometry of the interacting materials, rate of injection of the melt, and the degree of interpenetration of melt and water (Wohletz, 1983, 1986; Wohletz and McQueen, 1984; Zimanowski *et al.*, 1991; Fröhlich *et al.*, 1993). These are parameters than can be at least partly constrained by detailed field observations.

If we are to attempt to quantify the amount of water ice in the Martian regolith, it is critical to develop a good understanding of the relationship between cone characteristics and the explosion process under terrestrial conditions. Again, this can only be achieved with detailed field studies. Once this is achieved, theoretical models of the physics of rootless explosion dynamics can be developed to study cone formation under various planetary conditions.

Therefore, analog studies are needed to understand (i) how and under what circumstances these features form; (ii) the geometric relationship between the host lava and the explosion centers that produce the cones; (iii) how the lava and water interact to produce the explosion; (iv) what are the requirements for lava emplacement processes and substrate properties; and (v) what this means for the amount and spatial distribution of water on Mars.

In this chapter we will describe what can be learned from image analysis of the Martian cones and highlight how field studies of Icelandic rootless cones are critical both to developing models and to testing whether the Martian examples are indeed the products of rootless eruptions.

6.2 Martian volcanic cones: inferences from spacecraft data

Using high-resolution Viking images (~25–50 m/pixel), Frey and colleagues identified >2000 dome- and cone-like features occurring mainly in Acidalia, Utopia, Isidis, and Elysium Planitiae (Frey *et al.*, 1979; Frey and Jarosewich, 1982). Additional cone fields were identified north of Olympus Mons, in Amazonis Planitia, Hephaestus Fossae, and the Arrhenius region (Hodges and Moore, 1994) (see Figure 6.1 for location map). With the exception of the last region, all cone fields lie in the northern lowland plains, predominantly at mid to low latitudes.

In the Viking images, some of these features have a conical shape; others are more dome- or mound-like. Summit depressions are visible in many cones, but others are indistinct (Figure 6.2). Measurements of cone diameter range from ~300 to >1000 m, with summit craters varying from 0.25 to 0.65 times the basal diameter (Frey and Jarosewich, 1982).

Based on the conical expression, and apparently random spatial distribution, lack of structural control, and the relatively large ratio of crater diameter to cone diameter (larger than typical primary scoria cones, more similar to tuff rings), a rootless phreatomagmatic origin was assigned to these features (Frey and Jarosewich, 1982). However, for Mars, it was suggested that the lava interacted with ice confined within the regolith pore spaces (Frey and Jarosewich, 1982; Frey, 1987), rather than with a waterlogged substrate as in Iceland.

The origins of the cones remained somewhat equivocal, due to limitations in the resolution and coverage of the Viking data. However, recent MOC and THEMIS data have revealed numerous cone groups at much higher resolution, which allow for closer examination of cone morphology, as well as a better view of the geologic setting of the cone groups (Greeley and Fagents, 2001; Lanagan *et al.*, 2001; Fagents *et al.*, 2002). Figure 6.3 shows some clusters of cones lying on a remarkably youthful surface in Amazonis Planitia. These are clearly conical features with well-defined summit craters, and an apparent mantle of fine material surrounding the main cone. The surface on which they lie displays a platy, ridged appearance, similar to that observed on a section of the 1783–4 Laki flow field in Iceland. This surface texture is attributed to a mode of lava emplacement characterized by alternating flow stagnation and high effusion-rate surges, which disrupt the stationary pahoehoe crust into discrete plates (Keszthelyi *et al.*, 2000; Keszthelyi and McEwen, this volume). Lava flows in this region have been dated by crater-counting methods, and have ages as young as 10 Ma (Hartmann, 1999; Hartmann *et al.*, 1999). Figure 6.4 shows some closely spaced cones on a lava

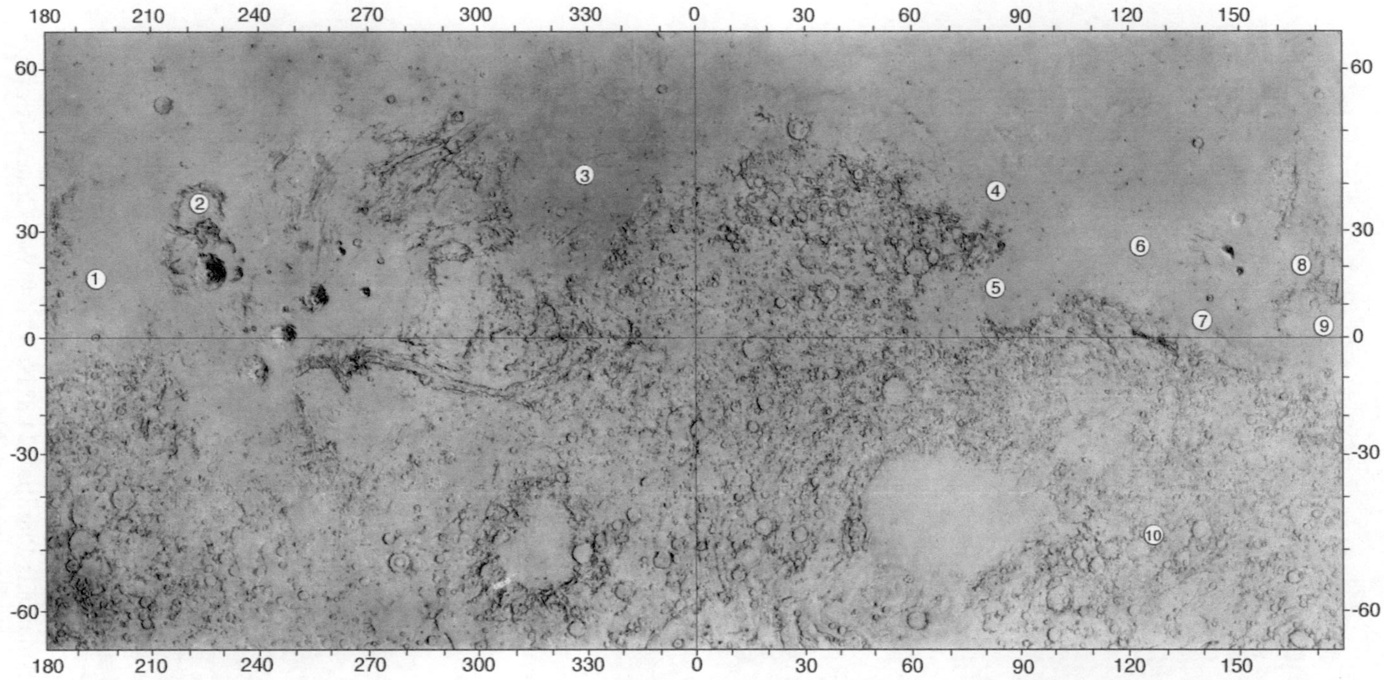

Figure 6.1. Map showing our current understanding of the distribution of candidate rootless volcanic cones on Mars, based on Viking, MOC, and THEMIS data. 1: Amazonis Planitia, 2: Olympus Mons aureole, 3: Acidalia Planitia, 4: S. Utopia Planitia, 5: Isidis Planitia, 6: Hephaestus Fossae, 7: Elysium Planitia, 8: Cerberus, 9: Marte Valles, 10: Arrhenius. Scale: 10° = 100 km at equator.

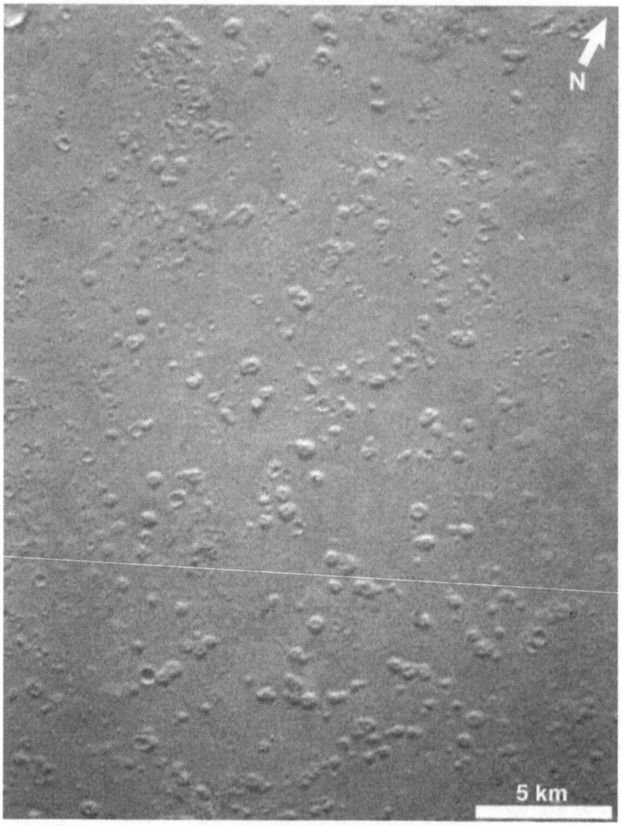

Figure 6.2. Portion of Viking Orbiter 1 image 38A11 (48 m/pixel) showing indistinct mounds located in eastern Acidalia Planitia. Some mounds appear to have summit depressions, others are too poorly resolved.

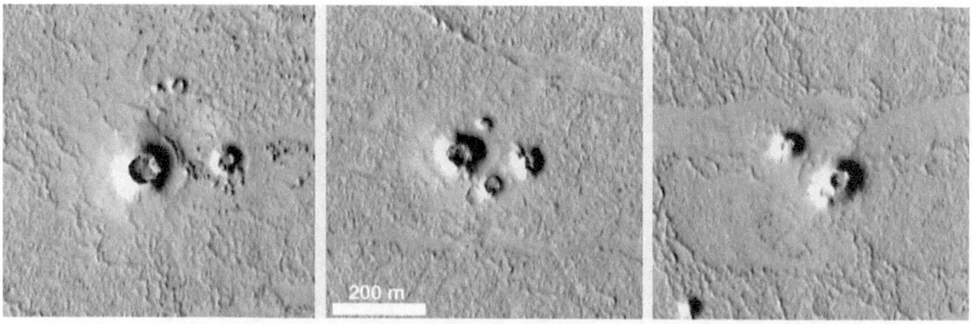

Figure 6.3. Portions of MOC image M0303958 (3.6 m/pixel) showing clusters of cones in western Amazonis Planitia (24.8° N, 171.4° W). Cones sit atop a remarkably youthful lava flow surface. North is up.

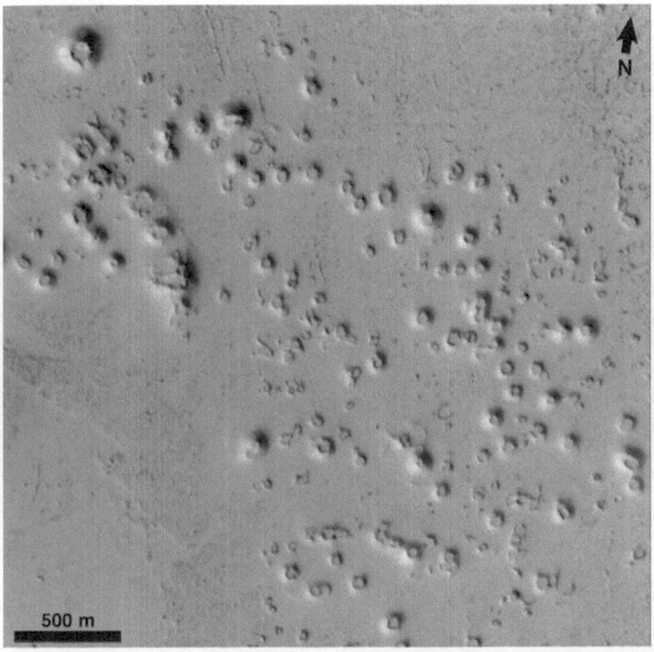

Figure 6.4. This portion of MOC image M0801962 (5.4 m/pixel), also in Amazonis Planitia (26° N, 189.7° W), shows a much denser distribution of cones than Figure 6.3. These cones appear to sit on an older flow surface, partially covered with (wind-blown?) sediment, and may perhaps be a source of some of that sediment in the form of pyroclastic ash.

surface (also in Amazonis) apparently partly covered with windblown sediment. In this instance, cones are more tightly clustered, with examples of multiple overlapping cones and coalesced craters. Figure 6.5 shows some more degraded cones in Acidalia Planitia; the presence of a significant number of impact craters testifies to the greater age of this surface. Whereas elsewhere the cones are apparently more or less randomly distributed, there is evidence of some alignment in this image.

Given the greater detail afforded by the MOC and THEMIS coverage, these constructs can be interpreted with more confidence as rootless volcanic cones, by analogy with Icelandic cone groups, based on a combination of (i) their distinct positive-relief, conical form, (ii) their location on lava flow surfaces, (iii) lack of association with eruptive fissures, (iv) location in areas of low regional slope, and (v) apparent haphazard clustering of cones (Greeley and Fagents, 2001; Lanagan *et al.*, 2001; Fagents *et al.*, 2002).

Other origins previously proposed for the Martian features include pingos, which are ice-cored sediment mounds produced in permafrost

Figure 6.5. Portion of THEMIS image (19 m/pixel), showing cone clusters in Acidalia Planitia (40.3° N, 6.9° W). Cones are morphologically quite distinct from impact craters, thus lending confidence to the interpretation of volcanic origins. Image release 20030404; NASA/JPL/Arizona State University.

zones by successive freeze–thaw cycles (Judson and Rossbacher, 1979; Theilig and Greeley, 1979; Lucchitta, 1981; Cabrol *et al.*, 2000); pedestal craters, which result from differential deflation of a surface armored by impact ejecta (McCauley, 1973; Arvidson *et al.*, 1976); and rampart craters, resulting from impacts into volatile-rich targets (Carr *et al.*, 1977; Gault and Greeley, 1978; Mouginis-Mark, 1987). However, the MOC and THEMIS data reveal that these cones are more regular and conical in morphology than would be the case for any of these other origins, and that they fulfill some or all of criteria (i)–(v). Impact craters are shown as morphologically distinct from the cones in Figure 6.5. Moreover, the association with clearly pristine lava flow surfaces in many cases (Figures 6.3 and 6.4), further strengthens the case for a volcanic origin (pingos require a sediment surface on which to form). The debate between a rooted and rootless origin relies on more subtle details. The number, spatial density and clustering of the cones argue against magmatic or hydrovolcanic eruptions fed by numerous separate volcanic conduits rooted deeper in the crust. Furthermore, the candidate rootless cones do not feed lava flows, as is commonly the case with primary cones built by strombolian eruptions. The postulated existence of a paleo-ocean on Mars (e.g., Parker *et al.*, 1993) suggests that littoral cone formation may have been possible. However, in all examples examined so far, the Martian cones do not exhibit typical littoral cone characteristics (e.g., truncation of cone flanks on the seaward side, or half-cones constructed on either side of a lava channel (Fisher, 1968)). Furthermore, the cones are distributed over a broad area, rather than narrower zones or breaks in slope that might represent paleo-shorelines. None of points (ii)–(v) alone point to a rootless origin for the Martian cones, but synthesis of a number of these lines of evidence strengthens this interpretation. Further Icelandic fieldwork on individual cone characteristics and the structure of cone fields will help to elucidate the origins of Martian cones.

Using the MOC images, the morphologies of candidate rootless volcanic cones have been characterized at resolutions of ~3–10 m/pixel (Greeley and Fagents, 2001; Lanagan *et al.*, 2001; Fagents *et al.*, 2002). Basal diameters range from ~20 to 1000 m. The very pristine cones in Amazonis Planitia (Figure 6.3) are the smallest, with a modal diameter around 100 m, whereas more ambiguous cones in Isidis have typical diameters in the 400–600 m range. Cone heights are difficult to determine as these features are at the limit of the Mars Orbiter Laser Altimeter shot spacing. However, some fortuitously placed shots provide lower limits of 25–60 m for the larger cones in Isidis Planitia (Fagents *et al.*, 2002).

There is increasing evidence that water has been present in some form at the surface and near-surface of Mars in the geologically recent past (Malin and Edgett, 2000), and may persist today (Boynton *et al.*, 2002; Feldman *et al.*, 2002; Mitrofanov *et al.*, 2002). This is supported by thermal models of regolith–atmosphere vapor exchange and ground ice stability (Paige, 1992; Mellon and Jakosky, 1995; Mellon *et al.*, 1997). Cyclic recharge of ground ice stores can arise due to outburst flooding episodes (Burr *et al.*, 2000; Lanagan *et al.*, 2001) or due to climate change caused by variations on Mars' orbital parameters (particularly orbital obliquity – the angle between the spin axis and orbital plane) (Mellon and Jakosky, 1995). It seems reasonable, therefore, to postulate the presence of ground ice at the time of emplacement of cone-hosting lava flows, even at low latitudes.

Therefore, mapping the locations of the cone fields, and crater-dating the host lava flows, constrains the temporal and spatial distribution of water, as well as providing indications of climatic conditions or flood events.

In order to determine the amount of water available to the lava to produce rootless cones, however, we must develop a model of the formation process using field observations both to guide our understanding of the physical mechanism, and to constrain the model.

6.3 Rootless cone groups in Iceland: field studies

The origin of rootless volcanic cones by explosive interaction of molten lava and surface water was first recognized in the late nineteenth century (Thoroddsen, 1894) and confirmed ~50 years later when Thorarinsson (1951, 1953) showed that the cones are constructional features formed when lava is emplaced over gently sloping waterlogged substrates such as shallow lakes, wetlands (bogs, marshes, etc.), braided streams, or glacial outwash plains.

Icelandic cone groups are ideally suited for analog studies since they are common in many recent flows (250–8000 years old; Table 6.1, Figure 6.6), and provide accessible high-quality outcrops to study interior structure and cone group architecture (Thordarson *et al.*, 1992; Thordarson, 2000). In addition, rare eyewitness accounts (e.g., observations of ~400 m tall lava fountains forming rootless cones in the 1783–4 Laki eruption (Steingrímsson and Ólafsson, 1783; Steingrímsson, 1788; Thordarson *et al.*, 1998)) provide constraints on eruption style and intensity. Explosive interactions between degassed lava have also been observed in coastal environments, where waves impacting on flowing lava, collapses of lava deltas resulting in sudden exposure of hot lava to seawater, or entrapment of water in lava tubes act to initiate explosive activity (Thorarinsson, 1965; Fisher, 1968; Mattox and Mangan, 1997). While the accessibility and opportunity to observe active littoral cone formation greatly enhances our understanding of rootless explosion dynamics, the Icelandic cone groups form in a somewhat different, lower-energy environment. While we cannot entirely rule out a littoral origin for some Martian cones, those identified to date possess characteristics indicating that Icelandic rootless cones represent the best analog to the Martian features (see discussion above). Detailed studies of the Icelandic cones will help distinguish between rootless and primary cones and volcanic and non-volcanic origins for Martian cones.

The following observations are pertinent to developing an understanding of rootless cone formation.

Table 6.1. *Rootless cone groups in Iceland*

Lava flow	Vol. (km^3)	Length (km)	Age (yr)	Name of cone group	Area (km^2)
Thjórsa	20	120	8500	Thjórsa	?
Leitin	3	35	4600	Raudhólar	1
Ketildyngja	4	84	4500	Goshólar	?
Nupar	7	25	3800	Unnamed	?
				Thjórsardalur	8–9
Búrfell	6	55	3300	Vadalda	4–5
				Tungnarkrókur	?
Y-Laxardalur	2–3	52	2300	Mývatn	30
				Raudhólar	0.15
				Adaldalur	28
Eldgjá AD 934	14.2	60	1060	Landbrotshólar	150
				Álftavershólar	35
				Atlaey	1.5
Laki AD 1783	15.1	55	210	Fljótsoddi	3
				Blængur	1
				Varmárdalur	1
				Hnúta	1
				Blágil	< 1
				Úlfarsdalur	3
				Grjótarhofud	1
				Leidólfsfell	12
				Skál	< 1

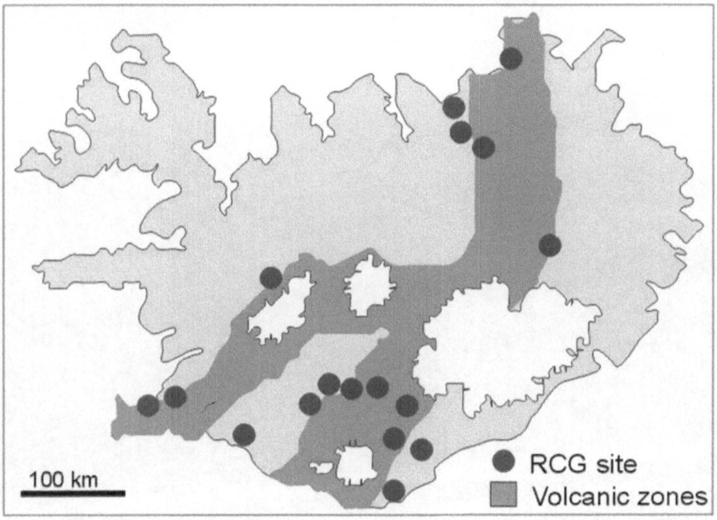

Figure 6.6. Distribution of rootless cone groups within the Holocene lava flows in Iceland.

6.3.1 Geologic setting

In Iceland, the rootless cone groups are found on Holocene lava flows underlain by impermeable units, such as the Plio-Pleistocene glacial diamictites that flank the active volcanic zones (Figure 6.6). These zones are characterized by high water tables, providing ample opportunity for interaction with the encroaching lava. Conversely, cones tend to be absent in lavas overlying the highly permeable Holocene lava flows, in which the water table is commonly tens of meters below the surface.

Rootless cone groups typically occur within tube-fed pahoehoe lava flow fields and are commonly found in association with tumuli (Figure 6.7). This suggests that rootless eruptions are related to insulated flow emplacement (i.e., endogenous growth via lava inflation beneath an insulating crust), and the development of internal lava pathways (i.e., tubes). The cones are not confined to a particular sector of the flow field; the flow extends beyond the margins of cone groups. Thus the lava fields are composed of compound flows, characterized by sustained flow through bifurcating feeder pathways.

6.3.2 Individual cone characteristics

Individual rootless cones in Iceland are roughly circular in plan form (Figure 6.8), and are constructed of fragmented degassed lava with varying

Figure 6.7. Photograph showing association of rootless cones (background) with a tumulus (foreground). Person for scale.

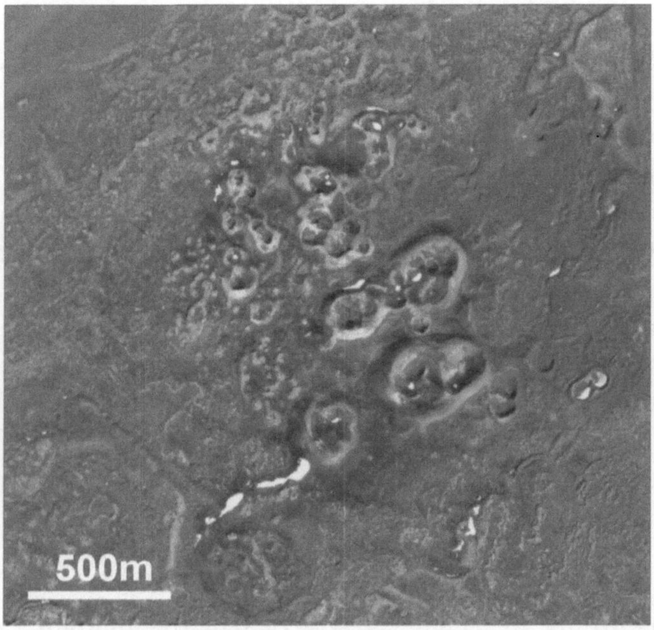

Figure 6.8. Aerial photograph of the Fljótsoddi rootless cone group within the 1783–4 Laki lava flow field, Iceland.

proportions of clasts derived from the substrate (Thordarson *et al.*, 1992). Cone morphology varies from narrow, steep-sided "hornitos" composed of fluidal spatter, to wider, broad-cratered cinder or tuff cones. Coexistence of different cone types within a group indicates a variation in the eruptive intensity, reflecting variations in the efficiency of lava–water interaction. Cone diameters range from 5 to 450 m and cone heights lie between 2 and 40 m (Thordarson, 2000), somewhat smaller on average than the Martian examples. Cone size is commonly greater towards the center of a cone group.

In all cases where dissected cones provide well-exposed cross-sections, a crudely funnel-shaped conduit extends up through the flow to form the crater. Individual cones exhibit internal layering; in a typical cone, a well-bedded, widely dispersed lower sequence with decimeter to meter thick beds of lapilli scoria alternates with thinner (<0.2 m) crudely laminated beds of excavated substrate sediment mixed with black ash. The lower sequence typically gives way to a crudely bedded upper sequence of larger (centimeter- to decimeter-sized) spatter bombs and agglutinates, which in turn is commonly capped by a 1–2 m thick unit of rheomorphic spatter (Thordarson *et al.*, 1992; Thordarson and Höskuldsson, 2002).

Substrate sediments are always present in the products of rootless eruptions, although in variable abundance and form. For example, at Raudholar, the lower sequence described above often contains distinct beds which are a mixture of red baked mud and ash-sized juvenile lava fragments and exhibit low-angle cross-bedding, indicating emplacement by low-concentration pyroclastic density currents (i.e., surges). In the upper sequence, the substrate sediment is present as individual centimeter- to decimeter-sized clasts (i.e., lumps of baked mud) dispersed throughout the succession but in significantly lower and upwardly decreasing abundances. It is also present as inclusions in armored bombs. The presence and distribution of substrate clasts in the rootless cone tephra indicates that physical mixing of molten lava and the substrate sediments took place during the explosive interaction.

6.3.3 Cone group architecture

Rootless cone groups in Iceland consist of tens to hundreds of constructs, superposed on and undeformed by their host lavas (Thorarinsson, 1951, 1953). Cone spacing varies from densely packed (Figures 6.8 and 6.9) to more sparse arrangements (Greeley and Fagents, 2001). Cones within a group

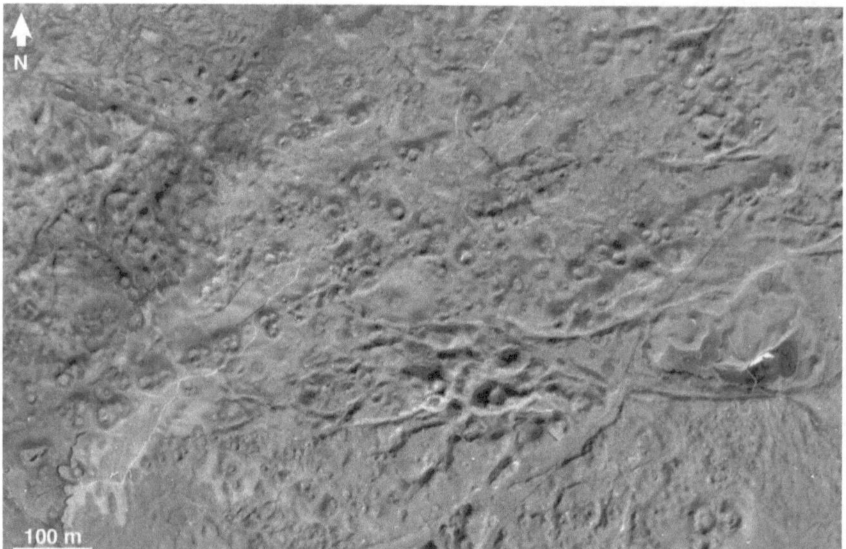

Figure 6.9. Aerial photograph of the Landbrotshólar rootless cone field within the 934–40 Eldgjá lava flow field, Iceland. Note alignment of cones in some areas.

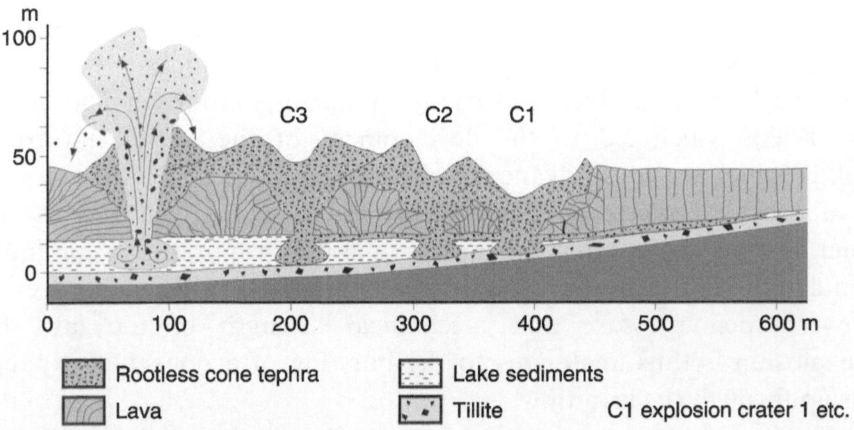

Figure 6.10. Reconstructed cross-section through the Raudhólar cone group, showing the conduit geometry and the onlapping arrangement of cones.

may form small clusters, and local alignments of several cones are observed in some groups (Figure 6.9), which are linked to eruptions along a lateral feeder system (i.e., an internal lava pathway or tube). Commonly, cones overlap and partly obscure one another (Figure 6.10). The pattern of this onlap is typically in the upflow direction, implying prolonged activity with upflow propagation of cone-building explosions along a lateral feeder segment. The formation of a cone at a given point in the flow will cut off the supply of lava downflow. Once explosions at a particular site cease (perhaps because the water supply becomes exhausted), further cone-building explosions can only be initiated where there is a continued supply of lava, i.e., in the upstream direction of the same tube segment or at sites along new tube segments elsewhere in the flow field (Thordarson, 2000; Thordarson and Höskuldsson, 2002).

6.4 Models of rootless cone formation

The goal of the Iceland analog study is to develop computational models of the cone formation process in order to quantify the amount of ice in the Martian regolith leading to cone-forming explosions. Such quantitative models must be based on sound conceptual models of the physical mechanisms of the lava–water interaction, and as such will draw heavily on field observations and past experimental studies.

Until recently the details of the cone-forming process have been far from clear, and much remains to be done. Two end member models have been proposed to describe the lava–water interaction, based on the mode of heat transfer: the static and dynamic heat transfer models.

Early work on the problem was based on the observations of Thorarinsson (1951, 1953), which led to the development of the static heat transfer model, in which the lava advances over the waterlogged substrate, trapping and vaporizing pockets of water beneath a solidified basal crust. Heat transfer is thus by conductive means, and the vapor must continue to accumulate until a critical pressure is reached that is sufficient to overcome the overburden pressure and mechanical strength of the lava flow. The explosion is thus analogous to the bursting of a pocket of expanding vapor up through the lava flow.

The few quantitative models that have addressed the lava–substrate heat transfer have been based on this static model. Calculations of conductive heat transfer to the substrate, demonstrated that sufficient vapor pressure could be generated only if the water or ice was within a depth from the surface equivalent to one-third to one-half the lava flow thickness (Allen, 1979; Squyres *et al.*, 1987). In theory, this condition should not be difficult to meet in Iceland, where surface water is ubiquitous. Under Martian conditions, this implies ice lying within a few meters to a few tens of meters of the surface, seemingly in conflict with early models of ground ice stability (Clifford and Hillel, 1983; Fanale *et al.*, 1986), but not unreasonable based on more recent modeling advances (Paige, 1992; Mellon and Jakosky, 1995; Mellon *et al.*, 1997) and results from spacecraft remote sensing (Boynton *et al.*, 2002; Feldman *et al.*, 2002; Mitrofanov *et al.*, 2002).

However, the static heat transfer model fails to explain all field observations. For example, the above models were based on simple cooling of an instantaneously emplaced flow unit, which ignores the evidence for continued input of lava to the explosion site via preferred pathways in the flow (e.g., lava tubes), leading to repeated explosions. Any accurate conceptual model of rootless cone formation must account for the following observations and inferences:

- Cones are constructed on top of a stationary, solidified lava crust, since they rest on top of, and are undeformed by, their host lava.
- Cones are constructed by multiple explosive events, as implied by the deposit bedding. This requires sustained activity characterized by discrete, sequential explosions.
- Sustained lava input (and the potential for multiple explosions) is provided by internal transport pathways (e.g., lava tubes), with which cones are commonly associated. A bifurcating tube system is also evidenced by the fact

that flow fields commonly continue to form downslope from the cones; flow advance may have been delayed as cone construction temporarily cut off supply, but was not halted completely.

- Decreasing explosion intensity with time is indicated by the upward decrease in dispersal and coarsening of fluidal clasts in the deposit, which in turn indicates decreasing efficiency of the lava–water interaction. This is best explained by a decrease in water available to the lava (Figure 6.11).
- Variable eruption intensity (and hence efficiency of lava–water heat transfer) across a given cone field is indicated by the range of cone morphologies observed.
- Significant mechanical mixing of the lava and substrate is indicated by the incorporation of substrate sediment clasts in the deposits. The amounts tend to decrease upwards through the deposits, again indicating a waning in explosivity.

From the above observations, a conceptual model of dynamic heat transfer is postulated (Figures 6.12 and 6.13) (Thordarson, 2000). Physical mixing of lava and substrate sediments allows for highly efficient heat transfer and rapid conversion of thermal to mechanical energy by explosive vaporization of the water, leading to thorough fragmentation and acceleration of the lava out of the explosion site. A mechanism for inducing and maintaining mixing is required; the lava must continue to move through the flow field and feed into the explosion site. This can be accomplished most easily if the lava is

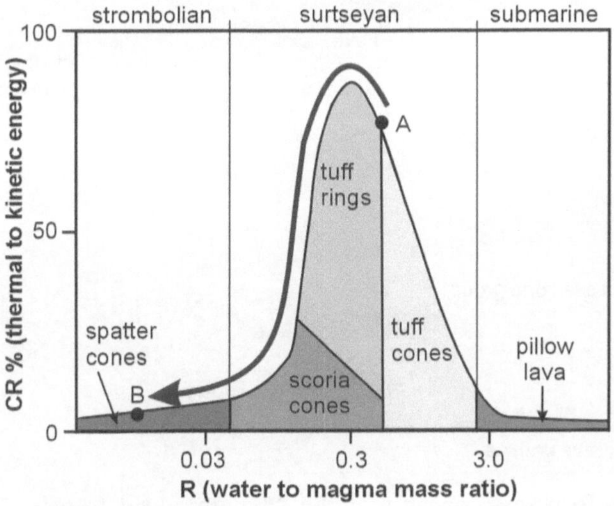

Figure 6.11. Inferred changes (using deposit properties) in water/lava mass ratio during a rootless eruption at Raudhólar cone group, SW Iceland. Point A represents high-intensity explosions at the onset of eruption, whereas point B represents low-intensity explosions towards the end of cone growth (T. Thordarson, unpublished data).

a tube-fed pahoehoe flow (Figure 6.12). Initially the lava enters a shallow lake basin or wetland as relatively small pahoehoe lobes from a set of tubes at the active lava fronts. An insulating crust seals the lobe interiors from the source of water, and lobes inflate and expand laterally in response to continued injection of lava. The internal lava tubes are thus extended and the process repeats itself, extending the lava further into the lake/wetland. Another consequence of this process is that, as the lava behind the active flow fronts increases in thickness by inflation, it begins to sink into soft sediment where it is thickest. However, the subsidence is not uniform, and cracks open in the base of the lava below the internal pathways, so that the hot lava flows into water-saturated mud and initiates steam explosions (Figure 6.13). If the explosions are powerful enough, they burst through the overlying lava to emerge as rootless eruptions that build cones around the vents. Continued influx of lava to the explosion site generates repeated cycles of vaporization, fragmentation, pressurization, and excavation, and hence multiple explosive pulses. Within particular sectors of the lava transport system, the explosion centers

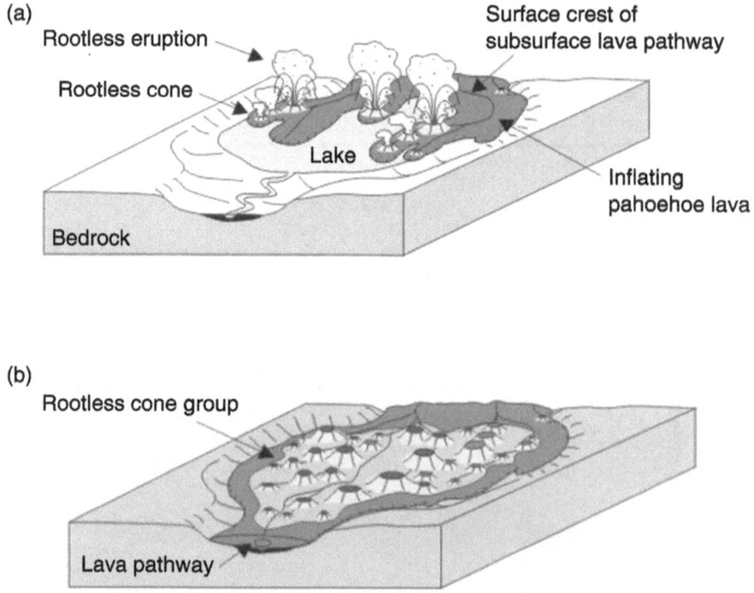

Figure 6.12. Principal concepts of the "dynamic heat transfer" model for rootless eruptions resulting in the construction of a cone group. (a) Tube-fed lava partly covers a basin and thickens by inflation. Cracks in the tube floors allow direct contact between the hot lava and waterlogged sediments, initiating series of rootless eruptions. (b) A fully developed cone group within a pahoehoe flow field.

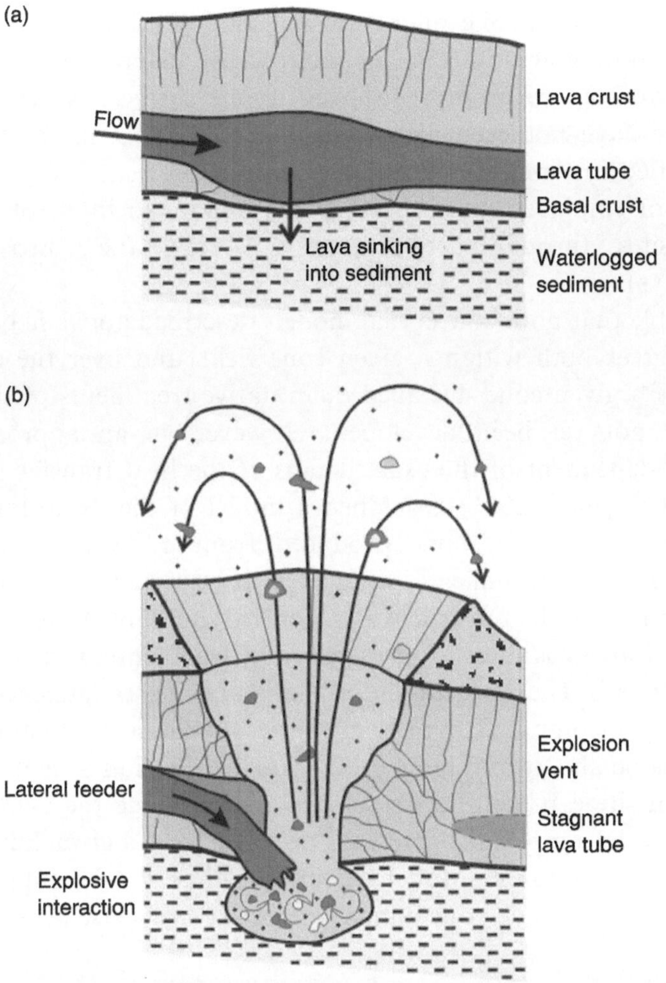

Figure 6.13. Inferred conditions for the initiation of a rootless eruption. (a) Internal lava pathway (i.e., lateral feeder) some distance upflow from the active flow front of pahoehoe lava that is advancing over and sinking into lower density sediments. (b) A crack through the basal crust of the lava allows the hot lava to come in direct contact and physically mix with waterlogged sediments, thus initiating a rootless eruption.

migrate in the upflow direction following the course of the pathway that feeds them. At any site, the eruption stops when the supply of water (from the substrate sediment) and/or fresh hot lava runs out. As the flow migrates across the lake/basin via bifurcating pathways, the locus of explosive vents follows and gradually builds a rootless cone group on top of the lava. When the lava eventually reaches the opposite margin, it continues its advance in the same manner as it did before its encounter with the basin.

In the Martian case, one might imagine that the lavas could encounter areas of ice-containing substrate, or areas where heterogeneities in regolith properties allow ice to be present closer to the surface. Although obvious tube systems have not been identified, it is likely that in the broad sheet flows identified in Amazonis Planitia (Keszthelyi *et al.*, 2000), the surging mechanism of emplacement would provide continued input of lava to the explosion sites through active zones of flow (i.e., broad internal lava pathways).

It is possible that both conceptual models described above may operate to different degrees both within a given cone field, and over the duration of formation of a given cone. Detailed quantitative treatments of the dynamic model have not yet been developed. However, in an approach that is somewhat independent of the exact details of the heat transfer mechanism, Greeley and Fagents (2001) developed a model of the explosion dynamics to determine the size of cone produced from a given set of eruption conditions. This model takes a volume of pressurized gas as the starting point (ignoring how the pressure was generated), and integrates the equation of motion of an expanding slug of gas and fragmenting lava to find the eruption velocity. The trajectories of individual clasts launched with this velocity are then computed, subject to the aerodynamic interactions with the moving volcanic and atmospheric gases. The distribution of material around the explosion site can then be synthesized to determine the diameter of the cone produced by a repeated explosions resulting from a given initial pressure and water vapor amount. Model constraints that can be supplied by the field observations include cone dimensions, clast size, clast density, and lava thickness.

Figure 6.14 shows an example of the results of this model: cone diameter is plotted as a function of starting pressure and initial gas mass under both terrestrial and Martian conditions. Comparisons of the results for Icelandic and Martian cones show that for equivalent starting conditions, Martian rootless cones should be 5–10 times wider than terrestrial cones, due to the lower atmospheric pressure allowing greater gas expansion (and hence greater eruption velocities) and lower particle drag forces, and the low gravity allowing clasts to stay aloft for longer (Greeley and Fagents, 2001; Fagents *et al.*, 2002). For the sizes of cones observed in the MOC images (e.g., typically 100 m in Amazonis), much less water vapor is required than their Icelandic counterparts: 2–350 kg per explosion for eruptions through 1–5 m of lava on Mars, compared with 500–3000 kg per explosion for a similar cone in Iceland. This is consistent with the likely low availability of water on Mars compared with Iceland: if the Mars climate allows ice stability at any given time

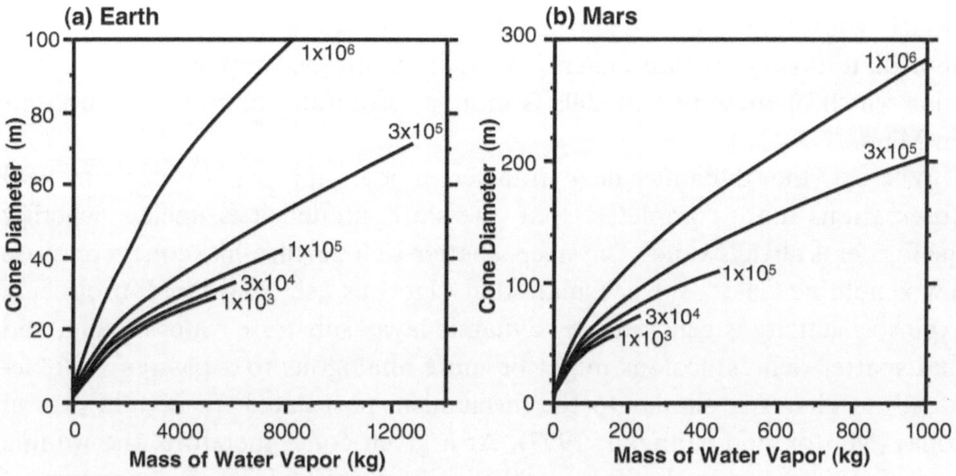

Figure 6.14. Computed cone diameter as a function of initial pressure and H_2O mass. Calculations were performed for a typical clast size of 10 cm, and density of 2500 kg m^{-3}, appropriate for degassed lava. Labeled curves showed the initial vapor pressure in Pa. Maximum clast ejection angle is assumed to be 70°.

(Paige, 1992; Mellon and Jakosky, 1995; Mellon *et al.*, 1997), the ice is likely to be contained within regolith pore spaces. Importantly, because the lava remains fluid and continues to be fed into the explosion site, the requirement for water/ice to be present within a few meters of the surface (Allen, 1979; Squyres *et al.*, 1987) is relaxed, because theory predicts that deeper levels of the substrate can be heated with the continued advective heat transfer (Fagents and Greeley, 2001).

6.5 Discussion

The two conceptual models outlined above are fundamentally different as regards heat transfer processes and energy conversion modes, and this has direct implications for the approach to numerical modeling of these eruptions as well as for quantitative assessment of water mass involved in such eruptions. Theoretical considerations of the static heat transfer model indicate that conductive heating by continuously flowing lava could lead to vaporization of a volatile phase present at depths up to ~20 m. However, because water vapor is a poor conductor of heat, generation and entrapment of a discrete vapor layer would act to reduce the penetration of the heating effects, which is not taken into account by this model. In the case of the dynamic heat transfer model, where the water and lava intermingle, the water must be present in the immediate lava substrate. Therefore, it is of utmost importance to determine which of these two models is more appropriate for rootless eruptions on Mars.

Whereas the dynamic heat transfer model appears to explain field observations more completely than the static model, it is unclear whether the former is always valid. The steep, spatter-rich hornito-like cones comprise few xenolithic clasts. We speculate that vigorous ash- and lapilli-producing explosive activity is generated by dynamic lava–substrate water mixing, and that spatter-rich explosions might be more analogous to a passive "bubble-burst" mechanism, similar to the mechanism postulated for certain littoral cones (Mattox and Mangan, 1997). At a given cone, therefore, the waning explosion intensity with time might signify a shift along the spectrum of activity from the dynamic to the static model. This transition is probably driven by a decreasing water/lava mass ratio as the eruption proceeds. However, we emphasize that more fieldwork is required to determine the relative roles of the two heat transfer models in Icelandic rootless cone-forming explosions.

There are a number of limitations in the use of Icelandic cones as analog for the Martian features. We are faced with the possibility that on Mars, because ice might be present below meters of dehydrated regolith, the ability of the lava to mix mechanically with water-rich substrate materials might be limited. If the dynamic heat transfer model is shown to explain all of the observed features of the Icelandic cones, then the Icelandic analog study might be less relevant to Mars. Moreover, if the water is present as ice, as it may also have been in some winter-time Icelandic eruptions, this configuration and the phase change energy requirements will affect the process. It seems likely, however, that the explosion onset will simply be delayed (as the ice first melts before the lava-sediment intermingling takes place), since the solid–liquid phase change only consumes ~7% of the energy required to drive a given explosion (Allen, 1979). Both dynamic and static heat transfer models will need to be addressed theoretically.

There remain several key aspects of this problem that need to be explained quantitatively if we are to understand rootless eruptions on Earth and their applicability to Mars, including:

- Does the process of explosive vaporization of the volatile phase and hence fragmentation of the lava always require physical mixing between lava and water-rich sediment or is conductive heat transfer across thickening basal crust adequate for driving rootless eruptions?

- If the former condition is required, then what are the conditions that allow physical mixing of molten lava and waterlogged substrate sediments? Some insights might be gleaned from peperite studies (e.g., Skilling *et al.*, 2002).
- What is the exact relationship between lava emplacement modes and the formation of rootless cones? Are they restricted to flow characterized by endogenous growth (i.e., insulated transport by internal pathways and flow inflation)?
- What is the mechanism that controls (a) the range in cone types within each cone group, (b) the random and partly overlapping arrangement of cones and craters, and (c) internal stratification of individual cones?
- What is the relationship between water/magma mass ratio and cone morphology, and how can we determine these ratios?

These issues are best addressed with a suite of field, laboratory, and computational tools. Key strategies include:

- Using ground- and image-based characterization of cone morphologies and cone group architecture to develop unambiguous ways of distinguishing rootless cones on Mars from primary cones or non-volcanic features.
- Analysis of deposit components (juvenile lava, solidified lava, and substrate sediments) and clast sizes (and vertical variations of both through the deposit) to determine fragmentation mechanism and degree of lava–substrate mixing.
- Development of theoretical treatments of the explosive pressures generated by mechanical mixing of lava and substrate water/ice.
- Use of clast size and density measurements, together with isopach and isopleth maps of deposits, to determine, via numerical models, the eruption velocities, confining pressures, and vapor amounts in rootless explosions.
- Application of models to Martian environmental conditions, to tie observed cone morphology to eruption conditions and substrate water amounts.

Together with crater-dating of Martian flow surfaces, the above approaches should ultimately yield estimates of spatial and temporal distribution of ground ice amounts and depths across the surface of Mars, wherever cone fields are present.

6.6 Concluding remarks

Studies of Icelandic rootless cones will continue to provide insights into the Martian examples since they remain the best analog, in terms of morphology, geologic setting, and eruption mechanism. Additional information can be gleaned from field observations of littoral cones and peperites,and laboratory simulations of explosive magma–water interactions. Thus far, we have been able to broadly quantify the amount of vapor involved in rootless explosions, and make inferences on the amount of water that must be held in the Martian regolith. However, since the dynamics of lava–water interactions are poorly understood, even on Earth, refinements to our understanding will continue to be made. In particular, if we can further develop our terrestrial suite of tools (field, laboratory analyses and computational modeling) to better understand the relationship of cone morphology and clast characteristics to water–lava mass ratio, then we can apply this knowledge to Mars and make inferences on water amounts based on the characteristics of features observed in the remote-sensing images.

Acknowledgments

This work was supported in part by NASA grants NAG5-11199 and NAG5-13458 to SAF and in part by a Nordic Volcanological Institute Research Fellowship to TT. This is HIGP contribution #1331, and SOEST contribution #6383.

References

Allen, C. C. (1979). Volcano–ice interactions on the Earth and Mars. Ph.D. thesis, University of Arizona, Tucson.

Arvidson, R. E., Coradini, M., Carusi, A. *et al.* (1976). Latitude variations of wind erosion of crater ejecta deposits on Mars. *Icarus*, **27**, 503–16.

Boynton, W. V., Feldman, W. C., Squyres, S. W. *et al.* (2002). Distribution of hydrogen in the near-surface of Mars: evidence for subsurface ice deposits. *Science*, **297**, 81–5.

Burr, D. M., Grier, J. A., McEwen, A. S., and Keszthelyi, L. P. (2000). Repeated aqueous flooding from the Cerberus Fossae: evidence for very recently extant, deep groundwater on Mars. *Icarus*, **159**, 53–73.

Büttner, R., Zimanowski, B., and Röder, H. (2000). Short-time electrical effects during volcanic eruption: experiments and field measurements. *Journal of Geophysical Research*, **105**, 2819–27.

Cabrol, N. A., Grin, E. A., and Pollard, W. H. (2000). Possible frost mounds in an ancient Martian lake bed. *Icarus*, **145**, 91–107.

Carr, M. H., Greeley, R., Blasius, K. R., Guest, J. E., and Murray, J. B. (1977). Some Martian volcanic features as viewed from the Viking Orbiters. *Journal of Geophysical Research*, **82**, 3985–4015.

Clifford, S. M. and Hillel, D. (1983). The stability of ground ice in the equatorial region of Mars. *Journal of Geophysical Research*, **88**, 2456–74.

Fagents, S. A. and Greeley, R. (2001). Factors influencing lava–substrate heat transfer and implications for thermomechanical erosion. *Bulletin of Volcanology*, **62**, 519–32.

Fagents, S. A., Lanagan, P. D., and Greeley, R. (2002). Rootless cones on Mars: a consequence of lava–ground ice interaction. In *Volcano–Ice Interaction on Earth and Mars*, ed. J. L. Smellie and M. G. Chapman. Geological Society of London Special Publication 202, pp. 295–317.

Fanale, F. P., Salvail, J. R., Zent, A. P., and Postawko, S. E. (1986). Global distribution and migration of subsurface ice on Mars. *Icarus*, **67**, 1–18.

Feldman, W. C., Gasnault, O., Squyres, S. W. *et al.* (2002). Global distribution of neutrons from Mars: results from Mars Odyssey. *Science*, **297**, 75–8.

Fisher, R. V. (1968). Pu'u Hou littoral cones, Hawaii. *Geol. Rundschau.*, **57**, 837–64.

Fröhlich, G., Zimanowski, B., and Lorenz, V. (1993). Explosive thermal interactions between molten lava and water. *Experimental Thermal and Fluid Science*, **7**, 319–32.

Frey, H. (1987). Pseudocraters as indicators of ground ice on Mars. *Reports of the Planetary Geology and Geophysics Program, NASA Technical Memorandum* 89810, pp. 18–19.

Frey, H. and Jarosewich, M. (1982). Subkilometer Martian volcanoes: properties and possible terrestrial analogs. *Journal of Geophysical Research*, **87**, 9867–79.

Frey, H., Lowry, B. L., and Chase, S. A. (1979). Pseudocraters on Mars. *Journal of Geophysical Research*, **84**, 8075–86.

Gault, D. E. and Greeley, R. (1978). Exploratory experiments of impact craters formed in viscous-liquid targets: analogs for Martian rampart craters? *Icarus*, **34**, 486–95.

Greeley, R. and Fagents, S. A. (2001). Icelandic pseudocraters as analogs to some volcanic cones on Mars. *Journal of Geophysical Research*, **106**, 20527–46.

Hartmann, W. K. (1999). Martian cratering VI: crater count isochrons and evidence for recent volcanism from Mars Global Surveyor. *Meteoritics & Planetary Science*, **34**, 167–78.

Hartmann, W. K., Malin, M. C., McEwen, A. S. *et al.* (1999). Evidence for recent volcanism on Mars from crater counts. *Nature*, **397**, 586–9.

Hodges, C. A. and Moore, H. J. (1994). Atlas of volcanic landforms on Mars. US Geological Survey Professional Paper 1534.

Judson, S. and Rossbacher, L. A. (1979). Geomorphic role of ground ice on Mars. *NASA Technical Memorandum* 80339, pp. 247–9.

Jurado-Chichay, Z., Rowland, S. K., and Walker, G. P. L. (1996). The formation of circular littoral cones from tube-fed pahoehoe; Mauna Loa, Hawaii. *Bulletin of Volcanology*, **57**, 471–82.

Keszthelyi, L. P., McEwen, A. S., and Thordarson, T. (2000). Terrestrial analogs and thermal models for Martian flood lavas. *Journal of Geophysical Research*, **105**, 15027–49.

Lanagan, P. D., McEwen, A. S., Keszthelyi, L. P., and Thordarson, T. (2001). Rootless cones on Mars indicating the presence of shallow equatorial ground ice in recent times. *Journal of Geophysical Research*, **28**, 2365–8.

Lucchitta, B. K. (1981). Mars and Earth: comparison of cold climate features. *Icarus*, **45**, 264–303.

Malin, M. C. and Edgett, K. S. (2000). Evidence for recent groundwater seepage and surface runoff on Mars. *Science*, **288**, 1927–37.

Mattox, T. N. and Mangan, M. T. (1997). Littoral hydrovolcanic explosions: a case study of lava–seawater interaction at Kilauea Volcano. *Journal of Volcanology & Geothermal Research*, **75**, 1–17.

McCauley, J. F. (1973). Mariner 9 evidence for wind erosion in the equatorial and mid-latitude regions of Mars. *Journal of Geophysical Research*, **78**, 4123–37.

Mellon, M. T. and Jakosky, B. M. (1995). The distribution and behavior of Martian ground ice during past and present epochs. *Journal of Geophysical Research*, **100**, 11781–99.
Mellon, M. T., Jakosky, B. M., and Postawko, S. E. (1997). The persistence of equatorial ground ice on Mars. *Journal of Geophysical Research*, **102**, 19357–69.
Mitrofanov, I., Sanin, A., Tret'yakov, V. *et al.* (2002). Maps of subsurface hydrogen from the High Energy Neutron Detector. Mars Odyssey, *Science*, **297**, 78–81.
Mouginis-Mark, P. J. (1987). Water or ice in the Martian regolith?: Clues from rampart craters seen at very high resolution. *Icarus*, **71**, 268–86.
Paige, D. A. (1992). The thermal stability of near-surface ground ice on Mars. *Nature*, **356**, 43–5.
Parker, T. J., Gorsline, D. S., Saunders, R. S., Pieri, D. C., and Schneeberger, D. M. (1993). Coastal geomorphology of the Martian northern plains. *Journal of Geophysical Research*, **98**, 11061–78.
Skilling, I. P., White, J. D. L., and McPhie, J. (2002). Peperite: a review of magma–sediment mingling. *Journal of Volcanology & Geothermal Research*, **114**, 1–17.
Squyres, S. W., Wilhelms, D. E., and Moosman, A. C. (1987). Large-scale volcano–ground ice interactions on Mars. *Icarus*, **70**, 385–408.
Steingrímsson, J. (1788). Fulkomid Skrif um Sídueld (A complete description of the Sída volcanic fire), *Safn til Sögu Íslands, IV* (Copenhagen 1907–1915), 58–69.
Steingrímsson, J. and Ólafsson, S. (1783). Einföld og sönn frásaga um jardeldshlaupid í Skaftafellssyslu árid 1783 (A simple, but true narrative of the eruption in Skaftafell county in the year 1783), *Safn til Sögu Íslands, IV* (Copenhagen, 1907–1915), 58–69.
Theilig, E. and Greeley, R. (1979). Plains and channels in the Lunae Planum–Chryse Planitia region of Mars. *Journal of Geophysical Research*, **84**, 7994–8010.
Thorarinsson, S. (1951) Laxargljufur and Laxarhraun: A tephrochronological study, *Geografiska Annaler*. **H1–2**, 1–89.
Thorarinsson, S. (1953). The crater groups in Iceland. *Bulletin of Volcanology*, **14**, 3–44.
Thorarinsson, S. (1965). The Surtsey eruption: course of events and the development of the new island. *Surtsey Research Progress Report*, **1**, 51–5.
Thordarson, T. (2000). Rootless eruptions and cone groups in Iceland: products of authentic explosive water to magma interactions. Abstract in *Volcano/Ice Interactions on Earth and Mars*, ed. V. C. Gulick and M. T. Gudmundsson. Reykjavík: University of Iceland, p. 48.
Thordarson, T. and Höskuldsson, Á. (2002). *Iceland. Classic Geology in Europe, 3*. Harpenden, UK: Terra Publishing.
Thordarson, T., Morrissey, M. M., Larsen, G., and Cyrusson, H. (1992). Origin of rootless cone complexes in S-Iceland. Abstract in *The 20th Nordic Geological Winter Meeting*, ed. A. Geirsdóttir, H. Norddahl, and G. Helgadóttir. Reykjavík: Icelandic Geoscience Society, p. 169.
Thordarson, T., Miller, D. J., and Larsen, G. (1998). New data on the Leidolfsfell cone group on south Iceland. *Jökull*, **46**, 3–15.
Thoroddsen, T. (1894). Ferd um Vestur Skaftafellssyslu sumarid 1893 (Travelogue from Western Skaftafellshire in the summer of 1893). *Andvari*, **19**, 44–161.

White, J. D. L. (1996). Impure coolants and interaction dynamics of phreatomagmatic eruptions. *Journal of Volcanology & Geothermal Research*, **74**, 155–70.

Wohletz, K. H. (1983). Mechanisms of hydrovolcanic pyroclast formation: grain-size, scanning electron microscopy, and experimental studies. *Journal of Volcanology & Geothermal Research*, **17**, 31–63.

Wohletz, K. H. (1986). Explosive magma–water interactions: thermodynamics, explosion mechanisms, and field studies. *Bulletin of Volcanology*, **48**, 245–64.

Wohletz, K. H. and McQueen, R. G. (1984). Experimental studies of hydromagmatic volcanism. In *Explosive Volcanism: Inception, Evolution, and Hazards*. Washington, DC: National Academy of Sciences, pp. 158–69.

Zimanowski, B. and Büttner, R. (2002). Dynamic mingling of magma and liquefied sediments. *Journal of Volcanology & Geothermal Research*, **114**, 37–44.

Zimanowski, B., Fröhlich, G., and Lorenz, V. (1991). Quantitative experiments on phreatomagmatic explosions. *Journal of Volcanology & Geothermal Research*, **48**, 341–58.

7

Mars interior layered deposits and terrestrial sub-ice volcanoes compared: observations and interpretations of similar geomorphic characteristics

Mary G. Chapman
US Geological Survey, Flagstaff, Arizona

and

John L. Smellie
British Antarctic Survey, Cambridge

7.1 Introduction

This chapter discusses our investigation of the morphologies of numerous Mars interior layered deposits (ILDs) and terrestrial sub-ice volcanoes. Our study reveals multiple strong similarities in both groups of edifices that are supportive of extending interpretations of a sub-ice volcanic origin to the Mars landforms. In particular, distinctive morphological features common to both include relatively tall edifices and wing-like projections. Whereas most terrestrial sub-ice volcanoes described so far are monogenetic and relatively small, the generally much greater dimensions of many Mars ILDs may be explained as a consequence of long-lived polygenetic sub-glacial volcanism. Terrestrial sub-ice volcanoes can be observed in several northern latitude locales and on Mars and the ILDs are found within Valles Marineris.

The Valles Marineris canyon system consists of isolated and interconnected elliptical and linear troughs or chasmata near the equator of Mars. This canyon is cut into the crest of a topographic ridge having a high end at the Tharsis volcanoes and a low end 3000 km away where the canyon connects to the catastrophic flood channels that drain into Chryse basin (Figure 7.1a). The rocks on Mars are broken into three age systems, from oldest to youngest: Noachian, Hesperian, or Amazonian, based on the

The Geology of Mars: Evidence from Earth-based Analogs, ed. Mary Chapman. Published by Cambridge University Press.

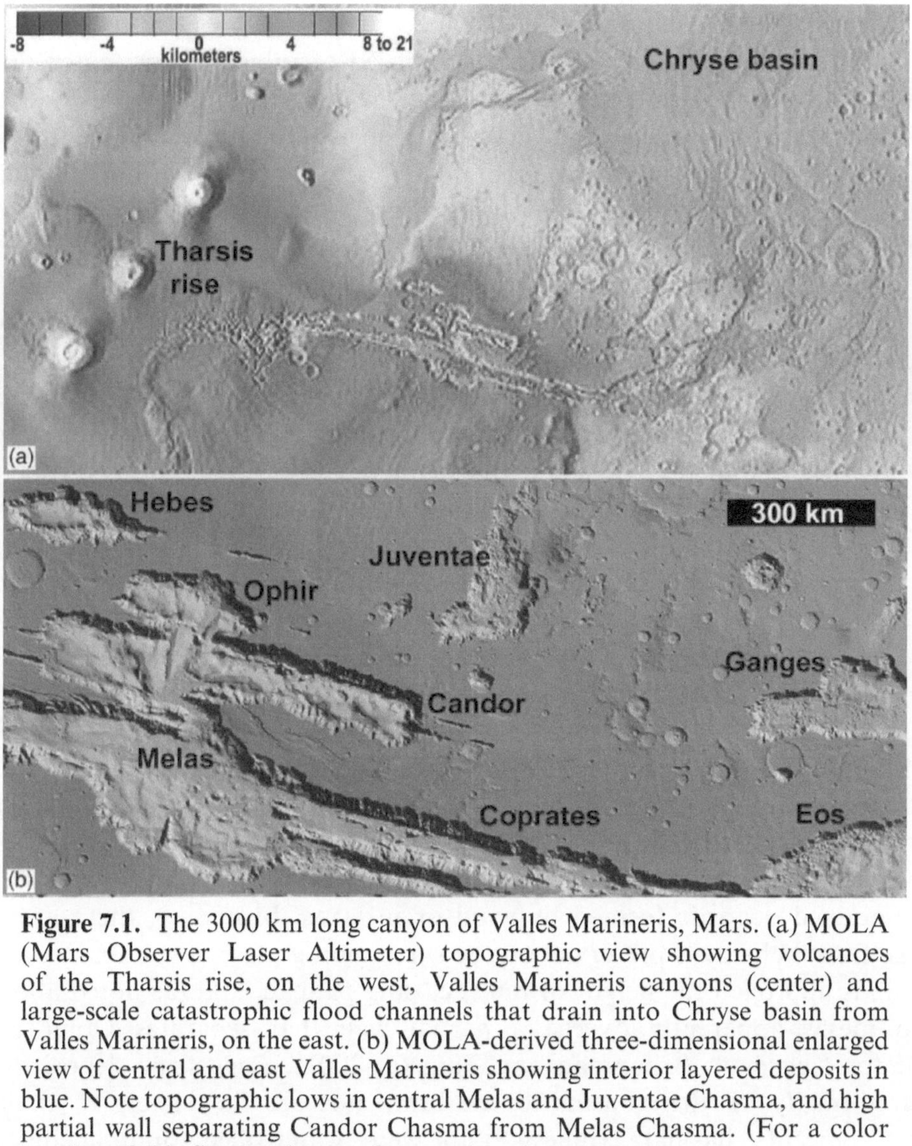

Figure 7.1. The 3000 km long canyon of Valles Marineris, Mars. (a) MOLA (Mars Observer Laser Altimeter) topographic view showing volcanoes of the Tharsis rise, on the west, Valles Marineris canyons (center) and large-scale catastrophic flood channels that drain into Chryse basin from Valles Marineris, on the east. (b) MOLA-derived three-dimensional enlarged view of central and east Valles Marineris showing interior layered deposits in blue. Note topographic lows in central Melas and Juventae Chasma, and high partial wall separating Candor Chasma from Melas Chasma. (For a color version of this figure, please refer to color plate section.)

number and size of impact craters on the surface. The chasmata have been generally interpreted as large grabens and (or) collapse structures, produced in association with tensional stresses generated by the Tharsis rise during the Late Noachian to Early Hesperian (Lucchitta *et al.*, 1992). On the adjacent plateau, these chasms cut "ridged plains" rocks that have a surface age of Early Hesperian and therefore Valles Marineris must postdate deposition of that unit. The layered plateau rock exposed in the walls of Valles Marineris (and underlying the plateau surface composed of ridged plains material(s)) is mostly Noachian, is likely basaltic, and in many troughs can extend up to 8 km from the plateau rim to the canyon floor (McEwen *et al.*, 1999). Flat-topped mesas and mounds of ILDs occur mainly within elliptical chasmata and are constructed of horizontal to angled beds of laterally continuous materials (Figure 7.1b).

There is abundant evidence that the ILDs were not formed by down-faulting of materials in the walls of Valles Marineris. (1) Two ILD mesas extend across the wall-like ridge separating Ophir and Candor Chasmata (Figure 7.1b) and show no signs of displacement; (2) the linear graben-type chasmata, the best candidates for likely downfaulting, do not contain ILD mounds; (3) erosional styles indicate that the ILDs and chasmata wall rock are composed of very different materials (Lucchitta *et al.*, 1992); and (4) most workers believe that the ILDs are no older than Hesperian based on the stratigraphic age of the chasmata and superposition of ILD outcrops on adjacent wall rock and chasma-floor chaotic blocks (Lucchitta *et al.*, 1992; Komatsu *et al.*, 1993; Lucchitta, 1999; Chapman and Tanaka, 2001; Chapman, 2002). (There are downdropped blocks of wall rock (horsts) in the Valles Marineris, but these are not ILDs.)

Mapping suggests that the ILDs formed at different times (Lucchitta, 1990). Adjacent outcrops have different morphologies and contain dissimilar sequences, suggesting variable depositional histories (Komatsu *et al.*, 1993). These characteristics are incompatible with an origin for ILDs as coeval fine-grained deposits formed by wind or in lakes. Moreover, there are no large fluvial channels preserved on the surrounding plateau surfaces, several mounds approach the height of the enclosing plateau, and (as noted above) two mesas apparently overlie a partial wall that separates Ophir and Candor Chasmata (Chapman and Tanaka, 2001). Lacustrine deposits do not normally fill their basins to the brim, nor are they likely to extend into adjacent chasmata (unless both were water-filled simultaneously). Most ILD mesas are free-standing (i.e., separated from each other and the chasma wall rock by moats), many show resistant horizontally layered caprocks above friable flank deposits (Lucchitta *et al.*, 1992), and some show layers in the flank deposits dipping steeply downslope (Chapman and Tanaka, 2001). The lacustrine and eolian hypotheses cannot explain the formation of moats that separate the deposits from the trough walls, nor the presence of angled beds on mound flanks beneath horizontally bedded caprock: additional mechanisms need to be evoked to explain these features.

Evidence in support of a volcanic origin for ILDs includes: (1) the variable lithologies, heights, and relative ages of ILDs; (2) their occurrence as free-standing massifs; (3) their compositions; (4) the volcano-tectonic setting; and (5) the low albedo, high competence and tuff-like weathering of some layers (Peterson, 1981; Lucchitta, 1990; Witbeck *et al.*, 1991). Many of the ILD mounds form flat-topped mesas. Sub-glacially erupted volcanoes also form free-standing flat-topped mesas, known as tuyas (Mathews, 1947; Smellie, 2000). The ILDs are composed of basalt and palagonite (Murchie *et al.*, 2000; Mustard and Murchie, 2001). Although there are some terrestrial tuya edifices with silicic compositions (e.g., Tuffen *et al.*, 2002), most sub-ice volcanoes are mafic and consist of horizontal layered basaltic lavas overlying friable flank deposits of steeply dipping (angled) hyaloclastite breccias composed of variably palagonite altered basaltic clasts (Figure 7.2). When terrestrial volcanoes erupt underneath ice, they melt a vault and generate a meltwater lake within which a tephra pile is constructed (Björnsson, 1988; Smellie, 2000). Following emergence above the lake, lava-fed deltas are formed. ILDs are compositionally and morphologically similar to many terrestrial basaltic tuyas and most ILD-containing chasmata lead to outflow channels. Because of the significant weaknesses inherent in alternative interpretations, many workers have suggested that the ILDs may be sub-ice volcanoes (Croft, 1990; Lucchitta *et al.*, 1994; Chapman and Tanaka, 2001; Komatsu *et al.*, 2004).

Although the hypothesis that the ILDs are sub-ice volcanoes is not new, there is a paucity of comparative planetary studies of sub-ice volcanoes using terrestrial examples and Mars Observer Camera (MOC) imagery of the ILDs. In order to test the subglacial volcanic hypothesis for ILDs, we have investigated terrestrial sub-ice volcanoes and compared them to characteristics of analog ILDs on Mars. In doing so, we have found several commonalities that suggest comparable volcanic processes. In addition,

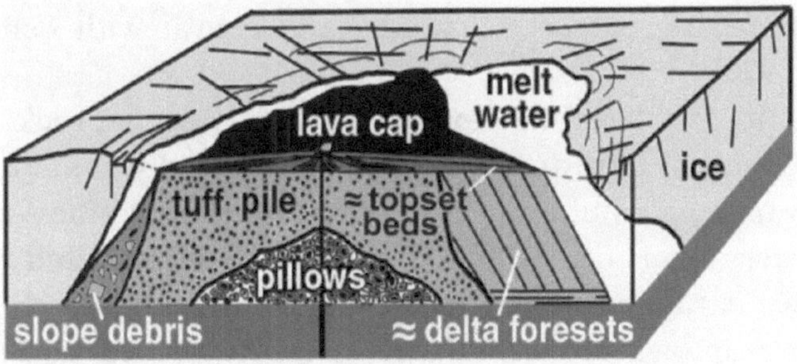

Figure 7.2. Schematic cross-sectional view of an ideal tuya.

in this chapter we describe the effects of enclosing ice sheets on terrestrial edifices to postulate possible consequences for the putative volcanic morphologies of the ILDs. This chapter discusses the initial findings of our analog studies.

7.2 Geomorphic commonalities between terrestrial Tuyas and Mars ILDs

Barring a returned sample or human/robotic science data from Valles Marineris, we are constrained to using remotely sensed spacecraft data for investigating possible volcanic origins of the ILDs. Imagery and derived topographic data allow us to examine only three ILD attributes: gross morphology, stratigraphy, and field relationships of the caprock and flank deposits. Comparative analog studies of these attributes on terrestrial sub-ice volcanoes reveal some unique and definitive morphologic characteristics.

7.2.1 Gross morphologic characteristics (observed mostly in medium-resolution datasets)

Planetary scientists generally think of a sub-ice volcano as the idealized flat-topped subcircular mesa or tuya (Figures 7.2 and 7.3a). Tuyas are defined by the presence of lava-fed deltas (Skilling, 2002). The deltas represent a late subaerial progradational phase of edifice-building (Smellie, 2000). Although the deltas are constructed on top of a subaqueous tuff cone and/or pillow mound, these earlier-formed constructs do not significantly affect the final shape of the edifice (Figure 7.2). The lavas are the volcanic equivalent of topset beds in a sedimentary delta (Figure 7.2). They overlie cogenetic coarse hyaloclastite breccias (equivalent to sedimentary delta foreset beds; Figure 7.2), which dip radially outward at steep repose angles (25–40°). The breccias formed by a combination of quench-induced shattering and avalanching at the delta brinkpoint (Skilling, 2002). The junction between the lavas and dipping breccias is a planar surface (known as a passage zone) that records the water level coeval with delta formation (Jones, 1969, 1970).

In addition to the ideal flat-topped tuya having caprock lying above friable flank deposits, terrestrial volcanic fields usually contain sub-ice edifices that show many permutations of geomorphic shapes. These permutations take several basic forms due to the idiosyncrasies of syn- and post-eruptive geologic events particular to an individual edifice and to the geographic location. For example, sub-ice volcanoes in Antarctica and Siberia may also show cuspate margins (concave out) and arm- or wing-like projections

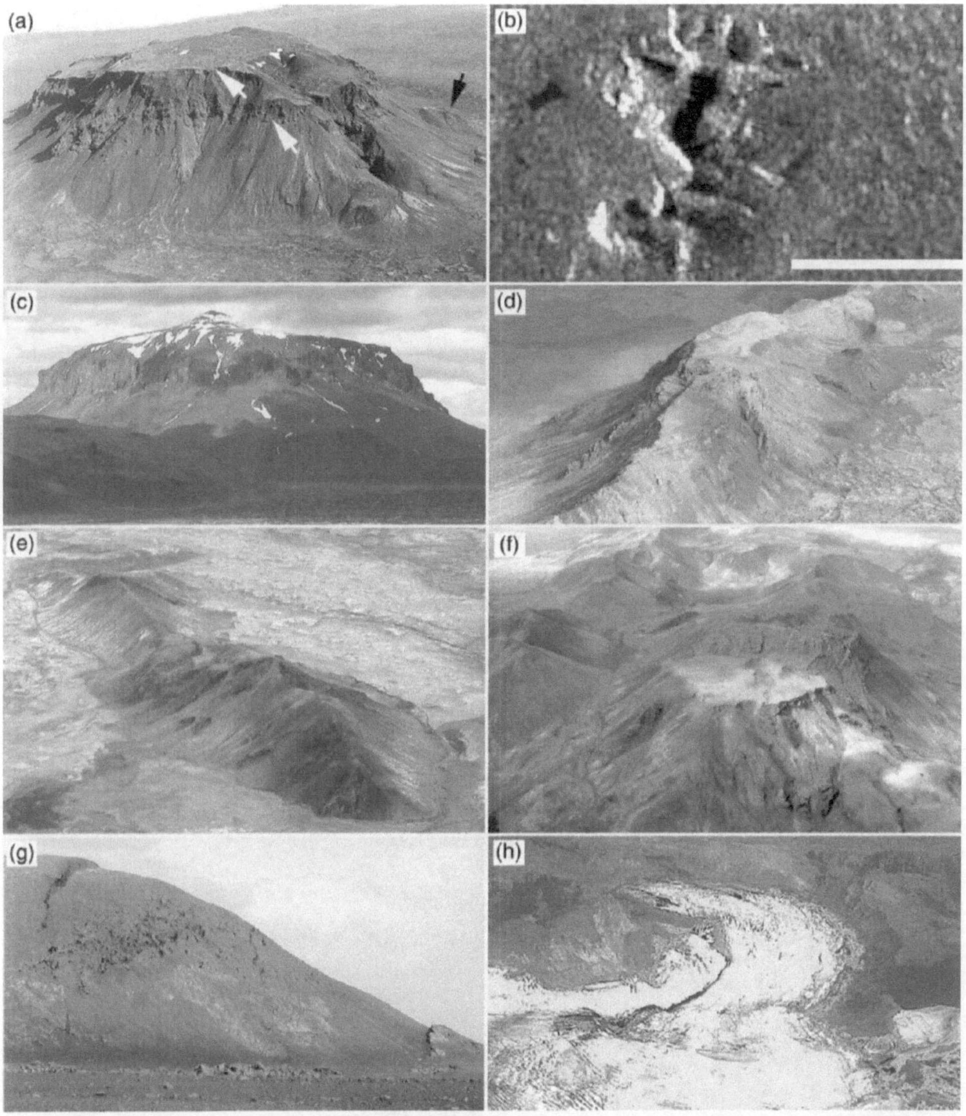

Figure 7.3. Views of terrestrial tindars and tuyas; (a) Hludufell, an Icelandic tuya with two horizons of layered caprock (white arrows) that embays older small hyaloclastic mound (black arrow); (b) Derbi-Taiga (Siberia), a tuya showing cuspate margins and wing-like flank protrusions; scale bar = 5 km (part of RADARSAT image R120050130U3S016 courtesy of Goro Komatsu); (c) Herdubried (Iceland), a tuya capped by a prominent scoria cone (central peak in view) that overlies resistant, horizontal lava layers; (d) Tindaskagi (Iceland), a capped tindar ridge with no visible vents; (e) Sodulholar (Iceland), a classic tindar; (f) Vifilsfell (Iceland), a tuya with a complex plateau summit; (g) down-dropped block on un-named 600 m high tindar mound, 20 km west of Askja (Iceland); (h) glacier cutting the north end of Tungnaárfjöll (Iceland; a tindar).

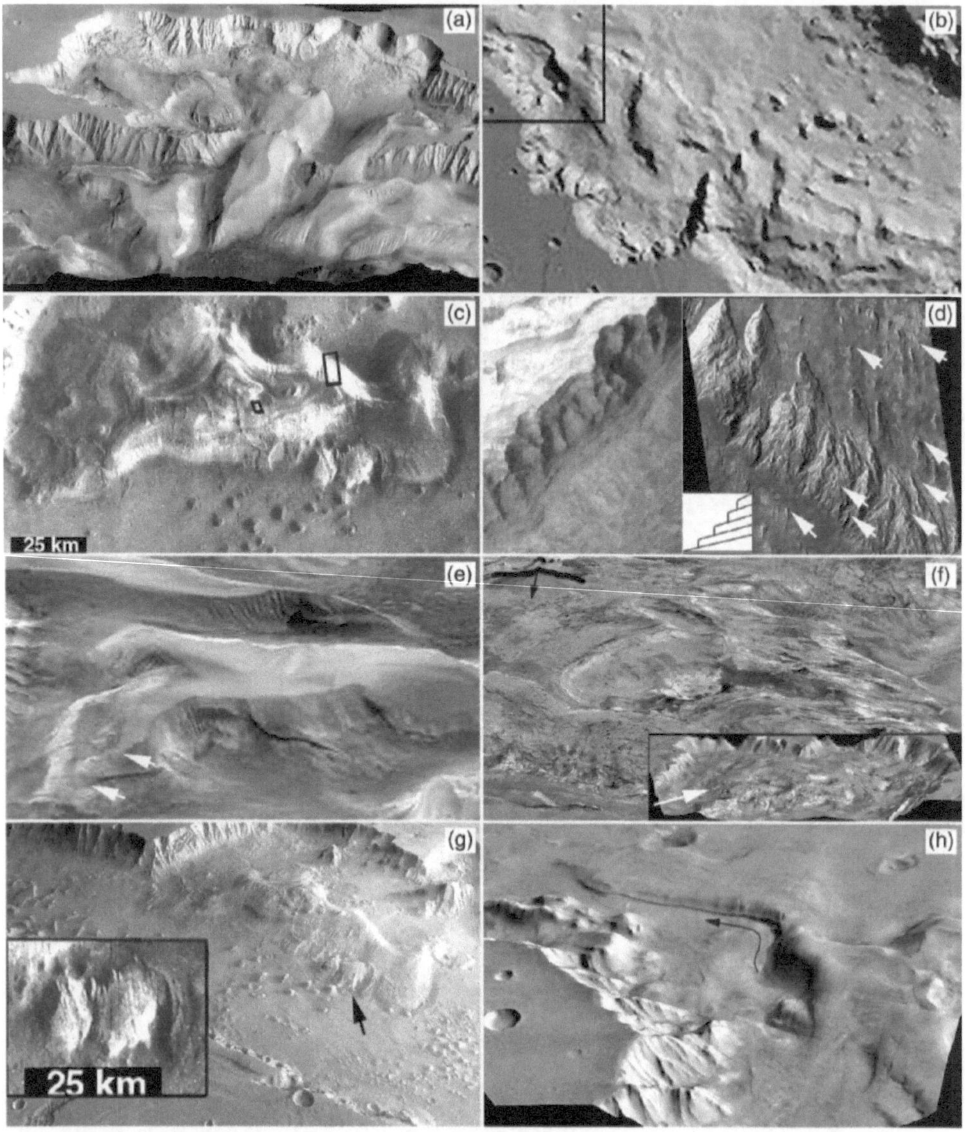

Figure 7.4. Views of Martian interior layered deposits (ILDs). (a) Viking three-dimensional image of ILD mounds and landslide materials on floors of Ophir (300 km wide) and central Candor Chasmata; (b) MOLA derived three-dimensional view of complex ILD mounds in Melas Chasma (90 km wide box denotes location of (h)); (c) Viking image of Ganges Mensa; note narrow linear ridges on top of mound that may be vent sites (boxes denote location of images on (d)); (d) parts of MOC images of Ganges Mensa: on left, horizontally layered resistant caprock (M804332, 2.86 m/pixel) and on right, arrows on flank denote alignment of ridge wedge points indicating steeply dipping layers schematically shown on illustration (M304405 and M401737 mosaic, 5.7 m/pixel); (e) Viking-imagery-draped MOLA topographic view

along the edifice base (Smellie and Skilling, 1994; Komatsu *et al.*, 2004; Figure 7.3b). Subaerial scoria cones formed concurrently with the layered lava cap. This scoria may be preserved on the mesa top to give edifices a more cone-like appearance (e.g., Herdubried (Iceland); Figure 7.3c), although scoria is friable and tends to be easily eroded if glacial ice reforms on top. Ridges composed of tephra, known as tindars or hyaloclastic ridges, are also common subglacial volcano forms and are formed by eruptions along volcanic fissures. The ridges are dominated by tephra but they may be capped locally by lava where they are beginning to transform to a ridge-shaped tuya (e.g., Figures 7.3d and e). The more familiar mesa-like tuyas are edifices that, over time, have become centralized over the erupting vents and have transformed to subaerial effusion and progradation of lava-fed deltas. Some of these edifices may be compound, formed of interlayered subaerial lava and subaqueous hyaloclastite breccia, indicating multiple eruptive episodes into successively formed meltwater lakes (arrows in Figure 7.3a). Some tuya plateaus are complexes with multiple volcanic sources and likely overlapping ages (e.g., Jones, 1969, 1970; Figure 7.3f). Because of ice confinement, most subglacial volcanoes are relatively tall. By contrast, pillow lava deposits typically form low mounds, because these volcanoes never reached a height at which magmatic volatiles could exsolve and form hydromagmatic tuff eruptions. In terrestrial volcanic fields, postglacial and interglacial subaerial volcanoes are commonly interspersed among sub-ice edifices. For example, since the last glacial maximum in Iceland, large subaerial volcanoes, pyroclastic cones, and embaying subaerial lava flows have been constructed above or adjacent to subglacial edifices. Tuyas and tindars are commonly faulted owing to seismic shock and flank collapse associated with eruptions (Van Bemmelen and Rutten, 1955; Smellie and Skilling, 1994; Smellie, 2001; Figure 7.3g). After an eruption ceases, reformed wet-based glaciers may carve U-shaped winding canyons and cirques into sub-ice volcanoes (Figure 7.3h).

If the ILDs on Mars are mostly sub-ice volcanoes, we should see similar permutations of shape. In fact, the shapes of the ILD mounds do vary

Caption for Figure 7.4. (cont.) of Candor Mensa (110 × 50 km) shows two resistant horizons (arrows); (f) Viking-imagery-draped MOLA topographic view of Ceti Mensa in west Candor Chasma showing horseshoe-shaped scarp and arrowed cone (inset shows arrowed direction of northeast view); (g) Viking draped MOLA view of Ganges Mensa showing two blocks (arrow, shown larger scale in Viking image inset) that may be have been down-dropped and wind eroded; (h) U-shaped, flat-floored canyon with sharp angled curved valley (arrow) cut into west Melas ILD (location in b).

considerably (Figures 7.1a and 7.4a). ILD mounds in Ganges and Candor Chasmata (Figures 7.4c, e, and g) and Hebes Mensa show cuspate, arm- or wing-like projections (cf. figure 19 in Komatsu *et al.*, 2004). Many ILDs are ridges that parallel the long axis of their enclosing chasmata; others are mounds, mesas, or complex plateaus (e.g., Melas Chasma; Figure 7.4b). Ganges Mensa is a typical free-standing ILD mesa (Figure 7.4c) with a horizontally bedded caprock and steeply dipping flank deposits (Chapman 2002; Figure 7.4d). Candor Mensa shows two horizons of caprock (arrows in Figure 7.4e). The western ILD plateau in Melas Chasma shows numerous horizontal layers that overlie and partly cover what appears to be a 55 km long linear trough, which is visually analogous to linear subaerial pyroclastic cones on Earth (figure 7.9 in Chapman, 2002). Ceti Mensa has a low dome shape with a superposed cone, and the entire mound is locally cut by huge horseshoe-shaped scarps with what appear to be massive outlying displaced blocks of material (Figure 7.4f). Ceti Mensa also has resistant layers that dip down at the side of the mound but is mostly composed of numerous layers of less-resistant materials with highly variable dips and folds. These characteristics and the dome-like overall shape of Ceti Mensa are more reminiscent of a subaerial volcano. In addition, Ceti Mensa also has two late-stage large blocky plate flows that may be inconsistent with a subglacial origin (Skilling *et al.*, 2002). Younger dark materials embay the ILDs. These materials may be some of the youngest basaltic volcanic pyroclastic eruptions on the planet (Geissler *et al.*, 1990; Lucchitta, 1990). Local vents have been observed in MOC images of this dark material (Chapman and Smellie, 2001; Lucchitta, 2001) and new THEMIS data suggests that some outcrops of the material may be olivine basalts (Christensen, 2003). Possible post-depositional faulting has been suggested for apparent structures and offsets on ILDs in Juventae (Komatsu *et al.*, 2004) and Hebes (Lucchitta, 2001) Chasmata. In addition, on its southeast side Ganges Mensa shows two small brighter mounds that may be have been down-dropped and eroded by wind (Figure 7.4g). Alternatively, these small mounds may be older hyaloclastic volcanoes that have been draped by the larger edifice. This relation may be analogous to that observed on the Hludufell tuya in Iceland which embays older small hyaloclastic mounds on its eastern side (Figure 7.3a; I. Skilling, personal communication). Similar to glacial erosion, the complex plateau mound of west Melas Chasma (Figure 7.4b) shows a semicircular, deep, steep-walled cirque heading a U-shaped canyon that displays a sharp right-angle bend (arrow in Figure 7.4h). High-standing glaciers can preferential cut cirques on south exposures of terrestrial mountains due to the sun seasonally melting ice only on that side (Figure 7.5a). The tops of certain ILDs in

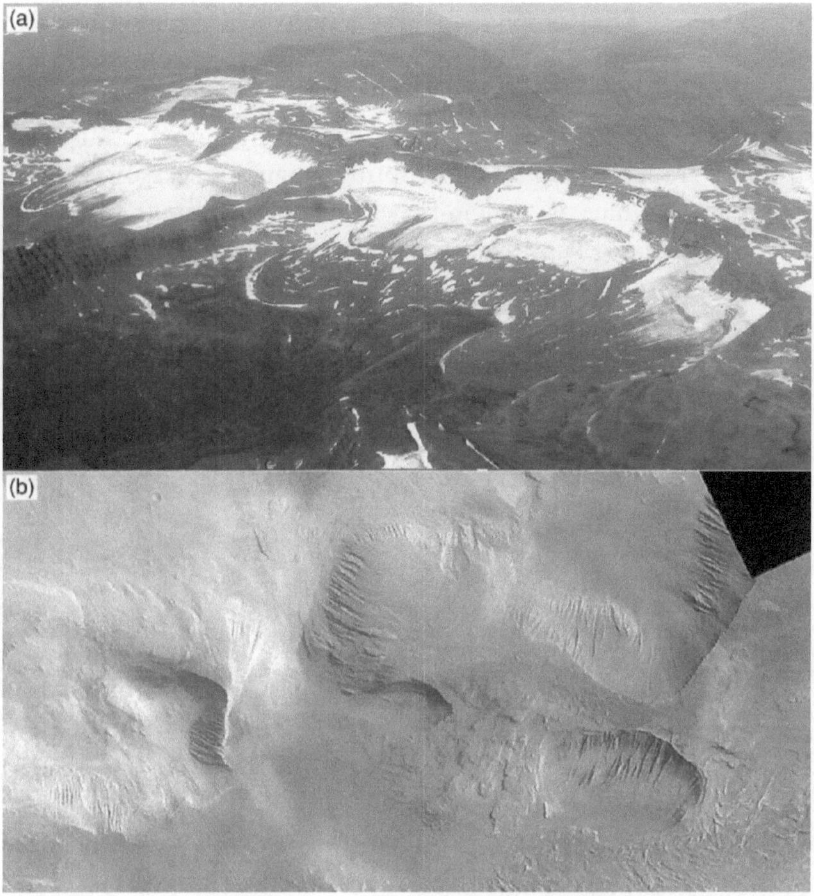

Figure 7.5. Views of hollows cut into Icelandic mountains and Martian mounds. (a) Aerial photograph showing south-facing chain of glacial cirques in Icelandic mountain (field of view approximately 30 km); (b) Viking mosaic showing chain of three south-facing, cirque-like hollows in ILDs of east Candor Chasma (field of view approximately 120 km).

East Candor Chasma also show steep cirque-like hollows mostly on southern exposures (Figure 7.5b).

7.2.2 Caprock and surface features

Terrestrial sub-ice volcanoes often form uncapped piles of tephra/pillow lavas (tindars) or can have caps of layered subaerial lava (tuyas). However, another distinct morphology occurs in Iceland just west of the Hekla volcano. Naefurholtsfjöll tindar is capped by deposits of bright white pumice and black lithics derived from Hekla that show unusual folds and drapes (Figure 7.6a). Similar to Naefurholtsfjöll tindar, Candor Mensa has a friable cap of mostly bright-albedo layers eroded into unusual folds and drapes (Figure 7.6b). On Naefurholtsfjöll, the Hekla ash lithics are heavier pre-existing dark volcanic rock fragments from the Hekla eruption that were dislodged by the erupting magma (accessory or accidental fragments). The pumice is vesicle-rich juvenile magma that is very light in weight. The Hekla 4 eruption that occurred 4000 years ago was lithic-rich (42 wt.% lithics; Larsen and Thorarinsson, 1977). Wind erosion preferentially winnows the lighter pumice, leaving the darker heavier lithics on the crests of dunes, creating an illusion of folds on the outcrop (Figure 7.6c). By contrast, the younger Hekla 3 eruption (3000 ka) is lithic-poor (Larsen and

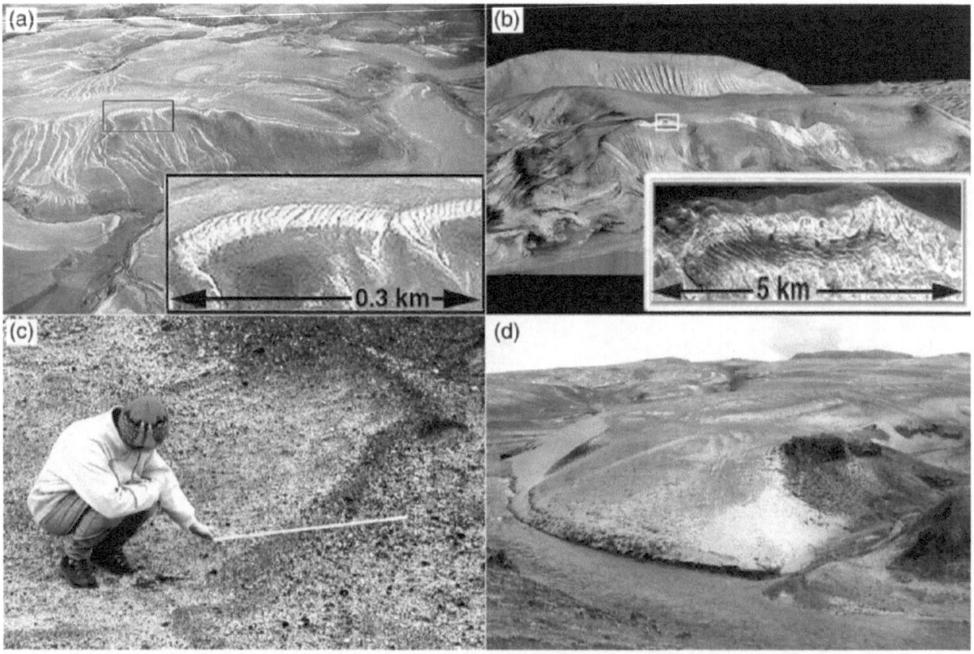

Figure 7.6. Views of friable mantles on Icelandic and Martian mounds. (a) Naefurholtsfjöll tindar capped by Hekla ash deposits of bright white pumice with black lithics that show unusual folds and drapes (inset); (b) Viking and MOC-imagery-draped MOLA topographic view of Candor Mensa ILD, showing a friable cap of mostly bright-albedo material eroded into unusual folds and drapes (inset); (c) recent tephra (Hekla 4 ash) on Naefurholtsfjöll (Iceland); the ash lithics are heavier and occur on the crests of dunes; lighter white pumice is preferentially removed and now occurs mainly on lee and stoss sides of ripple-like eolian bedforms; (d) lithic-poor, mostly pumiceous Hekla 3 ash eroded into drapes on Naefurholtsjöll.

Thorarinsson, 1977) and the wind erodes this pumiceous ash layer into drapes (Figure 7.6d).

Resistant layered lava caprock can show vents, including volcanic craters and feeder dikes, on both terrestrial tuyas (Figure 7.7a, b) and ILDs (Figure 7.7c; figure 7.12 of Chapman and Tanaka, 2001). However, many ILDs show no obvious vents (Figure 7.4a). Vents are also commonly eroded

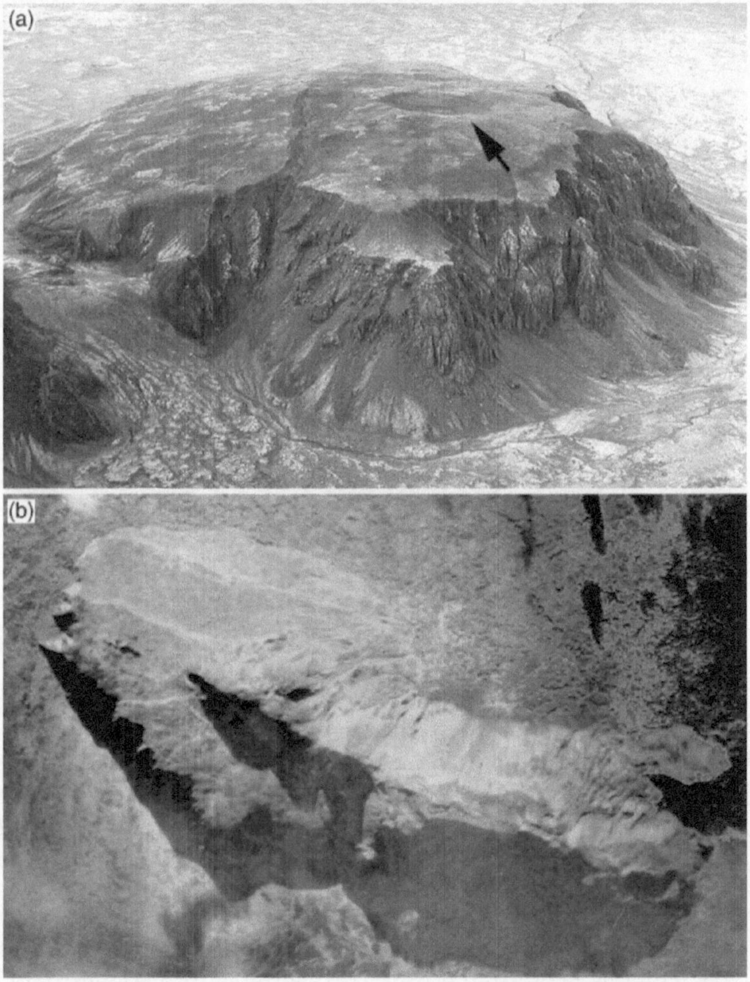

Figure 7.7. Vent sites on resistant layered lava caprock of tuyas in Iceland and possibly on Mars. (a) Circular volcanic crater (arrow) on 2.5 km diameter Hrafnabjörg; (b) narrow central fissure vent on 4 km wide Herdubreidartögl (adapted from Van Bemmelen and Rutten, 1955); (c) HRSC/Mars Express image of Central Ophir and Candor Chasmata showing possible circular volcanic vents (black arrows) and central vent (white arrows), box denotes location of Figure 7.14a, 30 km scale bar at top.

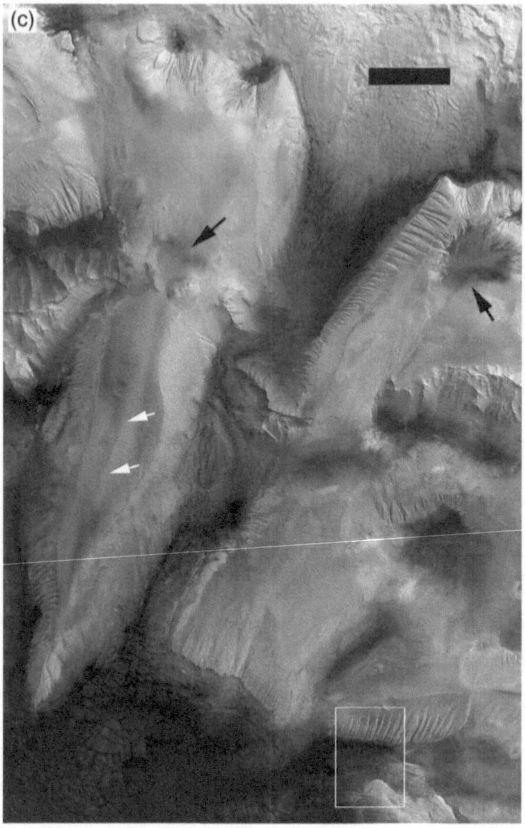

Figure 7.7. (cont.)

or obscured on many terrestrial sub-ice volcanoes (Figure 7.3d) as, after volcanism has subsided, the reformed ice can scour the caprock surfaces and mantle them with a layer of glacial till.

Resistant caprock layers with distinctive, comparatively small-scale scalloped edges are observed on Lagafell tuya in Iceland (Figure 7.8a). On some ILDs, resistant caprock layers show similar scalloped edges and a comparable origin is inferred (e.g., Ganges Chasma; Figure 7.8b). These scalloped margins are formed by preferential weathering at the terminations of lobate lava flow lobes (Figure 7.8c) and along edges of cooling cracks (Figure 7.8d), which is promoted if the caprock margins are slightly back-tilted during post-eruption subsidence in the vent area or uplift from younger magmatic intrusion.

In Iceland, Dyngjufjöll Ytri is an uncapped tephra mound with partly filled pits in random and linear distributions scattered on its surface (Figure 7.9a). These pits were interpreted as small maar craters formed when younger lavas intruded the tephra mound and interacted with

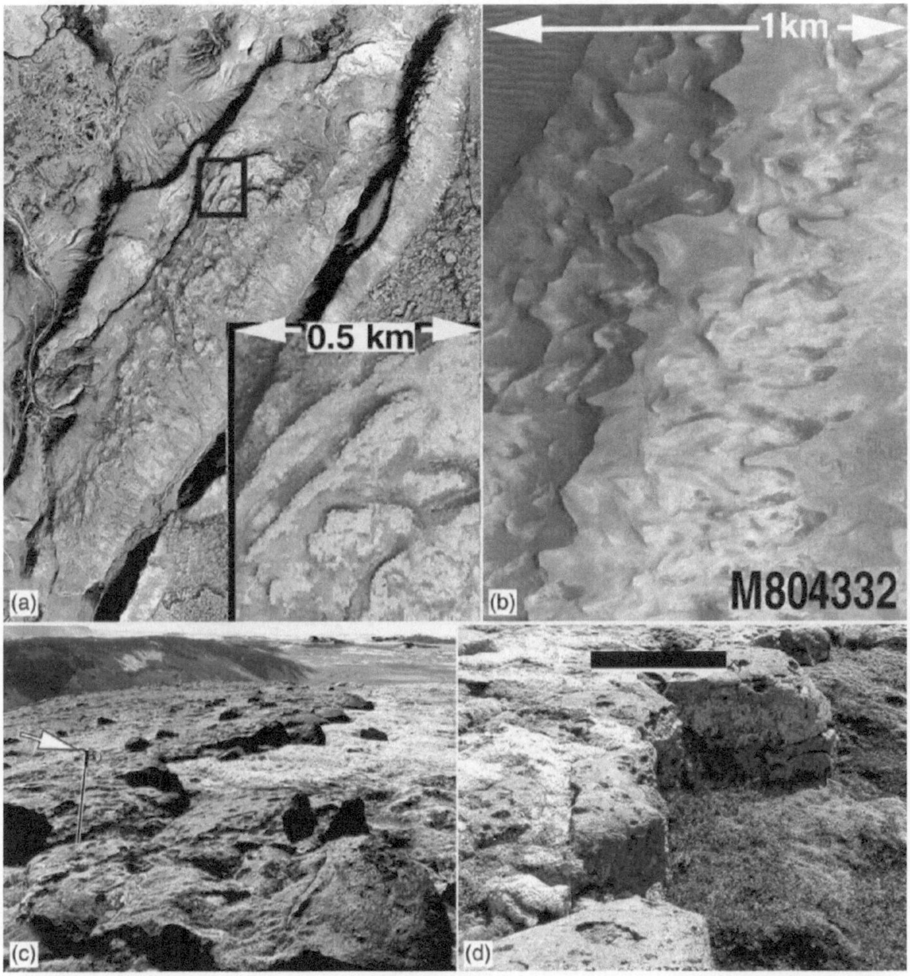

Figure 7.8. Resistant caprock layers with scalloped edges on Icelandic and Martian mounds. (a) Lagafell tuya; (b) MOC image of Ganges Chasma mound (rotated 180°); (c) ground view of back-tilted scalloped lava flow margin on Lagafell (arrow points to top of 3 m staff); (d) note straight edges of cooling cracks and protrusion of normal lobate weathered margin of Lagafell lava flow (3 m scale bar).

groundwater or ground ice (Sigvaldason, 1992). Cemented pipe-like structures that mark the centers of eroded pits are also present on the sides of Dyngjufjöll Ytri (Figure 7.9b). The pipes are hollow and consist of indurated tephra cemented by hot fluids during the phreatomagmatic eruptions. The friable surface of Candor Mensa (Figure 7.9c) also shows unusual pits, which are circular or irregularly shaped, may have double rims or appear filled, and are randomly or linearly distributed (Figure 7.9d).

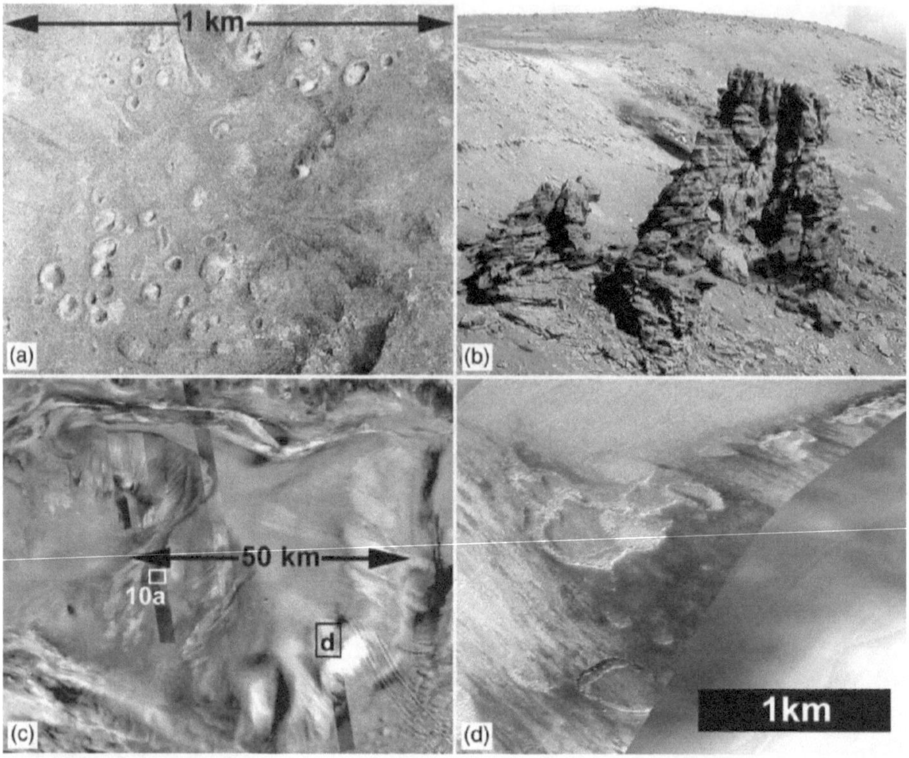

Figure 7.9. Pits on friable tops of Icelandic and Martian mounds. (a) Dyngjufjöll Ytri; (b) cemented pipe-like structure that marks the subsurface center of a now-eroded pit on east side of Dyngjufjöll Ytri; (c) Viking and MOC draped MOLA view of Candor Mensa (box indicates location of d); (d) irregularly shaped pits (now mostly filled with probable eolian material) on part of MOC image M1300507 (5.81 m/pixel).

They are broadly similar to those at Dyngjufjöll Ytri and may have had a similar origin.

7.2.3 Flank characteristics

Viking and MOC images of apparently wind-eroded friable flanks of many ILDs show several instances where the flank layers appear to consist of steeply dipping beds with different erosional resistances (Komatsu *et al.*, 1993; Chapman and Tanaka, 2001; Chapman, 2002; Figure 7.4d). However, several ILD mounds also have near-horizontal beds in their flanks (figure 7.13 in Chapman, 2002; Komatsu *et al.*, 2004). For example, cliffs at the base of mounds in Melas and Juventae Chasmata and near the summit region at Candor Mensa show nearly horizontal, thinly stratified to

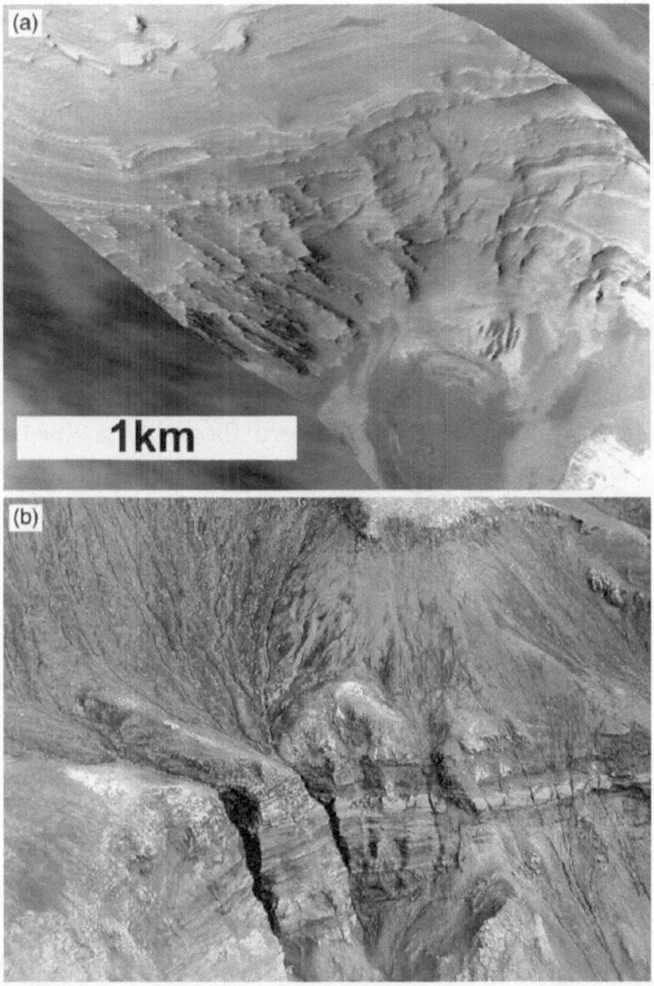

Figure 7.10. Nearly horizontal, alternating thinly stratified and thicker massive layers in Icelandic and Martian mounds. (a) Part of MOC image M1801893 (5.83 m/pixel) on west midpoint of Candor Mensa; (b) approximately 1 km wide aerial view of Tindaskagi.

massive layers (Figure 7.10a). Subglacial tindar outcrops (i.e., subaqueously deposited tephra mounds/ridges) also show flanking cliffs composed of alternating thinly stratified and massive layers (Figure 7.10b). This is a very characteristic feature of tindars (e.g., Smellie, 2000, 2001). The individual layers are due to repeated volcanic eruptions that generate large columns of pyroclastic material, which collapse onto the surrounding meltwater lake and cascade down submerged volcanic slopes, mainly as sediment gravity flows (Smellie, 2001; White *et al.*, 2003). Layer thickness is relative and is determined by the volume of each eruption (i.e., thicker massive layers result from large eruptions, whereas thinly stratified layers form from rapidly repeated series of smaller eruptions). The attitude of beds, in particular the presence of flat-lying beds, depends on the eruptive environment and morphology of the underlying edifice. In an ice-confined (englacial) setting, the meltwater lake has a relatively small size. It is rapidly infilled by tephra layers that abut the encircling ice walls and horizontal bedding is rapidly established. This has been demonstrated in some Icelandic tindars, in which bedding at higher levels in the volcanic pile may even be slightly back-tilted (Smellie, 2000). Similarly, a tindar at Helgafell may show morphological evidence of possible ice confinement on its east side (Figure 7.11a). The eastern flank has a linear termination interpreted as largely unmodified and caused by hyalotuff material banked against a former ice wall that has since melted away (Schopka *et al.*, 2003). Tephra cools down during its passage through the air and in a meltwater lake.

Figure 7.11. Views of 900 m wide Helgafell tindar, Iceland. (a) Aerial image shows abrupt linear east flank termination; (b) left, view of north Helgafell showing tephra flow lobe with a steep west wall, and right, gray shading marking possible location of a former ice wedge.

As a result, it contains insufficient energy to melt ice rapidly at the base of the edifice. Hence, the tephra accumulates more rapidly than the enclosing ice barrier can retreat. A few ILD mesas in Melas, Ganges, and Juventae Chasmata also show flank material with abrupt termination along a lineation that suggests terrace erosion or confinement against subsequently removed material (figure 16 in Chapman, 2002). Alternatively, the Helgafell tindar has been affected by faulting and the abrupt east flank termination of this terrestrial tuya may be due to faulting alone (T. Thardarson, personal communication 2005).

On Mars, the talus material eroded off the ILD shows no large blocks at the MOC scale of about 1.5 m/pixel per picture element (Malin and Edgett, 2000). This phenomenon also occurs on some terrestrial tuyas and tindars. The south end of Naefurholtsfjöll tindar (elsewhere capped by Hekla ash, see above) is covered by a veneer of pebbles to fist-size cobbles of rounded volcanic lavas (Figure 7.12a). The exposed and underlying tindar at this locale is capped by fluvio-glacial deposits, consisting of angled beds of rounded lava clasts embedded in a matrix of hyaloclastic ash presumably eroded off some upslope part of the tinder (Figure 7.12b). The ash matrix of the fluvio-glacial material weathers away leaving behind an inclined pediment of rounded rocks which armor and stabilize the flanks of the tindar. Another type of armoring can be observed at the Icelandic Armansfell tindar where the slopes show a lack of boulders at scales below that of the MOC (Figure 7.12c). Closer examination of Armansfell showed that armoring was due to coherent (lacking internal cooling cracks or joints), cemented fluvio-glacial and hyaloclastic tuff that were deposited at the angle of repose. The cement (probably zeolites and palagonite), the steep dip of the beds, and cohesiveness allow water to run off the edifice without undercutting materials and creating boulders.

The Mars ILDs show gullies cut into discrete, mostly low-lying flank deposits (Figure 7.13a). Gullies cut into discrete layers in volcanic rocks are common on Earth and are due to groundwater seepage that emerges on slopes along the base of less permeable horizons (Figure 7.13b). It is an essentially subaerial process and the gullies form after the ice has melted and the meltwater lake has drained. Gullies are eroded into all sides of the ILDs including their north-facing slopes, so solar heating alone likely did not generate the gullies, but rather some internal heating mechanism like intrusion of magma, like that which may have erupted to form the much younger and embaying dark material (Chapman and Smellie, 2001). This is common in active volcanic areas like Iceland, where many subice volcanoes have been eroded and altered by local hot springs.

Figure 7.12. Views of sediment armoring on Icelandic tindars. (a) South end of Naefurholtsfjöll (arrow points to location of b) covered by a veneer of pebbles to fist-size cobbles of rounded volcanic lavas (camera case and hammer for scale in right foreground); (b) Naefurholtsfjöll fluvio-glacial deposits, consisting of dipping and horizontal beds of rounded lava clasts embedded in a matrix of hyaloclastic ash; (c) steeply dipping tephra deposits on boulder-free slope on Armansfell showing that armoring is due to cohesive tephra mantle (15 m wide road on lower right for scale).

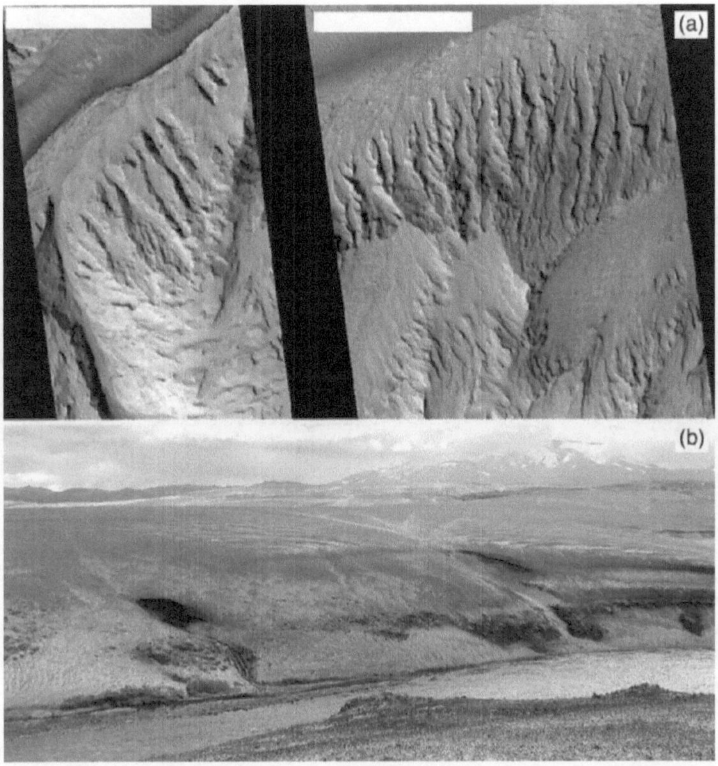

Figure 7.13. Low-lying flanks of Martian and Icelandic mounds, showing gullies (groundwater breakouts) associated with less permeable layers. (a) MOC images M1003480 on left and M1102514 on right (scale bars = 1 km) of ILDs in east Candor Chasma; (b) southwest side of Naefurholtsfjöll tindar; Hekla in background.

We may now be observing this same type of local hot spring alteration on Mars, as the OMEGA/Mars Express has locally detected sulfates on some ILDs (Gendrin *et al.*, 2005).

During and after the last glacial maximum, glaciers have continued to erode sub-ice volcanoes in Iceland. In central Candor Chasma new data from a Mars Express Mission anaglyph, THEMIS, and MOC images clearly show the lower flanks of two ILD mounds (downslope from the cirque) have been sheared off at level heights (Chapman *et al.*, 2004). The substance that sheared off the ILD flanks has subsequently been removed (Figure 7.14). Ice can shear off material in this level fashion and later melt away. These new images and some older Viking images also show a scarp (cirque?) and medial/terminal ridges (moraines?) in dark floor material in central Candor Chasma. The shearing in this area, associated with possible moraines and a cirque are strong

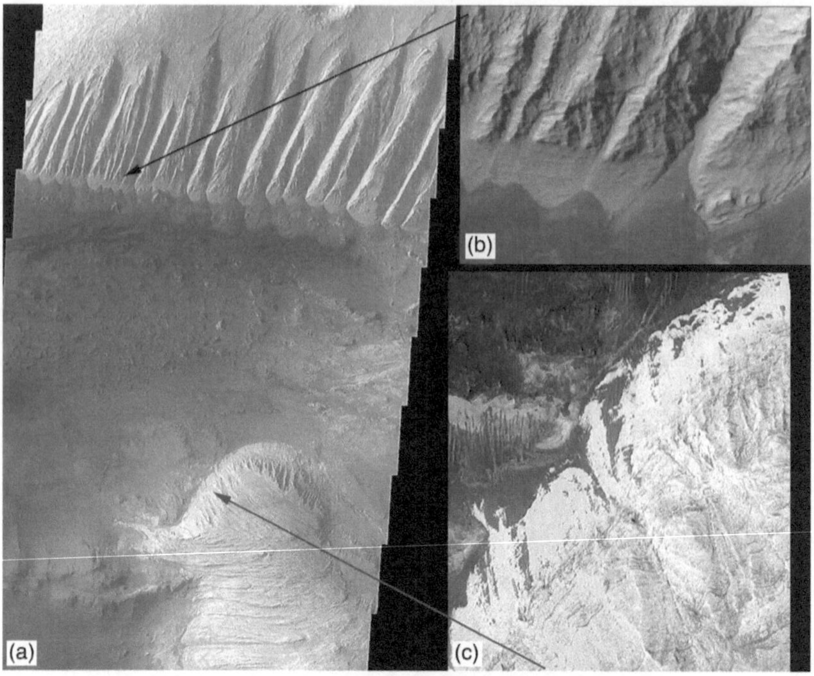

Figure 7.14. Cliff erosion or shearing along the base of ILDs flanks in central Candor Chasma shown in (a) approximately 30 km wide THEMIS image V10551002 (location shown in Figure 7.7c); (b) MOC image E1900200 (1.5 km wide); and (c) MOC image E1700142 (1.5 km wide).

evidence of glacial processes. The possible cirque and moraines mostly form in young dark floor materials (Figure 7.15) which likely indicates that the ancient glacier was very young (Chapman *et al.*, 2004). The low level of these possible glacial features on the ILD flanks may suggest the hypothetical paleoglacier to be some of the very last surface ice in central Ophir Chasma, as a much larger volume of ice would be required to fill the chasmata and form the ILDs if the mounds are indeed mostly sub-ice volcanoes (Chapman, 2003b).

7.3 The influence of ambient conditions on terrestrial subglacial eruptions

The most important effect of an enclosing ice sheet on an erupting volcano is to confine the eruptive products, whether meltwater or magmatic materials. Otherwise, eruptions will be indistinguishable from their submarine or lacustrine equivalents (Walker and Blake, 1966). Beyond that influence, the thermal regime, rheologic properties, and compositions of ice sheets are paramount in their influence on sub-ice eruptions.

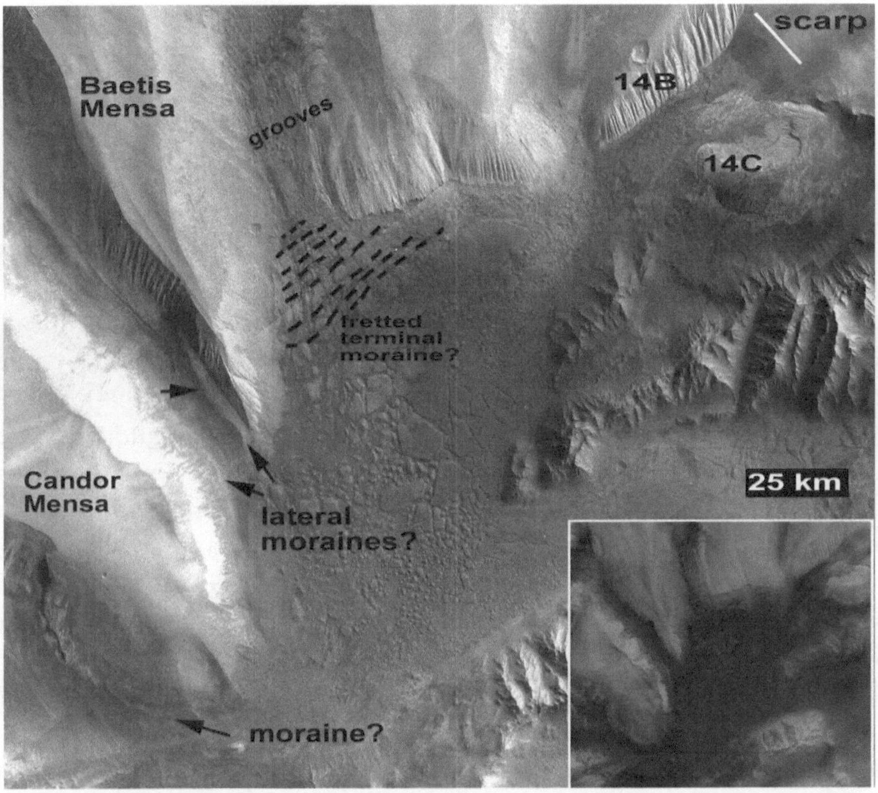

Figure 7.15. Lightened Viking image of central Candor Chasma (north toward upper left corner) showing possible moraines and scarp in dark floor material; unlightened inset image shows relative dark tone of floor material; Figures 7.14b and c show locations of these figures.

7.3.1 Thermal regime

The temperature distribution, or thermal regime, of a glacier or ice sheet is fundamentally important for the way a glacier deforms (i.e., rheology), and with regard to the role of meltwater (i.e., hydraulics). Glaciers may be classified empirically by their thermal regime. There are three main types (temperate, polar, and polythermal), although transitions may occur in the same glacier, as in the present Antarctic Ice Sheet (Wilch and Hughes, 2000). *Temperate* ice is relatively warm, i.e., at the melting point throughout except for a thin surface layer. A thin layer of meltwater intervenes between the glacier sole and underlying bedrock and the glacier moves by sliding. Englacial vaults in temperate ice are capable of having their meltwater drained suddenly, in jökulhlaups. In other cases, the vaults will maintain their meltwater and lava-fed deltas may develop within the ice cauldron lake. Being relatively warm, temperate ice will melt and deform (by flow) comparatively rapidly (cf. Gudmundsson, 2003). Englacial lakes will therefore be relatively wide and ice walls will collapse and recede (by melting) ahead of an advancing lava-fed delta. Rapid ice recession will repeatedly remove support from the volcano flanks, leading to frequent flank collapses and numerous slump scars that should be prominent, particularly in the early-formed tephra sequences (tindar/hyaloclastic stage). Any subglacial meltwater tunnels are likely to get repeatedly blocked by these collapses. Edifice morphologies generally should be identical to terrestrial examples, possibly modified to higher volumes caused by potentially greater magma discharges on Mars (Wilson and Head, 1994).

Polar ice is much colder and well below freezing throughout the year. The ice is frozen to the bed and movement is usually assumed to be accommodated by intra-ice shear. Polar ice lacks free water, and meltwater cannot escape subglacially because it is frozen to the base. Englacial vaults will form but will always overflow, hence there will be no morphological evidence for meltwater-derived landforms, such as channels and eskers. In its colder state, the ice will not deform so readily by flowing into the vault or lake. It is also less liable to collapse. The additional thermal requirements for melting will also ensure that any vault/lake will be narrower and will be rapidly back-filled by erupted magmatic products. Any lava-fed deltas will thus be very short and subaerial lava could overrun onto the surrounding ice sheet surface, which, following ablation, should form extensive fields of chaotic lava rubble surrounding tuya landforms. There is no limit to the maximum height of an edifice constructed englacially, being determined solely by volume of magma discharged and thickness of the ice. More stable ice walls will lead to fewer flank collapses and slump scars in the volcanic sequences. Edifices may have generally higher aspect ratios than those in temperate ice. Currently, Mars is a frigid planet and was likely so in the past; therefore polar ice might be the expected thermal regime.

Polythermal (or sub-polar) ice is also cold with temperatures well below freezing, but their bases are warmed to the melting point from beneath by geothermal heat and they are therefore wet-based. A major consequence of the variable thermal regimes of ice is that colder ice requires significantly more energy to melt compared with warmer deposits, since energy has to be used to warm the cold ice to its melting point. For example, every kilogram of ice at minus 50 °C at atmospheric pressure requires 209 kJ of energy simply to raise its temperature to the freezing point. Meltwater can escape subglacially in polythermal ice throughout some eruptions, forming channels and eskers extending radially away from subglacial volcanic centers, and relatively tall edifices compared with submarine and (non-glacial) lacustrine eruptions. The extended "wings" or "arms" may be a consequence of the progradation of multiple individual lava (delta) streams, probably at different times during an eruption, or else these shapes are erosional, formed by glacial downcutting, long after an eruption ceased.

7.3.2 Rheology

Ice below its pressure melting point is harder to deform than ice at its pressure melting point. For example, under a given stress (lithostatic stress $(\sigma) = \rho g h$; where ρ is hydrostatic pressure, g is gravity, and h is height of the ice), ice at 0 °C deforms at a rate 100 times faster than ice at minus 20 °C (Hambrey, 1994). The presence of solid (non-volatile) impurities also affects rheology, with debris–ice mixtures sharply increasing ice strength. Finely dispersed particulate concentrations less than 10% have a negligible effect, but they significantly strengthen ice (compared with pure ice) when they exceed 10%. The presence of solid impurities (tephra particles) is likely to be ubiquitous in glaciers in regions affected by volcanic activity. Their influence on ice rheology is relatively predictable, by increasing ice strength and, thus, strain response time, and these effects will be greatest for polar (colder) glaciers. If the ice is dirty, it will better resist flow into the vault or lake. The lower atmospheric pressure on Mars should yield higher eruption velocities and lower final densities of decompressed gas, which would cause eruption columns to rise much higher, have larger near-vent particles, and distribute fine particles much farther than those on Earth (Wilson and Head, 1983, 1994; Wilson and Heslop, 1990). In addition, regional and global dust storms are common events on Mars. Thus, dirty ice on Mars should be widespread and common, as has been suggested by several authors (e.g., Clifford, 1993; Carr, 1996; Ghatan and Head, 2002).

7.3.3 Composition

The presence of uniformly distributed impurities, such as air, salts, and CO_2, lowers the melting point relative to pure water ice, so more ice will melt per unit mass of magma forming broader, more voluminous, meltwater lakes. Sub-ice volcanoes may thus form broad-based edifices under these conditions. It has been suggested that any ice sequestered in the Martian subsurface would have eventually segregated to form a large volume of highly concentrated brines (Brass, 1980). The possibility of brines had been considered by others, but subsequently dismissed based on a long-held assumption that existence of near-surface fluids was unlikely throughout much the latter half of Mars' history (Clark and Van Hart, 1981; Jakosky, 1985; Zent *et al.*, 1990). However, the subject of brines has been revisited and convincingly supported on the argument that frozen eutectic brines could be possible if an initial saline hydrosphere underwent evapoconcentration and evolved by chemical interaction with mafic rocks, and this brine would be theoretically stable and could persist in local subsurface aquifers perched between ground ice and crystalline salts (Knauth and Burt, 2002). Recently, the casts of possible evaporitic minerals seen on "El Capitan" rock at the Mars Opportunity Rover Site now strongly suggest that salty fluids did occur in the subsurface of Mars (Squyres *et al.*, 2004). Without precipitation, surface ice on Mars is fed solely from subsurface sources. In the subsurface, aquifers eventually segregate into ice above brine and salts beneath; the brine being the last to freeze is also the first to melt and drain out of country rock (Knauth and Burt, 2002). However, as there are no comparable terrestrial volcanoes that have formed in brines more saline than ocean water, we are unable to make informed hypotheses about the resulting possible end morphological shapes. Potentially, volcanic eruption into brines could form sulfate alteration minerals and may perhaps account for the OMEGA detected sulfates on ILDs, rather than local hydrothermal spring alteration.

7.4 Discussion

Although the origins of ILDs are unknown, an origin by sub-ice volcanic processes makes sense on many levels. For example, one aspect of the ILDs that has been consistently overlooked is their occurrence only in elliptical chasmata. The linear chasmata are noticeably devoid of ILDs, which suggests a causal relationship between the origin of elliptical troughs and their internal mounds (or an unusual amount of erosion within only linear troughs). The elliptical chasmata are not eroded and filled impact craters, as they are (1) not circular, like impact craters; (2) are in all cases completely rimless; (3) are much too deep for impacts of this size; and (4) have long axes aligned with the tectonic trend of Valles Marineris. Assuming a subglacial or subaqueous setting for the ILDs, a simple tectonic control (and associated rupture of a confined aquifer) alone seems an unlikely trigger for ILD volcanism, or we should also observe ILDs in the linear chasmata, for which a tectonic setting is most likely and in which ILDs are absent. We suggest that the method of formation of the elliptical chasmata and the ILDs may be genetically related. Rise of plutons within the Mars crust could have generated the elliptically shaped chasms as very large explosive calderas above the magmatic foci (linear chasma could be tensional structures related to deeper magma chambers). On Earth, such super-calderas generate huge ash blankets and on Mars we would expect even more widespread ash blankets with much finer-grained particles (Wilson and Head, 1994). Some of the fine-grained layered sediments (Malin and Edgett, 2000) that occur directly east of Valles Marineris may be the eroded remnants of hypothetical ash blankets. The putative magma sources may have provided enhanced geothermal heat to drive a subsurface circulation that was ultimately responsible for filling the volcano-tectonic depressions with water/brine. These magma chambers would then be responsible for providing the lava that erupted to form the ILDs.

Another observation consistent with eruption of ILDs beneath ice is that the Valles Marineris canyon and Juventae Chasma terminate in topographic barriers immediately prior to emergence of associated outflow channels. The negative topography of the chasmata could have trapped any extant ice/water (Smith *et al.*, 1999; Chapman *et al.*, 2003). However, the presence of those channels indicates that water in large volumes had to have escaped from the chasmata, thus requiring some mechanism (like geothermal/volcanic heating) to melt the ice (Chapman *et al.*, 2003). The uphill flow of water at the chasmata terminations is strongly suggestive of meltwater flow at the base of a glacier(s) whose surface dipped out towards its termination, since the direction of basal water flow is strongly influenced by surface gradient and meltwater can flow up a basement surface (Björnsson, 1988). Such a scenario relies, however, on that ice being temperate (or polythermal), not polar, and we should also expect to see evidence for subglacial meltwater in the form of eskers (like the wing-like forms of Ganges Mensa) and/or channels carved in the chasmata floors (cf. Head and Pratt, 2001). Alternatively, eruption of the ILDs could have provided ample heat to melt any ice in the chasmata (Chapman *et al.*, 2003). Meltwater escaping supraglacially (cf. Smellie, 2000) could also run out over the plateau surface adjoining chasmata if the ice sheet and plateau surfaces were of similar elevations. Such eruption-related supraglacial flooding may occur initially as a widespread sheet of meltwater (Smellie, 2002) that will not cause significant erosional modification of the surrounding plateau landscape, although small-scale erosional features are likely. Later flooding is typically confined to supraglacial channels (Smellie, 2002), and is associated with a reduced volume of meltwater. The volume may not be sufficient, nor the event of long enough duration, to carve large fluvial water courses on the plateau, which may explain their absence on Mars.

An original water-rich environment in several chasmata is suggested by the high proportions of combined water predicted from epithermal neutron rates measured by the Neutron Spectrometer on the Mars Odyssey spacecraft. For example, parts of Hebes, Ophir, west Candor, and west Coprates Chasmata show 6 mass% H_2O and <2 mass% in East Candor, Melas, east Coprates, Juventae, and Ganges Chasmata. Volcanism in the chasmata could also explain the differing amounts of detected H_2O. After gradual sublimation of unstable surface ice, continued late-stage volcanism (signaled by presence of young dark material and associated vents; Chapman and Smellie, 2001; Lucchitta, 2001) could have extended the geothermal/hydrologic cycle and driven any remaining subsurface ice out of the more volcanically active chasmata (Chapman, 2003a).

Several morphological commonalities exist between terrestrial analogs and the ILDs that appear to suggest comparable volcanic processes and may also lend support to interpretations of a sub-ice eruptive environment on parts of Mars. For example, like terrestrial deposits, the ILDs vary widely in overall shape. Some ILDs show the ideal flat-topped tuya shape, formed of resistant horizontally layered caprock over friable steeply dipping slopes (e.g., ILDs in Ophir Chasma). Other ILDs show morphologies similar to terrestrial tindars, i.e., they are friable ridges with local occurrences of resistant horizontally layered caps (e.g., Baetis Mensa, central and east Candor Chasma, and the large mound in Juventae Chasma). The smaller mounds in Juventae Chasma and Hebes Mensa resemble uncapped tindars. In addition, some ILDs closely resemble volcanoes formed by subaerial eruptions (e.g., Ceti Mensa and parts of Melas Chasma troughs). The geographical association of subglacial and subaerial edifices is also a feature of terrestrial volcanic fields. Other ILDs show angular unconformities and displaced blocks that suggest unstable edifices possibly affected by seismic shocks. ILD mounds in Ganges and Candor Chasmata and Hebes Mensa also show the unusual cuspate, wing-like projections that are found in some terrestrial subglacial volcanoes. Many smaller-scale caprock features observed on terrestrial edifices can also be seen in the Martian analogs, such as resistant and soft wind-eroded (ash?) caps, a few vent sites, scalloped lava flow edges, and partly filled pits on soft or uncapped mounds. Finally, like some terrestrial sub-ice volcanoes, the flanks of the ILDs show fine to massive layers having both steep and nearly horizontal dips, confined flank flows at the base, lack of flank boulders, gullies confined to discreet layers, and possible glacial erosion.

Comparing observations of the ILD morphologies with our interpretations of ambient ice conditions and their influence on the resulting edifices leads to some interesting speculations. For example, as discussed above, eruptions underneath ice on Mars should lead to edifice morphologies generally identical to terrestrial examples, possibly modified to higher volumes caused by potentially greater magma discharges on Mars. ILDs that have the idealized tuya/tindar forms are observed in Ophir and central and east Candor Chasmata. Although cold-based ice would have been likely on the surface of paleo-Mars (e.g., Head and Marchant, 2003), closely associated ILDs erupting at approximately the same times may create conditions of basal melting on a regional scale comparable to ice with a temperate thermal regime (see Ghatan and Head, 2002). The height of isolated Hebes Mensa rivals that of the surrounding plateau. The plateau surface also lacks significant morphological evidence of overflowing meltwater. Perhaps, similar to polar ice, water in the englacial vault overflowed across the surface of the ice and out of Hebes Chasma onto the surrounding plateau. Finally, edifices formed in temperate ice may have wing-like channels, eskers, and may be relatively tall. ILDs in Ganges and Candor Chasmata, and even Hebes Mensa, have similar cuspate, arm- or wing-like projections, possibly indicating ancient temperate or polytherma ice conditions.

A common observation about Mars is that many of its geologic features are huge. Similarly, the ILDs, if they are sub-ice volcanoes, have higher aspect ratios (J.L. Smellie, 2003, unpublished data) and are 10 orders of magnitude larger in size (Chapman and Tanaka, 2001) than terrestrial edifices. There are some differences between the planets that could explain the deviations in size and form. First, of the studied terrestrial volcanoes erupted beneath glaciers and ice sheets, none have formed in ice thicknesses comparable to those postulated to have occurred in Valles Marineris. If the Earth's ocean froze, then eruptions that produced the Hawaiian volcanoes could produce tuyas/tindars on the same scale or larger than those of Mars. For example, the big island of Hawaii (Earth's largest volcano) is 135 km long and 9 km high, comparable to the Ganges Mensa ILD (105 km long and 4 km high). Second, many volcanoes on Mars grow much larger than those on Earth, because magma discharges are potentially much greater (Wilson and Head, 1994) and there are no tectonic plates to shift and displace the volcanism. Third, stress ($\sigma = \rho g h$) is less on Mars, because the surface gravity for Mars is 3.71 m/s^2 and is 9.8 m/s^2 for Earth, so Martian ice should deform more slowly and perhaps support taller edifices. Finally, the exceptionally large volumes of the Martian edifices (compared to Earth) would also be easier to explain if many are polygenetic volcanoes formed over a longer period of time. Polygenetic volcanoes are constructed during eruptions from multiple overlapping vents at different times. These types of volcanoes could construct an aerially large edifice through a long-lived thick ice sheet. Eruptive stages could be a combination of entirely sub-ice and (ultimately) subglacial-to-emergent. In the final emergent stages, laterally extensive lava-fed deltas could form creating gently dipping shield-like volcano summit regions. Conversely, the edifices might be built of multiple stacked lava-fed deltas constructed through a much thinner but continuously replenished ice sheet (as for the terrestrial James Ross Island volcano (Antarctica); cf. Nelson, 1975; Smellie, 1999; J.L. Smellie, unpublished data). Certain attributes of the ILDs may suggest polygenetic eruptions. For example, Ganges Chasma shows several aligned linear features on the ILD caprock that may be multiple feeder vents (figure 7.11 in Chapman and Tanaka, 2001). That ILD also shows possible embayment of older hyaloclastic mound (compare Figures 7.3a and 7.4g) and post-depositional back-tilting of what could have been horizontally deposited lava volcanic topset beds (Figure 7.8). Tilting may be due to intrusion by late-stage magma bodies. Candor Mensa has a pitted surface that can be interpreted as multiple small maar craters formed by later intrusion of magma. In addition, the Ceti Chasma ILD has two large late-stage blocky plate flows, superposed cones, scarps, and jumbled layers that may suggest multiple-vent effects and perhaps subaerial deposition.

7.5 Conclusions

In order to investigate the hypothesis that ILDs on Mars are sub-ice volcanoes, we have examined analog terrestrial sub-ice volcanoes and compared the characteristics of these to the Mars examples. In addition, we have compared the influence and effects of enclosing ice sheets on terrestrial edifices to postulate possible consequences for the putative volcanic morphologies of the ILDs, such as their wing- and arm-like projections, heights, volumes, and aspect ratios. Polygenetic volcanism may have also been responsible for the much greater heights, volumes, and different shapes of putative volcanic edifices on Mars compared with terrestrial analogs, since our knowledge of the latter is based mainly on small monogenetic volcanoes. Several morphological commonalities exist between terrestrial analogs and the Mars ILDs and appear to suggest comparable volcanic processes, lending support to interpretations of the latter as sub-ice volcanic edifices. As in any terrestrial volcanic field, if the ILDs are mostly sub-ice volcanoes, we should expect an association with subaerial volcanoes, ash beds, lava flows, fluvio-glacial materials, and eolian deposits. That such features do occur in association with ILDs strengthens our general thesis that Mars ILDs are indeed likely sub-ice volcanoes. However, while current hypotheses are strongly permissive, they require confirmation. Accordingly, we look forward with eager anticipation to the successful acquisition of return samples or human/robotic science data obtained *in situ* from the Valles Marineris region during future Mars missions.

Acknowledgments

We wish to thank Lazlo Keszthelyi and Paul Geissler for their helpful comments and suggestions. This work was funded under a grant from the NASA Mars Data Analysis Program.

References

Björnsson, H. (1988). *Hydrology of Ice Caps in Volcanic Regions.* Reykjavik: Societas Scientarium Islandica, pp. 1–139.

Brass, G. W. (1980). Stability of brines on Mars. *Icarus*, **42**, 20–8.

Carr, M. H. (1996). *Water on Mars.* Oxford: Oxford University Press.

Chapman, M. G. (2002). Layered, massive, and thin sediments on Mars: possible Late Noachian to Late Amazonian tephra? In *Volcano–Ice Interaction on Earth and Mars*, ed. J. L. Smellie and M. G. Chapman. Geological Society of London Special Publication 202, pp. 273–93.

Chapman, M. G. (2003a). Geologic evidence for late-stage equatorial surface and ground ice on Mars. *EOS Transactions, American Geophysical Union Fall Meeting*, **84** (no. 46), Abstract P22C-063, F959.

Chapman, M. G. (2003b). Sub-ice volcanoes and ancient oceans/lakes: a Martian challenge. *Global and Planetary Change*, **35**, 185–98.

Chapman, M. G. and Smellie, J. L. (2001). Putative large and small volcanic edifices in Valles Marineris, Mars and evidence of ground water/ice. *EOS Transactions, American Geophysical Union Fall Meeting*, **82** (no. 47), Abstract P21C-11, F698.

Chapman, M. G. and Tanaka, K. L. (2001). Interior trough deposits on Mars: sub-ice volcanoes? *Journal of Geophysical Research*, **106** (E5), 10,087–10,000.

Chapman, M. G., Gudmundsson, M. T., Russell, A. J., Hare T. M. (2003). Possible Juventae Chasma sub-ice volcanic eruptions and Maja Valles ice outburst floods, Mars: implications of MGS crater densities, geomorphology, and topography. *Journal of Geophysical Research*, **108** (E10), doi:10.1029/2002JE002009.

Chapman, M. G., Soderblom, L. A., and Cushing, G. (2004). Active mission data show convincing evidence of amazingly young glacial processes and variable ages of ILD formation in an equatorial chasm of Valles Marineris. *Geological Society of America Abstacts with Programs*, **36** (5), 139, abstract 52–5.

Christensen, P. R. (2003). Mars as seen from the 2001 Mars Odyssey Thermal Emission Imaging System Experiment. *EOS Transactions, American Geophysical Union Fall Meeting*, **84** (no. 46), Abstract P21A-02, F944.

Clark, B. C. and Van Hart, D. C. (1981). The salts of Mars. *Icarus*, **45**, 370–8.

Clifford, S. M. (1993). A model for the hydrologic and climatic behavior of water on Mars. *Journal of Geophysical Research*, **98**, 10,973–11,016.

Croft, S. K. (1990). Geologic map of the Hebes Chasma quadrangle, VM 500K 00077, (abs.), *NASA Technical Memorandum*, 4210, pp. 539–41.

Geissler, P. E., Singer, R. B., and Lucchitta, B. K. (1990). Dark materials in Valles Marineris: indications of the style of volcanism and magmatism on Mars. *Journal of Geophysical Research*, **95**, 14399–413.

Gendrin, A., Mangold, N., Bibring, J.-P. *et al.* (2005). Sulfates in Martian layered terrains: the OMEGA/Mars Express View. *Science*, **307**, 1587–91.

Ghatan, G. J. and Head, J. W. (2002). Candidate subglacial volcanoes in the south polar region of Mars: morphology, morphometry, and eruption condition. *Journal of Geophysical Research*, **107** (E7), doi:10.1029/2001JE001519.

Gudmundsson, M. T. (2003). Melting if ice by magma-ice-water interactions during subglacial eruptions as an indicator of heat transfer in subaqueous eruptions. In *Explosive Subaqueous Volcanism*, ed. J. D. L. White, J. L. Smellie, and D. A. Clague. American Geophysical Union Geophysical Monograph 140, pp. 61–72.

Hambrey, M. J. (1994). *Glacial Environments*. London: UCL Press.

Head, J. W. and Marchant, D. R. (2003). Cold-based mountain glaciers on Mars: western Arsia Mons. *Geology*, **31**, 641–4.

Head, J. W. and Pratt, S. (2001). Extensive Hesperian-aged south polar ice sheet on Mars: evidence for massive melting and retreat, and lateral flow and ponding of meltwater. *Journal of Geophysical Research*, **106**, 12275–99.

Jakosky, B. M. (1985). The seasonal cycle of water on Mars. *Space Science Reviews*, **41**, 131–200.

Jones, J. G. (1969). Intraglacial volcanoes of the Laugarvatn region, south-west Iceland. *International Quarterly Journal of the Geological Society of London*, **124**, 197–211.

Jones, J. G. (1970). Intraglacial volcanoes of the Laugarvatn region, south-west Iceland, II. *Journal of Geology*, **78**, 127–40.

Knauth, L. P. and Burt, D. M. (2002). Eutectic brines on Mars: origin and possible relation to young seepage features. *Icarus*, **158**, 267–71.

Komatsu, G., Geissler, P. E., Strom, R. G., and Singer, R. B. (1993). Stratigraphy and erosional landforms of layered deposits in Valles Marineris, Mars. *Journal of Geophysical Research*, **98**, 11105–21.

Komatsu, G., Ori, G. G., Ciarcelluti, P., and Yuri, Y. D. (2004). Interior layered deposits of Valles Marineris, Mars: analogous subice volcanism related to Baikal Rifting, Southern Siberia. *Planetary and Space Science*, **52**, 167–87.

Larsen, G. and Thorarinsson, S. (1977). H-4 and other acid Hekla tephra layers. *Jökull*, **27**, 28–46.

Lucchitta, B. K. (1990). Young volcanic deposits in the Valles Marineris, Mars? *Icarus*, **86**, 476–509.

Lucchitta, B. K. (1999). Geologic map of Ophir and Central Candor Chasmata (MTM -05072) of Mars. *US Geological Survey Miscellaneous Investigation Series Map*. I-2568, scale 1:500,000.

Lucchitta, B. K. (2001). Late mafic volcanism in Valles Marineris, Mars. *EOS Transactions, American Geophysical Union Fall Meeting*, **82** (no. 47), Abstract P32C-0563, F713.

Lucchitta, B. K., Clow, G. D., Geissler, P. E. *et al.* (1992). The canyon system on Mars. In *Mars*, ed. H. H. Kieffer, B. M. Jakosky, C. W. Snyder, and M. S. Matthews. Tucson: University of Arizona Press, pp. 453–92.

Lucchitta, B. K., Isbell, N. K., and Howington-Kraus, A. (1994). Topography of Valles Marineris: implications for erosional and structural history. *Journal of Geophysical Research*, **99**, 3783–98.

Malin, M. C. and Edgett, K. S. (2000). Sedimentary rocks of Mars. *Science*, **290**, 1927–37.

Mathews, W. H. (1947). "Tuyas": flat-topped volcanoes in northern British Columbia. *American Journal of Science*, **245**, 560–70.

McEwen, A. S., Malin, M. C., Carr, M. H., and Hartmann, W. K. (1999). Voluminous volcanism on early Mars revealed in Valles Marineris, *Nature*, **397**, 584–6.

Murchie, S. L., Kirkland, L., Erad, S., Mustard, J. F., and Robinson, M. (2000). Near-infrared spectral variations of Martian surface materials from ISM imaging spectrometer data. *Icarus*, **147**, 444–71.

Mustard, J. F. and Murchie, S. L. (2001). Unusual spectral properties observed in Valles Marineris. *Abstracts of Papers Submitted to the 32nd Lunar and Planetary Science Conference*. Houston: Lunar and Planetary Institute, CD 32, Abstract 2194.

Nelson, P. H. H. (1975). The James Ross Island Volcanic Group of north-east Graham Land. *British Antarctic Survey Scientific Reports*, 54, 62 pp.

Peterson, C. (1981). A secondary origin for the central plateau of Hebes Chasma. *Proceedings of the 12th Lunar and Planetary Science Conference*. Houston: Lunar and Planetary Institute, 1459–71.

Schopka, H. H., Gudmundsson, M. T., and Jakobsson, S. P. (2003). Formation of Helgafell, SW-Iceland: ice–volcano interaction during a fissure eruption under glacier and the effect of ice cap geometry on distribution of erupted products (In icelandic). The Geoscience Society of Iceland Spring Meeting 2003, abstracts, pp. 36–7.

Sigvaldason, G. E. (1992). Recent hydrothermal explosion craters in an old hyaloclastite flow, central Iceland. *Journal of Volcanology and Geothermal Research*, **54**, 53–63.

Skilling, I. P. (2002). Basaltic pahoehoe lava-fed deltas: large-scale characteristics, clast generation, emplacement processes and environmental discrimination. In *Volcano–Ice Interaction on Earth and Mars*, ed. J. L. Smellie and M. G. Chapman. Geological Society of London Special Publication 202, pp. 91–115.

Skilling, I. P., Chapman, M. G., and Lucchitta, B. K. (2002). Young, blocky flows in east Ius/west Melas and west Candor Chasmata, Mars: debris avalanche deposits derived from interior layered deposits (ILD) mounds? *Abstracts of Papers Submitted to the 33rd Lunar and Planetary Science Conference*. Houston: Lunar and Planetary Institute, CD 33, abstract 1361.

Smellie, J. L. (1999). Lithostratigraphy of Miocene – recent, alkaline volcanic fields in the Antarctic Peninsula and eastern Ellsworth Land. *Antarctic Science*. **11**, 362–78.

Smellie, J. L. (2000). Subglacial eruptions. In *Encyclopaedia of Volcanoes*, ed. H. Sigurdsson Academic Press, pp. 403–18.

Smellie, J. L. (2001). Lithofacies architecture and construction of volcanoes: in englacial lakes: Icefall Nunatak, Mount Murphy, eastern Marie Byrd Land, Antarctica. In *Lacustrine Volcaniclastic Sedimentation*, ed. J. D. L. White and

N. Riggs. International Association of Sedimentology Special Publication 30, pp. 73–98.

Smellie, J. L. (2002). The 1969 subglacial eruption on Deception Island (Antarctica): events, and processes during an eruption beneath a thin glacier and implications for volcanic hazards. In *Volcano–Ice Interaction on Earth and Mars*, ed. J. L. Smellie and M. G. Chapman. Geological Society of London Special Publication 202, pp. 59–79.

Smellie, J. L. (unpublished) Sub-ice volcanism: edifice morphologies and implications for Mars examples. In *Preservation of random mega-scale events on Mars and Earth: Influence on geologic history*, ed. M. G. Chapman and L. Keszthely Geological Society of America Special paper.

Smellie, J. L. and Skilling, I. P. (1994). Products of subglacial volcanic eruptions under different ice thickness: two examples from Antarctica. *Sedimentary Geology*, **91**, 115–29.

Smith, D. E., Zuber, M. T., Solomon, S. C. *et al.* (1999). The global topography of Mars and implication for surface evolution. *Science*, **284**, 1495–503.

Squyres, S. W., Grotzinger, J. P., Arvidson, R. E. *et al.* (2004). In situ evidence for an ancient aqueous environment at Meridiani Planum, Mars. *Science*, **306** (5702), 1709–14.

Tuffen, H., McGarvie, D. W., Gilbert, J. S., and Pinkerton, H. (2002). Physical volcanology of a subglacial-to-emergent rhyolite tuya at Rauđufossafjöll, Torfajökull, Iceland. In *Volcano–Ice Interaction on Earth and Mars*, ed. J. L. Smellie and M. G. Chapman. Geological Society of London Special Publication 202, pp. 213–36.

Van Bemmelen, R. W. and Rutten, M. G. (1955). *Tablemountains of Northern Iceland.* Leiden: E. J. Brill.

Walker, G. P. L. and Blake, D. H. (1966). The formation of a palagonite breccia mass beneath a valley glacier in Iceland. *Journal of the Geological Society of London*, **122**, 45–61.

White, J. D. L., Smellie, J. L., and Clague, D. A. (2003). Introduction: a deductive outline and topical overview of subaqueous explosive volcanism. In *Explosive Subaqueous Volcanism*, ed. J. D. L. White, J. L. Smellie, and D. A. Clague. American Geophysical Union Geophysical Monograph 140, pp. 1–23.

Wilch, E. and Hughes, T. J. (2000). Calculating basal thermal regimes beneath the Antarctic ice sheet. *Journal of Glaciology*, **46**, 297–310.

Wilson, L. and Head, J. W. (1983). A comparison of volcanic eruption processes on Earth, Moon, Mars, Io, and Venus. *Nature*, **302**, 663–9.

Wilson, L. and Head, J. W. (1994). Mars: review and analysis of volcanic eruption theory and relationships to observed landforms. *Reviews of Geophysics*, **32**, 221–64.

Wilson, L. and Heslop, S. E. (1990). Clast sizes in terrestrial and Martian ignimbrite lag deposits. *Journal of Geophysical Research*, **95**, 17309–14.

Witbeck, N. E., Tanaka, K. L., and Scott, D. H. (1991). The geologic map of the Valles Marineris region, Mars. *US Geological Survey Miscellaneous Investigation Series Map*. I-2010, scale 1:2,000,000.

Zent, A. P., Fanale, F. P., and Roth, L. E. (1990). Possible Martian brines: radar observations and models. *Journal of Geophysical Research*, **95**, 14531–42.

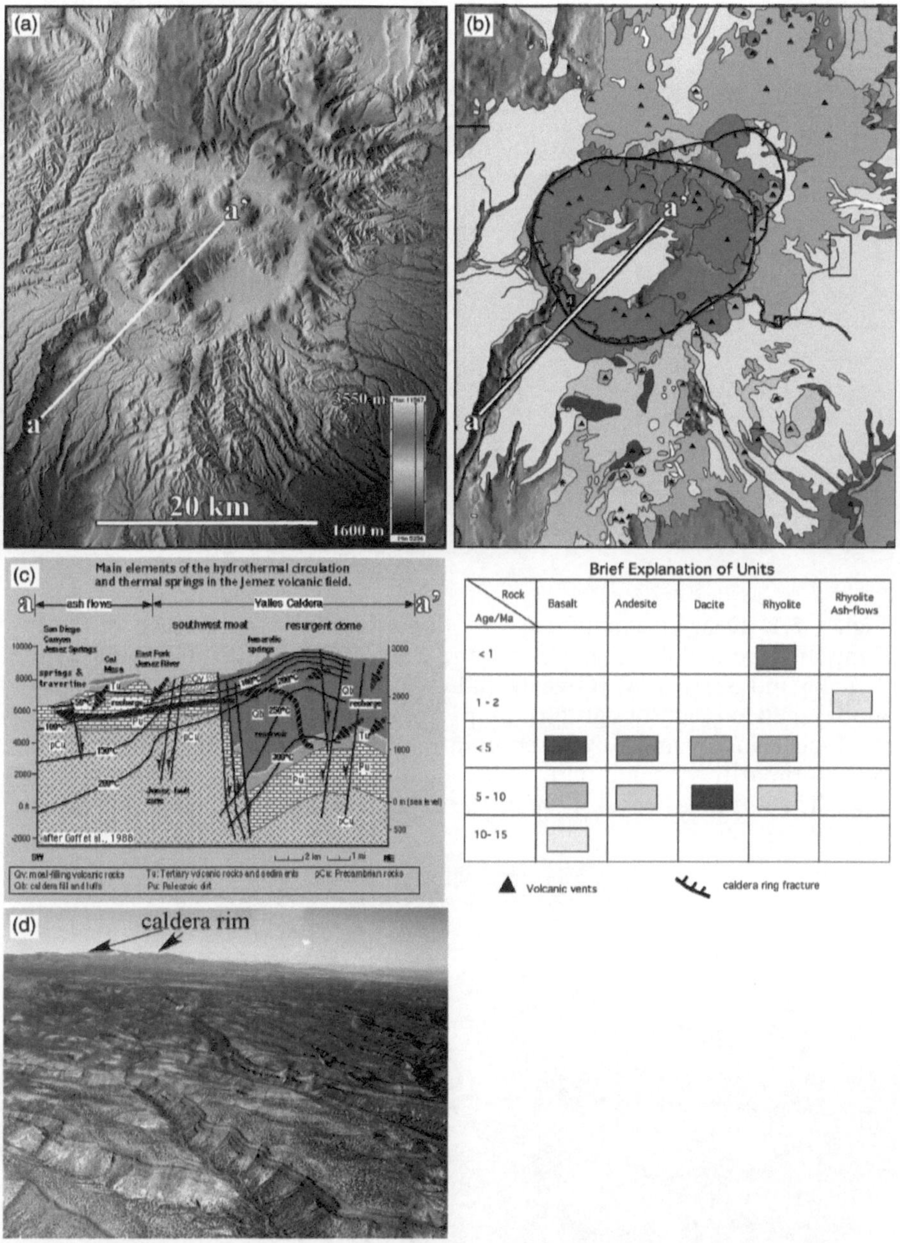

Figure 4.5. (a) Digital shaded relief map of the Valles Calderas, near Los Alamos, New Mexico. (b) Generalized geologic map of the Valles Caldera. After Luedke and Smith (1978) and Smith *et al.* (1970). (c) Schematic section (after Goff *et al.*, 1989) from the caldera center through the south margin and down the prominent valley on the southwest flank illustrating the principal characteristics of hydrothermal circulation and relationship to down-gradient springs and spring deposits. Similar circulation patterns may have existed shortly after formation of the highland paterae type calderas on Mars (see, for example, Figure 4.4a). (d) Oblique air photo of valleys cut into the extensive layered sheets of Bandelier ash flow tuffs that were erupted during caldera formation; view west from vicinity of the Rio Grande; note distance from the caldera rim.

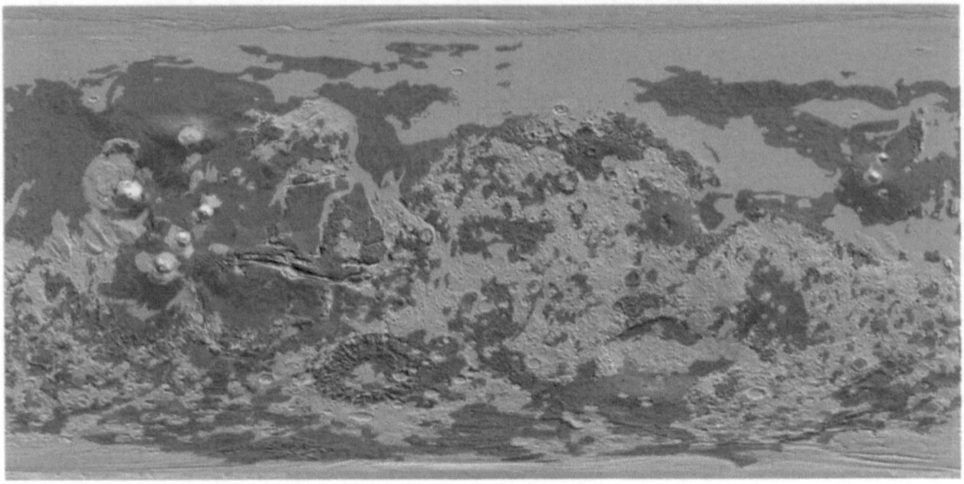

Figure 5.1. Distribution of possible flood lavas on the surface of Mars. Simplified from global maps (Scott and Carr, 1978; Scott and Tanaka, 1986; Tanaka and Scott, 1987; Greeley and Guest, 1987) and placed over MOLA global shaded relief topographic map. Areas in red have been either mapped as flood lavas or as likely to contain modified flood lavas. Yellow shows other primarily volcanic units. All remaining materials are shown in light blue. The red areas cover 47% of the surface of Mars, which is marginally

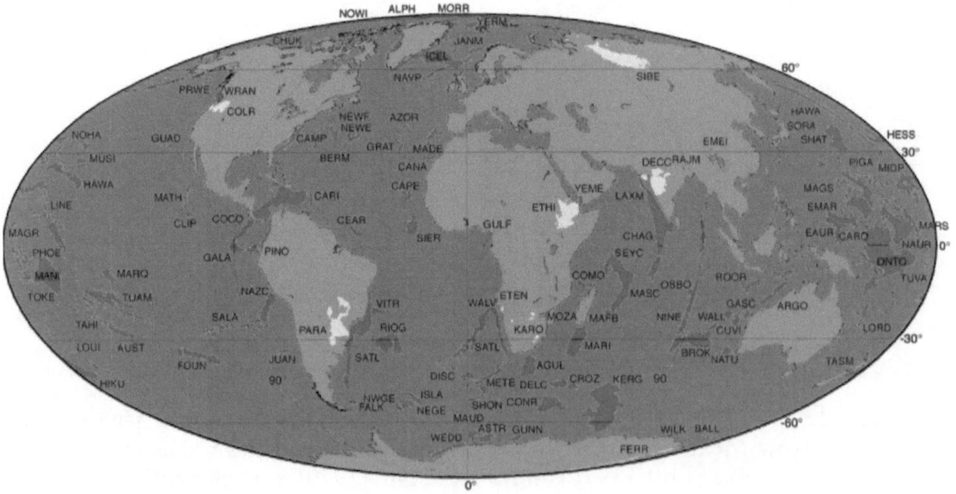

Figure 5.5. Map of Recent (≤250 Ma) Large Igneous Provinces (LIPs) on Earth. Well-preserved continental flood basalt provinces listed in Table 5.2 are depicted in yellow, other LIPs in red. While recent LIPs only cover a few percent of Earth's surface, it is important to note that this only represents a few percent of Earth's geologic history. Map after Coffin and Eldholm (1994). Abbreviated names listed in full in Coffin and Eldholm (1994).

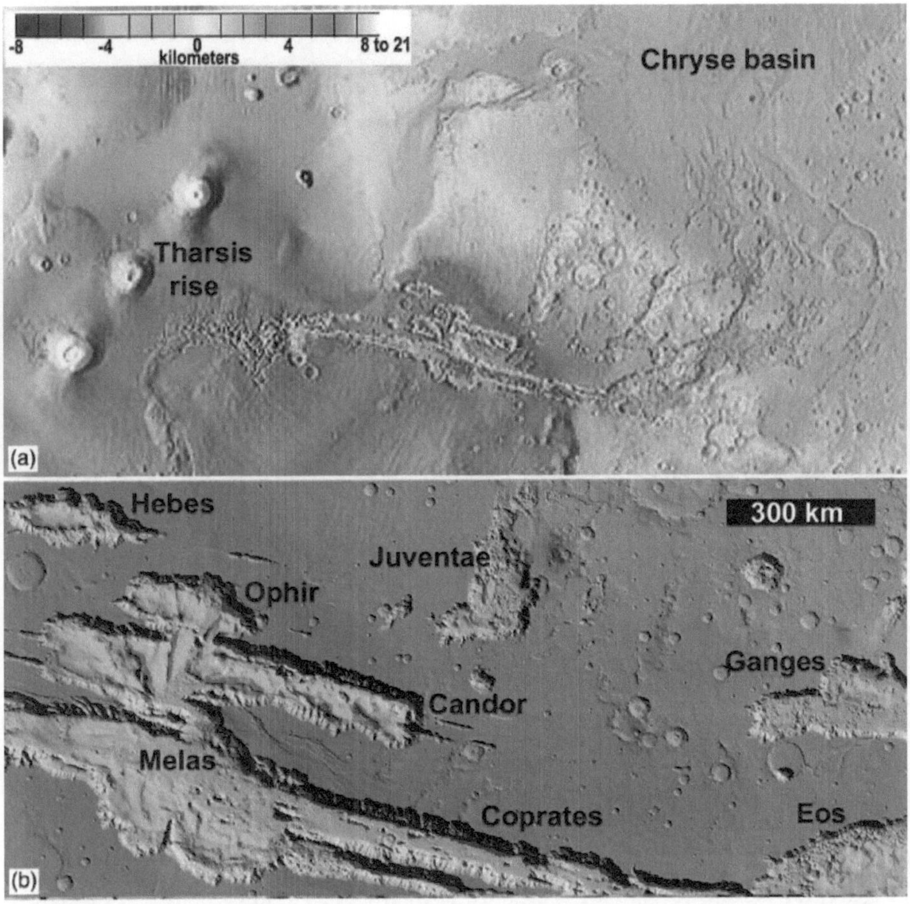

Figure 7.1. The 3000 km long canyon of Valles Marineris, Mars. (a) MOLA (Mars Observer Laser Altimeter) topographic view showing volcanoes of the Tharsis rise, on the west, Valles Marineris canyons (center) and large-scale catastrophic flood channels that drain into Chryse basin from Valles Marineris, on the east. (b) MOLA-derived three-dimensional enlarged view of central and east Valles Marineris showing interior layered deposits in blue. Note topographic lows in central Melas and Juventae Chasma, and high partial wall separating Candor Chasma from Melas Chasma.

Figure 9.4. Oblique view of a star dune in the Ibex Dunes, California. The star dune is ~30 m tall. The dark patches at the base of the dune consist of dark pebbles and granules derived from the nearby mountains, some of which have accumulated into large ripples of 2 to 8 m wavelength, over a sand substrate, possibly analogous to some ripple-like features on Mars (see Figure 9.8). JRZ, 2/03.

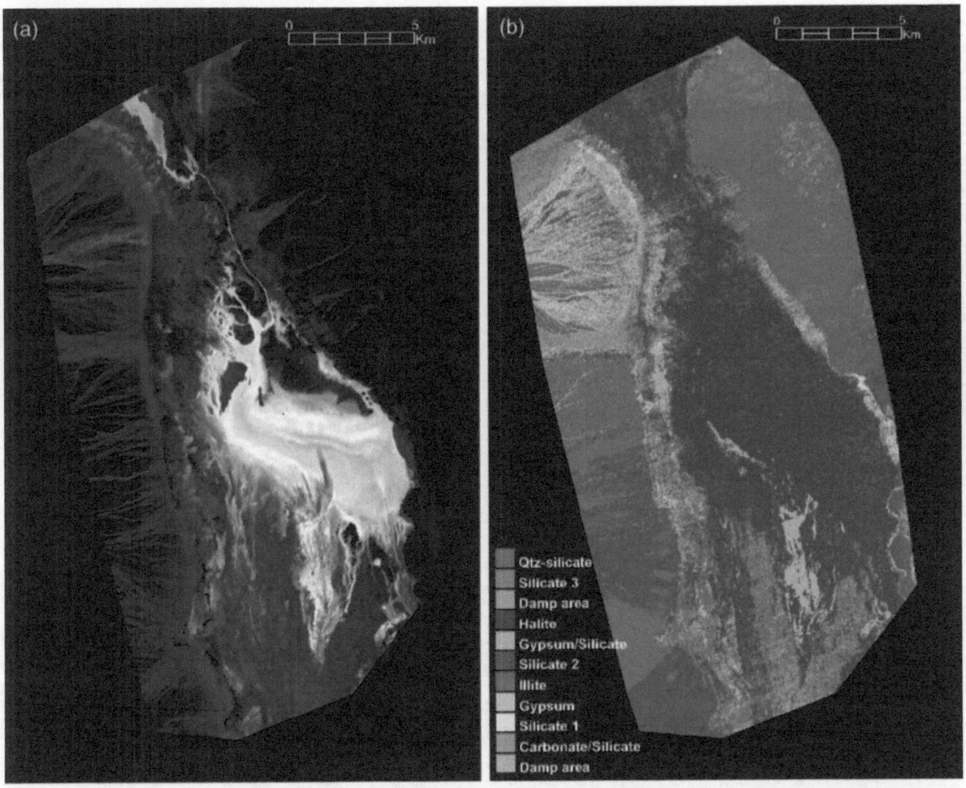

Figure 13.12. MASTER views of Badwater Basin, Death Valley, California. (a) Approximate true color image created using the MASTER 0.654, 0.542, and 0.460 μm reflectance bands. (b) Spectral classification map of MASTER thermal infrared bands.

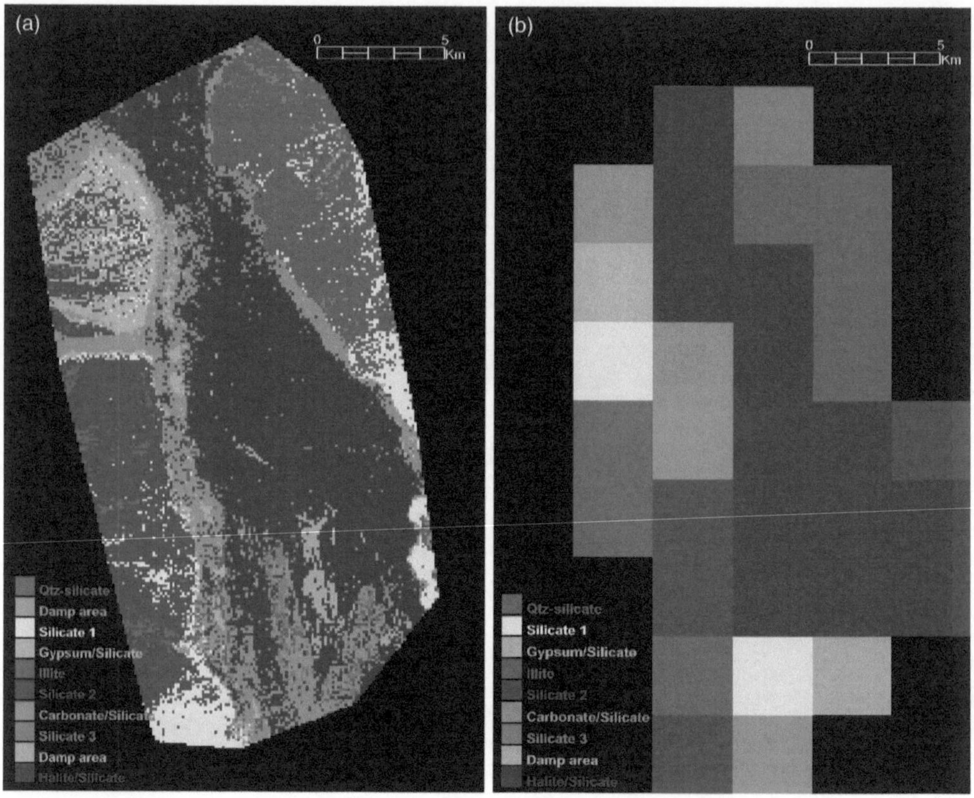

Figure 13.13. The effects of spatial resolution on spectral classification mapping of Badwater Basin. (a) MASTER thermal infrared emissivity data were degraded to the 100 m/pixel spatial resolution of the THEMIS instrument and then spectrally classified as in Figure 13.12b. (b) The same data were degraded to 3 km spatial resolution (approximately equal to TES resolution) and then spectrally classified. Spectral endmembers used to classify the TES-resolution data were taken from THEMIS-resolution data because there were insufficient pixels available at TES resolution to provide useful endmembers. Note that the typical spatial/spectral evaporite patterns on the western margin of the basin are clearly visible at THEMIS resolution, but are lost at TES resolution.

Figure 14.2. (a) Licancabur lake 100 m below the summit rim. Paleoshorelines are visible. The lake currently ~100 × 90 m and possibly up to 10 m deep, may have reached 65 m and ~200 m in diameter at its peak. (b) Laguna Verde with Licancabur in the background. (c) Hydrothermal springs in Laguna Blanca and algal mat. (d) Oxygen producing algae in the "Thermales" hot spring. Algae abound in the +36 °C water. Credit photographs: Brian H. Grigsby and Nathalie A. Cabrol.

Figure 16.1. The Martian meteorite EETA 79001 was found in Elephant Moraine, Antarctica in 1979. This meteorite is a basaltic rock of the Shergottite class of Martian meteorites. The inset shows the original sample with its fusion crust, while the larger image shows a sawn face and the igneous texture. The dark areas on the cut face are impact melt pockets that contain trapped Martian atmosphere whose composition is the strongest evidence for a Martian origin for this meteorite.

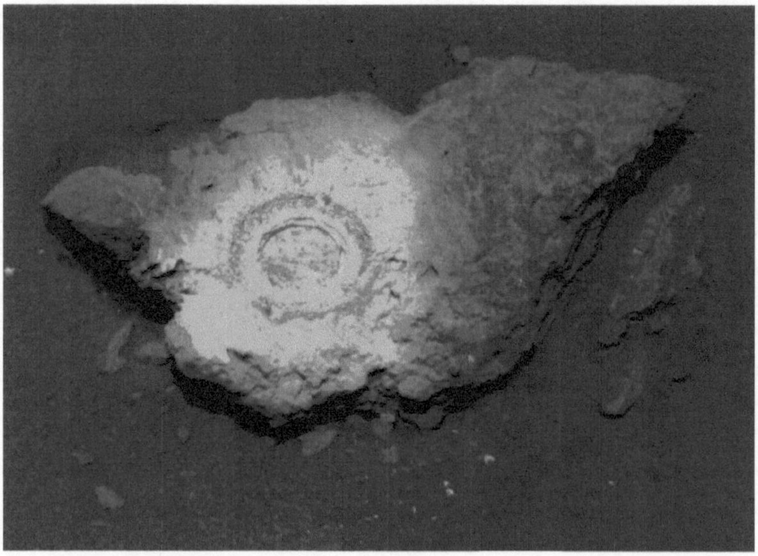

Figure 16.2. The Mars Exploration Rover Opportunity studied a rock dubbed 'Bounce' as shown in this false-color composite taken on sol 68. The 40 cm long rock was drilled to a depth of 7 mm by the rover's rock abrasion tool. The chemical composition of this sample measured by the Rover's Alpha Particle X-ray Spectrometer is nearly identical to the Martian meteorite EETA 79001 illustrated in Figure 16.1.

8

Lava–sediment interactions on Mars: evidence and consequences

Tracy K. P. Gregg
The University at Buffalo, New York

8.1 Introduction

Solidified lava flow morphologies are a consequence of complex interactions between the moving, cooling lava and its environment. Because no active Martian lava flow has been observed, eruption and emplacement parameters must be determined from the resulting volcanic morphologies. Griffiths and Fink (1992a, b) demonstrated the effects that ambient conditions exert on the gross morphology of lava flows with Newtonian rheologies. Through the use of analog experiments, they concluded that typical lava flow morphologies are created by a balance between the rate at which heat is advected within the flow and the cooling rate – a ratio they quantified with the dimensionless parameter Ψ (Fink and Griffiths, 1990). Gregg and Fink (2000) examined the effect of underlying slope on lava flow morphologies, and concluded that increasing slope has a similar effect to increasing effusion rate. However, Gregg and Smith (2003) show that this relationship breaks down somewhat on slopes steeper than about 20°. Griffiths and Fink (1997) and Fink and Griffiths (1998) examined the effect of ambient conditions on laboratory flows with a Bingham rheology, and observed a similar dependence of morphology with Ψ.

Thus, the main parameters that appear to control lava flow morphologies for lavas with Newtonian or Bingham rheologies are effusion rate, eruption temperature, lava viscosity, underlying slope, and ambient conditions (e.g., Fink and Griffiths, 1990, 1998; Gregg and Fink, 2000). Quantitatively constraining any of these characteristics for Martian lavas would allow greater insight into the generation, rise, storage, and eventual eruption of magma on Mars, thereby revealing important information about the thermal, physical,

The Geology of Mars: Evidence from Earth-based Analogs, ed. Mary Chapman. Published by Cambridge University Press.

and chemical evolution of that planet. Data collected from Viking Orbiter (VO), Mars Global Surveyor (MGS), and Mars Odyssey (MO) allow for some of these parameters to be directly measured, and narrows the range of reasonable possibilities for the remaining characteristics.

Data returned from the Thermal Imaging System (THEMIS) has revealed that much of Mars is covered with dust (e.g., McSween and Keil, 2000; Christensen *et al.*, 2003); global dust storms attest to the ubiquitous nature of Martian dust. Mars – particularly the northern hemisphere – is also covered with volcanoes and lava flows. Similarly, there is a great deal of morphologic evidence suggesting that liquid water once flowed on the Martian surface (e.g., Carr, 1996) although there are also indications that water-ice existed on the surface at low latitudes (e.g., Chapman, 2002) and at higher latitudes (Ghatan and Head, 2002; Head and Wilson, 2002) in the past. It seems likely, then, that dust, lava, and water (as liquid or solid) have interacted on the Martian surface. Lava–ice interactions generate signature morphologies and volcaniclastics. Lava–water interactions also create unique landforms (such as psuedocraters, tuff rings, tuff cones, and maars; see Fagents *et al.* (2002), and references therein). Peperites are generated when lavas (or magmas) come in contact with wet sediment, resulting in a mixture of sediment and fragmented, quenched lava (see Skilling *et al.* (2002) for review.)

What is less well studied on Earth is the interaction of lavas with dry, unconsolidated sediments. On Earth, most sediments are wet because of the ubiquitous nature of water. However, Jerram and Stollhofen (2002) report on basaltic lavas flowing over sand dunes, and sand being blown onto actively flowing lavas. It is probable that Mars, with its currently arid surface conditions, abundance of dust, and numerous volcanoes, contains examples of lava flows interacting with dust (and possibly sand) deposits.

A field of lava flow lobes, covering $< 10^6$ km^2, is located southwest of Arsia Mons, Mars, within Daedalia Planum (Warner and Gregg, 2003) (Figure 8.1). The unique surface morphologies of the lobes comprising this flow field were first identified in high-resolution ($<$45 m/pixel) VO images and were presented as possible examples of large-volume evolved lavas on Mars (Moore *et al.*, 1978; Fink, 1980; Theilig and Greeley, 1986). The flow lobes within this flow field display a ridged surface morphology that can be identified on MOC narrow-angle, high-resolution VO images and THEMIS images (both visible and infrared) (Figure 8.2), but not low-resolution ($>$150 m/pixel) VO images. These flows are characterized by steep ($>30°$), thick (~65 m) margins, and surface ridges (Warner and Gregg, 2003). Near the flow margins, surface ridges are oriented with their axes roughly perpendicular to the inferred flow direction, and have been interpretted to be folds

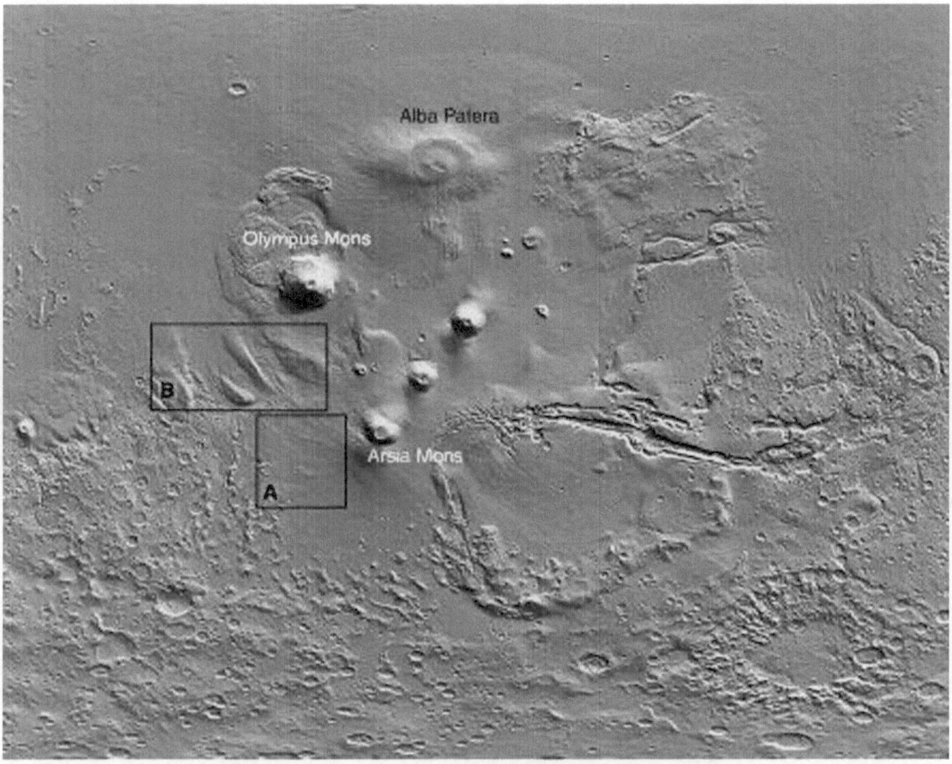

Figure 8.1. MOLA grid of the eastern hemisphere of Mars. Box "A" shows the approximate location of the flow field containing flow lobes with ridged surface textures and thick flow fronts. Box "B" indicates the approximate extent of the Medusa Fosse Formation. Field of view is ~6000 km.

(cf. Fink, 1980; Gregg *et al.*, 1998). The precise areal extent of these unique flows cannot be determined because the flow-field is covered by the Medussae Fossae formation to the north and west (Figure 8.1). Keszthelyi and others (2000) suggested that these lava morphologies may be caused by a surge-and-stasis type of pahoehoe emplacement, similar to what they interpret for the 1783 eruption of Laki, Iceland. Alternatively, the lavas may achieve that morphology simply by flowing for hundreds of kilometers over slopes of $< 1°$ (cf. Warner and Gregg, 2003). But it is also possible – given the proximity of the Medussae Fossae formation to these flows – that at least some of the lavas may have interacted with a fine-grained (possibly dust-sized), unconsolidated deposit during there emplacement, and this may be at least partly responsible for their unique morphologies (Porter and Schultz, 1990; Gregg and Schultz, 1997).

Emplacement beneath this deposit may have affected the thermodynamics and kinetics of these flows, thereby generating the observed large surface

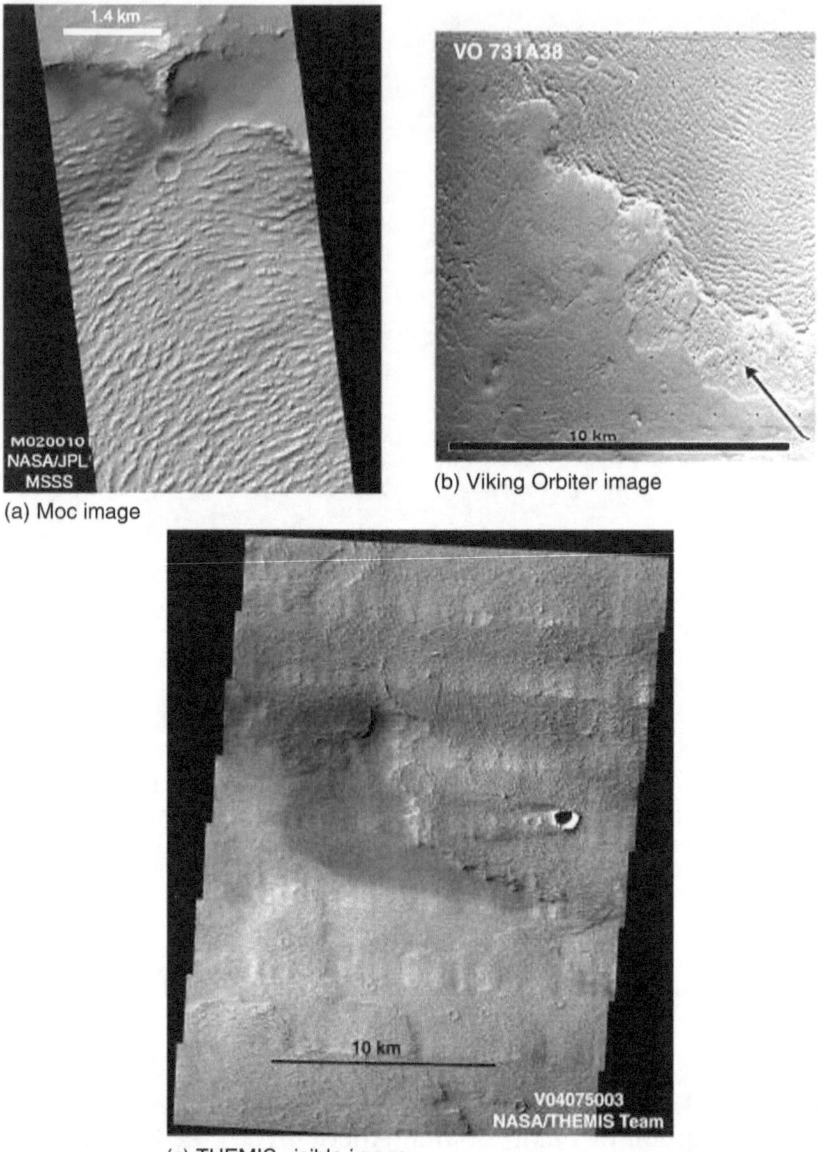

Figure 8.2. Ridged lavas shown in MOC (a), Viking Orbiter (b) and THEMIS (c) images. Ridged flows are only visible in high- to moderate-resolution (< 100 m/pixel) VO images, but can be seen in most MOC and THEMIS images within the region outlined in box "A" in Figure 8.1. North is at the top of images (a) and (c); arrow indicates north in (b).

ridges and thick flow fronts. Similar morphologic features are observed on Earth where lavas have burrowed beneath wet sediments, or flowed into shallow lakes (e.g., Schminke, 1967; Walker and Francis, 1987; Godchaux *et al.*, 1992). Observations made using available data indicate that these flows lie unconformably beneath the Medussae Fossae formation (Figure 8.3). The presence of pedestal craters on top of these ridged flows strongly suggest that the lavas are currently being exhumed rather than buried. Pedestal craters are interpreted to be the remnants of craters whose ejecta blanket have protected the underlying material from erosion (primarily eolian erosion) (e.g., Arvidson *et al.*, 1976). Here, I explore the possibility and implications that lava flows were emplaced beneath this deposit.

Identifying the sub-surface emplacement, and subsequent exhumation, of a large lava flow field on Mars would have profound implications for our understanding of the present and past behavior of the surface. Mars is a dynamic planet, and it would be simplistic to assume that the surface observed today is the same surface that was present throughout Martian history. Recent data collected by MGS and MO demonstrate a complexity and rich history of the Martian surface that was only hinted at in Viking Orbiter images.

8.2 Background and previous work

Investigation of lava-flow morphology is a primary method for constraining the temporal and spatial evolution of magma generation, storage, and final emplacement on any planetary surface. The change in ambient conditions as lava entered the sea off the Hawaiian coast was drastically reflected in lava flow morphology: subaerial pahoehoe flows became submarine bulbous, striated pillowed flows (Tepley and Moore, 1978; Sansone, 1990). Similarly, although sub-glacial lavas have a very different appearance from their subaerial counterparts (Lescinsky and Fink, 1996; Lescinsky and Sisson, 1998), an active subglacial lava flow has not yet been observed. Thus, virtually all of the information we have for basaltic lava flow behavior – on land, on sea, and on the other planets – is based on observations of active subaerial terrestrial lavas (e.g., Greeley, 1971, 1972; Pinkerton and Sparks, 1978; Guest *et al.*, 1987; Keszthelyi and Delinger, 1996). However, it is unlikely that observed Martian lavas experienced ambient conditions similar to those encountered by the active flows studied today on Earth.

The effects of planet surface temperature (e.g., Crisp and Baloga, 1990, 1994), atmospheric conditions (e.g., Head and Wilson, 1986; Gregg and Fink, 1996; Bridges, 1997), and gravitational acceleration (e.g., Wilson and

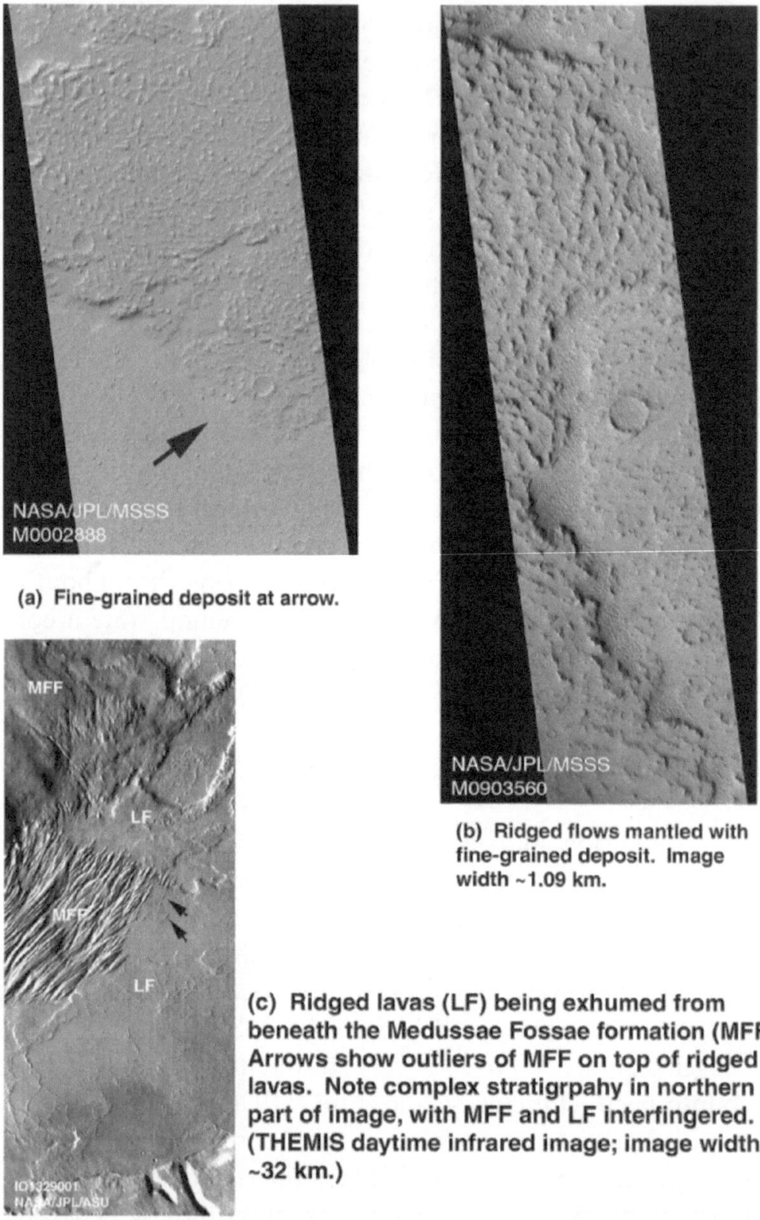

(a) Fine-grained deposit at arrow.

(b) Ridged flows mantled with fine-grained deposit. Image width ~1.09 km.

(c) Ridged lavas (LF) being exhumed from beneath the Medussae Fossae formation (MFF). Arrows show outliers of MFF on top of ridged lavas. Note complex stratigrpahy in northern part of image, with MFF and LF interfingered. (THEMIS daytime infrared image; image width ~32 km.)

Figure 8.3. MOC images (a) and (b) and THEMIS image (c) showing association of ridged flows with fine-grained deposit. Image (c) reveals pedestal craters on top of the ridged flow, indicating that substantial material has been removed from above the flow since the formation of the impact.

Head, 1981; Glaze *et al.*, 1997; Sakimoto *et al.*, 1997) have been incorporated into numerical models designed to interpret extraterrestrial volcanic morphologies. General petrologic models (e.g., McSween, 1985, 1995; Hess and Head, 1990; Longhi *et al.*, 1992; Kargel *et al.*, 1994; McSween and Treiman, 1998) provide constraints on possible lava compositions – and therefore rheologies – for those planets from which we have no returned samples. Remote sensing of the Martian surface using the Thermal Emission Spectrometer (TES) (e.g., Bandfield, 2000) and THEMIS suggest that the surface of Mars can be generally characterized as basaltic to basaltic andesite (Bandfield *et al.*, 2000).

Studies into the controls on final Martian lava flow morphology typically assume that the lavas were emplaced subaerially on a Mars with a climate similar to that of today's (e.g., Griffiths and Fink, 1992b; Gregg and Fink, 1996; Glaze and Baloga, 1998). However, the outflow channels strongly suggest that the surface of Mars supported large amounts of liquid water in the past (e.g., Baker *et al.*, 1992), and Parker and others (Parker *et al.*, 1993; Parker and Currey, 2001) and Head and others (Head *et al.*, 1998, 1999) cite evidence for the possible existence of a vast ocean once covering the northern plains on Mars. Similarly, some investigators have identified evidence of past widespread glaciation on Mars (Baker *et al.*, 1991; Kargel and Strom, 1992; Kargel *et al.*, 1995; Ghatan and Head, 2002). Allen (1979) and Hodges and Moore (1979) point to Martian features with gross morphologic similarities to terrestrial moberg and table mountains as products of subglacial eruptions; Chapman and others (Chapman and Tanaka, 2001; Chapman, 2002) have identified possible subglacial volcanic deposits within Valles Marinaris. High-resolution images collected using the Mars Orbiter Camera (MOC) reveal young (possibly active) gullies within crater and canyon walls (Malin and Edgett, 2001); Christensen *et al.* (2003) suggest that a present-day snow cover may be responsible for small gully formation. Moreover, the surface records evidence for different epochs of eolian deposition and removal leaving easily identified relict unconformable deposits in certain locations (Schultz and Lutz, 1988; Grizzaffi and Schultz, 1989; Grant and Schultz, 1993; Edgett and Malin, 1999; Malin and Edgett, 1999). Global dust storms, layered polar deposits, thick sedimentary sequences and sand dunes on Mars attest to the dynamic ability of the Martian atmosphere to transport dust- and sand-sized particles around the globe (e.g., Christensen and Moore, 1992; Greeley *et al.*, 1992; Thomas *et al.*, 1992). Mustard and others (2001) have identified the presence of a fine-grained surface layer that they argue is in the process of being removed – probably through sublimation of ice that cements the material together. Finally, data collected from MO reveal large quantities of hydrogen (presumably in the form of water) in the northern plains of Mars. All of these observations strongly suggest that Mars is dynamic and the surface we observe today is not necessarily the same one exposed in the past. This must be considered when interpreting Martian volcanic morphology.

The thick, textured lava flows on Mars identified southwest of Arsia Mons have briefly caught the attention of previous investigators (Fink, 1980; Schaber, 1980; Theilig and Greeley, 1986; Porter and Schultz, 1990; Gregg and Fink, 1994, 1996; Gregg and Schultz, 1997; Keszthelyi *et al.*, 2000; Warner and Gregg, 2003). Schaber (1980) identifies textured lavas in the Memnonia Fossae region, and characterizes their surfaces as including "various combinations of channels and knobs, random ridges and troughs, and curvilinear ridges and troughs" and refers to this as "ridge-and-trough" texture. The ridged lavas are ~65 m thick, with pseudo-parallel curvilinear ridges approximately 27 m ($\pm$11 m) tall and spaced 100 $\pm$ 34 m apart (Warner and Gregg, 2003). Assuming that these features are folds formed during flow emplacement, their dimensions are more similar to those seen on terrestrial evolved lavas (andesites, dacites and rhyolites) than typical basalts (Fink, 1980; Schaber, 1980; Gregg *et al.*, 1998). Theilig and Greeley (1986) measured amplitudes of 12 m and spacing of 30–120 m for these features (they appear to have investigated different flows from those studied by Schaber (1980). Using folding analysis (Fink and Fletcher, 1978), Theilig and Greeley (1986) determined that these flows were composed of a lava with an interior viscosity of 10^8 to 10^{10} Pa – approaching the viscosity at which a basalt becomes too viscous to flow (e.g., Marsh, 1981; Moore, 1987; Pinkerton and Stevenson, 1992). Schaber (1980) noted that the lack of small impact craters on these textured flows suggests a recent eolian mantle, but did not mention that if these flows are traced upstream, they appear to emerge from beneath a fine-grained deposit. Gregg and others (Gregg and Fink, 1996; Gregg *et al.*, 1998) identified two generations of folding on portions of these textured flows, and used the ratio of the second-generation wavelength to the first-generation wavelength to suggest that the surface texture could have been generated by ductile deformation of a lava with a viscosity consistent with that of an andesite or basaltic andesite. Alternatively, Porter and Schultz (1990) and Gregg and Schultz (1997) proposed that a "typical" basalt (viscosity = 10^2–10^4 Pa s) could generate large-scale surface folding (cf. Fink and Fletcher, 1978) if the appropriate balance between surface deformation and crust growth is obtained, and suggest that this balance could be most easily obtained by emplacing the lavas beneath an ice-containing dust mantle. Keszthelyi and others (2000) refer to these types of flows as "platey-ridged" pahoehoe lavas. By comparing the Martian Daedalia Planum flows with basalts erupted in 1783–4 from Laki, Iceland, Keszthelyi and others (2000) suggest that the Martian flows formed by long-lived pahoehoe flows punctuated by episodes of lava surges. Most recently, Warner and Gregg (2003) applied a series of rheologic models to these Martian ridged flows, and compared model results with those obtained from terrestrial lavas with known basaltic and trachyandesitic/andesitic compositions. They concluded that the rheology of the Arsia Mons lavas is more consistent with a basaltic composition than that of a more-evolved lava.

Although the process of relatively dense lava flowing into lighter, unconsolidated sediments has not heretofore been unequivocally documented on Mars, it is common on Earth wherever volcanic vents are juxtaposed with sediments (such as sedimentary basins, lake beds, and the ocean floor) (e.g., Gifkins *et al.*, 2002; Lavine and Aalto, 2002; Skilling *et al.*, 2002). There are clearly places on Mars where volcanics and sediments have had the opportunity to interact, so it is reasonable to expect that this process may have occurred there as well – and may have profoundly influenced the resulting lava flow morphology.

8.3 Terrestrial analogs: quiescent peperites

When lava intrudes into a solid host rock, the resulting feature is either a "dike" or a "sill," depending on the orientation of the intrusion. However, when lava flows into relatively unconsolidated, wet sediments, the outer portion of the flow in contact with the sediments may catastrophically quench and fracture, and the host sediments deform around the shallow intrusion, resulting in quenched lava fragments peppered throughout the surrounding sediments – hence their name of "peperites." While peperites, *sensu strictu*, refers only to this mixture of lava fragments and host sediment surrounding the intrusive lava/magma body (Skilling *et al.*, 2002), the term was broadened in the literature to refer to any lavas/magmas that came in contact with wet sediments during their emplacement (Fisher and Schminke, 1984).

Most commonly, terrestrial peperites form when basaltic lava (or, less frequently, more evolved compositions Hanson and Wilson, 1993) intrudes unconsolidated wet sediments (e.g., Brooks *et al.*, 1982; Kokelaar, 1982; Busby-Spera and White, 1987; Walker and Francis, 1987; White and Busby-Spera, 1987; Kano, 1989), although there are examples of basaltic lava flowing on top of, and subsequently "burrowing into" soft sediments (Schminke, 1967). Additionally, evidence from submarine investigations suggests that basaltic peperites may be a common occurrence on the sea floor, where sufficient thicknesses of wet sediment exist (Einsele *et al.*, 1978; Shen *et al.*, 1997). Field investigations clearly demonstrate that both the gross and fine-scale flow morphology of a peperite is distinct from the morphology the lava would have formed had it flowed on the surface. For example, investigation of a rhyolite peperite in Chile (Hanson and Wilson, 1993) reveals that flow is disrupted into a series of large, roughly cylindrical pod-like bodies, ~15–20 m wide, of coherent rhyolite, separated by dense zones of quench-fractured rhyolite. Hanson and Wilson (1993) state that this indicates that the lava advanced as a series of lobes or buds, similar to submarine pillow lavas but on a much larger scale. This is a vastly different morphology from other, similar-sized rhyolite flows, such as the Inyo Domes in California, which are thick, blocky flows with a rough radial symmetry about a central vent (e.g., Sampson, 1987). Similarly, the surface of basaltic peperites may exhibit tall ridges (>3 m) and troughs, whereas the subaerial portion of the flow is more commonly smooth (Schminke, 1967; Busby-Spera and White, 1987).

The morphologic expression of basaltic peperites (and probably for more evolved lavas as well) depends on the properties of the host sediment (e.g., Busby-Spera and White, 1987). A fine-grained, unconsolidated deposit is more likely to deform in a ductile manner and generate little resistance to lava advance; in contrast, a coarse-grained deposit may contain large pockets of water that could result in steam explosions during emplacement (Busby-Spera and White, 1987). Whereas the lower margins of peperites tend to be planar and smooth, their upper surfaces are more typically irregular on the scale of centimeters to meters (Schminke, 1967; Busby-Spera and White, 1987). Schminke (1967) provides an excellent description of peperites in the Pomona flow of the Columbia River Basalts, and his observations are summarized here.

Field relations, geochemical and petrologic analyses confirm that the Pomona flow was a surface basalt flow that locally burrowed into at least 4.5 m of fluffy, unconsolidated sediments. Where the lava was emplaced subaerially, it has a relatively constant thickness of ~30 m. In contrast, lavas that flowed under the sediments are irregular: basalt stringers, a meter or more thick, protrude into the overlying sediment, and locally envelop it. Like the subaerially emplaced portions of the flow, the peperite is highly vesicular to scoriacious, suggesting that the overlie sediment did not exert sufficient pressure on the flow to keep volatiles from exsolving. Schminke (1967) envisions that the basalt initially advanced as a series of thin sheets, which were broken as they were overridden by larger pods of basaltic lava. Eventually, the main body of the peperites apparently advanced similar to a subaerial

Hawaiian inflated flow (Hon *et al.*, 1994). Importantly, large portions of the surface of this peperite are currently exposed where the overlying tuff has been eroded (Schminke, 1967), allowing for good investigation of the lava flow surface morphology.

In most documented cases of terrestrial basaltic peperites, there is evidence that liquid water, with varying amounts of suspended sediments, has been ejected from the host sediment (Einsele *et al.*, 1978; Busby-Spera and White, 1987; White and Busby-Spera, 1987; Shen *et al.*, 1997; Wohletz, 2002; Zimanowski and Büttner, 2002). When the intruding lava contacts the pore water, the thermal energy generates a thin boundary layer of steam surrounding the flow (Wohletz, 2002). This steam layer fluidizes the host sediment, which then is easily transported away from the flow (Kokelaar, 1982, 1986). In contrast, Jerram and Stollhofen (2002) report on the interaction between basaltic lava and dry, eolian quartz sand dunes in the Etendeka Flood Basalt Province, Namibia. Locally, the lavas both flowed over the dry sand with minimal interactions and elsewhere burrowed into the sand and generated zones of dynamic mixing between sand and lava.

8.4 Evidence for lava emplacement beneath dust

Images collected by VO, THEMIS, and MOC reveal an intimate relationship between dust and lavas west and north of Arsia Mons. Deposits of Medussae Fossae lay unconformably on top of the thick, ridged flows (Figure 8.3). THEMIS night infrared images (Figure 8.4) reveal that the ridged flows are dust-covered, and this is also shown in MOC images of the flows (Figure 8.3). Most significantly, pedestal craters on top of the ridged flows (Figure 8.5) – particularly abundant near the current margin of the Medussae Fossae formation – indicate that the deposit in which the craters formed has been at least partially removed from the flow surface.

These observations indicate that the ridged flows were once covered with a relatively unconsolidated, fine-grained (as suggested by the low thermal inertia displayed in the THEMIS images) deposit that has be removed by eolian processes. Jerram and Stollhofen (2002) describe a host of features (mostly at scales beneath available Martian resolutions) that are indicative of dry sediment–lava interactions. These include: preserved dunes layered between successive lava flows; sand-filled cracks and fissures within the flows; and flow-top breccias that are 1.5–4.0 m thick, composed of sand and bits of quenched lava. However, available Mars images cannot reveal whether the Daedalia Planum lavas were emplaced beneath the observed deposit, nor

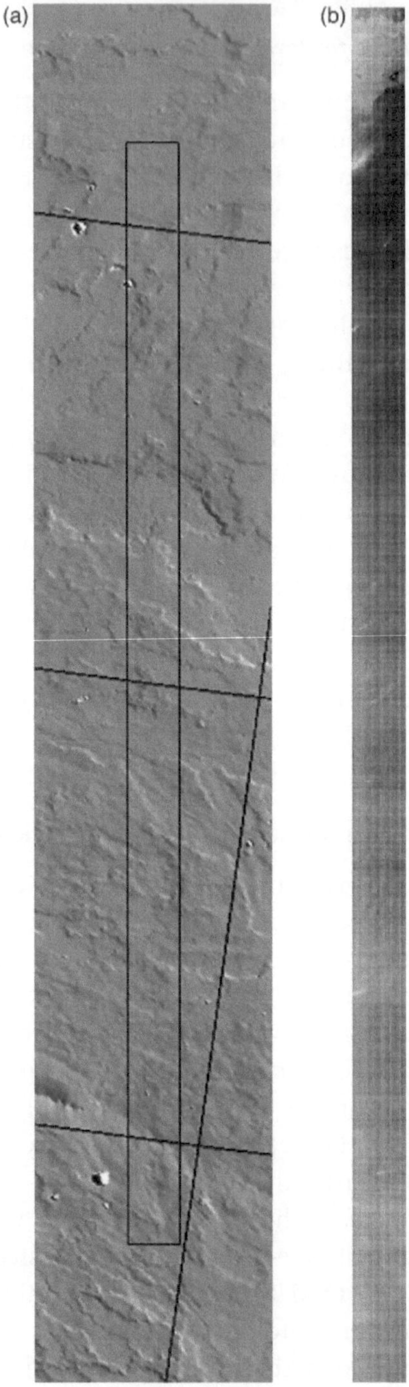

Figure 8.4. Nightime thermal infrared image of the ridged lavas flow field, demonstrating that the dust cover is relatively uniform across the flows and too thick to be penetrated by THEMIS. The image on the right (a) shows the context of image (b); context image is MOLA shaed relief with rectangular outline showing location of (b). Grid in (a) has 5° spacing. Image I02408002, courtesy NASA/JPL/ASU. Width of image (b) is ~32 km.

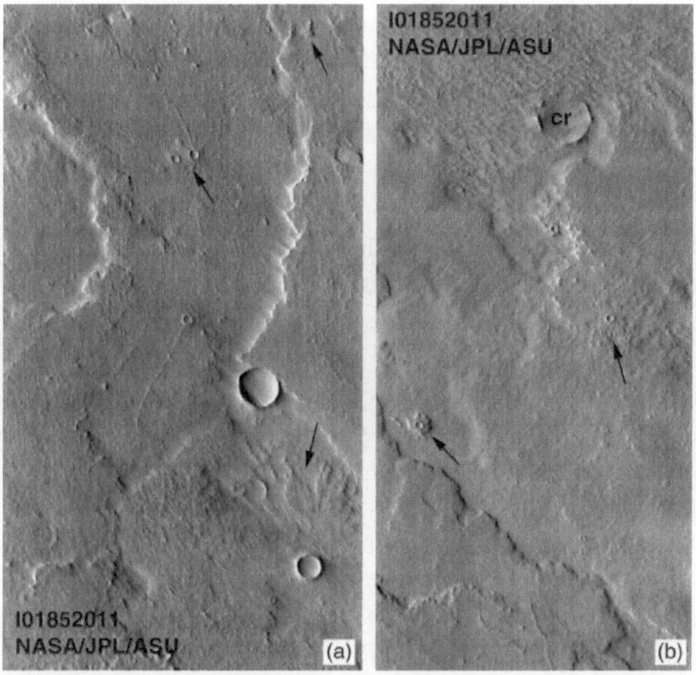

Figure 8.5. (a) Daytime THEMIS infrared image showing pedestal craters (arrows) on top of ridged flows west of Arsia Mons. Image width is ~32 km. (b) Daytime THEMIS infrared imaged showing pedestal craters (arrows) on top of ridged flows west of Arsia Mons. Note crater (cr) in northern part of image being exhumed from beneath Medussae Fossae formation. Image width ~32 km. Courtesy NASA/JPL/ASU.

for how long these flows may have been buried. Modeling of flow emplacement processes may yield some insight.

8.5 Implications for lava flow emplacement

Lavas emplaced beneath an unconsolidated deposit may be similar to terrestrial sills, or to terrestrial subaqueous flows, depending on the nature of the deposit. If the deposit is fine-grained (dust-sized particles), unconsolidated, and unconfined, the lava does not need to do much work to push the deposit aside, but the deposit does provide a certain resistance to flow (a buoyancy force) that is related to the difference in density between the flow and the deposit. On Earth, for example, submarine lavas essentially advance within a reduced gravity environment, in which the gravitational acceleration (g') is given by (Fink and Griffiths, 1990):

$$g' = g(\rho_a/\rho_l) \tag{1}$$

where g is the planetary gravitational acceleration (9.78 m/s^2 on Earth; 3.67 m/s^2 on Mars), ρ_a is the density of the ambient material (e.g., atmosphere, water, or dust), and ρ_l is lava density. Thus, any process that is directly correlated with gravity (such as flow velocity, and flow length for gravity-driven flows) would be reduced. We therefore might expect slower, shorter, and thicker flows emplaced beneath an unconsolidated deposit compared with a identical lava emplaced on the planet's surface.

Emplacement beneath a dry, dusty deposit would act to insulate the flow surface by reducing the rate at which heat is able from the flow both by radiant heat and by atmospheric convection. A dust cover would allow heat to leave the flow by conduction through the individual dust grains, and radiation between the grains. Such insulation would, in theory, allow a lava flow to travel farther beneath a dust cover than would an identical, uncovered flow.

If the dusty deposit were dense, with closely packed and larger (sand-sized) grains, it is less likely that a gravity-driven flow would have the ability to burrow into it, or to push it aside. However, Jerram and Stollhofen (2002) report that most commonly the basaltic lava did not disturb the underlying, active eolian sand deposits during emplacement, but locally was able to plough through the sand like a bulldozer.

Finally, if the deposit – regardless of grain size – were filled with interstitial ice, it would act more like solid country rock, or like a glacier. Lava would either be emplaced on top of the deposit – and, by comparison with hot lahars in New Zealand, would likely stay on top rather than melt through (although admittedly that depends on eruption duration) – or be forced to intrude into the material like a sill (Cronin *et al.*, 1996; Head and Wilson, 2002; Wilson and Head, 2002). In the latter case, we would predict the formation of Martian peperites. Depending on the degree of induration, the peperites could be preferentially eroded from the flow surface through time, or contribute to the preserved flow morphology. We would also expect to see evidence of water–lava interaction (such as pseudo-craters (Fagents *et al.*, 2002) and of dewatering (such as channels sourced stratigraphically above the flows) in this case. All of these features would be difficult to unequivocally identify on Mars. Dewatering could be supported by identification of channels that appear to be sourced beneath a deposit or appear, fully formed, from the toe of a lava flow (which would require assuming that the channel originated in the now-removed sediments stratigraphically above the lava flow). Such channels have been observed to emanate from beneath the Medussae Fossae formation (Shockey *et al.*, 2004).

8.6 Discussion and conclusions

Intrusion of lavas into unconsolidated sediments is common on Earth, wherever volcanoes erupt near (or in) sedimentary basins, and the intruded lava has a different morphologic expression than the same surface flow. Mars, replete with volcanoes, abundant large, thick sequences of fine-grained deposits (Scott and Tanaka, 1982; Schultz and Lutz, 1988; Thomas *et al.*, 1992; Zimbelman *et al.*, 1996, 1997), and dry stream channels, must have experienced this type of shallow intrusion in the past. There is evidence on Mars for substantial reworking of fine-grained deposits (Greeley *et al.*, 1992), so it is reasonable to expect that these Martian peperites may be subsequently exposed.

Identification of exhumed Martian lavas would force the generation of a new relative and absolute chronology for the Tharsis region. For example, crater size–frequency distributions from crater counts obtained from the flow surfaces would reveal an *exposure* age, rather than an *emplacement* age. In other words, lavas may appear fresh because they have been protected from erosion by an overlying deposit, not because they are relatively youthful. The stratigraphy here would not represent the chronology, which is vital to recognize if we are to correctly interpret Martian volcanic evolution through time.

Unequivocal identification of peperites on Mars will require detailed analyses at a hand-sample as well as outcrop scale. The unique texture of peperites, particularly in conjunction with their parent flow, would be difficult to misinterpret. More difficult to identify would be the interaction of dry, unconsolidated dust deposits with martial lavas. Our current understanding of Martian geology almost requires that such interactions occurred in the past; it is less clear if we will be able to readily identify it when we see it. Currently, the thick, ridged flows southwest and west of Arsia Mons that are presently being exhumed from beneath the Medussae Fossae formation, are the best place to look for lava–sediment interactions on Mars.

References

Allen, C. C. (1979). Volcano–ice interactions on Mars. *Journal of Geophysical Research*, **84**, 8048–59.

Arvidson, R. E., Coradini, M., Carusi A. *et al.* (1976). Latitudinal variation of wind erosion of crater ejecta deposits on Mars. *Icarus*, **27**, 503–16.

Baker, V. R., Strom, R. G., Gulick, V. C. *et al.* (1991). Ancient oceans, ice sheets, and the hydrological cycle on Mars. *Nature*, **352**, 589–94.

Baker, V. R., Carr, M. H., Gulick, V. C., Williams, C. R., and Marley, M. S. (1992). Channels and valley networks. In *Mars*, ed. H. H. Kieffer, B. M. Jakosky, C. W. Snyder, and M. S. Matthews. Tucson: University of Arizona Press, pp. 493–521.

Bandfield, J. L., Hamilton, V. E., and Christensen, P. R. (2000). A global view of Martian surface compositions from MGS-TES. *Science*, **287**, 1626–30.

Bridges, N. T. (1997). Ambient affects on basalt and rhyolite lavas under venusian, subaerial and subaqueous conditions. *Journal of Geophysical Research*, **102**, 9243–56.

Brooks, E. R., Wood, M. M., and Garbutt, P. L. (1982). Origin and metamorphism of peperite and associated rocks in the Devonian Elwell Formation, northern Sierra Nevada, California. *Geological Society of America Bulletin*, **93**, 1208–31.

Busby-Spera, C. J. and White, J. D. L. (1987). Variation in peperite textures associated with differing host-sediment properties. *Bulletin of Volcanology*, **49**, 765–75.

Carr, M. H. (1996). *Water on Mars*. Oxford: Oxford University Press.

Chapman, M. G. (2002). Layered, massive, and thin sediments on Mars: possible Late Noachian to Late Amazonian tephra? In *Volcano–Ice Interaction on Earth and Mars*, ed. J. L. Smellie and M. G. Chapman. Geological Society of London Special Publication 202, pp. 273–93.

Chapman, M. G. and Tanaka, K. L. (2001). Interior trough deposits on Mars: sub-ice volcanoes? *Journal of Geophysical Research*, **106** (E5), 10,087–10,000.

Christensen, P. R. and Moore, H. J. (1992). The Martian surface layer. In *Mars*, ed. H. H. Kieffer, B. M. Jakosky, C. W. Snyder, and M. S. Matthews. Tucson: University of Arizona Press, pp. 686–29.

Christensen, P. R., Bandfield, J. L., and Bell III, J. F. (2003). Morphology and composition of the surface of Mars, Mars Odyssey THEMIS results. *Science*, **300**, 2056–61.

Crisp, J. and Baloga, S. (1990). A model for lava flows with two thermal components. *Journal of Geophysical Research*, **95**, 1255–70.

Crisp, J. and Baloga, S. (1994). Influence of crystallization and entrainment of cooler material on the emplacement of basaltic aa lava flows. *Journal of Geophysical Research*, **99**, 11819–31.

Cronin, S J., Neall, V. E., Lecointre, J. A., and Palmer, A. S. (1996). Unusual "snow slurry" lahars from Ruapeu volcano, New Zealand, September 1995. *Geology*, **24**, 1107–10.

Edgett, K. S. and Malin, M. C. (1999). MGS MOC the first year: sedimentary materials and relationships. *Abstracts of Papers Submitted to the 30th Lunar and Planetary Science Conference*. Houston: Lunar and Planetary Institute, CD 30, Abstract 1029.

Einsele, G., Gieskes, J. M., Curray, J. *et al.* (1978). Intrusion of basaltic sills into highly porous sediments, and resulting hydrothermal activity. *Nature*, **283**, 441–5.

Fagents, S. A., Lanagan, P. D., and Greeley, R. (2002). Rootless cones on Mars: a consequence of lava–ground ice interaction. In *Volcano–Ice Interaction on Earth and Mars*, ed. J. L. Smellie and M. G. Chapman. Geological Society of London Special Publication 202, pp. 295–317.

Fink, J. H. (1980). Surface folding and viscosity of rhyolite flows. *Geology*, **8**, 250–4.

Fink, J. H. and Fletcher, R. C. (1978). Ropy pahoehoe: surface folding of a viscous fluid. *Journal of Geophysical Research*, **4**, 151–70.

Fink, J. H. and Griffiths, R. W. (1990). Radial spreading of viscous-gravity currents with solidifying crust. *Journal of Fluid Mechanics*, **221**, 485–509.

Fink, J. H. and Griffiths, R. W. (1998). Morphology, eruption rates, and rheology of lava domes: insights from laboratory models. *Journal of Geophysical Research*, **103**, 527–45.

Fisher, R. V. and Schminke, H.-U. (1984). *Pyroclastic Rocks*. New York: Springer-Verlag.

Ghatan, G. J. and Head, J. W. (2002). Candidate subglacial volcanoes in the south polar region of Mars: morphology, morphometry, and eruption conditions. *Journal of Geophysical Research*, **107**, 5048.

Gifkins, C. C., McPhie, J., and Allen, R. L. (2002). Pumiceous rhyolitic peperite in ancient submarine volcanic successions. *Journal of Geophysical Research*, **114**, 181–203.

Glaze, L. S. and Baloga, S. M. (1998). Dimensions of Pu'u O'o lava flows on Mars. *Journal of Geophysical Research*, **103**, 13659–66.

Glaze, L S., Baloga, S. M., and Wilson, L. (1997). Transport of atmospheric water vapor by volcanic eruption columns. *Journal of Geophysical Research*, **102**, 6099–108.

Godchaux, M. M., Bonnichson, B., and Jenks, M. D. (1992). Types of phreatomagmatic volcanoes in the western Snake River Plain, Idaho, USA. *Journal of Volcanology & Geothermal Research*, **52**, 1–25.

Grant, J. A. and Schultz, P. H. (1993). Degradation of selected terrestrial and Martian impact craters. *Journal of Volcanology and Geothermal Research*, **98**, 11025–42.

Greeley, R. (1971). Observations of actively forming lava tubes and associated structures, Hawaii. *Modern Geology*, **2**, 207–23.

Greeley, R. (1972). Additional observations of actively forming lava tubes and associated structures, Hawaii. *Modern Geology*, **3**, 157–60.

Greeley, R., Lancaster, N., Lee, S., and Thomas, P. (1992). Martian aeolian processes, sediments, and features. In *Mars*, ed. H. H. Kieffer, B. M. Jakosky, C. W. Snyder, and M. S. Matthews. Tucson: University of Arizona Press, pp. 730–66.

Gregg, T. K. P., and Chadwick, Jr. W. W. (1996). Submarine lava flow inflation: a model for the formation of lava pillars. *Geology*, **24**, 981–4.

Gregg, T. K. P. and Fink, J. H. (1994). Ratio of first and second generation fold wavelengths on lavas may indicate flow composition. *Abstracts of Papers Submitted to the 25th Lunar and Planetary Science Conference*. Houston: Lunar and Planetary Institute, pp. 473–4.

Gregg, T. K. P. and Fink, J. H. (1996). Quantification of extraterrestrial lava flow effusion rates through laboratory simulations. *Journal of Geophysical Research*, **101**, 16891–900.

Gregg, T. K. P. and Fink, J. H. (2000). A laboratory investigation into the effects of slope on lava flow morphology. *Journal of Volcanology and Geothermal Research*, **96**, 145–59.

Gregg, T. K. P., Fink, J. H., and Griffiths, R. W. (1998). Formation of multiple fold generations on lava flow surfaces: influence of strain rate, cooling rate, and lava composition. *Journal of Volcanology and Geothermal Research*, **80**, 281–92.

Gregg, T. K. P. and Schultz, P. H. (1997). Ridged Martian lava flows: intrusions or Extrusions? *Abstracts of Papers Submitted to the 28th Lunar and Planetary Science Conference*. Houston: Lunar and Planetary Institute, pp. 463–4.

Gregg, T. K. P and Smith, D. K. (2003). Volcanic investigations of the Puna Ridge, Hawai'i: relations of lava flow morphologies and underlying slopes. *Journal of Volcanology and Geothermal Research*, **126**, 63–77.

Griffiths, R. W. and Fink, J. H. (1992a). The morphology of lava flows in planetary environments: predictions from analog experiments. *Journal of Geophysical Research*, **97**, 19739–48.

Griffiths, R. W. and Fink, J. H. (1992b). Solidification and morphology of submarine lavas: a dependence on extrusion rate. *Journal of Geophysical Research*, **97**, 19729–37.

Griffiths, R. W. and Fink, J. H. (1997). Solidifying Bingham extrusions: a model for the growth of silicic lava domes. *Journal of Fluid Mechanics*, **347**, 13–36.

Grizzaffi, P. and Schultz, P. H. (1989). Isidis basin: site of ancient volatile-rich debris layer. *Icarus*, **77**, 358–81.

Guest, J. E., Kilburn, C. R. J., and Duncan, A. M. (1987). The evolution of lava flow-fields: observations of the 1981 and 1983 eruptions of Mount Etna, Sicily. *Bulletin of Volcanology*, **49**, 527–40.

Hanson, R. E. and Wilson, T. J. (1993). Large-scale rhyolite peperites (Jurassic), southern Chile. *Journal of Volcanology and Geothermal Research*, **54**, 247–64.

Head, J. W. and Wilson, L. (1986). Volcanic processes and landforms on Venus: theory, predictions and observations. *Journal of Geophysical Research*, **91**, 9407–46.

Head, J. W. and Wilson, L. (2002). Mars: a review and synthesis of general environments and geologcial settings of magma–H_2O interactions. In *Volcano–Ice Interaction on Earth and Mars*, ed. J. L. Smellie and M. G. Chapman. Geological Society of London Special Publication 202, pp. 27–58.

Head J. W., III, Hiesinger, H., Ivanov, M., and Kreslavsky, M. (1998). Ancient northern seas and oceans on Mars: analysis of evidence using MOLA data. *Geological Society of America Abstracts with Programs*, **30**, 402.

Head J. W., III, Hiesinger, H., Ivanov, M., Kreslavsky, M., Pratt, S., and Thomson, B. J. (1999). Northern seas and oceans in the past history of Mars: new evidence from Mars Orbiter Laser Altimeter (MOLA) data. *Geological Society of America Abstracts with Programs*, **31**, 132.

Hess, P. C. and Head, J. W. (1990). Derivation of primary magmas and melting of crustal materials on Venus: some preliminary petrogenetic considerations. *Earth Moon* & Planets, **50/51**, 57–80.

Hodges, C. A. and Moore, J. H. (1979). The subglacial birth of Olympus Mons and its aureoles. *Journal of Geophysical Research*, **84**, 8061–74.

Hon, K., Kauahikaua, J., Denlinger, R., and Mackay, K. (1994). Emplacement and inflation of pahoehoe sheet flows: observations and measurements of active lava flows on Kilauea Volcano, Hawaii. *Geological Society of America Bulletin*, **106**, 351–70.

Jerram, D. A. and Stollhofen, H. (2002). Lava–sediment interaction in desert settings: are all peperite-like textures the result of magma–water interaction? *Journal of Volcanology & Geothermal Research*, **114**, 231–49.

Kano, K. (1989). Interactions between andesitic magma and poorly consolidated sediments: examples in the Neogene Shirahama Group, South Izu, Japan. *Journal of Volcanology & Geothermal Research*, **37**, 39–75.

Kargel, J. S. and Strom, R. G. (1992). Ancient glaciation on Mars. *Geology*, **20**, 3–7.

Kargel, J. S., Kirk, R. L., Fegley, Jr. B. and Treiman, A. H. (1994). Carbonate-sulfate volcanism on Venus? *Icarus*, **112**, 219–52.

Kargel, J. S., Baker, V. R., Beget, J. E. *et al.* (1995). Evidence of ancient continental glaciation in the Martian northern plains. *Journal of Volcanology and Geothermal Research*, **100**, 5351–68.

Keszthelyi, L. P. and Denlinger, R. (1996). The initial cooling of pahoehoe flow lobes. *Bulletin of Volcanology*, **58**, 5–23.

Keszthelyi, L., McEwen, A. S., and Thordarson, T. (2000). Terrestrial analogs and thermal models for Martian flood lavas. *Journal of Geophysical Research*, **105**, 15027–49.

Kokelaar, B. P. (1982). Fluidization of wet sediments during the emplacement and cooling of various igneous bodies. *Journal of the Geological Society of London*, **139**, 21–33.

Kokelaar, B. P. (1986). Magma–water interaction in subaqueous and emergent basaltic volcanism. *Bulletin of Volcanology*, **48**, 275–89.

Lavine, A. and Aalto, K. R. (2002). Morphology of a crater-filling lava lake margin, the Peninsula tuff cone, Tule Lake National Wildlife Refuge, California: implications for formation of peperite textures. *Journal of Volcanology & Geothermal Research*, **114**, 147–63.

Lescinsky, D. T. and Fink, J. H. (1996). Lava and ice interaction: controls on lava flow morphology and texture. In *Glaciers, Ice Sheets and Volcanoes: A Tribute to Mark F. Meier*, ed. S. C. Colbeck. *Cold Regions Research and Engineering Laboratory Special Report*, **96** (27); 81–8.

Lescinsky, D. T. and Sisson, T. W. (1998). Ridge-forming lava flows at Mount Rainier, Washington. *Geology*, **26**, 351–4.

Longhi, J., Knittle, E., Holloway, J. R., and Wanke, H. (1992). The bulk composition, mineralogy and internal structure of Mars. In *Mars*, ed. H. H. Kieffer, B. M. Jakosky, C. W. Snyder, and M. S. Matthews. Tucson: University of Arizona Press, pp. 184–208.

Malin, M. C. and Edgett, K. S. (1999). MGS MOC the first year: geomorphic processes and landforms. *Abstracts of Papers Submitted to the 30th Lunar and Planetary Science Conference*. Houston: Lunar and Planetary Institute, CD 30, Abstract *1028*.

Malin, M. C. and Edgett, K. S. (2001). Mars Global Surveyor Mars Orbiter Camera: interplanetary cruise through primary mission. *Journal of Geophysical Research*, **106**, 23429–570.

Marsh, B. D. (1981). On the crystallinity, probability of occurrence, and rheology of lava and magma, *Contributions to Mineralogy and Petrology*, **78**, 85–98.

McSween, H. Y. (1985). SNC meteorites: clues to Martian petrologic evolution? *Reviews of Geophysics*, **23**, 391–416.

McSween, H. Y., Jr. (1995). Mars, on the rocks with a squirt of seltzer. *Meteoritics*, **3**, 241–2.

McSween, H. Y., Jr. and Keil, K. (2000). Mixing relationships in the Martian regolith and the composition of globally homogeneous dust. *Geochimica et Cosmochimica Acta*, **64**, 2155–66.

McSween, H. Y., Jr. and Treiman, A. H. (1998). Martian meteorites, in Papike. *Reviews in Mineralog*, **36**, 6.1–6.53.

Moore, H. J. (1987). Preliminary estimates of the rheological properties of 1984 Mauna Loa lava. In *Volcanism in Hawaii*, ed. R. W. Decker, T. L. Wright, and P. H. Stauffer. US Geological. Survey Professional Paper 1350, pp. 1569–88.

Moore, H. J., Arthur, D. W. G., and Schaber, G. G. (1978). Yield strengths of flows on the Earth, Mars and Moon. *Proceedings of the 9th Lunar and Planetary Science Conference*. Houston: Lunar and Planetary Institute, pp. 3351–78.

Mustard, J. F., Cooper, C. D., and Rifkin, M. K. (2001). Evidence for recent climate change on Mars from the identification of youthful near-surface ground ice. *Nature*, **412**, 411–14.

Parker, T. J. and Currey, D. R. (2001). Extraterrestrial coastal geomorphology. *Geomorphology*, **37**, 303–28.

Parker, T. J., Gorsline, D. S., Saunders, R. S., Pieri, D. C., and Schneeberger, D. M. (1993). Coastal geomorphology of the Martian northern plains. *Journal of Geophysical Research*, **98**, 11061–78.

Pinkerton, H. and Sparks, R. S. J. (1978). The 1975 sub-terminal lavas, Mount Etna: a case history of the formation of a compound lava field. *Journal of Volcanology and Geothermal Research*, **1**, 167–82.

Pinkerton, H. and Stevenson, R. J. (1992). Methods of determining the rheological properties of magmas at sub-liquidus temperatures. *Journal of Volcanology & Geothermal Research*, **53**, 47–66.

Porter, T. K. and Schultz, P. H. (1990). Formation of rhyolitic ridges on Martian basalts. *Abstracts of Papers Submitted to the 21st Lunar and Planetary Science Conference*. Houston: Lunar and Planetary Institute, pp. 973–4.

Sakimoto, S. E. H., Crisp, J. A., and Baloga, S. M. (1997). Eruption constraints on tube-fed planetary lava flows. *Journal of Geophysical Research*, **102**, 6597–613.

Samspon, D. E. (1987). Textural heterogeneities and vent area structures in the 600-year-old lavas of the Inyo volcanic chain, eastern California. In *The Emplacement of Silicic Domes and Lava Flows*, ed. J. H. Fink. Geological Society of America Special Paper 212, pp. 89–102.

Sansone, F. (1990). *Lava Meets the Sea*. Honolulu: LavaVideo Productions, 27 minutes.

Schaber, G. G. (1980). Radar, visual and thermal characteristics of Mars: rough planar surfaces. *Icarus*, **42**, 159–84.

Schminke, H.-U. (1967). Fused tuff and peperites in south-central Washington. *Bulletin of the Geological Society of America*, **78**, 319–30.

Schultz, P. H. and Lutz, A. B. (1988). Polar wandering of Mars. *Icarus*, **73**, 91–141.

Scott, D. H. and Tanaka, K. L. (1982). Ignimbrites of Amazonis Planitia region of Mars. *Journal of Geophysical Research*, **87**, 1179–90.

Shen, Y., Forsythe, D. W., and Dorman, L. M. (1997). Investigation of microearthquake activity following an intraplate teleseismic swarm on the west flank of the Southern East pacific Rise. *Journal of Geophysical Research*, **102**, 459–75.

Shockey, K. M., Zimbelman, J. R., Friedmann, S. J., and Irwin, R. P. (2004). Geologic mapping of the Medusae Fossae Formation on Mars. *US Geological Survey Open-File Report*, OF 2004–1289.

Skilling, I., White, J. D. L., and McPhie, J. (2002). Peperite: a review of magma-sediment mingling. *Journal of Volcanology and Geothermal Research*, **114**, 1–17.

Tepley, L. and Moore, J. G. (1978). *Fire Under the Sea: The Origin of Pillow Lava*. Mountain View, CA: Moonlight Productions, 20 minutes.

Theilig, E. and Greeley, R. (1986). Lava flows on Mars: analysis of small surface features and comparisons with terrestrial analogs. *Journal of Geophysical Research*, **90**, E193–206.

Thomas, P., Squyres, S. W., Herdenhoff, K., Howard, A., and Murray, B. (1992). Polar deposits of Mars. In *Mars*, ed. H. H. Kieffer, B. M. Jakosky, C. W. Snyder, and M. S. Matthews. Tucson: University of Arizona Press, pp. 767–98.

Walker, B. H. and Francis, E. H. (1987). High-level emplacement of an olivine-dolerite sill into Namurain sediments near Cardenden, Fife. *Transactions of the Royal Society of Edinburgh: Earth Science*, **77**, 295–307.

Warner, N. H. and Gregg, T. K. P. (2003). Evolved lavas on Mars? Observations from Southwest Arsia Mons and Sabancaya Volcano, Peru. *Journal of Geophysical Research*, doi:10.129/2002JE001969.

White, J. D. L. and Busby-Spera, C. J. (1987). Deep marine arc apron deposits and syndepositional magmatism in the Alisitos Group at Punta Cono, Baja California, Mexico. *Sedimentology*, **34**, 911–27.

Wilson, L. and Head, J. W. (1981). Ascent and eruption of basaltic magma on the Earth and Moon. *Journal of Geophysical Research*, **86**, 2987–3001.

Wilson, L. and Head, J. W. (2002). Tharsis-radial graben systems as the surface manifestation of plume-related dike intrusion complexes: models and implications. *Journal of Geophysical Research (Planets)*, **107** (E08), doi: 10.1029/2001JE001593.

Wohletz, K. (2002). Water/magma interaction: some theory and experiments on peperite formation. In *Volcano–Ice Interaction on Earth and Mars*, ed. J. L. Smellie and M. G. Chapman. Geological Society of London Special Publication 202, pp. 19–35.

Zimbelman, J. R., Johnston, A. K., and Lovett, C. G. (1996). Geologic map of the Medusae Fossae formation within MTM Quadrangle 05142 on Mars. *Geological Society of America Abstracts with Programs*, **28** (7), A–128.

Zimbelman, J. R., Crown, D. A., Grant, J. A., and Hooper,D. M. (1997). The Medusae Fossae Formation, Amazonis Planitia, Mars: evaluation of proposed hypotheses of origin. *Abstracts of Papers Submitted to the 28th Lunar and Planetary Science Conference*. Houston: Lunar and Planetary Institute, pp. 1623–4.

Zimanowski, B. and Buttner, R. (2002). Dynamic mingling of magma and liquified sediments. In *Volcano–Ice Interaction on Earth and Mars*, ed. J. L. Smellie and M. G. Chapman. Geological Society of London Special Publication 202, pp. 37–43.

9

Eolian dunes and deposits in the western United States as analogs to wind-related features on Mars

James R. Zimbelman and Steven H. Williams
National Air and Space Museum, Smithsonian Institution, Washington

9.1 Introduction

Eolian processes produce distinctive features and deposits on planetary surfaces where the atmosphere is sufficiently dense to allow interactions between the wind and sediments on the surface (Greeley and Iversen, 1985). Arid and semi-arid regions on Earth contain abundant evidence of wind–surface interactions (e.g., Lancaster, 1995a; Thomas, 1997), and the Martian surface shows a diverse array of eolian features across the planet (e.g., Greeley *et al.*, 1992). The characteristics of several eolian localities (primarily sand dunes) in the western part of the United States have been used previously as analogs to features seen on Mars in data obtained from several spacecraft (e.g., Greeley *et al.*, 1978; Greeley and Iversen, 1987; Golombek *et al.*, 1995), yet the analog potential of other western eolian sites is relatively underutilized. Rather than attempting a comprehensive survey of all eolian features in the United States, this chapter will focus on several examples illustrative of a variety of dune forms and their potential applicability as analogs to eolian features observed on Mars. Dunes in the Great Plains, east of the Rocky Mountains, and all coastal dunes are excluded from this survey in order to concentrate on discrete sand accumulations in arid or semi-arid environments. Both traditional publications and selected internet sites (cited here as W#) are referenced throughout the text.

Eolian features in the western United States reflect varying climatic and drainage conditions that have directly contributed to the formation of the individual deposits. Cold conditions during the last glacial period gave way to warmer and more arid conditions during the ensuing interglacial

The Geology of Mars: Evidence from Earth-based Analogs, ed. Mary Chapman. Published by Cambridge University Press.

(early Holocene) period (e.g., Spaulding, 1990; Hamblin and Christiansen, 1998), which in turn contributed directly to the development of numerous isolated sand deposits in North America (Smith, 1982; Tchakerian, 1997, W1). Both rivers and glacial pluvial lakes represent sources of sand-sized sediment that can be mobilized by the wind (see Kocurek and Lancaster, 1999); similar sand sources may exist on Mars (see Chapters 11, 12, and 13). Arid to semi-arid conditions in the western United States at present should not lead to an underestimation of the crucial role of water in the development of eolian deposits on Earth, and a similar caution is appropriate when analyzing eolian features on Mars. The current emphasis by NASA to "follow the water" in the design of the Mars Exploration Program and its associated spacecraft missions (W2) makes it particularly important to recognize the strong link between fluvial and lacustrine processes in generating the sand-sized sediments that subsequently are redistributed by the wind across the Martian surface.

Twenty-two sites throughout the western United States that possess significant eolian deposits were selected for discussion here, based on the abundance of eolian sediments, ease of access, and potential relevance for analog studies. Where applicable, nearby eolian deposits are mentioned along with the listed primary location. The sites are next discussed within the context of a simplified category basis, relating them to probable sediment sources and the conditions that may have contributed to their present location. The selected sites contain common types of sand deposits and associated dune forms, wind streaks, wind-eroded features, and large eolian ripples; this list is representative but by no means comprehensive. A review of the current view of eolian features on Mars emphasizes the new insights gained through the on-going Mars Global Surveyor (MGS) and Mars Odyssey spacecraft missions, including current working hypotheses derived from the new data. The similarities and differences between conditions on Earth and Mars as they relate to eolian features, along with prospects for study of eolian features from missions soon to be launched to Mars are presented in Section 9.5.

9.2 Selected eolian sediment locations

Twenty-two sites in the western United States (Figure 9.1) are briefly described below, chosen to illustrate the major types of dunes and associated eolian deposits. The selected sites are by no means the only eolian sediments present in the western United States, but they do represent the diversity of landforms developed by the interaction of the wind with sand-sized particles.

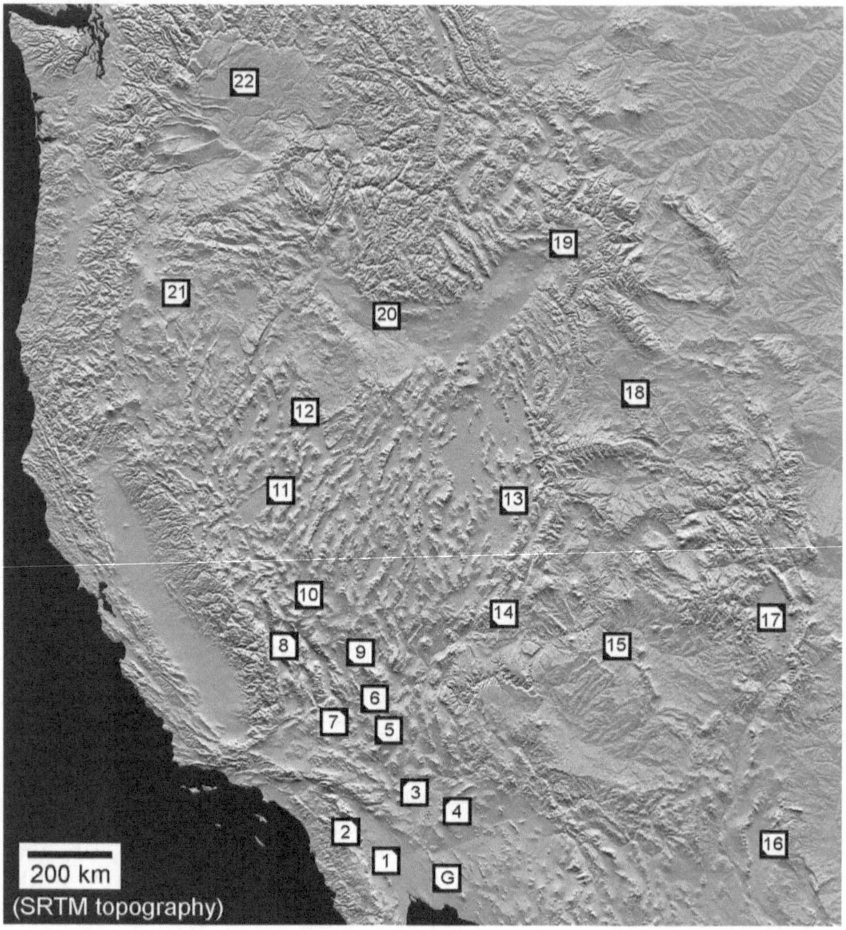

Figure 9.1. Map of eolian dune localities in the western United States discussed in this chapter. Numbers indicate the location of selected eolian sites discussed in the text, and listed in Table 12.1. G indicates the Gran Desierto in northern Mexico. Shaded relief base map uses data from the Shuttle Radar Topography Mission (SRTM) collected in February 2000, scaled to ~1 km/pixel horizontal resolution, obtained from one channel of an anaglyph product (W34). Area shown corresponds approximately to 32° to 48.5° N latitude, 103° to 125° W longitude, Mercator projection.

Smith (1982) provides an extensive treatment of sand dune locations throughout western North America (see also W1), and both McKee (1979) and Breed *et al.* (1979) present a global perspective of sand deposits derived from aircraft and spacecraft imaging. The reader interested in more detailed information about terrestrial sand deposits than what is provided below is encouraged to consult the references cited, as well as several excellent textbooks about eolian processes (e.g., Cooke and Warren, 1973; Greeley and Iversen, 1985; Pye and Tsoar, 1990; Lancaster, 1995a; Tchakerian, 1995; Thomas, 1997).

Use of off-highway vehicles (OHVs) at a site can be either a benefit (increased range of access) or a detriment (strong modification of original morphology), depending on the goals of the particular study. OHV accessibility is described below for many sites to aid the reader in identifying those sites where either the presence or absence of OHVs might prove critical to the potential investigation being envisioned. Locations within national or state parks can limit OHV access, but sometimes this does not include all of the deposits at a site. Websites maintained by the Bureau of Land Management (BLM) and other government-sponsored agencies usually include specific information about OHV use and restrictions.

The sites are arranged by the general physiographic providence in which they occur: the Mojave Desert, the Great Basin, the High Deserts, and locations in the Northwest. Each site description includes information on the general setting of the deposits, the principal dune type present, restrictions on use of the site, and an assessment with respect to four categories related to deposit emplacement conditions and sediment source (Table 9.1). Note that the existing literature is quite disparate among the sites discussed, with some having an extensive publication record while others have almost no citations. The following summaries are not intended to be exhaustive treatments of the eolian literature, but the references cited should point the reader to important publications from which a more detailed investigation can be carried out.

9.2.1 Mojave Desert

Algodones Dunes. The Algodones Dunes cover ~2600 km^2, making them one of the largest dune complexes in North America (Smith, 1978; W3). The Algodones Dunes primarily consist of barchanoid ridges (broad linear to sinuous sand accumulations) up to 90 m high with superposed barchan (crescentic, horns pointing downwind) undulations, both of which indicate that the main sand-driving winds here are to the southeast (Norris and Norris, 1961; McCoy *et al.*, 1967; Smith, 1978; Sharp, 1979; Kocurek and Nielson, 1986; Havholm and Kocurek, 1988; Sweet *et al.*, 1988). The dunes occur on the eastern margin of pluvial Lake Cahuilla, but detailed sedimentological studies indicate that the sand was derived from Colorado River sediments deposited in the basin that contained the lake rather than

Table 9.1. *Selected western United States eolian sites*

				Category of eolian feature			
		Name	Dune type	Topo-graphic	Pathway	Lake-related	River-related
Mojave Desert	1	Algodones	BR			X	
	2	Salton Sea	B			X	
	3	Bristol/Palen	–		X		
	4	Cactus Plain	TR				X
	5	Cady Mtns./Kelso	–	(x)	X		X
Great Basin	6	Dumont	TR				(x)
	7	Ibex	C	(x)			(x)
	8	Eureka	C	X		(x)	
	9	Big Dune	S			X	
	10	Clayton Valley	TR			X	
	11	Sand Mountain	TR	(x)	X	X	(x)
	12	Winnemucca	P		X	X	
	13	Little Sahara	TR			X	
	14	Coral Pink	BR	(x)			
High Deserts	15	Navajo	L	(x)			
	16	White Sands	BR			(x)	
	17	Great Sand Dunes	C	X			
	18	Killpecker	BR				
Northwest	19	St. Anthony	P		X		
	20	Bruneau	C	X			X
	21	Christmas Valley	P			(x)	
	22	Moses Lake	P				

B = barchan; BR = barchanoid ridge; C = complex; L = longitudinal; P = parabolic; S = star; TR = transverse ridge; X = significant aspect of site; (x) = secondary aspect of site.

from the surrounding mountains (Muhs *et al.*, 1995). North of California Highway 78, which crosses the dune complex, much of the dunes are a wilderness area with restricted access, but south of the highway and the very northern part of the dunes are designated as the Imperial Sand Dunes Recreational Area and are extensively used by OHVs.

Immediately south-southeast of the Algodones Dunes is the Gran Desierto (G in Figure 9.1), where ~5700 km^2 of northern Mexico is covered by sand as part of the Sonoran desert (Lancaster *et al.*, 1987; Lancaster and Blount, 1990; Lancaster, 1995a, b). The sand has accumulated into large (80 to >100 m relief) star dunes (three or more arms branching from a central deposit, formed under multiple wind regimes), crescentic dunes 5 to 20 m high that merge into complex crescentic mounds 10 to 80 m high (analogous to the hybrid forms present in the Algodones), as well as both linear and parabolic dunes and sand sheets along the margins (Lancaster, 1992, 1995b). Similar to the Algodones Dunes, sand in the Gran Desierto was transported to the region by the Colorado River where arid conditions allowed remobilization by the wind of the sand-sized sediments (Lancaster and Blount, 1990). OHVs are essential to access this broad sand deposit where only very limited logistical support is available in close proximity to the dunes.

Salton Sea Barchan Dunes. Barchans are crescent-shaped dunes whose horns point away from the primary wind direction (the opposite of the stabilized horns of parabolic dunes). Barchans form where the sand supply is less abundant than in other dune fields (McKee, 1979). The classic location in the United States for barchans is on the western side of the Salton Sea (Rempel, 1936; Norris, 1966; van de Kamp, 1973; Theilig *et al.*, 1978; Abbott, 1980). The barchans have been used to monitor dune migration over many decades (see Haff and Presti, 1995), confirming the inverse relationship noted between barchan slip face height and dune advance rate (Bagnold, 1941; Long and Sharp, 1964). The sand source is thought to be Lake Cahuilla sediments, similar to the nearby Algodones Dunes (site #1). Also nearby are longitudinal dunes that cover the southern portion of the Superstition Mountains, spawning barchans that migrate east toward the Salton Sea (Smith, 1982). The Salton Sea barchans are readily accessible from nearby highways but OHV use is restricted.

Bristol Trough/Palen. Sand accumulations in the southeastern Mojave Desert have morphologic indicators of eolian transport of sand across low drainage divides, resulting in two roughly parallel transport pathways through the Bristol Trough and through the Palen playa (Figure 9.2; Zimbelman *et al.*, 1995; W4). Sand deposits along these paths consist of a combination of sand sheets, sand ramps (see Lancaster and Tchakerian, 1996), and stabilized transverse ridges. Sediment studies show the sands are immature and geochemically distinct from the mature Colorado River sands (Zimbelman and Williams, 2002), with indications of varied local sediment sources (Pease and Tchakerian, 2003). OHV use is restricted throughout this portion of the eastern Mojave Desert, and local inquiries are required.

Cactus/LaPosa Plains. Stabilized transverse sand dunes of 1 to 3 m height and 100 to 200 m wavelength (Figure 9.3) cover the Cactus and LaPosa Plains east of the Colorado River (W5). The sand in the dunes has strong chemical affinities to sand from the Colorado River, unlike the Mojave sands west of the river (Muhs, 2002; Zimbelman and Williams, 2002).

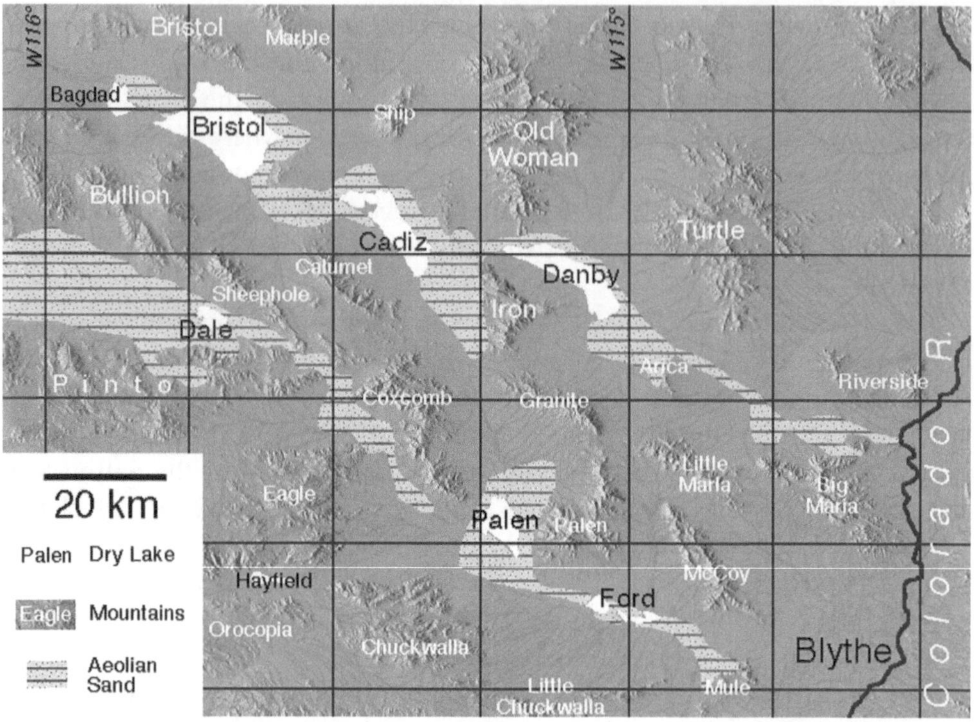

Figure 9.2. The Bristol trough and Palen transport paths. Lined pattern shows sand deposits that the wind has blown across several low topographic divides (modified from figure 9.1 of Zimbelman and Williams, 2002).

OHVs are permitted at some locations around the dunes; enquire locally for permission.

Cady Mountains/Kelso Dunes. Sand from the Mojave River has blown across portions of the Cady Mountains and forms thick ramps against the mountains on their western (windward) side (Lancaster and Tchakerian, 1996). Mojave River sand also exits out of Silver Lake playa through the Devil's Playground and into the 170 m tall transverse ridges of the Kelso Dunes (Sharp, 1966, 1978; Smith, 1982, 1984; Lancaster, 1993; W6; W7). Heavy mineral concentrations within the Kelso sands (MacDonald, 1970; Earl, 1981) led to commercial mining of the dunes for a short time. A variety of sand sheet deposits and stabilized transverse dunes populate these pathways, similar to the Bristol and Palen pathways (#3) discussed above. Thermal infrared remote sensing has proved to be a valuable tool for mapping and documenting the sand sources and transport pathways leading to the Kelso Dunes (Ramsey *et al.*, 1999). Restriction of both cattle grazing and OHV use has led to extensive stabilization of the dunes.

Figure 9.3. Oblique aerial photograph of stabilized transverse dunes on Cactus Plain near Parker, Arizona. The dunes are 2 to 3 m high with 100 to 200 m spacing. JRZ, 2/99.

The Devil's Playground and the Kelso Dunes are within the recently established Mojave National Preserve, restricting all OHV and cattle grazing activity in these areas, resulting in significant vegetation growth and dune stability over the last decade.

9.2.2 Great Basin

Dumont Dunes. A core of star and complex dunes up to 120 m tall occurs along a broad transverse ridge of sand, surrounded by smaller transverse and barchan dunes (Smith, 1982). Nielson and Kocurek (1987) provide an excellent overview of these dunes, including evidence supporting the possible origin of the dunes being concentrated at this location. This area is a major OHV site for southern California (W8), so very little detailed surface morphology remains undisturbed. Dune migration studies have been carried out, however, with recent aerial photography (W9).

Ibex Dunes. A small field of star dunes is located in the southeastern corner of Death Valley National Park (Figure 9.4). The highest of the dunes is ~50 m tall and the field extends ~3 km along the southwestern slope of the Saddle Peak Hills (Smith, 1982). Located within a Wilderness section of the park, no vehicular traffic is allowed within ~1.5 km of the dunes, making them among the most pristine but still accessible sand dunes in the United States.

Figure 9.4. Oblique view of a star dune in the Ibex Dunes, California. The star dune is ~30 m tall. The dark patches at the base of the dune consist of dark pebbles and granules derived from the nearby mountains, some of which have accumulated into large ripples of 2 to 8 m wavelength, over a sand substrate, possibly analogous to some ripple-like features on Mars (see Figure 9.8). JRZ, 2/03. (For a color vision of this figure, please refer to color plate section.)

Eureka Dunes. This complex dune mass includes small star dunes around a broad ridge of sand, the highest portion of which has 208 m of relief, making this the tallest dune complex in California (Smith, 1982). A plant found only on these dunes (Pavlik, 1980) makes the Eureka Dunes one of the most protected areas within Death Valley National Park, so that not only are OHVs restricted, but even walking on the dunes is only allowed along a designated trail (W10). Although not part of the Eureka Dunes, in nearby Panamint Valley are two star dune complexes, 50 to 60 m tall, surrounded by low transverse ridges, and the Stovepipe Wells star dunes rise to ~40 m above the floor of the center of Death Valley (Smith, 1982).

Big Dune. Big Dune is located in the Amargosa Desert of southwestern Nevada, and is a popular OHV site. Star dunes, with relief up to 80 m, are surrounded transverse ridges over an area of ~13 km^2, derived from nearby playa deposits (Swadley and Carr, 1987). The dune complex is misnamed in that total relief considerably less than the relief at Sand Mountain (#11), Eureka (#8), and Kelso (#5), but it is a locality where sand avalanches have been observed to emit audible acoustic energy (Trexler and Melhorn, 1986).

Clayton Valley. Also known as the Silver Peak dunes, sand is collected into a series of transverse dunes in the southern part of Clayton Valley, about 7 miles south of the Nevada town of Silver Peak (Trexler and Melhorn, 1986). These dunes were the site of a recent field study of sand grain deposition in the lee of transverse dunes (Nickling *et al.*, 2002). The dune sand is of local origin, derived from deflation of playa and alluvial fan deposits that cover ~310 km^2 of Clayton Valley, comparable to the origin of the nearby Crescent Dunes, located northwest of Tonopah, Nevada (Trexler and Melhorn, 1986).

Sand Mountain. The largest dune in Nevada, with ~120 m of relief, is Sand Mountain (W11), consisting of a complex pattern of transverse ridges superposed on a terminal seif dune (Trexler and Melhorn, 1986), surrounded by transverse dunes that extend into the nearby hills (Figure 9.7). The sand at Sand Mountain has the distinctive properties that allow sand avalanches to produce audible sounds, and it has been the site of several investigations of this phenomenon (Criswell *et al.*, 1975; Lindsay *et al.*, 1976; Maloney, 1982; Trexler and Melhorn, 1986). Sand Mountain sediments are ultimately derived from materials deposited within nearby glacial Lake Lahontan (see Morrison, 1964, 1991; Benson *et al.*, 1990), similar to the origin of several other Nevada

Figure 9.5. Great Sand Dunes National Monument, Colorado. The dune mass, more than 200 m tall, is trapped against the western margin of the lofty Sangre De Christo Mountains. JRZ,9/02.

dune sites like Big Dune (#9) and Clayton Valley (#10). Sand Mountain is a major OHV destination, so it is often difficult to find pristine sand surfaces, and increasingly difficult to find undisturbed slip faces for generating audible sounds.

Winnemucca. Dunes cover an area of ~900 km^2 north of Winnemucca, Nevada, as first described by Russell (1885). Stabilized parabolic dunes are the most common landform, but isolated barchans and transverse ridges occur scattered throughout the dune field. The dune morphologies and locations indicate sand transport has occurred to the east (Smith, 1975), crossing two mountain ranges that separate Paradise Valley, Silver State Valley, and Desert Valley; this leads to the dune field also being known as the Silver State Dune Complex (Eissmann, 1990; Epps *et al.*, 2000). The sand source ultimately was Lake Lahontan sediments (Morrison, 1964, 1991; Benson *et al.*, 1990; Epps *et al.*, 2000), as at Sand Mountain (#11). OHVs are allowed on portions of the dune field; enquire locally for permission.

Little Sahara Dunes. The largest field of active dunes in Utah is now a major OHV site (W12). Transverse ridges 100 to 200 m high are surmounted by active transverse dunes 5 to 15 m tall, all surrounded by stabilized parabolic dunes (Smith, 1982). Most of the sand comes from deposits left by the Sevier River, which used to flow into glacial Lake Bonneville (W13). A link was established between the beetle *Eusattus muricatus* found on the Little Sahara dunes and those found on dunes in Silver State Valley (Winnemucca, #12), many hundreds of kilometers to the west (Epps *et al.*, 2000) (see 12 and 13 in Figure 9.1).

Coral Pink Sand Dunes. Sand grains eroded from nearby Navajo sandstone outcrops provide the distinctive color of the barchanoid ridge dunes within the ~15 km^2 of this Utah state park (Ford and Gilman, 2000; W14; W15). Much of the dune field is still active, although some portions of the field have become fully stabilized by vegetation. OHVs are allowed on only selected portions of the dunes.

9.2.3 High Desert

Navajo Dunes. Longitudinal dunes (aligned parallel to the major wind direction) 2 to 9 m high and ~5 km long dominate a large portion of northern Arizona, along with isolated barchans, transverse dunes, and climbing and falling dunes (banked against bedrock obstacles) (Hack, 1941; Smith, 1982; Breed *et al.*, 1984; Billingsley, 1987; Stokes and Breed, 1993). The dune field includes the largest collection of longitudinal dunes in the western

United States, but any research effort in the area must be coordinated through the Navajo nation.

White Sands. Barchanoid ridges of very white sand make up a compositionally unique dune field in New Mexico. The whiteness of the sand results from being composed of gypsum (deflated from nearby Lake Lucero, but initially leached from rocks in the surrounding mountains), which accumulates through eolian processes into barchan-like dune forms (Murbarger, 1950; McKee, 1966; McKee and Douglass, 1971; McKee and Moiola, 1975; W16). Interaction of the gypsum sand grains with water results in some very unique modifications of the dune sedimentary structures (Simpson and Loope, 1985; Schenk and Fryberger, 1988). White Sands is now protected as a National Monument (W17), but the dunes are easily accessible by regular automobiles.

Great Sand Dunes. National Monument status also protects this large mass of sand trapped against the western margin of the Sangre De Christo Mountains of central Colorado (Figure 9.5; Wegemann, 1939; Johnson, 1967, 1968; Burford and Hutchinson, 1968; Freyberger *et al.*, 1979; Andrews, 1981; W18). The dunes are the tallest of the selected sites discussed here (215 m of relief), as well as at the highest elevation in the United States

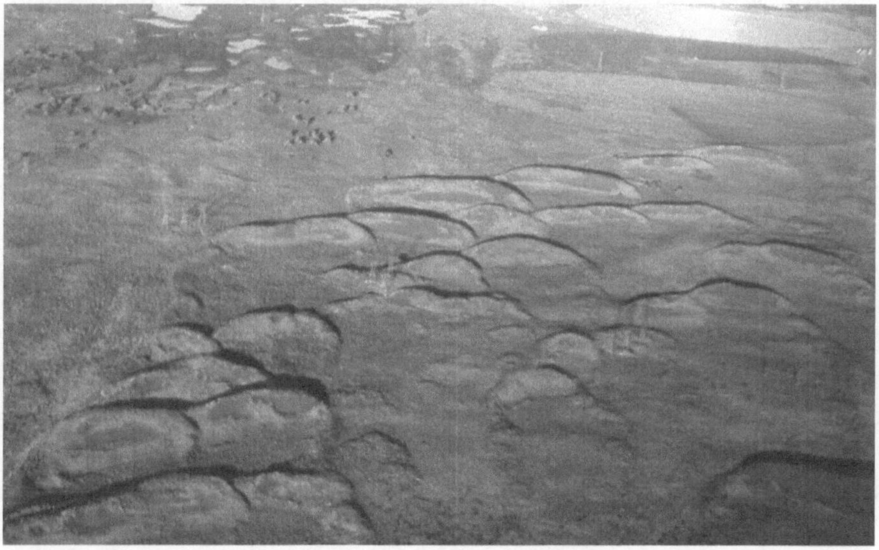

Figure 9.6. Oblique aerial photograph of stabilized parabolic dunes near Moses Lake, Washington. The prevailing wind is from the lower left; parabolic dunes have horns stabilized by vegetation (a condition highly unlikely on Mars!) that point into the driving wind direction. Individual dunes are ~5 m tall and ~30 to 60 m wide. The dune sand has a large basaltic component. JRZ, 9/98.

(2400 m; USGS, 1982). Temporal changes in Landsat images of the dunes document movement of sand but reveal no systematic migration trends (Janke, 2002). However, satellite observations do document extensive dust plumes originating from Great Sand Dunes (figure 9.3 of Edgett, 2002).

Killpecker Dunes. The largest and most active dune field in Wyoming covers ~274 km^2 and extends over a length of ~100 km (Smith, 1982; W19). The dunes consist of barchanoid ridges, with some transverse ridges up to 45 m tall and parabolic dunes up to 15 m tall (Ahlbrandt, 1973, 1974a, b, 1975). The sand was derived from the Eocene Green River Formation instead of local glacio-fluvial deposits (Ahlbrandt, 1973, 1974a) and its mineralogical maturity is quite low, only slightly more mature than the sands in the Bristol pathway (#3) (Muhs, 2002). OHVs are allowed on the dunes not within designated Wilderness areas.

9.2.4 Northwest

St. Anthony Dunes. Both climbing and falling parabolic dunes cover much of the Juniper Buttes near the town of St. Anthony in eastern Idaho (Greeley, 1977; W20). The sand is most likely derived from local sediments washed into nearby Mud Lake and subsequently blown northeastward to form the dunes (Dort, 1959). A modern migration rate of 3 m per year has been documented for these dunes (Chadwick and Dalke, 1965). OHVs are allowed on the dunes.

Bruneau Dunes. The Bruneau Dunes have collected in an abandoned cut-off meander of the Snake River, carved into the basalts of the Snake River Plains, and they include the tallest single-structured sand dune (> 140 m relief) in North America (Murphy, 1973, 1975; W21). Basaltic (iron-rich) particles comprise 12% (by volume) of the sand grains in the Bruneau Dunes, likely derived from the surrounding basalts (Murphy, 1973). The Bruneau Dunes were proposed as analogs to features on Mars shortly after Mariner 9 first provided confirmation of sand dunes on Mars (Murphy and Greeley, 1972). OHVs are not allowed on the dunes, located within a state park preserve.

Christmas Valley. Southern central Oregon was the site of several lakes during the last glaciation; the demise of Lake Chewaucan led to sand accumulations at Christmas Valley (W22), while other isolated lakes led to sand-sized sediments in (now dry) Alkali Lake and Alvord Valley (Orr and Orr, 2000, p.88). The dunes are primarily parabolic, possessing long horns stabilized by vegetation that connect to an active crescentic main body,

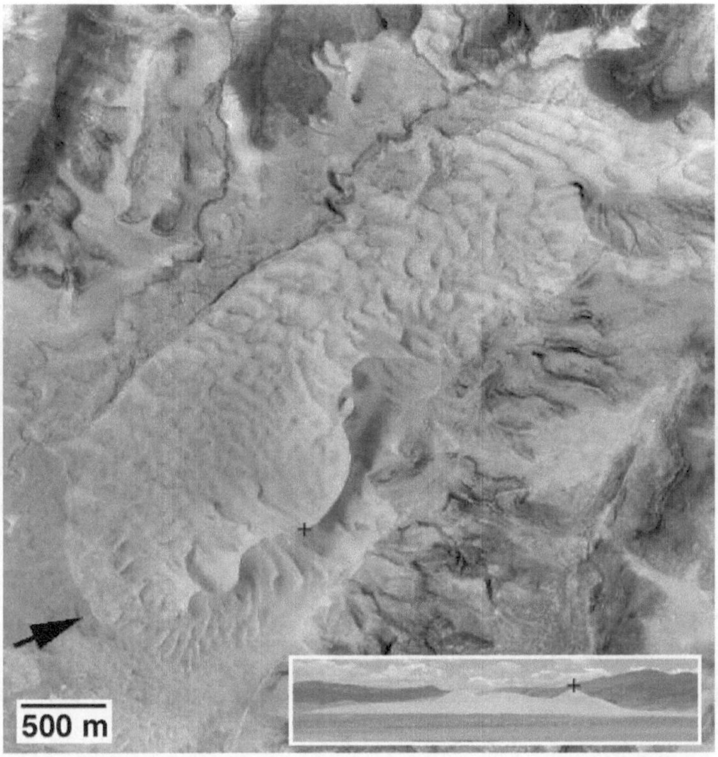

Figure 9.7. Sand Mountain, Nevada. A sinuous transverse ridge contains the highest portions of the dune complex, with transverse ridge dunes extending into the surrounding mountains. Inset shows a horizontal view, looking in the direction of the black arrow; "+" marks the same high point in both views. Portion of digital orthophoto mosaic of the Four Mile quadrangle. Inset, JRZ, 4/03.

and the sand is essentially basaltic in composition (Edgett, 1994). OHVs are permitted on these dunes.

Moses Lake Dunes. A broad field of stabilized parabolic dunes (Figure 9.6) is found southwest of the Moses Lake (Washington) city limits (W23), portions of which are flooded by water from the Potholes Reservoir. The dune sand consists of significant amounts of basalt and granodiorite fragments, as well as clay weathering products; all of these components can be detected remotely using hyperspectral mapping techniques similar to those provided by recent spacecraft instruments (Bandfield *et al.*, 2002). Basalt and granodiorite outcrops within the Channeled Scablands, caused by massive floods from glacial Lake Missoula (e.g., Baker, 1973; Alt, 2001), are the likely source of the sand and not nearby Columbia River (Bandfield *et al.*, 2002). The Juniper Dunes Wilderness is a similar nearby collection of stabilized parabolic dunes, south of Moses Lake, but the Juniper Dunes lack the potential fluvial–eolian interactions of the Moses Lake dunes. OHVs are permitted on most of the Moses Lake dune field, although some dunes are on private land requiring local approval for access.

9.3 A process-based view of the deposits

Regional sand transport by the wind can be semi-independent of topography, unlike water transport which must at all times go downhill. In most places on Earth, the geomorphic effects of running water mask those of wind; however, in some desert regions, the interaction between eolian transport and topography can be clearly seen. Sand can be transported between basins over considerable distances, surmounting substantial topographic barriers without significant obstruction of sand movement (Zimbelman *et al.*, 1995). Pathways of active eolian sands result, and may extend many for many tens of kilometers. Sand will form ramps at topographic barriers along such a sand pathway (see Lancaster and Tchakerian, 1996); Table 9.1 includes two sites (#3 and #5) in such transport pathways.

Layering and/or soil horizons in those ramps provide dates and other clues to the changing climate prevailing while the ramps were created. For example, interbasin transport of sand from the Dale Dry Lake area of Pinto Valley eastward via Clarks Pass to Palen Dry Lake (site #3) has built a ramp incised by drainage from the mountains at the margin of the valley. Layers and soil horizons within the ramp can be correlated to Holocene variations in the level of Dale Lake (Tchakerian, 1991; Lancaster and Tchakerian, 1996). A similar sand ramp has been built by the wind at Soldier Pass along the Mojave River, part of the Devil's Playground regional sand transport path that feeds sediments into the Kelso Dunes (#5). Quarrying activities have revealed layering similar to that at Clarks Pass (Rendell and Sheffer, 1996). The nearby Bristol Trough (#3) contains a sand pathway connecting Bristol, Cadiz, Danby, and Rice Valleys (Figure 9.2; Zimbelman *et al.*, 1995).

Once sand ramps are fully developed, sand motion over the topographic barrier proceeds apace, at least in theory. The Bristol Trough and Palen sand paths (#3) in California are examples where geomorphic evidence clearly indicates that some interbasin transport of sand has occurred (Zimbelman *et al.*, 1995). Chemical studies of samples from along transport paths confirm the geomorphic observations (Zimbelman and Williams, 2002); however, detailed analysis also shows that local contribution of sediments can be important (Pease and Tchakerian, 2003).

The other end-member case is when sand transport between basins does not occur, probably due to some combination of weak or multi-directional winds, vigorous destruction of sand ramps by running water, and steep boundaries between basins. Wind may trap sand near such a topographic barrier, producing a trapped dune mass. Overland flow of water down the steep gradient of the topographic obstruction may assist the trapping process. Examples include Great Sand Dunes in Colorado (#17) and the Eureka Dunes in California (#8). In the case of the former, prevailing winds transport sediments from the Rio Grande drainage across the San Luis Valley to the front of the Sangre de Christo Mountains. The wind is funneled up steep Medano and Mosca Passes, but the gradient is too steep and stream action from those passes is too vigorous for sand ramps to form. Similarly for the latter, prevailing northwesterly winds trap eolian sediments in a pocket in the Last Chance Range, but a ramp has not formed, presumably because of a combination of wind regime and stream action.

The intermediate case is a topographic barrier that is sufficiently large to cause a significant obstruction to sand flow and the formation of a large dune mass, but not so large that all interbasin flow is prevented. An example is Sand Mountain (#11) east of Fallon, Nevada, where sand from pluvial Lake Lahontan has blown against the mountains east of Fallon. The sand transport path surmounts the Blow Sand Mountains to the southwest, and the Sand Mountain dune complex extends over the topographic trap, but much of the sand in motion resides at Sand Mountain.

Both interbasin sand transport and the formation of dune masses trapped by topography require an ultimate source of the sand supply. On Earth, that source is typically one or more of the following: a paleo-lake, a paleo-river, erosion of local sandstone outcrops, paleo-flood deposits, and volcanic deposits. Each terrestrial source type is potentially analogous to sand deposits on Mars.

Tectonic activity has produced many enclosed drainage basins throughout much of the western United States. During the last glacial epoch, many pluvial lakes formed in those basins. Some were small and occupied but a single basin; others filled many interconnected basins with very large bodies of water (e.g., lakes Bonneville and Lahontan). Rivers and streams feeding these lakes supplied large quantities of sand, and when the lakes dried up with the changing of the climate, that sand was then available for mobilization by the wind. The sand supply is not renewed unless the lake returns, and the subsequent transport of lacustrine sands diminishes with time as a consequence. The Algodones Dunes (#1) are composed of sand collected in Lake Cahuilla, a precursor to the modern Salton Sea. The barchan dunes on the west side of the Salton Sea (#2) similarly were derived from paleo-shoreline sediments, and the dunes at Sand Mountain (#11) and north of Winnemucca (#12) are formed from sands from pluvial Lake Lahontan.

Some lake-related dunefields were associated with temporary accumulations of runoff rather than a pluvial lake that might have persisted for many centuries. In many cases, wind deflates part of the playa surface to form lunette dunes, unusually high in silt content (parna) (44 to 61 wt.% clay fraction for dunes in Grass Valley, Nevada; Greeley and Williams, 1994). Not enough sand was delivered to the parent playa to generate a large dune or dunefield. The dunes of Alkali Lake, Christmas Valley (#21), Big Dune (#9), St. Anthony (#19), and Moses Lake (#22) may include examples of this type.

Rivers and streams can, in some cases, provide sufficient sands to build dunes, without a significant lake having been present. Examples include the sand of the Devil's Playground/Kelso Dunes and Cady Mountains pathways (#5), related to the Mojave River, and the Dumont (#6) and Ibex (#7) Dunes (California) are related to the Amargosa River. The sands of the Great Sand Dunes National Monument (#17) come from the Rio Grande and its tributaries, and the transverse dunes of the Cactus and La Posa Plains (#4) in western Arizona are built of sands derived from the Colorado River (Zimbelman and Williams, 2002). The large-scale floods that formed the Channeled Scablands of Washington State produced large sand deposits that are the source of some of the dunes in the Pacific Northwest (e.g., #22). In some cases, the massive floods eroded volcanic materials into the size range amenable to subsequent eolian transport into dunes and pathways.

In a few cases, the primary source of the sand is local erosion of sandstone outcrops or older alluvium; overland water flow plays only a minor role in the sand source. The beautifully colored sands of the Coral Pink Sand Dunes (#14) in southern Utah are derived from local erosion of the Navajo Sandstone deposits that ring the valley in which the dunes reside. The sands of the Killpecker Dunes (#18) in Wyoming are derived from the erosion of the Laney Member of the Eocene-age Green River Formation, which underlies the dune area (Smith, 1982). The dunes of the Cactus and La Posa Plains of Arizona (#4), the Algodones Dunes (#1) of California, and the Gran Desierto (G in Fig. 12.1) are composed of sand transported by the Colorado River; however, those sands may have resided in local alluvial deposits in between the time they were transported by the river and freed to form the present-day dunes.

9.4 A post-MGS perspective of eolian deposits on Mars

Eolian features were first clearly documented on Mars in images taken by the Mariner 9 spacecraft, and the Viking orbiters revealed details of the global distribution of sand and dust around the planet (see reviews in Greeley and Iversen, 1985; Greeley *et al.*, 1992). The instruments on the MGS spacecraft (see MGS at W2) are providing data that continue to produce revisions to many of the Viking-based concepts for Mars (see Chapter 1). In particular, the 2 to 6 m/pixel images from the Mars Orbiter Camera (MOC; W24) on MGS provide views of eolian features that are comparable to aerial photographs of Earth (Malin *et al.*, 1998; Edgett and Malin, 2000; Malin and Edgett, 2001; Edgett, 2002). The MOC images have not revealed "new"

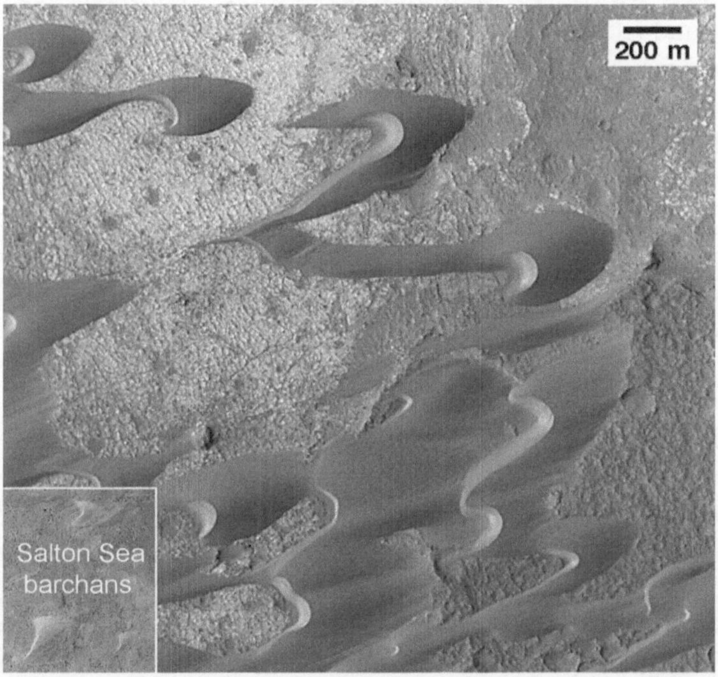

Figure 9.8. Large barchans and barchanoid ridges on the floor of the caldera of Nili Patera, Mars. Prevailing wind was from the upper right; barchans have horns that point away from the prevailing wind direction. Slip faces on the downwind side of individual dunes indicates the dunes are ~70 m tall. TES data indicate the dark sands are basaltic in composition. Portion of MOC image E03-02016, centered at 8.8° N, 292.9° W, 3.22 m/pixel, NASA/JPL/MSSS (W25). Inset (lower left) shows three of the Salton Sea barchans (2) at the same scale as the Mars barchans, rotated here 180° to match the orientation of the Martian features; scaled subset of a US Geological Survey Digital Orthophoto Mosaic.

eolian features as much as they have clarified the perception of specific details for Martian sand dunes and dust deposits, including instances of eolian deposition and erosion that are comparable in scale to features on Earth, a change from the Viking view (Edgett and Malin, 2000). Comparison of MOC and the best Viking and Mariner 9 images have yet to reveal any observable movement of dune-forms on Mars (Zimbelman, 2000; Malin and Edgett, 2001). Classification of Martian dunes is now becoming feasible, although barchans (Figure 9.8; W25) and transverse dunes are still by far the most abundant Martian dune forms (Malin and Edgett, 2001). Martian dune assemblages include an enormous erg (sand sea) surrounding the north polar cap, covering ~680 000 km^2 between 75° and 80° N (Tsoar *et al.*, 1979). Numerous sand deposits are trapped within large impact craters in the southern hemisphere, the largest deposits in the craters Kaiser, Proctor, and Rabe between 42° to 48° S and 320° to 345° W (Greeley *et al.*, 1992), potentially analogous to topographically trapped sand on Earth (Table 9.1).

MOC images are providing tests of hypotheses for the growth and development of dune fields trapped within craters like Proctor (e.g., Fenton *et al.*, 2002; Williams *et al.*, 2003).

While the thin Martian atmosphere can clearly set sand in motion when winds are sufficiently strong, once the sand falls within a topographic depression with steep walls (like a crater or valley), it is effectively trapped. Some material is blown away from the trapped sediments (Edgett, 2002), but MOC images have yet to reveal sufficient sand to form sand ramps or climbing dunes, as occur along terrestrial transport pathways. MOC images have revealed abundant eolian transverse ridges over much of the Martian surface, with wavelengths of tens to one hundred meters (Figure 9.9; W26; W27); these features could either be small dunes or large ripples, which form by different transport mechanisms (Malin *et al.*, 1998; Malin and Edgett, 2001; Zimbelman and Wilson, 2002). Large eolian ripples on Earth have wavelengths of several to tens of meters and heights of 0.3 to 2 m and always involve a bimodal particle distribution of sand and granules to pebbles (Williams *et al.*, 2002); efforts are underway to test whether the smallest Martian transverse ridges are more likely ripples or dunes (Wilson *et al.*, 2003).

MGS includes non-imaging instruments that also have greatly altered the Viking view of Mars, most notably its topography and the composition of its surface materials. The Mars Orbiter Laser Altimeter (MOLA; W28) has measured topography to a vertical precision of better than 1 m over spatial scales of hundreds of meters, providing the first quantitative assessment of regional and local slopes across the entire planet (Smith *et al.*, 1999).

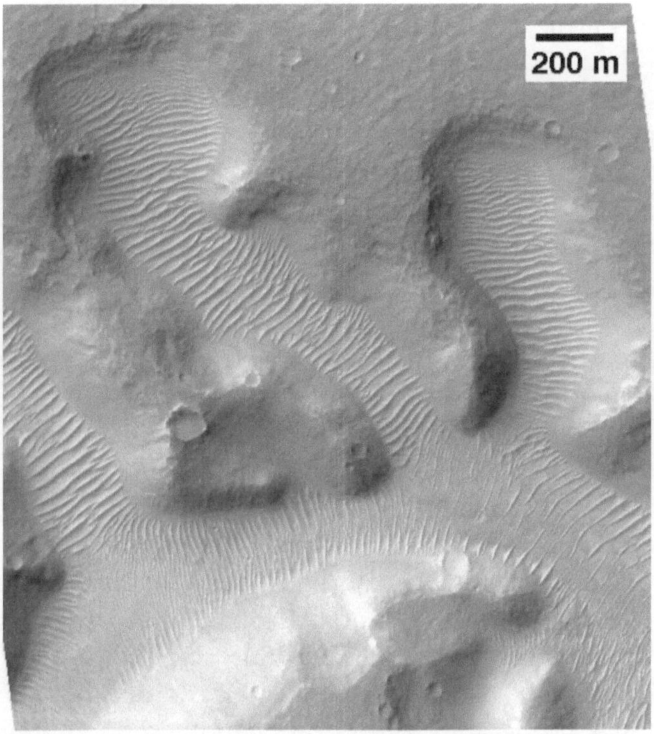

Figure 9.9. Transverse ripple-like features confined to the floor of Nirgal Vallis, Mars. Features have wavelengths of 30 to 100 m, and could be formed by either dune or ripple processes. Portion of MOC image E02-02651, centered at 27.8° S, 43.3° W, 2.8 m/pixel, NASA/JPL/MSSS (W26).

Unfortunately, MOLA does not resolve individual dunes, except for some exceptional transverse ridges in the northern erg, and thus will not provide morphometric information on Martian dune masses. In spite of this, MOLA has established precise control for the slopes across which sand is transported by the wind. The Thermal Emission Spectrometer (TES; W29) on MGS has provided compositional constraints for Martian surface materials over areas as small as 3 by 3 km (Christensen *et al.*, 1998, 2001). TES spectra have revealed two distinct volcanic compositions abundant on the Martian surface: a typical basaltic composition and a slightly more chemically evolved basaltic andesite composition (Bandfield *et al.*, 2000) that is comparable to several rocks at the Mars Pathfinder landing site (Golombek *et al.*, 1999). The TES basalt component has a high concentration in Syrtis Major (Bandfield *et al.*, 2000; Ruff and Christensen, 2002), a classic low-albedo feature that includes large barchan dunes (Figure 9.8). The presence of basaltic dunes on Mars increases the importance of studies of basaltic sand dunes on Earth, like those at Moses Lake (#22) (e.g., Bandfield *et al.*, 2002). TIMS data (Ramsey *et al.*, 1999) have been used to document the transport of specific compositions of sand at the Kelso Dunes (#5), but it is not yet clear how severely the dust on Mars will interfere with a similar analysis of Martian sediments. TES has thus far failed to detect much quartz, the most common sand component on Earth, even in regions that contain abundant dunes (Christensen *et al.*, 2001). Edgett and Lancaster (1993) discuss the role of composition for many sand deposits, particularly those derived from volcanic products.

MGS data have also intensified the investigation of hypothesized oceans and isolated lakes on Mars, first proposed based on analysis of Viking images (e.g., Parker *et al.*, 1989; Baker *et al.*, 1991; Forsythe and Zimbelman, 1995; Edgett and Parker, 1997). The ocean hypothesis continues to be controversial, with MOLA data providing topographic evidence in support of at least one northern ocean level (Head *et al.*, 1998) but with inconclusive MOC evidence of diagnostic shoreline morphologies (Malin and Edgett, 2001). Detailed topography using Differential Global Positioning System surveys of shorelines from glacial Lake Lahontan (Nevada), when scaled for gravity and the shallow slopes of the northern lowlands of Mars, compare very well with MOLA topography across one hypothesized shoreline on Mars (Zimbelman *et al.*, 2004); additional field data are being collected to test this correlation further. MOLA data revealed the presence of large lakes ponded within the southern highlands, one of which helps to explain the origin of Ma'adim Vallis at a kilometer-wide channel that empties into Gusev crater (Irwin *et al.*, 2002). Water has played a crucial role throughout Martian history (e.g., Carr, 1996), but the lack of a recently active hydrosphere on Mars is a substantial difference from the environment of all sand deposits on Earth, and this situation must be considered during analog eolian field investigations.

9.5 Discussion

The ubiquitous nature of eolian features on Mars has been revealed by spacecraft missions with progressively more capable remote sensors: Mariner 9, Viking, Mars Global Surveyor, and Mars Odyssey. Mars is obviously enough like Earth to support a similarity in eolian processes and products; however, the Martian eolian environment is sufficiently different to account for significant differences between the two planets.

Mars is like the Earth when it comes to eolian transport in general, sediment sources, bedform formation, and interbasin sediment transport.

The basic process of eolian transport, especially by saltation, is the same on both planets, as are the features produced by saltating sand. Fluvial processes on both planets create and transport sediment later subject to wind action. Viking and MOC data show abundant examples of eolian materials trapped by Martian topography. Many craters contain fields of barchans/barchanoid ridges that cover significant portions of their floors (e.g., Rabe and Proctor craters). Valleys such as Nirgal Vallis (Figure 9.9) also trap considerable quantities of eolian sediments.

But Mars is not like the Earth in other ways. The thin atmosphere of Mars, its surface geologic history, and lack of an active, terrestrial-style, hydrologic cycle affects the source and evolution of eolian sediments, the details of the sediment transportation process, and the location and durability of the resulting bedforms. The thinness of the Martian atmosphere requires wind speeds an order of magnitude larger than on Earth to initiate and sustain eolian transport (Greeley *et al.*, 1982, 1985). Viking and Pathfinder observations suggest that winds capable of moving loose sediments are rare. The rocks at the Viking landing sites show some evidence of eolian sandblasting (Binder *et al.*, 1977; Mutch *et al.*, 1977), and many rocks at Pathfinder show considerable sculpting by wind-borne sand (Bridges *et al.*, 1999; Greeley *et al.*, 2002). The crater populations of these sites indicate that they are relatively old, from the Hesperian (middle) period of Martian history (Scott and Tanaka, 1986). Calculations of potential eolian abrasion rates at those sites show that only a very small amount of the abrasion that potentially could have occurred has, in fact, actually occurred. The most likely explanations are: competent sand-sized material is very scarce on Mars, sand-sized sediments are there in abundance, but its motion is inhibited by some mechanism; or the surfaces we have landed on to date have been buried (protected from abrasion) for a substantial proportion of their existence (Greeley *et al.*, 1982, 1985).

Sand-sized material must be abundant in the Martian surface environment, as evidenced by the presence of ergs, dunefields, isolated dunes, and large-wavelength bedforms, all of which require saltation for their formation (by terrestrial analogy), which in turn implies particle sizes in the few hundred micrometer diameter range. The Algodones Dunes (#1) show a mix of many dune types (longitudinal, transverse, and barchan), and parts of it strongly resemble dune fields imaged by MOC, for example, the dunes within Proctor crater.

Just how mobile the sand-sized material is, however, is less easy to determine from orbital images. Comparisons of Viking and MOC images of dunes, for example, show zero movement to within the spatial resolution of the Viking images (8 m/pixel, Zimbelman, 2000; 17 m/pixel, figure 40 of Malin and Edgett, 2001) over periods of 10 Mars years (20 Earth years). Further, there are numerous examples where large-wavelength bedform features, which by terrestrial analogy should be highly mobile, have persisted in their positions long enough to have been subject to impact cratering and be overriden by debris flows and channel deposits, events that should have obliterated bedforms composed of loose sand.

Neither theory nor observation has revealed widespread granitic rocks on the surface of Mars. (Bandfield *et al.*, (2004) note that the Thermal Emission System (TES) on the Mars Odyssey Spacecraft recently detected only local outcrops of granite-like material in the central peaks of two impact craters.) Consequently, particulate materials are more likely to be composed of basalt rather than quartz. Erosion of basalt by fluvial/eolian processes could be the source of much of the sand-sized sediments seen on the Martian surface, which would make the dunes at Christmas (#21) and Moses (#22) lakes particularly good analogs (Edgett, 1994; Bandfield *et al.*, 2002).

Another consequence of the difference in terrestrial and Martian surface environments is that Mars does not have parabolic dunes analogous to those at Alkali Lake, Christmas Valley (#21), Juniper Flats, Moses Lake (#22), St. Anthony (#19), and Winnemucca (#12). Parabolic dunes like these owe their distinctive shape in large part to the stabilization of part of the sand mass by vegetation, which of course is not present on Mars. However, perhaps other agents (thick dust cover?) may stabilize Martian dunes within the dusty low thermal inertia regions of Mars (Zimbelman, 2000).

Recent ideas of Martian gullies notwithstanding (e.g., Christensen, 2003), lakes and streams are unlikely to have played a significant role in the formation of the eolian bedforms that are nearly ubiquitous in MOC images. However, postulated oceans, paleo-lakes, and the emplacement of dendritic drainage patterns in ancient terrains (see Chapters 1 and 13), could have been an older source of at least some of the sediments now residing in dunes and ripple-like bedforms. Large outflow channels on Mars are analogous to (and larger than) the flooding caused by breaching of glacial dams like that which created the Channeled Scablands (see Chapter 12). Outflow channels may have produced significant quantities of eolian material, much of which has been blowing around the Martian surface for a long time (see Chapters 11 and 12).

New data from current and future Mars missions should help to refine the strength of some of the analog sites discussed above. The Mars Odyssey spacecraft (W30) now in orbit includes the Thermal Emission Imaging

System (THEMIS; W31), a camera that shows the Martian surface at both thermal infrared and visual wavelengths. THEMIS images have already led to new hypotheses about features such as the enigmatic gullies (Christensen, 2003), and the capability is just as great for similar advances for eolian features. The Mars Exploration Rovers (MER) are providing detailed imaging and compositional information from two new landing sites on Mars; while not specifically targeted at eolian features, new insights have been gained for wind-blown materials and eolian deposits (W32; Greeley *et al.*, 2005). The 2005 Mars Reconnaissance Orbiter should provide sub-meter-scale images, which will likely change many of our current ideas about the Martian surface and provide new opportunities for investigating analog eolian sites here on Earth (W33).

9.6 Summary

Twenty locations in the western United States are discussed as potential analog sites for eolian features and deposits on Mars. The sites represent sand that is topographically trapped (a common occurrence on Mars), sand deposits along transport pathways (potentially present on Mars, but not reported to date), and sand that is derived from either lake or river sediments (again potentially present on Mars, but not specifically identified as such), although not all eolian sediments remain confined to topographic traps (Edgett, 2002). MGS data have greatly improved the level of detail available for Martian eolian features. On-going MGS and MER data collection, combined with prospects from future missions to Mars, holds great potential for more specific analog studies of eolian features and deposits in the future.

Acknowledgments

The authors are very grateful for the detailed review comments of A.Wesley Ward and Nicholas Lancaster, which substantially improved revision of the manuscript. This work was supported by NASA MDAP grant NAG5-11075 and funds from the Walcott and Becker endowments to the Smithsonian Institution.

References

Abbott, P. L. (1980). Provenance of Salton Dunes, southwest of the Salton Sea, California. In *Geology and Mineral Wealth of the California Desert*, ed. D. Fife. Santa Ana, CA: South Coast Geological Association, pp. 409–13.

Ahlbrandt, T. S. (1973). Sand dunes, geomorphology and geology, Killpecker Creek area, northern Sweetwater county, Wyoming. Ph.D. dissertation, University of Wyoming.

Ahlbrandt, T. S. (1974a). The source and sand for the Killpecker sand dune field, southwestern Wyoming. *Sedimentary Geology*, **11**, 39–57.

Ahlbrandt, T. S. (1974b). Dune stratigraphy, archaeology, and the chronology of the Killpecker dune field. In *Applied Geology and Archeology – The Holocene History of Wyoming*, ed. M. Wilson. *Wyoming Geological Survey Report of Investigations*, **10**, 51–60.

Ahlbrandt, T. S. (1975). Comparison of textures and structures to distinguish aeolian environments, Killpecker dune field, Wyoming. *Mountain Geologist*, **12** (2), 61–73.

Alt, D. (2001). *Glacial Lake Missoula and its Humongous Floods*. Missoula, MT: Mountain Press.

Andrews, S. (1981). Sedimentology of Great Sand Dunes, Colorado. In *Recent and Ancient Non-marine Depositional Environments: Models for Exploration*, ed. F. P. Ethridge and R. M. Flores. Society of Economic Paleontologists and Mineralogists Special Publication 31, pp. 279–91.

Bagnold, R. A. (1941). *The Physics of Blown Sand and Desert Dunes*. London: Chapman and Hall (reprinted in 1973).

Baker, V. R. (1973). Paleohydrology and sedimentology of Lake Missoula flooding in eastern Washington. Geological Society of America Special Paper 144.

Baker, V. R., Strom, R. G., Croft, S. K. *et al.* (1991). Ancient oceans, ice sheets and the hydrological cycle on Mars. *Nature*, **352**, 89–594.

Bandfield, J. L., Hamilton, V. E., and Christensen, P. R. (2000). A global view of Martian surface compositions from MGS-TES. *Science*, **287**, 1626–30.

Bandfield, J. L., Edgett, K. S., and Christensen, P. R. (2002). Spectroscopic study of the Moses Loake dune field, Washington: determination of compositional distributions and source lithologies. *Journal of Geophysical Research*, **107** (E11), doi:10.1029/2000JE001469.

Bandfield, J. L., Hamilton, V. E., Christensen, P. R., and McSween, Jr. H. Y. (2004). Identification of quartzofeldspathic materials on Mars. *Journal of Geophysical Research*, **109**, doi:10.1029/2004JE002290.

Benson, L. V., Currey, D. R., Dorn, R. I. *et al.* (1990). Chronology of expansion and contraction of four Great Basin lake systems during the past 35,000 years. *Paleogeography, Paleoclimatology, Paleoecology*, **78**, 241–86.

Billingsley, G. H. (1987). Geologic map of the southwestern Moenkopi Plateau and southern Ward terrace, Coconino county, Arizona. *US Geological Survey Miscellaneous Investigations Series Map* I-1793, scale 1:31,680.

Binder, A. B., Arvidson, R. E., Guinness, E. A. *et al.* (1977). The geology of the Viking Lander 1 site. *Journal of Geophysical Research*, **82** (28), 4439–51.

Breed, C. S., Fryberger, S. G., Andrews, S. *et al.* (1979). Regional studies of sand seas, using Landsat (ERTS) imagery. *US Geological Survey Professional Paper* 1052, pp. 305–97.

Breed, C. S., Grolier, M., McCauley, J. *et al.* (1984). Eolian (wind formed) landscapes. In *Landscapes of Arizona, the Geological Story*, ed. T. L. Smiley, J. D. Nations, and T. L. Pewe. University Presses of America, pp. 359–413.

Bridges, N. T., Greeley, R., Haldemann, A. F. C. *et al.* (1999). Ventifacts at the Pathfinder landing site. *Journal of Geophysical Research*, **104** (E4), 8595–616.

Burford, A. E. and Hutchinson, D. M. (1968). Insight into origin of the Great Sand Dunes, Colorado. *Geological Society of America Special Paper* 115.

Carr, M. H. (1996). *Water on Mars*. Oxford: Oxford University Press.

Chadwick, H. W. and Dalke, P. D. (1965). Plant succession on dune sands in Fremont County, Idaho. *Ecology*, **46**, 675–780.

Christensen, P. R. (2003). Formation of recent Martian gullies through melting of extensive water-rich snow deposits. *Nature*, **422**, 45–8.

Christensen, P. R., Anderson, D. L., Chase, S. C. *et al.* (1998). Results from the Mars Global Surveyor Thermal Emission Spectrometer. *Science*, **279**, 1692–8.

Christensen, P. R., Bandfield, J. L., Hamilton, V. E. *et al.* (2001). Mars Global Surveyor Thermal Emission Spectrometer experiment: investigation description and surface science results. *Journal of Geophysical Research*, **106** (E10), 23823–71.

Cooke, R. U. and Warren, A. (1973). *Geomorphology in Deserts*. Berkeley: University of California Press.

Criswell, D. R., Lindsay, J. F., and Reasoner, D. L. (1975). Seismic and acoustic emissions of a booming dune. *Journal of Geophysical Research*, **80** (35), 4963–74.

Dort, W. (1959). Sand dunes of northeastern Snake river Plain, Idaho. *Geological Society of America Bulletin*, **69**, 1555 (abstract).

Earl, P. I. (1981). Sand mountain: dune of mystery. *California Mining Journal*, **51**, 67–8.

Edgett, K. S. (1994). The volcaniclastic shifting sand dunes of Christmas Lake Valley, Oregon. In The sand component of the modern Martian aeolian sedimentary system. Ph.D. dissertation, Arizona State University, Tempe, pp. 145–201.

Edgett, K. S. (2002). Low-albedo surfaces and eolian sediment: Mars Orbiter Camera views of western Arabia Terra craters and wind streaks. *Journal of Geophysical Research – Planets*, **107**, doi: 10.1029/2001JE001587.

Edgett, K. S. and Lancaster, N. (1993). Volcanoclastic aeolian dunes. *Journal of Arid Environments*, **25**, 271–97.

Edgett, K. S. and Parker, T. J. (1997). Water on early Mars: possible subaqueous sedimentary deposits covering ancient cratered terrain in western Arabia Terra and Sinus Meridiani. *Geophysical Research Letters*, **24**, 2897–900.

Edgett, K. S. and Malin, M. C. (2000). New views of Mars eolian activity, materials, and surface properties: three vignettes from the Mars Global Surveyor Mars Orbiter Camera. *Journal of Geophysical Research – Planets*, **105**, 1623–50.

Eissmann, L. J. (1990). Eolian sand transport in western Nevada, M.S. thesis, University of Nevada, Reno.

Epps, T. M., Britten, H. B., and Rust, R. W. (2000). Allozyme variability, isolation, and dispersal of Eusattus Muricatus (Coleoptera: Tenebrionidae) within Silver State dune complex, Great Basin, western North America. *Western North American Naturalist*, **60**, 281–90.

Fenton, L. K., Richardson, M. I., and Toigo, A. D. (2002). Sand transport in Proctor crater on Mars based on dune morphology and mesoscale modeling. *Lunar and Planetary Science XXXIII*. Houston; TX: Lunar and Planetary Institute, Abstract 1953.

Ford, R. L. and Gilman, S. L. (2000). Geology of Coral Pink Sand Dunes. In *Geology of Utah's Parks and Monuments*, ed. D. A. Sprinkel, T. C. J. Chidsey, and P. B. Anderson. Salt Lake City: Utah Geological Association, pp. 365–89.

Forsythe, R. D. and Zimbelman, J. R. (1995). A case for ancient evaporite basins on Mars. *Journal of Geophysical Research*, **100**, 5553–63.

Freyberger, S. G., Ahlbrandt, T. S., and Andrews, S. (1979). Origin, sedimentary features and significance of low-angle eolian "sand sheet" deposits, Great Sand Dunes National Monument and vicinity, Colorado. *Journal of Sedimentary Petrology*, **49**, 733–46.

Golombek, M. P., Edgett, K. S., and Rice, J. W. (eds.) (1995). Mars Pathfinder landing site workshop II: Characteristics of the Ares Vallis region and field trips in the channeled scabland, Washington. *Lunar and Planetary Institute Technical Report 95–01*, Part 1 and Part 2.

Golombek, M. P. *et al.* (1999). Overview of the Mars Pathfinder mission: launch through landing, surface opertations, data sets, and science results. *Journal of Geophysical Research – Planets*, **104**, 8523–54.

Greeley, R. (1977). Aerial guide to the geology of the central and eastern Snake River Plain. Volcanism of the eastern Snake River Plain, Idaho: a comparative planetary geology guidebook, NASA Conference Report CR-154621, pp. 60–111 (St. Anthony dunes, pp. 101–8).

Greeley, R. and Iversen, J. D. (1985). *Wind as a Geological Process on Earth, Mars, Venus and Titan*. New York: Cambridge University Press.

Greeley, R. and Iversen, J. D. (1987). Measurements of wind friction speeds over lava surfaces and assessment of sediment transport. *Geophysical Research Letters*, **14**, 925–8.

Greeley, R. and Williams, S. H. (1994). Dust deposits on Mars: the "parna" analog. *Icarus*, **110**, 165–77.

Greeley, R., Papson, R. P., and Spudis, P. D. (eds.) (1978). *Aeolian Features of Southern California: A Comparative Planetary Geology Guidebook*. US Government Printing Office, Stock No. 033–000–00706–0.

Greeley, R., Leach, R. N., Williams, S. H. *et al.* (1982). Rate of wind abrasion on Mars. *Journal of Geophysical Research*, **87**, 10009–24.

Greeley, R., Williams, S. H., White, B. R., Pollack, J. B., and Marshall, J. R. (1985). Wind abrasion on Earth and Mars. In *Models in geomorphology*, ed. M. J. Woldenburg. Boston: Allen and Unwin, pp. 373–422.

Greeley, R., Lancaster, N., Lee, S., and Thomas, P. (1992). Martian aeolian processes, sediments, and features. In *Mars*, ed. H. H. Kieffer, B. M. Jakosky, C. W. Snyder, and M. S. Mathews. Tucson: University of Arizona Press, pp. 730–66.

Greeley, R., Bridges, N. T., Kuzmin, R. O., and Laity, J. E. (2002). Terrestrial analogs to wind-related features at the Viking and Pathfinder landing sites on Mars. *Journal of Geophysical Research*, **107**, 10129–50.

Greeley, R., Arvidson, R., Bell, J. F. III. *et al.* (2005). Martian variable features: new insight from the Mars Express Orbiter and the Mars Exploration Rover Spirit. *Journal of Geophysical Research*, **110**, E06002, doi:10.1029/2005JE002403.

Hack, J. T. (1941). Dunes of the western Navajo country. *Geographical Journal*, **31**, 240–63.

Haff, P. K. and Presti, D. E. (1995). Barchan dunes of the Salton Sea region, California. In *Desert Aeolian Processes*, ed. V. P. Tchakerian. New York: Chapman and Hall, pp. 153–77.

Hamblin, W. K. and Christiansen, E. H. (1998). The effects of the Pleistocene glaciation. In *Earth's Dynamic Systems: A Textbook in Physical Geology*, 8th edn. Upper Saddle River, NJ: Macmillan, pp. 370–80.

Havholm, K. G. and Kocurek, G. (1988). A preliminary study of the dynamics of a modern draa, Algodones, southeastern California, USA. *Sedimentology*, **35**, 649–69.

Head, J. W., Kreslavsky, M., Heisinger, H. *et al.* (1998). Oceans in the past history of Mars: tests for their presence using Mars Orbiter Laser Altimeter (MOLA) data. *Geophysical Research Letters*, **25**, 4401–4.

Irwin, R. P., Maxwell, T. A., Howard, A. D., Craddock, R. A., and Leverington, D. W. (2002). A large paleolake basin at the head of Ma'adim Vallis, Mars. *Science*, **296**, 2209–12.

Janke, J. R. (2002). An analysis of the current stability of the dune field at Great Sand Dunes National Monument using temporal TM imagery (1984–1998). *Remote Sensing of the Environment*, **83**, 488–97.

Johnson, R. B. (1967). The Great Sand Dunes of southern Colorado. US Geological Survey Professional Paper 757-C, pp. 177–83.

Johnson, R. B. (1968). The Great Sand Dunes of southern Colorado. *Rocky Mountain Geologist*, **5**, 23–39.

Kocurek, G. and Nielson, J. (1986). Climbing zibars of the Algodones. *Sedimentary Geology*, **48**, 1–15.

Kocurek, G. and Lancaster, N. (1999). Aeolian sediment states: theory and Mojave Desert Kelso dunefield example. *Sedimentology*, **46**, 505–16.

Lancaster, N. (1992). Relationships between dune generations in the Gran Desierto, Mexico. *Sedimentology*, **39**, 631–44.

Lancaster, N. (1993). Development of Kelso Dunes, Mojave Desert, California. *National Geographic Research and Explorations*, **9**, 444–59.

Lancaster, N. (1995a). *Geomorphology of Desert Dunes*. New York: Routledge.

Lancaster, N. (1995b). Origin of the Gran Desierto sand sea, Sonora, Mexico: evidence from dune morphology and sedimentology. In *Desert Aeolian Processes*, ed. V. P. Tchakerian. London: Chapman and Hall, pp. 11–35.

Lancaster, N. and Blount, G. (1990). Development of the Gran Desierto sand sea. *Geology*, **18**, 724–8.

Lancaster, N. and Tchakerian, V. P. (1996). Geomorphology and sediments of sand ramps in the Mojave Desert. *Geomorphology*, **17**, 151–66.

Lancaster, N., Greeley, R., and Christensen, P. R. (1987). Dunes of the Gran Desierto sand sea, Sonora, Mexico. *Earth Surface Processes and Landforms*, **12**, 277–88.

Lindsay, J. F., Criswell, D. R., Criswell, T. L., and Criswell, B. S. (1976). Sound-producing dune and beach sands. *Geological Society of America Bulletin*, **87**, 463–73.

Long, J. T. and Sharp, R. P. (1964). Barchan dune movement in Imperial Valley, California. *Geological Society of America Bulletin*, **75**, 149–56.

MacDonald, A. A. (1970). The northern Mojave's little Sahara. *Mineral Information Service*, **23**, 3–6.

Maloney, J. F. (1982). Sand Mountain: the dune that booms. *Nevada*, **42**, 8–11.

Malin, M. C. and Edgett, K. S. (2001). Mars Global Surveyor Mars Orbiter Camera: interplanetary cruise through primary mission. *Journal of Geophysical Research – Planets*, **106**, E10, 23429–570.

Malin, M. C., Carr, M. H., Danielson, G. E. *et al.* (1998). Early views of the Martian surface from the Mars Orbiter Camera of Mars Global Surveyor. *Science*, **279**, 1681–5.

McCoy, F. M., Nokleberg, W. J., and Norris, R. M. (1967). Speculations on the origin of the Algodones dunes, California. *Geological Society of America Bulletin*, **78**, 1039–44.

McKee, E. D. (1966). Structures of dunes at White Sands National Monument, New Mexico (and a comparison with structures of dunes from other selected areas). *Sedimentology*, **7**, 1–69.

McKee, E. D. (1979). Introduction to a study of global sand seas. US Geological Survey Professional Paper 1052, pp. 1–19.

McKee, E. D. and Douglas, J. R. (1971). Growth and movement of dunes at White Sands National Monument. US Geological Survey Professional Paper 750-D, pp. 108–14.

McKee, E. D. and Moiola, R. J. (1975). Geometry and growth of the White Sands dune field, New Mexico. *US Geological Survey Journal of Research*, **3**, 59–66.

Morrison, R. B. (1964). Lake Lahontan: geology of the southern Carson Desert, Nevada. US Geological Survey Professional Paper 401, p. 156.

Morrison, R. B. (1991). Quaternary stratigraphic, hydrologic, and climatic history of the Great Basin, with emphasis on lakes Lahontan, Bonneville, and Tecopa. In *The Geology of North America*, vol. K-2, ed. R. B. Morrison. Boulder, CO: Geological Society of America, pp. 283–320.

Muhs, D. R. (2002). The concept of mineralogical maturity and the origin and evolution of dune fields in the western United States. In *Proceedings of ICAR5/GCTE-SEN Joint Conference*, ed. J. A. Lee and T. M. Zobeck. International Center for Arid and Semiarid Lands Studies, Texas Tech University, Lubbock, TX, Publication 02–2, pp. 408–9.

Muhs, D. R., Bush, C. A., Cowherd, S. D., and Mahan, S. (1995). Geomorphic and geochemical evidence for the source of sand in the Algodones dunes, Colorado Desert, Southeastern California. In *Desert Aeolian Processes*. ed. V. P. Tchakerian. New York: Chapman and Hall, pp. 37–74.

Murbarger, N. (1950). The great White Sands. *Natural History*, **59**, 228–35.

Murphy, J. D. (1973). The geology of Eagle Cove at Bruneau, Idaho, M.A. thesis, State University of New York at Buffalo.

Murphy, J. D. (1975). The geology of Bruneau Dunes State Park, Idaho. *Geological Society of America Abstracts with Program*, **7**, 633 (abstract).

Murphy, J. D. and Greeley, R. (1972). Sand dunes at Eagle Cove (Bruneau), Idaho: possible analogs to Martian aeolian features. *Eos, Transactions of the American Geophysical Union*, **53**, 1037 (abstract).

Mutch, T. A., Arvidson, R. E., Binder, A. B., Guinness, E. A., and Morris, E. C. (1977). The geology of the Viking Lander 2 site. *Journal of Geophysical Research*, **82**, 4452–67.

Nickling, W. G., McKenna Neuman, C., and Lancaster, N. (2002). Grainfall processes in the lee of transverse dunes, Silver Peak, Nevada. *Sedimentology*, **49**, 191–209.

Nielson, J. and Kocurek, G. (1987). Surface processes, deposits, and development of star dunes: Dumont dune field, California. *Geological Society of America Bulletin*, **99**, 177–86.

Norris, R. M. (1966). Barchan dunes of Imperial Valley, California. *Journal of Geology*, **74**, 292–306.

Norris, R. M. and Norris, K. S. (1961). Algodones dunes of southeastern California. *Geological Society of America Bulletin*, **72**, 605–20.

Orr, E. L. and Orr, W. N. (2000). *Geology of Oregon*, 5th edn. Dubuque, IA: Kendall/Hunt Publishing Company.

Parker, T. J., Saunders, R. S., and Schneeberger, D. M. (1989). Transitional morphology in the west Deuteronilus Mensae region of Mars: implications for modification of the lowland/upland boundary. *Icarus*, **82**, 111–45.

Pavlik, B. M. (1980). Patterns of water potential and photosynthesis of desert dune plants, Eureka Valley, California. *Oecologia*, **46**, 147–54.

Pease, P. P. and Tchakerian, V. P. (2003). Geochemistry of sediments from Quaternary sand ramps in the southeastern Mojave Desert, California. *Quaternary International*, **104**, 19–29.

Pye, K. and Tsoar, J. (1990). *Aeolian Sand and Sand Dunes*. London: Unwin Hyman.

Ramsey, M. S., Christensen, P. R., Lancaster, N., and Howard, D. A. (1999). Identification of sand sources and transport pathways at the Kelso Dunes, California using thermal infrared remote sensing. *Geological Society of America Bulletin*, **111**, 646–62.

Rempel, P. (1936). The crescentic sand dunes of the Salton Sea and their relation to vegetation. *Ecology*, **17**, 347–58.

Rendell, H. M. and Sheffer, N. L. (1996). Luminescence dating of sand ramps in the eastern Mojave Desert. *Geomorphology*, **17**, 187–97.

Ruff, S. W. and Christensen, P. R. (2002). Bright and dark regions on Mars: particle size and mineralogical characteristics based on Thermal Emission Spectrometer data. *Journal of Geophysical Research – Planets*, **107**, E12, doi: 10.1029/2001JE001580.

Russell, I. C. (1885). Geological history of Lake Lahontan, a Quaternary lake of northeastern Nevada. *U.S. Geological Survey Monograph 11*.

Schenk, C. J. and Fryberger, S. C. (1988). Early diagenesis of eolian dune sand and interdune sands at White Sands, New Mexico. *Sedimentary Geology*, **55**, 109–20.

Scott, D. H. and Tanaka, K. L. (1986). Geologic map of the western equatorial region of Mars. *US Geological Survey Miscellaneous Investigations Series map I-1802-A*, scale 1:15, 000, 000.

Sharp, R. P. (1966). Kelso dunes, Mojave Desert, California. *Geological Society of America Bulletin*, **77**, 1045–74.

Sharp, R. P. (1978). The Kelso dune complex. In *Aeolian Features of Southern California: A Comparative Planetary Geology Guidebook*, ed. R. Greeley, M. B. Womer, R. P. Papson, and P. D. Spudis. US Government Printing Office, pp. 54–63.

Sharp, R. P. (1979). Intradune flats of the Algodones chain, Imperial Valley, California. *Geological Society of America Bulletin*, **90**, 908–16.

Simpson, E. L. and Loope, D. B. (1985). Amalgamated interdune deposits, White Sands, New Mexico. *Journal of Sedimentary Petrology*, **55**, 361–5.

Smith, D. E., Zuber, M. T., Soloman, S. C. *et al.* (1999). The global topography of Mars and implications for surface evolution. *Science*, **284**, 1495–503.

Smith, R. S. U. (1975). Eolian transport of sand on actively accreting slip face of a sand dune northwest of Winnemucca, Nevada. *Geological Society of America Abstracts with Program*, **9**, 502 (abstract).

Smith, R. S. U. (1978). Guide to selected features of aeolian geomorphology in the Algodones dune chain, Imperial county, California. In *Aeolian Features of Southern California: A Comparative Planetary Geology Guidebook*, ed. R. Greeley, M. B. Womer, R. P. Papson, and P. D. Spudis. US Government Printing Office, pp. 74–98.

Smith, R. S. U. (1982). Sand dunes in the North American deserts. In *Reference Handbook of the Deserts of North America*, ed. G. L. Bender. Westport, CT: Greenwood Press, pp. 481–524.

Smith, R. S. U. (1984). Aeolian geomorphology of the Devil's Playground, Kelso Dunes and Silurian Valley, California. In *Surficial Geology of the Eastern Mojave Desert*, ed. J. C. Dohnenwend. Boulder, CO: Geological Society of America, pp. 162–73.

Spaulding, W. G. (1990). Late Quaternary vegetation and climate of the Mojave Desert: the last Glacial Maximum to the present. In *Packrat Middens: The Last 40,000 Years of Biotic Change*, ed. J. L. Betancourt, T. R. VanDevender, and P. S. Martin. Tucson: University of Arizona Press, pp. 166–99.

Stokes, S. and Breed, C. S. (1993). A chronostratigraphic re-evaluation of the Tusayan Dunes, Moenkopi Plateau and Ward Terrace, northeastern Arizona. In *The Dynamics and Environmental Context of Aeolian Sedimentary Systems*, ed. K. Pye. London: Geological Society, pp. 75–90.

Swadley, W. C. and Carr, W. J. (1987). Geologic map of the quaternary and tertiary deposits of the Big Dune quadrangle, Nye county, Nevada, and Inyo county, California. *US Geological Survey Miscellaneous Investigations Series I-1767*, scale 1:24,000.

Sweet, M. L., Nielson, J., Havholm, K., and Farrelly, J. (1988). Algodones dune field of southern California: case history of a migrating modern dune field. *Sedimentology*, **35**, 939–52.

Tchakarian, V. P. (1991). Late Quaternary aeolian geomorphology of the Dale Lake sand sheet, southern Mojave Desert, California. *Physical Geography*, **12**, 347–69.

Tchakerian, V. P. (ed.)(1995). *Desert Aeolian Processes*. New York: Chapman and Hall.

Tchakerian, V. P. (1997). North America. In *Arid Zone Geomorphology*, 2nd edn, ed. D. S. G. Thomas. New York: John Wiley & Sons, pp. 523–41.

Theilig, E., Womer, M., and Papson, R. (1978). Geological field guide to the Salton Trough. In *Aeolian Features of Southern California: A Comparative Planetary Geology Guidebook*, ed. R. Greeley, M. B. Womer, R. P. Papson, and P. D. Spudis. US Government Printing Office, pp. 100–59 (Salton Sea barchans are on pp. 119–24).

Thomas, D. S. G. (ed.)(1997). *Arid Zone Geomorphology*, 2nd edn. New York: John Wiley & Sons.

Trexler, D. T. and Melhorn, W. N. (1986). Singing, and booming sand dunes of California and Nevada. *California Geology*, **39**, 147–52.

Tsoar, H., Greeley, R., and Peterfreund, A. R. (1979). Mars: the north polar sand sea and related wind patterns. *Journal of Geophysical Research*, **84**, 8167–80.

US Geological Survey (1982). Zapata Ranch quadrangle, Colorado, 7.5 minute series (topographic) map, scale 1:24,000.

van de Kamp, P. C. (1973). Holocene continental sedimentation in the Salton basin, California: a reconnaissance. *Geological Society of America Bulletin*, **84**, 827–48.

Wegemann, C. H. (1939). Great Sand Dunes of Colorado. *Mines Magazine*, **29**, 445–8.

Williams, K. K., Greeley, R., and Zimbelman, J. R. (2003). Using overlapping MOC images to search for dune movement and to measure dune heights. *Lunar and Planetary Science XXIV*. Houston: Lunar and Planetary Institute, Abstract 1639.

Williams, S. H., Zimbelman, J. R., and Ward, A. W. (2002). Large ripples on Earth and Mars. *Lunar and Planetary Science XXIII*. Houston: Lunar and Planetary Institute, Abstract 1508.

Wilson, S. A., Zimbelman, J. R., and Williams, S. H. (2003). Large aeolian ripples: extrapolations from Earth to Mars. *Lunar and Planetary Science XXIV*. Houston: Lunar and Planetary Institute, Abstract 1862.

Zimbelman, J. R. (2000). Non-active dunes in the Acheron Fossae region of Mars between the Viking and Mars Global Surveyor eras. *Geophysical Research Letters*, **27**, 1069–72.

Zimbelman, J. R. and Williams, S. H. (2002). Geochemical indicators of separate sources for eolian sands in the eastern Mojave Desert, California, and western Arizona. *Geological Society of America Bulletin*, **114**, 490–6.

Zimbelman, J. R. and Wilson, S. (2002). Ripples and dunes in the Syrtis Major region of Mars, as revealed in MOC images. *Lunar and Planetary Science XXXIII*. Houston: Lunar and Planetary Institute, Abstract 1514.

Zimbelman, J. R., Williams, S. H., and Tchakerian, V. P. (1995). Sand transport paths in the Mojave Desert, southwestern United States. In *Desert Aeolian Processes*, ed. V. P. Tchakerian. New York: Chapman and Hall, pp. 101–29.

Zimbelman, J. R., Williams, S. H., Johnston, A. K., and Head, J. W. (2004). Lake shorelines: Earth analogs for hypothesized Martian coastal features. *Lunar and Planetary Science XXXV*. Houston: Lunar and Planetary Institute, Abstract 1683.

Website addresses cited in Chapter 9

W1: Eolian sand, SW US http://geochange.er.usgs.gov/sw/impacts/geology/sand/swsand.html
W2: Mars Exploration Program http://mars.jpl.nasa.gov
W3: Algodones http://www.ca.blm.gov/elcentro/algodones.html
W4: Bristol/Palen http://www.shef.ac.uk/~igcp413/pdf/abstracts_volume.pdf
W5: Cactus Plain http://azwww.az.blm.gov/rec/ECACTUS.HTM
W6: Kelso http://wrgis.wr.usgs.gov/docs/parks/mojave/kelso1.html
W7: Kelso http://www.blm.gov/nhp/Preservation/wilderness/kelsod.html
W8: Dumont http://www.ca.blm.gov/barstow/dumont.html
W9: Dumont http://www-wmc.wr.usgs.gov/home-esic/exhibit/dumont.html
W10: Eureka http://www.nature.nps.gov/nnl/Registry/USA_Map/States/California/nnl/ed
W11: Sand Mountain http://www.nv.blm.gov/carson/Recreation/Rec_SandMtn.html
W12: Little Sahara http://www.ut.blm.gov/recsite/little.html
W13: Little Sahara http://www.utah.com/schmerker/1999/little_sahara.html
W14: Coral Pink http://www.stateparks.utah.gov/park_pages/parkpage.php?id=cpsp
W15: Coral Pink http://www.utah.com/stateparks/coral_pink.html
W16: White Sands http://www2.nature.nps.gov/geology/parks/whsa
W17: White Sands http://www.nps.gov/whsa/index.html
W18: Great Sand Dunes http://www.nps.gov/grsa/index.html
W19: Killpecker http://www.wy.blm.gov/rsfo/rec/dunes.html
W20: St. Anthony http://www.id.blm.gov/spec_places/other_stant.html
W21: Bruneau http://www.idahoparks.org/parks/bruneaudunes.html
W22: Christmas Valley http://atv.prd.state.or.us/site_detail.php?UID=4
W23: Moses Lake http://www.tourgrantcounty.com/rec_offroad.asp
W24: MOC http://www.msss.com/mars_images/index.html

W25: MOC image http://www.msss.com/moc_gallery/e01_e06/images/E03/E0302016.html
W26: MOC image http://www.msss.com/moc_gallery/e01_e06/images/E02/E0202651.html
W27: MOC Press Rel. http://www.msss.com/mars_images/moc/E01_E06_sampler2002/nirgal
W28: MOLA http://ltpwww.gsfc.nasa.gov/tharsis/mola.html
W29: TES http://emma.la.asu.edu
W30: Mars Odyssey http://mars.jpl.nasa.gov/odyssey
W31: THEMIS http://themis.la.asu.edu
W32: MER http://marsrovers.jpl.nasa.gov
W33: MRO http://mars.jpl.nasa.gov/mro
W34: SRTM shaded relief image http://photojournal.jpl.nasa.gov/catalog/PIA03378

10

Debris flows in Greenland and on Mars

François Costard
UMR 8148 IDES, Université Paris-Sud

François Forget
Laboratory for Dynamic Meteorology, Paris

Vincent Jomelli
CNRS UMR 8591, Laboratoire de Géographie Physique, Meudon

Nicolas Mangold and Jean-Pierre Peulvast
UMR 8148 IDES, Université Paris-Sud

10.1 Introduction: Martian gullies and terrestrial debris flows

The observation of small gullies on Mars was one of the more unexpected discoveries of the Mars Observer Camera (MOC) aboard the Mars Global Surveyor (MGS) spacecraft (Malin and Edgett, 2000). Gullies are the flutes and narrow troughs formed by the debris flows process and not the process itself. They mostly occur in a latitudinal band higher than 30°. The upper parts of the slopes (mostly south facing slopes in the southern hemisphere) exhibit alcoves, with generally broad and deep channels. They are characterized by their distinct V-shaped channels with well-defined levees. Individual channels exhibit low sinuosity and deep erosion down to the fans that bury the lower parts of the crater walls (Figure 10.1). These debris fans correspond to one or several lobes.

The characteristics of these gullies suggest that they were formed by flowing water mixed with soil and rocks transported by these flows. They appeared to be surprisingly young, as if they had formed in the last few million years or even more recently. In their initial analysis, the MGS Camera investigators Mike Malin and Ken Edgett (2000) proposed a scenario involving ground water seepage from a sub-surface liquid water reservoir located a few hundred meters or less below the surface. However, the process capable of maintaining such a shallow aquifer at temperatures above the freezing point of water remains unclear. Several other hypotheses have been proposed taking into account the geothermal activity (Gaidos, 2001; Hartmann, 2001)

The Geology of Mars: Evidence from Earth-based Analogs, ed. Mary Chapman. Published by Cambridge University Press.

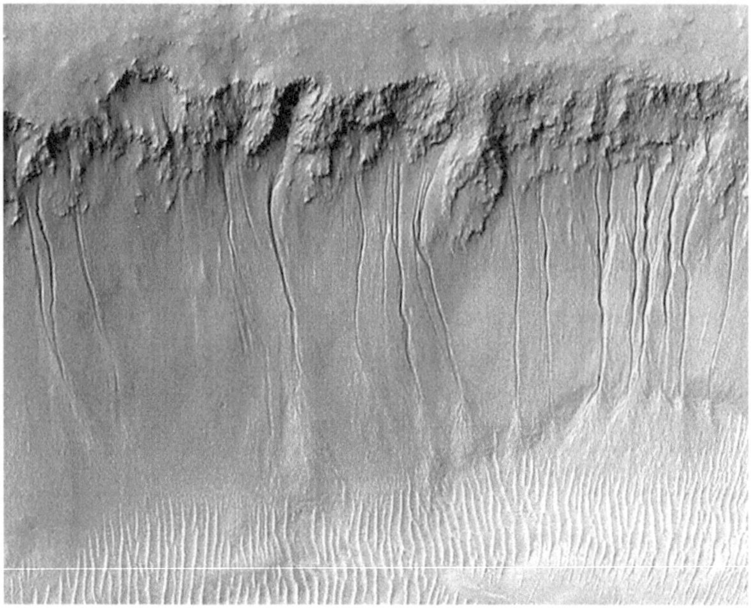

Figure 10.1. Debris flows on Mars in Nirgal Vallis (MOC image M03–02290) (29° S and 39° W). MOC image, MSSS. Total length of debris flows is 900 m.

or the presence of brines (Knauth *et al.*, 2000; Knauth and Burt, 2003), snowmelt in more recent periods (Christensen, 2003), or liquid CO_2 breakout (Musselwhite *et al.*, 2001). Through analysis of the MGS Camera data archive, some examples have been discovered of gullies originating from the top of isolated peaks and from dune crests (Baker, 2001; Mellon and Phillips, 2001; Mangold *et al.*, 2003). In these cases, the involvement of a subsurface aquifer is unlikely. The morphometric characteristics of Martian gullies, especially the existence of levees, strictly focus the formation of gullies by quick flows of rock–water mixture with high rock proportion inconsistent with the slow erosional process described by Christensen (2003).

On Earth, hillslope debris flow can be defined as a rapid mass movement of a flowing mixture of sediment and water (Phillips and Davies, 1991). Two types of debris flows can be defined: valley confined debris flow (or lave torrentielle in the French literature; Coussot, 1994) and hillslope debris flows (which may correspond to the Martian gullies) triggered at the base of a cliff which is only partially channelized on the high part of the slope. Geomorphologically the debris flow system typically consists of the following parts: (1) a funnel-shaped debris source area consisting of a ravine (or series of deep gullies) below steep rocky slopes; (2) a zone where debris levees are found on both sides of the flow track; and (3) a terminal part where the levees join to form a frontal or a combination of several lobes (Van Steijn *et al.*, 1988; Van Steijn, 1991; Major, 1997). A channel meandering pattern is possible especially in sections were slope angles are relatively low. From a physical point of view, these debris flows consist of a mixture of unsorted rocky material with low water content that forms a muddy slurry moving downslope. Of particular interest are the periglacial debris flows induced by the seasonal temperature change in the polar regions with continuous permafrost (i.e., the sub-surface is permanently frozen). Debris flows occur when soils begin to be saturated with water after the melting of the snow cover and/or the ground ice (French, 1988). Pure surface runoff cannot explain the formation of debris flows, because the existence of levees implies a significant thickness of material to initiate the debris flows (Mangold *et al.*, 2003).

10.2 The Greenland analogy

With the existence of a global permafrost and from a thermal point of view, Mars is considered as a periglacial planet. In order to get a more precise idea of the formation of these debris flows, we therefore proposed to study a typical terrestrial analog in a dry periglacial area. Martian gullies were strikingly similar to some "debris flows" that were observed by two of the authors in East Greenland (Jameson Land) during two field trips in the summer seasons of 1987 and 1989 (Peulvast, 1988). Both features exhibit the same scale and the same morphology (Figure 10.2). Like Mars, East Greenland has a dry periglacial climate. Jameson Land (70 to 71° N) is located north of the Scoresby Sund fjord and comprises wide plateaus 600 to 1000 m high mainly composed of clastic sediments (sandstones, mudstones, and shales from Permian to Lower Cretaceous age). The materials are relatively cohesionless, coarse, and easily removable. This is a particularly favorable condition for the initiation of debris flows. The annual mean temperature is −6.7 to −8 °C at sea level. Mean monthly temperatures are −18 °C in March and +4.5 °C in July at sea level; and winter minimum and summer maximum temperatures are, −43 °C and +20 °C respectively. Annual precipitations, mainly snowfalls, are 428 mm on the coast, with maxima in October and December; they are probably much less inland, in Jameson Land, as shown by the steep gradient of the glaciation limit, from 300 m at the coast to more than 900 m at a distance of 30 km to the northwest and even 1500 m in Jameson Land. The low temperatures favor a thick continuous permafrost

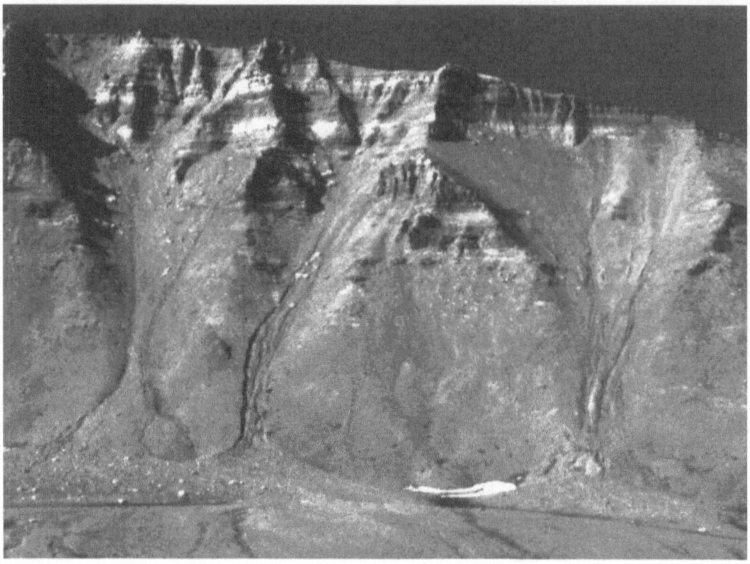

Figure 10.2. Debris flows in Jameson Land (East Greenland). The mean length of the transport zone is about 500 m.

of 80 to 220 m, and the active layer (part of the frozen ground thawing in summer) is about 1 m thick.

The source areas of the flows correspond to an outcropping in a high-walled cuesta (a ridge with one steep slope) near Harris Fjeld, which exhibits deep alcoves and funnels of about 15 to 60 m high and 50 to 100 m wide, carved into highly jointed shales and sandstones of Lower Jurassic age (Figure 10.2). Some of them originate in the upper parts of talus aprons below the rock cornice. Single or multiple linear channels with lateral ridges (levees) characterize the transport zone, as in many Martian gullies (Figure 10.1). The Greenland levees are observed lateral to both sides of the internal channels as continuous and narrow ridges reaching heights up to 10 to 50 cm above the surrounding slope (Figure 10.3). These levees are deposits of coarse and poorly sorted materials. Their mean length is a about 300–500 m, with slope angles between 25 and 30° and their size are comparable to gullies observed on several MOC images. For most periglacial areas, levees are typical of a particular kind of flows with a yield strength (Allen, 1997). The yield strength corresponds to the minimal shear stress the material needs to reach before to flow. They are typically associated with flows that contain 50 to 90% clastic materials (silt to pebble size; Allen, 1997). The existence of levees implies the incorporation of meltwater in the debris over a significant thickness of material (Innes, 1983; Boelhouwers *et al.*, 2000). The flow of such mixtures

Figure 10.3. Lateral levees of 10 to 50 cm height along the debris flows. Location: Jameson Land.

usually follows mechanical behavior of a Bingham fluid. Therefore, the flow stops rapidly when the slope decreases at the foot of the hillslope due to a reduction of the shear stress to values below the critical stress needed for such a mixture to flow.

On the hillsides below simple or multiple channels, their deposits form fans whose top is found up to 60 m above the base of the slope. Radial channels, with simple or anastomosing patterns, are observed on the distal cones or fans, whose slope angles are more gentle (25–10° or less). Distal-lobes are found lower down at the end of the channels at the foot of hill slopes. Half buried or exhumed sandstone blocks are found scattered on the fans, which are made of roughly stratified material.

10.3 Triggering factors in debris flow occurrence

In periglacial environments, two main conditions are needed for debris flow to be triggered: (1) the existence of a permafrost with an active layer, and (2) a large volume of weathered materials (Caine, 1980; Johnson and Rodine, 1984; Iverson, 1997).

10.3.1 The permafrost influence

Field studies showed that the periglacial debris flows in Greenland are not formed by ground water seepage, but that they result from the thawing of the near-surface ice (melting of the active layer), which becomes impregnated by liquid water when the snow and the interstitial ice melt (Harris and Gustafson, 1993). Seasonal snow covers occur in the alcoves and interstitial ice is usually found under the surface and can be seen to a depth of several meters. The permafrost limits the infiltration and permits debris flows triggering. In this case, debris flow triggering can be linked either to a rain event on an active layer already saturated with water or can be due simply to snow/ice melt without former precipitation (French, 1988; Wenshou and Cunhai, 1992). Generally speaking, the presence of a continuous permafrost explains that no water sources were observed in the study area in Greenland. Similarly, because this region is subject to a dry and cold climate, showers are not frequent. Although summer rains can contribute to the erosion of the plateau (shallow hortonian gullying on bare slopes), direct run-off from rains is not the main process creating the debris flows. Some heavy and continuous rainfalls (one or two days) may occur in summertime, but we could not observe any debris flow triggered by them. On the contrary, the most recent debris flows that were observed had occurred while snow was still present on the lower slopes, as shown by the preservation of residual cones of snow protected until the end of July by thin layers of mud and debris (Figure 10.4). In conclusion, the melting of the active layer represents the main source of liquid water. Debris flows are usually initiated when the critical shear stress is reached after the increase of fluid pressure within the layer of weathered debris (Johnson and Rodine, 1984; Iverson, 1997).

10.3.2 Weathered debris and debris flow occurrence

After a flow has occurred, the time needed for the store of rock debris to be reconstituted depends on climatic and geomorphological parameters.

Figure 10.4. Recent debris flows in Jameson Land (Greenland) which occurred by the infiltration of melting snow during the summer season.

The climatic parameters are linked to the temperatures and precipitations (Jomelli *et al.*, 2004). Generally, the rate of this type of weathering depends on the frequency of freeze/thaw alternations inside the rock-wall, their duration, and their intensity. However, fragmentation of the rock by freezing occurs only when the rock has a certain water content. This moisture can be related to rain or the melting of the snow cover. However, the time necessary for the reconstitution of the stock also depends on lithology, the nature of the contributive surface, and the altitude of the rock-wall (since temperatures vary according to altitude).

Generally speaking, large volume of debris can be due either to morainic accumulations on a slope (Haeberli *et al.*, 1990; Rickenman and Zimmermann, 1993) or to debris accumulated in a couloir (or deep, narrow, steep-sided valley) or at the apex of a slope deposit by frost weathering (Pech and Jomelli, 2001). In Greenland, this process is helped by the action of freeze–thaw cycles which fracture rocks and form many debris-covered hillslopes (Innes, 1983), especially in the sandstones and shales that outcrop in this cuesta. Debris (slabs, pebbles, and abundant fine products) accumulate on slopes with angles of repose approaching 30°.

On Mars, detailed analyses of MOC images show some correlation between large-scale debris flows and geological context. In particular, two interior walls of impact craters, a Gorgonum impact crater (37.4° S, 168° W) and a Terra Cimmeria crater (204° W, 41° S), exhibit unusually well-developed flow channels and large end-lobes (Figures 10.5–10.8). Both craters have internal channels that are about 3 to 6 km long and are several tens of meters wide, and both exhibit a highly dissected crater rim and unusual large-scale debris flows at the intersection between a wrinkle ridge and the crater rim. The diameter of the Gorgonum impact crater is 14 km. Its morphology is relatively fresh and exhibits lobate ejecta and a central peak. High-resolution images of this crater indicate a general stratigraphic uplift of the layered terrains along the crater rim. The western side of this crater crosscuts the eastern flank of a north trending wrinkle ridge (Figure 10.5) without any change in the morphology of the crater rim or floor topography.

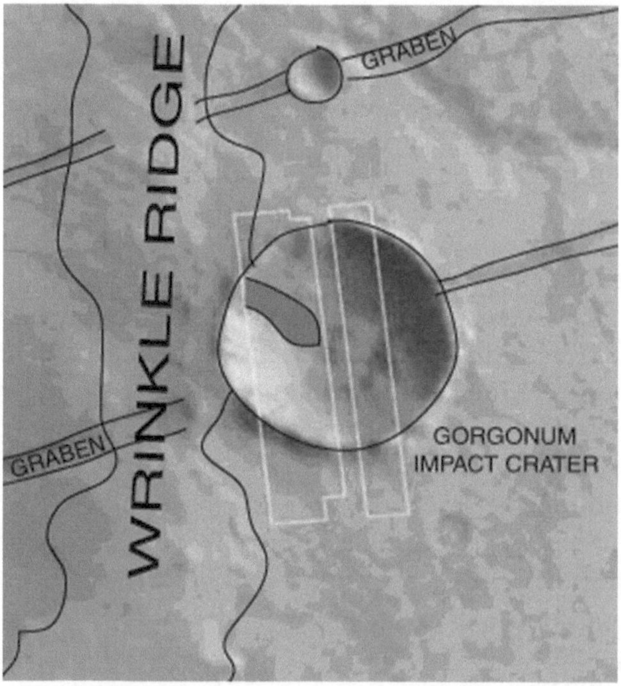

Figure 10.5. Geomorphological context of the Gorgonum impact crater (37.4° S, 168° W). Red area indicates the localization of large-scale debris flows. MOC image MSSS. North is at the top of the image.

Figure 10.6. Large-scale debris flows near 37.4° S and 168° W. The greatest distance travelled, measured from the head of the debris flow to its distal deposit, is 6 km. MOC2–241 image, MSSS.

The other impact crater located at Terra Cimmeria displays unusually large debris flows. A wrinkle ridge crosscuts the eastern part of the crater rim (Figure 10.7). This crater is 20 km in diameter. The mean width of the ridge is 3 km and its height is 216 m (from MOLA data). The elevation of the crater rim on the surrounding supposed lava plain is 670 m.

In both impact craters, the upper part of the walls have steep slopes that are dissected by channels. The debris flow deposits bury the lower parts of the crater walls. The morphology of these debris flows suggests several rapid mass movements of debris. They are characterized by their distinct steep V-shaped channels emanating from the crater rim and by their large-scale debris fans on the crater floor (Figures 10.6 and 10.8). These large-scale debris flows mostly originate from faults with thick accumulations of debris at the base of escarpments. In both cases, there is a distinct correlation between these wrinkle ridges and large-scale debris flows. Impact events crater, disturb, and break up the upper crust to a depth of the same order as the crater diameter (Melosh, 1989). The broken material near the crater is weaker

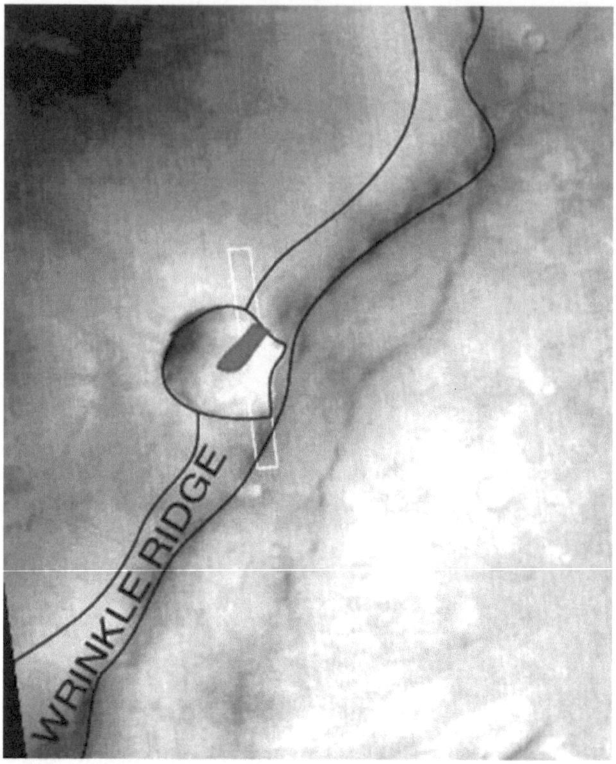

Figure 10.7. Geomorphological context of the impact crater at Terra Cimmeria (204.82° W and 41.25° S). MOC image. Red surface shows the localization of the large scale debris flow. North is at the top of the image.

than the material from the surrounding zone by the reduction of its cohesion (Allemand and Thomas, 1995). By analogy with terrestrial analogs, one hypothesis is that these large-scale debris flows are due to the relatively high porous structure of the highly fractured zone. Underneath the ejecta deposits probably lies a fractured regolith of significant porosity with various fault structures in relation to wrinkle ridges. Under the present cold environment, water or ice may have been trapped in pores in the breccia. Fractures or faults controlled preferential condensation of atmospheric H_2O and its accumulation in the layers.

10.4 The obliquity variation scenario: the Martian case

On the basis of the Greenland analogy and the near-surface melting scenario, it was tempting to assess if such a process involving the melting of the near-surface could have occurred on Mars. Using a state-of-the-art model

Figure 10.8. The Terra Cimmeria impact crater exhibits exactly the same kind of large-scale debris flows (MOC image M10–01143, MSSS) as those shown in Figure 10.6. Scene 2.8 × 5 km.

of the current Martian climate used to analyze current mission observations, the temperature of the surface and sub-surface on various locations on Mars and for various obliquities has been calculated (Forget *et al.*, 1999). Obliquity is the inclination of the planet rotation axis on its orbit (currently about 25°). Mars orbit specialists believe that its obliquity has varied a lot in the past million years (from 0° to 60°), and these variations have strongly affected the climate (Laskar and Robutel, 1993; Touma and Wisdom, 1993).

Our calculations revealed that the only places on Mars where the daily mean temperature has been above the melting point of water during the past obliquity cycles are the mid and high latitudes above 30°, especially on poleward-facing slopes (Costard *et al.*, 2002). The corresponding thermal wave could have melted the ground ice over several tens of centimeters. The fact that poleward-facing slopes receive more sunlight and get warmer at high obliquity in the summer is due to the pole being tilted toward the Sun.

This preferential orientation and the latitudinal distribution of the warmest near-surface temperature coincide with the location of the observed Martian gullies, suggesting a link between near-surface warming and debris flows. In fact, to further test this hypothesis, a more detailed statistical analysis of the observed gully orientations has been performed (Forget *et al.*, 2003). The statistical data show an almost perfect agreement between the variations of the orientation of the gullies with latitude observed on Mars and those predicted by the model. On this basis, and since Mars at high obliquity is thought to have had a water-rich atmosphere (Toon *et al.*, 1980; Mellon and Jakosky, 1995; Richardson and Wilson, 2002), much thicker than today (so that liquid water could sometime flow on the surface), it is believed that the Mars gullies result from the melting of the near-surface ground ice and interstitial ice at high obliquity (Costard *et al.*, 2002).

10.5 Conclusion

Examination of the Martian conditions (low temperature and global permafrost) leads to the idea that analogies may be considered between the Martian and terrestrial periglacial climates. The morphological characteristics of most Martian gullies, with levees and accumulation of debris at their base, do suggest that they are formed by debris flows rather than a slow erosion by a continuous surface runoff. Debris flows occur as short and repetitive events. Some large-scale debris flows on Mars occur on highly fractured zones like wrinkle ridges. On the basis of the Greenland analogy and the near-surface melting scenario, it is tempting to assess if such a process involving the melting of the near-surface could have occurred on Mars. Calculations and preferential orientations of Martian gullies suggest that they result from the melting of the near-surface ground ice and interstitial ice at high obliquity.

Acknowledgments

The authors thanks D. Mercier, P. Coussot, and R. M. Haberle for helpful discussions. The expedition to Scoresby Sund was organized by the Institut Polaire Français and was supported by the Laboratoire de Géographie Physique, Meudon. The Mars climate model has been developed in collaboration with the University of Oxford with the support of CNRS, ESA (TRP contract 11369/95/NL/JG(SC)), and CNES. This work is supported by Programme National de Planétologie, INSU, CNRS, France. Helpful comments were made by K. Tanaka and D. Milton.

References

Allemand, P. and Thomas, P. G. (1995). Localization of Martian ridges by impact craters: mechanical and chronological implications. *Journal of Geophysical Research*, **100**, 3251–62.

Allen, P. A. (1997). *Earth Surface Processes*. Osney Mead: Blackwell Science.

Baker, V. R. (2001). Water and the Martian landscape. *Nature*, **412**, 228–36.

Boelhouwers, J., Holness, S., and Sumner, P. (2000). Geomorphological characteristics of small debris flows on Junior's Kop, Marion Island, maritime sub-Antarctic. *Earth Surface Processes and Landforms*, **25** (4), 341–52.

Caine, N. (1980). The rainfall intensity-duration control of shallow landslides and debris flow. *Geografiska Annaler*, **62A**, 23–8.

Costard, F., Forget, F., Mangold, N., and Peulvast, J. P. (2002). Formation of recent Martian debris flows by melting of near-surface ground ice at high obliquity. *Science*, **295**, 110–13.

Coussot, P. (1994). Rhéologie des laves torrentielles. *La Houille Blanche*, **3**, 32–7.

Christensen, P. R. (2003). Formation of recent Martian gullies through melting of extensive water-rich snow deposits. *Nature*, **422**, 45–8.

Forget, F., Hourdin, F., Fournier, R., Hourdin, C., and Talagrand, O. (1999). Improved general circulation models of the Martian atmosphere from the surface to above 80 km. *Journal of Geophysical Research*, **104** (E10), 24155–76.

Forget, F., Mangold, N., and Costard, F. (2003). Scenarios to explain the formation of gullies on Mars: numerical simulation with a climate model. 6th International Conference on Mars, Pasadena, July 20–25, 2003, LPI CD: CB–1164, abstract 3192.

French, H. M. (1988). The periglacial phenomena: ancient and modern. *Journal of Quaternary Science*, **3** (1), 110 pp.

Gaidos, E. J. (2001). Cryovolcanism and the recent flow of liquid water on Mars. *Icarus*, **153**, 218–23.

Haeberli, W., Rickenmann, D., and Zimmerman, M. (1990). Investigation of 1987 debris flows in the Swiss Alps: general concept and geophysical soundings. Hydrology in Mountainous regions: II. Artificial reservoirs; Water and slopes. Proceedings of two Lausanne Symposia. *International Association of Hydrological Sciences Publication*, **194**, 303–10.

Hartmann, W. K. (2001). Martian seeps and their relation to youthful geothermal activity. *Space Science Reviews*, **96** (1–4), 405–10.

Harris, S. A. and Gustafson, C. A. (1993). Debris flows characteristics in an area of continuous permafrost, St Elias Range, Yukon territory. *Zeitschrift fur Geomorphologie*, **37**, 41–56.

Innes, J. L. (1983). Debris flows. *Progress in Physical Geography*, **7** (4), 469–501.

Iverson, R. (1997). The physics of debris flows. *Reviews of Geophysics*, **35**, 245–96.

Johnson, A. and Rodine, J. R. (1984). Debris flow. In *Slope Instability*, ed. D. Brunsden and D. B. Prior. New York: Wiley, pp. 257–361.

Jomelli, V., Pech, P., Chochillon, C., and Brunstein, D. (2004). Geomorphic variations of debris flows and recent climatic change in the French Alps. *Climatic Change*, **64** (1–2), 77–102.

Knauth, L. P. and Burt, D. (2003). Electrically conducting, Ca-rich brines, rather than water, expected in the Martian subsurface. *Geological Society of America Bulletin*, **115** (5), 566–80.

Knauth, L. P., Klonowski, S., and Burt, D. (2000). Ideas about the surface runoff features on Mars (letter). *Science*, **290**, 711–12.
Laskar, J. and Robutel, P. (1993). The chaotic obliquity of the planets. *Nature*, **361** (6413), 608–12.
Malin, M. C. and Edgett, K. E. (2000). Evidence for recent ground water seepage and surface runoff on Mars. *Science*, **288**, 2330–5.
Major, J. J. (1997). Depositional processes in large-scale debris flow experiments. *Journal of Geology*, **105**, 345–68.
Mangold, N., Costard, F., and Forget, F. (2003). Debris flows over sand dunes on Mars: evidence for liquid water. *Journal of Geophysical Research*, **108** (E4), doi: 10.1029/2002JE001958.
Mellon, M. T. and Jakosky, B. M. (1995). The distribution and behavior of Martian ground ice during past and present epochs. *Journal of Geophysical Research*, **100**, 11781–9.
Mellon, M. T. and Phillips, R. J. (2001). Recent gullies on Mars and the source of liquid water. *Abstracts of Papers Submitted to the 32nd Lunar and Planetary Science Conference*. Houston: Lunar and Planetary Institute, CD 32, Abstract 1182.
Melosh, H. J. (1989). *Impact Cratering: A Geologic Process*. New York: Oxford University Press.
Musselwhite, D. S., Swindle, T. D., and Lunine, J. I. (2001). Liquid CO_2 breakout and the formation of recent small gullies on Mars. *Geophysical Research Letters*, **28**, 1283–5.
Pech, P. and Jomelli, V. (2001). Rôle du cône apical dans le déclenchement des coulées de débris alpines du massif du Dévoluy, Hautes-Alpes, France. *Géographie Physique et Quaternaire*, **1**, 47–61.
Phillips, C. J. and Davies, T. R. M. (1991). Determining rheological parameters of debris flow material. *Geomorphology*, **4**, 101–10.
Peulvast, J. P. (1988). Pre-glacial landform evolution in two coastal high latitude mountains: Lofoten-Vesteralen (Norway) and Scoresby Sund area (Greenland). *Geografiska Annaler*, **70A**, 351–60.
Richardson, M. I. and Wilson, R. J. (2002). Investigation of the nature and stability of the Martian seasonal water cycle with a general circulation model. *Journal of Geophysical Research*, **107** (E5), doi: 10.1029/2001JE001536.
Rickenmann, D. and Zimmermann, M. (1993). The 1987 debris flow in Switzerland: documentation and analysis. *Geomorphology*, **8**, 175–89.
Toon, O. B., Pollack, J. B., Ward, W., Burns, J. A., and Bilski, K. (1980). The astronomical theory of climatic change on Mars. *Icarus*, **44**, 552.
Touma, J. and Wisdom, J. (1993). The chaotic obliquity of Mars. *Science*, **259**, 1294–7.
Van Steijn, H. (1991). Frequency of hillslope debris flows in a part of the French Alps. *Jeomorfoloji Dergisi (Geomorphology Bulletin)*, **19**, 83–90.
Van Steijn, H., De Ruig, J., and Hoozemans, F. (1988). Morphological and mechanical aspects of debris flows in parts of the French Alps. *Zeitschrift für Geomorphologie*, **32**, 143–61.
Wenshou, W. and Cunhai, G. (1992). Studies of ice-snow melt debris flows in the western Tian Shan mountains, China. Erosion, debris flows and environment in mountain regions. Proceedings of the Chengdu Symposium, *International Association of Hydrological Sciences Publication*, **209**, 329–36.

11

Siberian rivers and Martian outflow channels: an analogy

François Costard
UMR 8148 IDES, Université Paris-Sud

E. Gautier and D. Brunstein
CNRS UMR 8591, Laboratoire de Géographie Physique, Meudon

11.1 Introduction

The origins of Martian outflow channels have been the subject of considerable debate. Our examination of Martian conditions has led us to the idea that many commonalities exist between the Martian and terrestrial periglacial climates. Particularly, the formation processes for Martian outflow channels and Siberian valleys seem to be similar on both planets. Therefore in this chapter we propose an analogy between the Lena River and the Martian outflow channel Ares Vallis. The fluctuating flows of the Lena River might also afford an analog to the Martian channel-forming flows. Both hydrosystems were or are associated with a periglacial environment characterized by a deep and continuous permafrost.

11.2 Periglacial environments in Yakutia and on Mars

Central Yakutia (Figure 11.1) is a very specific morphoclimatic zone, bound on the east by the Verkhoyansk Mountains (elevation 2900 m) and on the west by the Siberian shield of (200 m to 1000 m in elevation). The extreme continental Yakutian climate in Siberia is characterized by long and cold periods with minimum surface temperatures of −72 °C. In summer, the maximum surface temperature can reach 38 °C. Low precipitation (less than 200 mm per year), and high evaporation and sublimation rates are characteristics of a very dry climate (Katasonow and Soloviev, 1969).

The low mean annual surface temperature together with the thin snow cover favor the existence of a deep permafrost layer. With a permanent freezing

The Geology of Mars: Evidence from Earth-based Analogs, ed. Mary Chapman. Published by Cambridge University Press.

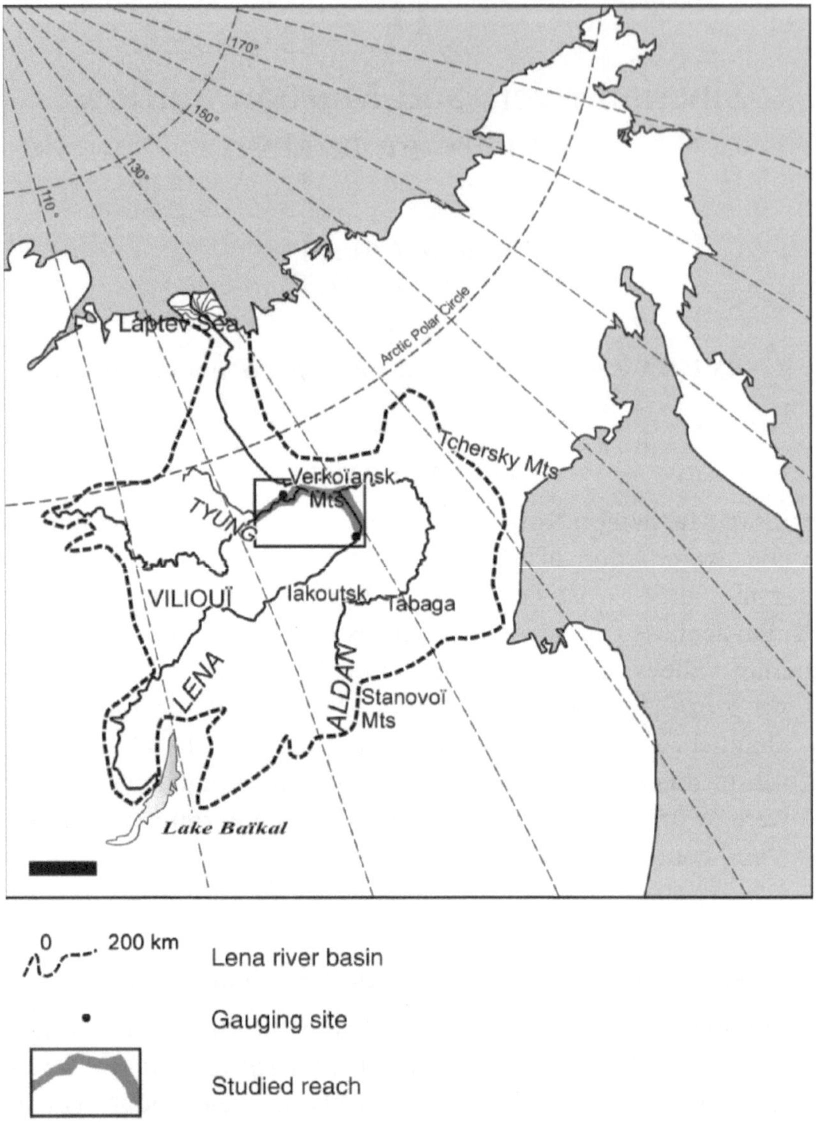

Figure 11.1. Location map of the Lena River in Siberia.

temperatures and a continuous permafrost, Central Yakutia is considered as a periglacial environment. The maximum depth of frozen soil in Oïmiakon (Yakutia) is about 1500 m and the thickness of the average Siberian permafrost is about 350 m (Anisimova *et al.*, 1973). A drilling down to 1450 m indicated the existence of a permafrost whose conditions of pressures and temperatures were favorable to the formation of clathrates (Makogon *et al.*, 2005). The temperature of the Siberian ground ice at the depth of minimum annual seasonal change varies from −5 °C to −12 °C (Péwé, 1991). Locally in Siberia, the temperature of ground ice decreases with depth down to −12 °C, indicating residual colder past climate during the Pleistocene (French, 1988). This permafrost contains 50% of interstitial ice, but segregation ice is 80% of ice content (Anasimova *et al.*, 1973). Massive icy beds are common along the Lena River banks and may reach an ice content as high as 250% (by mass). The thickness of the active layer varies from 1.5 to 2 m in the silts and 4 m in sands (Anisimova *et al.*, 1973).

From a thermal point of view, Mars looks like Siberia for many reasons. Mars is a dry cold planet with extensive permafrost. The mean annual temperature of the Martian surface is estimated to be −50 °C to −60 °C (Kieffer *et al.*, 1973), and the atmospheric surface pressure is about 600 Pa. At a latitude poleward of ±40°, the surface temperature remains under 0 °C all year. Because of these conditions, permafrost exists over the entire planet. Estimates of the permafrost thickness range from 3 to 7 km near the poles to 1 to 3 km near the equator (Fanale *et al.*, 1986; Clifford, 1993). The presence of ice in the ground was confirmed with the detection of near-surface ice poleward of 55° by the Neutron Spectrometer aboard Mars Odyssey (Boynton *et al.*, 2002; Feldman *et al.*, 2002). The thermal conditions of the Martian ground-ice (temperature and thickness) may be similar to those of the present terrestrial permafrost.

11.3 Comparative approach of hydrogeomorphology: the Lena River and Ares Vallis

It has been demonstrated that the fluvial dynamics is dependent on the main variables of the fluvial basin (mainly morphostructural and bioclimatic conditions), that determine semi-dependant variables (hydrological functioning: intensity and frequency of morphogenic discharges, and characteristics of the sediment load; Leopold and Wolman, 1957; Schumm, 1977; Knighton, 1999). The geometry (longitudinal gradient, and channel cross-section, width, and depth) and the fluvial pattern (fluvial plan forms defined by the number and the sinuosity of the channels) are adapted to spatial and temporal variations of water and sediment discharges (Leopold and Wolman, 1957; Schumm, 1977; Knighton, 1999). Therefore, the fluvial landforms express the influence of the whole system parameters, and their analysis is commonly used to understand the fluvial dynamics. The Lena River study provides a better understanding of the fluvial forms for Martian outflow channels.

The total length of the Lena River exceeds 4000 km with a main south–north course and the river receives few tributaries (Kirenga, Vitim,

Oliekma, Aldan, Viliouy rivers). The lower Lena River forms a delta into the Laptev Sea. During the major floods, for which the peak discharge can reach 20 000–35 000 $m^3 s^{-1}$, the inundation covers a 25 km width in the floodplain downstream from Yakutsk (62° N). The Lena River is composed of several wide and shallow channels, their width varying between several hundred meters and three kilometres; very long (1–5 km) sandy bars and forested islands constitute the main deposition zones (Figures 11.2 and 11.3). In the floodplain and islands, another type of channel can be observed: narrower branches (a few tens of meters wide), often sinuous, and frequently discontinuous (without outlet). Because of the very gentle slope of the Central Siberia rivers (0.0001), the specific power stream is very weak (less than 10 W m^{-2}; Figure 11.4).

The Lena River is comparable in scale to most Martian outflow channels (Ares, Tiu, Simud, and Kasei Valles) which are 10 to 30 km wide and more than 1000 km in length. With a total length of 1500 km, Ares Vallis is one of the largest Martian outflow channels (Figure 11.5). Ares Vallis consists of a valley with a few tributaries, averaging at least 25 km in width along a mostly rectilinear course. The channel has a gradient of 0.0001 – more or less the

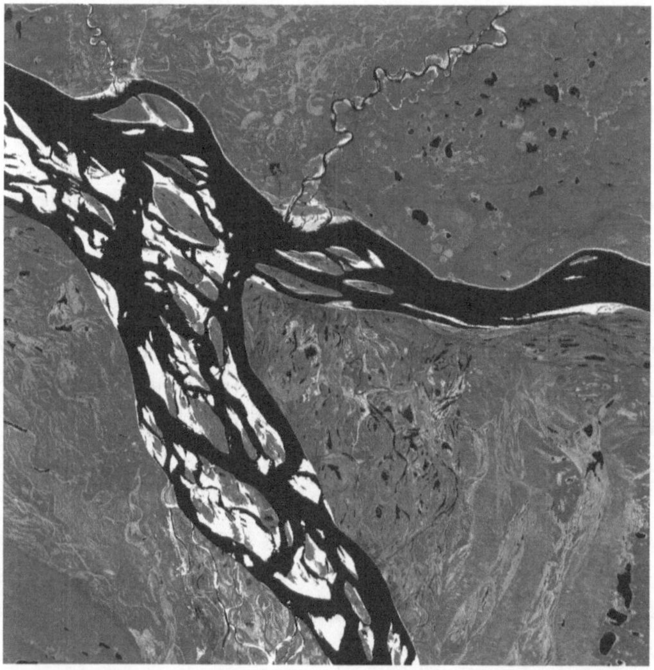

Figure 11.2. The Lena–Aldan Rivers junction (Siberia). Landsat image. One can note the presence of thermokarst lakes in the flood plain.

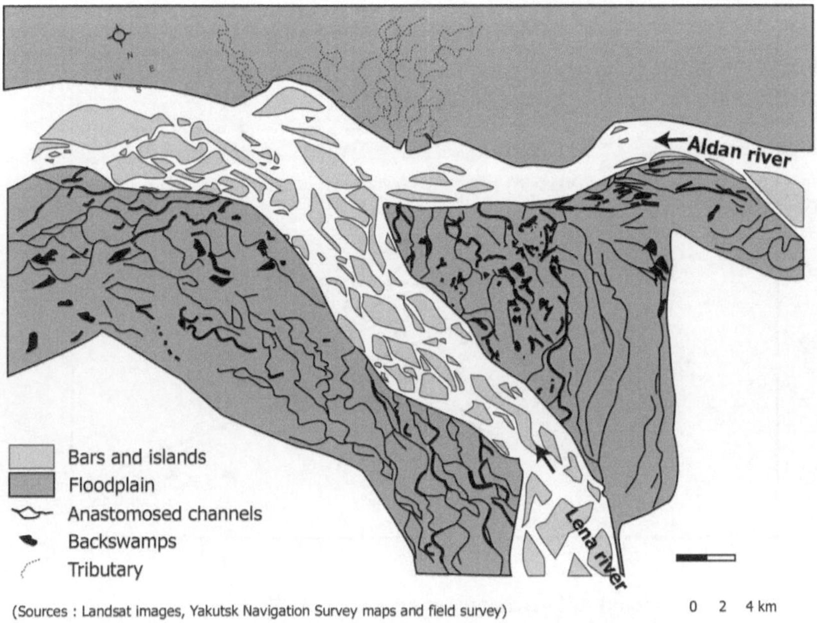

Figure 11.3. Flood plain of the Lena River. From Gautier and Costard (2000).

same as for the Lena River (0.0001; Figure 11.4). It generally is incised between 1 and 2 km into the adjacent plateau surface, and the source areas correspond to chaotic terrains which are collapse zones.

Both high-resolution Mars Orbital Camera (MOC) images (2 m/pixel) and Themis visible images (18 m/pixel) of Ares Vallis floodplain reveal a pattern of multiple channels. Streamlined islands and long and narrow forms can be interpreted as median bars of fluvial deposits (Figure 11.6). These supposed median bars are very similar to those observed along the Siberian rivers. The morphometric data for the Ares Vallis streamlined forms compare closely to the width/length ratio of the Lena islands (Table 11.1). However, there is a slight increase of the elongated shape for Mars indicating a higher turbulence of the flow (Baker, 1982), and so relatively higher velocity water flows for Ares Vallis than for Siberian rivers.

11.4 Specific hydrosystems dominated by short and intense outburst floods

Due to its large catchment (2.49×10^6 km^2), the Lena River brings annually 525 km^3 of water to the Laptev Sea (Antonov, 1960). The hydrology of the Lena and its tributaries is characterized by high discharge variations

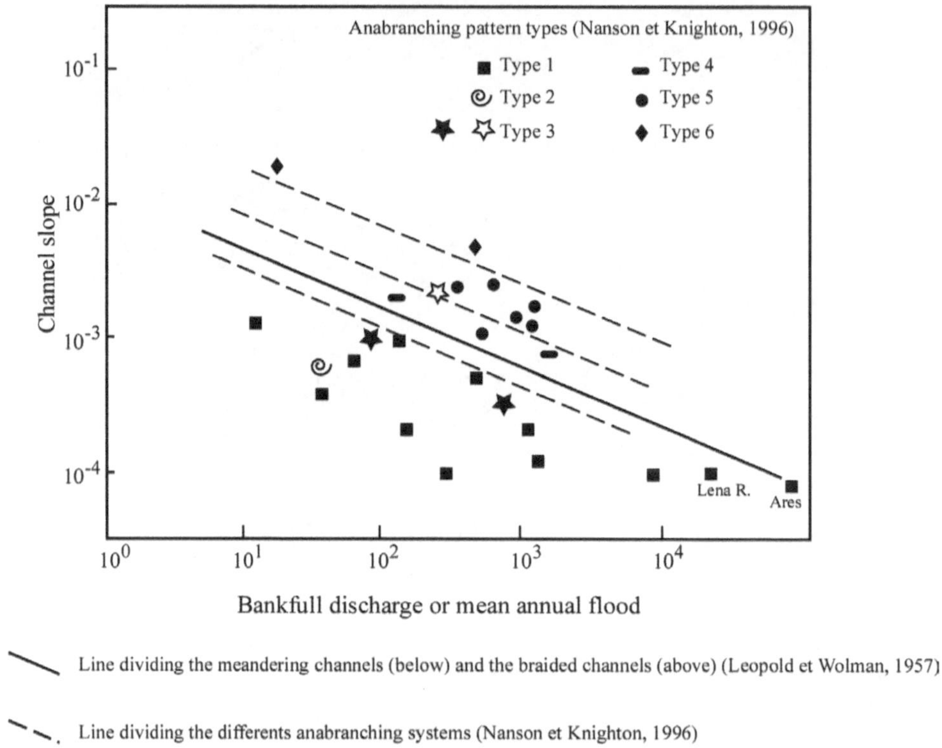

Figure 11.4. Position of the Lena River and Ares Vallis on a slope-discharge plot and distribution of the six types of anabranching rivers. From Leopold and Wolman (1957), Nanson and Knighton, (1996), modified. Ares Vallis (Mars) and Lena River (Siberia) are anabranching rivers which maximize bed-sediment transport under very low gradient and short efficient discharges.

(Gordeev and Sidorov, 1993; Gautier and Costard, 2000). The long drought during winter is strongly influenced by the Yakutian climate: under a carapace of thick ice, the river discharge is then extremely reduced (900–1500 $m^3 s^{-1}$ at Tabaga gauging site, near Yakutsk, and 4–12 $m^3 s^{-1}$ for the Vilioui River, a left side tributary), and its lowest level occurs at the end of winter, a few weeks before the flood. Figure 11.7 shows the discharge versus time and suggests that the majority of the discharge occurs during the short break-up period of about 4–6 weeks. The maximum flood peak of the Lena River can be 50 000 $m^3 s^{-1}$ near Yakutsk and reaches 100 000 $m^3 s^{-1}$ downstream from the Aldan junction, its main tributary (Gautier and Costard, 2000). The south–north flow of most Siberian rivers involves a fluvial outburst delay of 30 to 50 days between the upper basins and the middle valleys (Yamskikh *et al.*, 1999). Coming from the south, the flood arrives in Central Siberia before

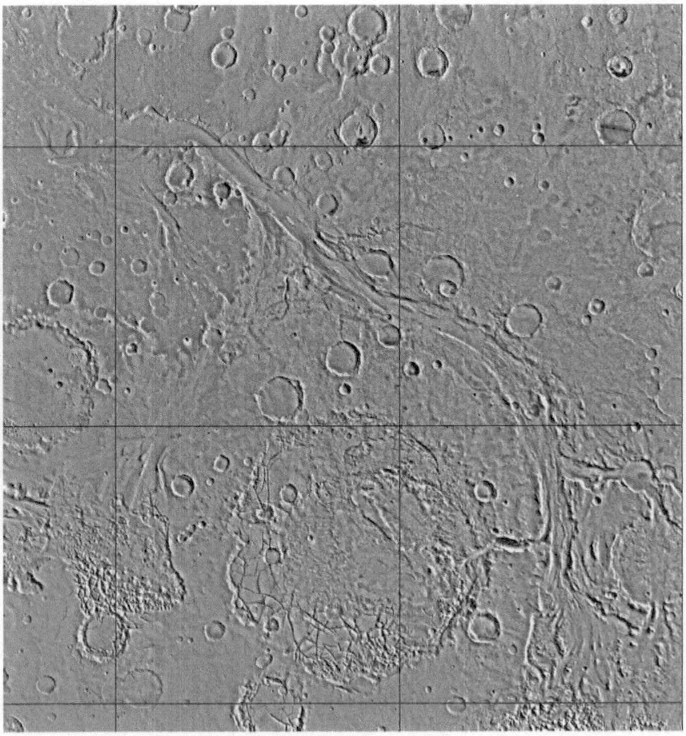

Figure 11.5. Part of the Martian outflow channel Ares Vallis. The image is 1000 km wide. Viking mosaic, NASA.

the increase of atmospheric temperature. In the middle Yenisseï valley, Yamskikh *et al.* (1999) also observed that the break-up of the river ice precedes (by 20 to 25 days) the increase of the atmospheric temperatures.

For this reason, the "warm" water provokes a massive break-up of the ice covering the river. During the outburst flooding, the Lena River carries large quantities of ice blocks and woody debris that accumulate locally. These ice- and log-dams provoke a water level increase, and thereby reroute the water in the floodplain and islands (Gautier and Costard, 2000). This specific mechanism is probably the origin of the small sinuous channels developed in the floodplain and islands. These anastomosed branches are rapidly dewatered at the end of the flood and the brevity of their submersion could explain why a great number of them disappear gradually downstream and do not have an outlet.

The extensive inundation can also be explained by the presence of a permafrost which limits the infiltration. The discharge decreases rapidly at the middle of July, because of the lack of precipitation and

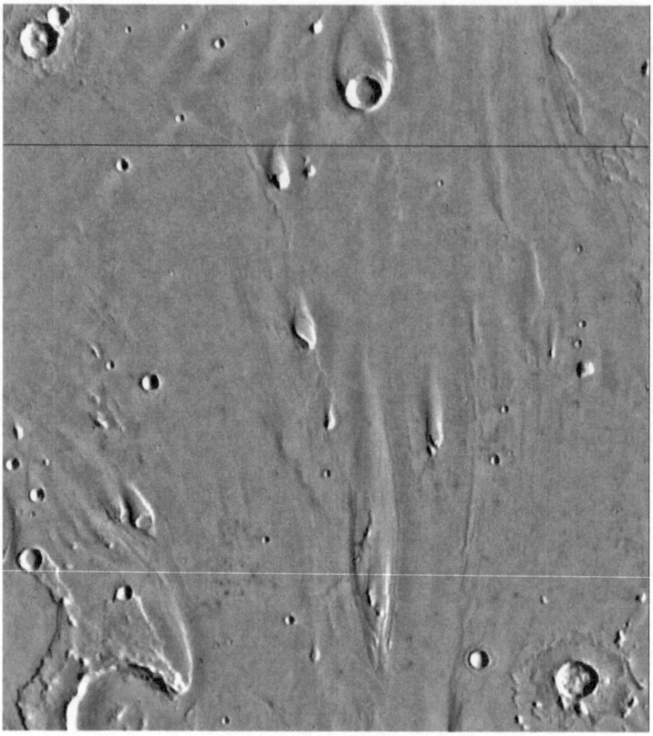

Figure 11.6. Fluvial bars and islands on the Ares Vallis outflow channel (Mars). Viking mosaic, NASA. The image is 120 km wide.

Table 11.1. *Comparative morphometric and hydrologic characteristics of the Lena River and Ares Vallis*

	Width/length ratio (islands, bars)	Length (km)	Width (km)	Gradient	Discharge ($m^3 s^{-1}$)
Lena River (Siberia)	2.8 to 3.1	4000	Up to 25	0.0001	Up to 50 000
Ares Vallis (Mars)	1.8 to 2.5	1500	Up to 25	0.0001	Up to 10^8

of ground-waters. A secondary peak of discharge occurs at the end of summer, which is related to storms.

The morphometric characteristics of the Lena River can be compared to the Martian outflow channels. Different interpretations of the Martian outflow channels have been proposed. Lucchitta (1982) considered the channels as possible glacial valleys. Carr (1979) suggested that these valleys were formed by sudden release of water from confined aquifers producing catastrophic

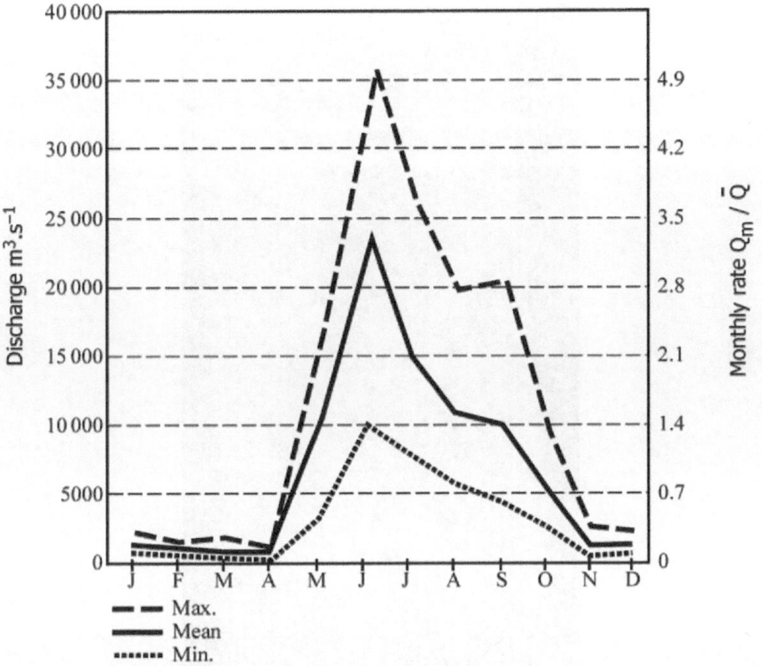

Figure 11.7. Regime of the Lena River. From Gautier and Costard (2000).

floods and Baker (1982) also proposed highly turbulent catastrophic floods. According to Wallace and Sagan (1979), the prevailing cold, dry Martian environment would have generated an ice cover.

Previous studies suggested the Martian floods would have been ice-rich (Lucchitta, 2001) and perhaps highly fluctuating, similar to the Siberian rivers. High-level fluid erosion on the upland surface indicates high discharge variations for Ares Vallis. High-resolution images of braided and crosscut channels indicate discharge variations (Figure 11.8). For example, Carr (1979) proposed discharge rates from 28 000 $m^3\ s^{-1}$ to 110 000 $m^3\ s^{-1}$ for Martian outflow channels and Baker (1982) gave discharge values ranging from 10^6 to almost 10^8 for Ares Vallis. These discharge estimations indicate catastrophic floods quite similar to the highly turbulent regime of the flow for the Lena River.

11.5 Erosional processes: evaluation and impacts of thermal erosion on sediment processes

Three main factors induce strong interactions between the "warm" water, on the one hand, and, on the other hand, the frozen banks and frozen

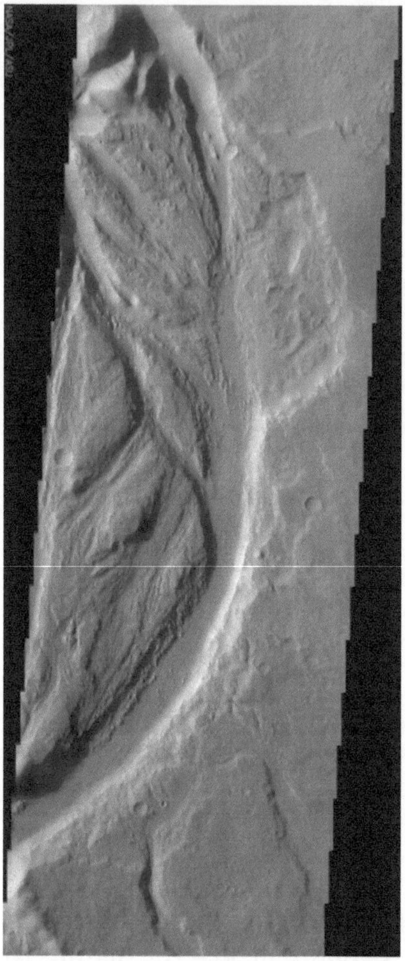

Figure 11.8. High-resolution Themis image of braided and crosscut channels indicating discharge variations near Lunai Planum. Resolution: 19 m/pixel. Image size: 57.8 km × 23.5 km. Image Credit: NASA/JPL/Arizona State University.

ground of floodplain and islands: (1) the marked water level variations in the fluvial bed, (2) the delay between the flood and the air temperature increase, and (3) the rapid increase of water temperature. In Siberia, ice dams and local high variations of the water level locally accelerate the erosion of the river banks. Active thermal erosion processes on river banks and immediate removal of sediments seem to be particularly effective on periglacial rivers (Walker and Arnborg, 1963; Jahn, 1975; Are, 1983; Walker, 1983), especially with non-cohesive sediments (sand, silt). Lateral thermal erosion produces specific forms, such as thermoerosive niches which favor collapsed blocks (Jahn, 1975). This type of thermokarstic subsidence exhibits degraded terraces and large slumps along the river banks. When massive ice (in ice wedges, thick ice lenses, etc.) is present, slumping and preferential bank erosion may occur more rapidly than would otherwise be the case (Jahn, 1975). Thawing of the ice reduces the strength of the thawed sediments and produces easily removable uncemented ground (Costard *et al.*, 2003). Individual pyramidal masses of soil appear separated from the bank by deep and narrow channels where a process of wash out must take place (Caveliev, 1958).

Are (1983) calculated a mean annual erosion of the Lena River banks of 6.5 m per year; on the main channels banks, the erosion can be 19–24 m per year, and on islands heads, a maximal rate of 40 m per year is observed. Gautier *et al.* (2003) conducted a 35-year diachronic study of the fluvial forms of the Middle Lena River. The joint study of aerial pictures (Corona satellite, 1967) and satellite images (Landsat 4 to 7: 1992, 1999, and 2002) allowed us to perform a follow-up of the temporal evolution of Lena River fluvial landforms (Figure 11.9). On the basis of a GIS, a delineation of the main fluvial units (channels and islands) was proposed to quantify their mobility. The analysis reveals a spatial variability of the efficiency of the thermal erosion on the fluvial form mobility. For example, on channels banks, a mean annual bank retreat of 3–4 m per year was calculated, and fast – but very local – retreats can be observed. With regards to the width of the channels, the bank retreat remain low; it represents 0.05–0.1% of the channel width (mean value) and a maximum of 0.45%. For this reason, the main channels remain stable and no channel migration was noticed during the proposed period. Concerning the small narrow channels, no change occurred. Contrarily, the islands heads undergo a fast erosion, with an average of 12 m per year and a maximal retreat of 40 m. The islands also migrate relatively rapidly downstream by erosion on the head and accretion downstream, especially in three zones: upstream and near Yakutsk and at the Aldan River junction (Figure 11.9).

The presence of ground-ice and wide rivers observed on Mars and in Siberia, suggests the occurrence of thermal erosion during the flooding event. The arguments for fluvio-thermal action on Mars as the erosive agent are based on analogy with Siberian rivers (Aguirre-Puente *et al.*, 1994). On Mars, relatively warm floods could have widened channels by a combination of thermal and mechanical erosion along frozen river banks. Thermomechanical action probably resulted from the thawing of the ground and the subsequent transport of thawed sediments (Costard *et al.*, 1999).

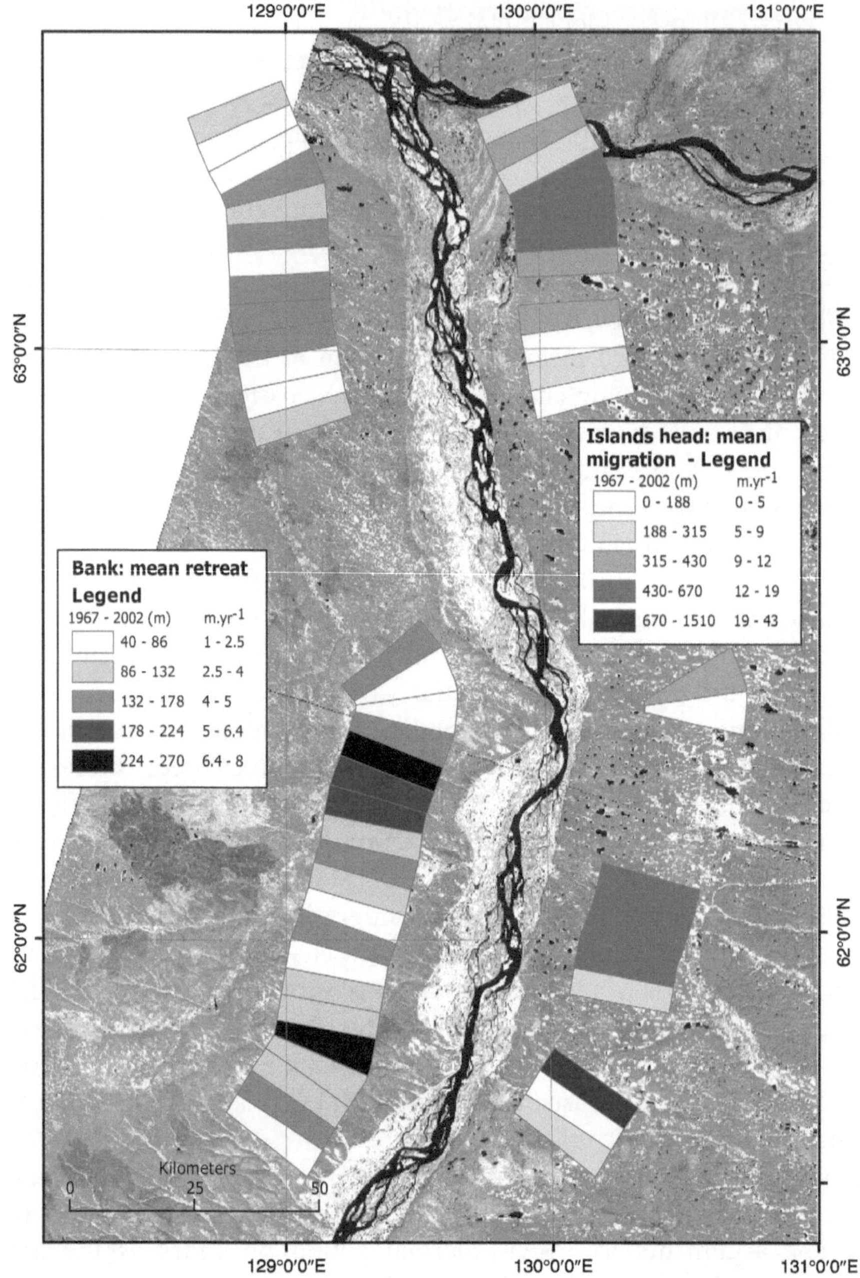

Figure 11.9. Mean and maximum retreats of channel and island banks on the middle Lena River.

11.6 Hydrodynamics of anabranching rivers: application to the Lena River and Ares Vallis

The fluvial pattern of the Siberian rivers (stable multiple channels, presence anastomosed branches in plains and islands, and forested islands) dependent on the specific hydrodynamic functioning and on the limited stream power allow us to compare those rivers with anabranching river patterns. Anabranching systems are composed of multiple channels separated by alluvial islands, generally associated with a low gradient (Smith, 1986; Knighton and Nanson, 1993; Nanson and Knighton, 1996). According to Nanson and Knighton (1996), anabranching forms are characterized by maximized bed-sediment transport under very low gradient and short efficient discharges (Figure 11.4).

On the Lena River, the fluvial thermal erosion on banks is one of the causes of aggradation. During the annual fluvial outburst, local bank retreats provide an abundant sediment supply. The annual suspended load is approximately 7 million tons for the Middle Lena River near Tabaga (Antonov, 1960) and 12–17 million tons (5–7 t km^{-2} yr^{-1}) for the lower Lena River (Lopatin, 1952; Lvovitch, 1971; Gordeev and Sidorov, 1993). The bars of the middle Lena River and of its tributaries are mainly formed by sands (D_{50}: 200–300 μm) and ancient eolian deposits constitute the greatest part of the present-day sediment load of the Siberian rivers (Gautier and Costard, 2000). Because of the very gentle gradient of the river, thermal and mechanical erosion provides a sediment load which exceeds the stream power (flow capacity). Therefore, in the case of the Lena River, the sediment load does not probably migrate over long distances downstream and is accumulated on wide bars and long islands.

Thermal inertia from both IRTM (Betts and Murray, 1994) and IR THEMIS images (Christensen *et al.*, 2003) is related to properties of the surface (particle size, abundance of rocks) and to the proportion of fine and rocks. From these data, the floor materials of Ares Vallis along its entire course have a much higher thermal inertia surface than surrounding upland (Jakosky *et al.*, 2000, Nowicki and Christensen, 2001). The interpretation is that the material is probably sand-sized (200–1000 μm) and may have constituted a abundant sediment supply (Nowicki and Christensen, 2001; Christensen *et al.*, 2003).

11.7 Thermokarst

The Lena valley forms a wide alluvial plain filled with Tertiary and Quaternary deposits as thick as 500 m. These Quaternary terraces contain loess, fluvial sands, and lake deposits of the middle Quaternary age (Katasonov and Soloviev, 1969; Journaux and Dresch, 1972). The height of the Lena terraces is several tens of meters, but can reach more than 100 m close to the Mammouth moutains. The annual flood carries sediments and spreads them along the lower terraces, submerged pools, and topographic depressions. The extreme Yakutia climate favors an intense development of thermokarst, and may be the most spectacular of the planet from its scale and its development (Figure 11.2). In the intense cold of Yakutia, large amounts of ice are incorporated into the accumulating alluvial deposits. Most of these terraces are pitted with round to irregularly shaped pits of kilometer size, termed alasses. They are formed when a large quantity of ground ice thaws (up to 80% by volume) in permafrost. Alasses are characteristics of the thermal degradation stages of permafrost. Their subsequent thaw induces a lowering of the ground surface with collapse and slumping.

The development of alasses are in close connection with the presence of polygonals nets and wedges. When individual alasses join each other, the lateral development of alass basins proceeds to form an alass valley which constitutes the mature development of the thermokarst evolution.

On Mars, Acidalia and Utopia Planitiae received extraordinarily large amounts of water and sediments from the Chryse and Elysium outflow channels (Baker, 1982). That possibility supposes an outwash plain in Acidalia and Utopia Planitiae with volatile rich sediments deposited directly from outflow channels (Lucchitta and Ferguson, 1983). Under the periglacial environment of Mars, discharge water at the mouth of most outflow channels presumably froze to form pore ice or massive icy beds (Costard and Kargel, 1995). This is confirmed with the analysis of Viking and MOC images showing a concentration of rampart craters as well as periglacial-like landforms (Costard, 1989; Barlow and Bradley, 1990; Chapman, 1994).

At the mouth of Ares Vallis, the valley floor is noticeably characterized by irregular hummocky topography and by a large number of interconnected linear ridges that enclose pits (Figure 11.10). Many of these pits are irregular and they lack rims and ejecta. The scale and morphology and these Martian pits are about the same as thermokarst in periglacial areas. In the lower Ares Vallis, these pits were previously hypothesized to be thermokarstic depressions by melting or sublimation of ground ice and subsequent collapse of the surface (Costard and Baker, 2001). On Earth, these kinds of pits are mostly found in periglacial areas where ground ice is widespread. These features suggest the presence of ice-rich sediments with possible massive icy beds in the subsurface Martian megaregolith. In Ares Vallis, intensive thermokarst development seems to be associated with local widening of the channel.

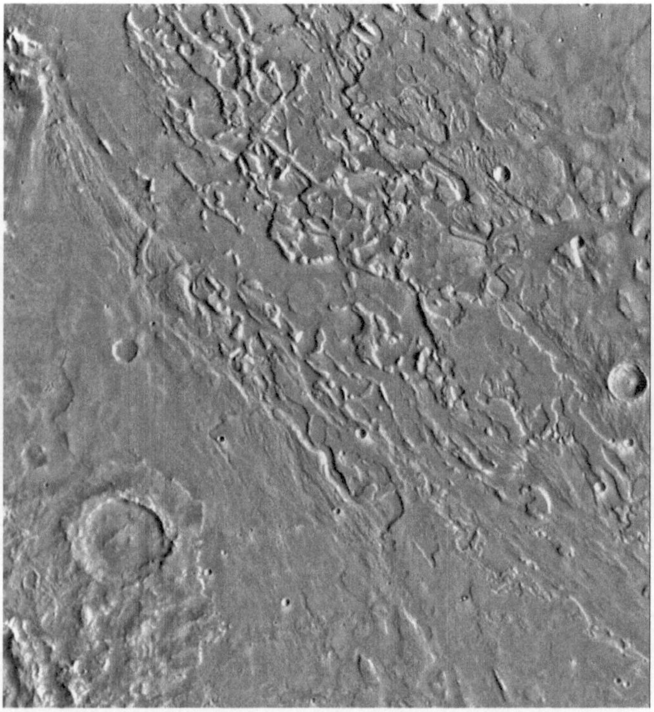

Figure 11.10. High resolution Viking image of interpreted thermokarst reach on Ares Vallis (Mars). Viking image, NASA. Size: 100 km × 120 km. Illumination from right.

Widening of the channel would be associated with decreased fluvial discharges favoring deposition of an alluvial plain (Costard and Baker, 2001). For the extremely cold conditions of Mars, processes related to ice and sediment accumulation would have been likely, as observed in northern Siberia.

11.8 Conclusions

Ares Vallis and the Lena River are anabranching rivers with high width to depth ratio channels. They are comparable in scale and exhibit the same hydrodynamic characteristics (high and variable discharge rate, very low longitudinal profile, sand-size materials). The comparison of several fluvial variables (slope, specific stream power, width to depth ratio, number of channels, bars and islands, sinuosity) between Siberian rivers and Martian outflow channels reveals similarities which suggest that similar processes have been active on both fluvial systems. Both hydrosystems were or are associated with a periglacial environment characterized by a deep and continuous permafrost. Flow regimes of Martian outflow channels would have been short and highly fluctuating, similar to those of Siberian rivers. The presence of ground-ice and wide rivers observed both on Mars and in Siberia suggests the occurrence of thermal erosion during the flooding event.

Acknowledgments

The authors are grateful to Devon Burr and Mary Chapman for helpful reviews. This work is supported by Programme National de Planétologie de l'Institut National des Sciences de l'Univers (CNRS, France).

References

Aguirre-Puente, J., Costard, F., and Posado-Cano, R. (1994). Contribution to the study of thermal erosion on Mars. *Journal of Geophysical Research*, **99** (E3), 5657–67.

Anisimova, N. P., Nikitina, N. M., Piguzova, U. M., and Shepelyev, V. V. (1973). Water sources of Central Yakutia-Guidebook. In *Proceedings, II International Conference on Permafrost. Yakutsk, URSS.* Washington, DC: National Academy of Sciences, pp. 1–47.

Antonov, S. (1960). Delta reki Leny. *Trudy Okeanographitcheskiy Comissyi. Ak. Nauk. SSSR*, **6**, 25–34.

Are, F. E. (1983). Thermal abrasion on coasts. In *Proceedings, Fourth International Conference on Permafrost, Fairbanks, Alaska.* Washington, DC: National Academy Press, pp. 24–8.

Baker, V. R. (1982). *The Channels of Mars.* Austin: University of Texas Press.

Barlow, N. G. and Bradley, T. L. (1990). Martian impact craters: correlations of ejecta and interior morphologies with diameter, latitude and terrain. *Icarus*, **87**, 156–79.

Betts, B. H. and Murray, B. C. (1994). Thermal studies of Martian channels and valleys using Termoskan data. *Journal of Geophysical Research*, **99**, 1983–96.

Boynton, W. V., Feldman, W. C., Squyers, S. W. *et al.* (2002). Distribution of hydrogen in the near-surface of Mars: evidence for subsurface ice deposits. *Science*, **297**, 81–5.

Carr, M. H. (1979). Formation of Martian flood features by release of water from confined aquifer. *Journal of Geophysical Research*, **84**, 2995–3007.

Caveliev, B. A. (1958). Particularités des processus de fusion des glaces dans la couverture glaciaire et les roches gelées. *Problemy Severa*, **1**, 149–55.

Chapman, M. G. (1994). Evidence, age, and thickness of a frozen paleolake in Utopia Planitia, Mars. *Icarus*, **109**, 393–406.

Christensen, P. R., Anderson, D. L., Chase, S. C. *et al.* (2003). Results from the Mars Global Surveyor Thermal Emission Spectrometer. *Science*, **200**, 2056–60.

Clifford, S. M. (1993). A model for the hydrologic and climatic behavior of water on Mars. *Journal of Geophysical Research*, **98**, 10973–11016.

Costard, F. (1989). The spatial distribution of volatiles in the Martian hydrolithosphere. *Earth Moon Planets*, **45**, 265–90.

Costard, F. and Baker, V. (2001). Thermokarst landforms and processes in Ares Vallis, Mars. *Geomorphology*, **37** (3–4), 289–301.

Costard, F. and Kargel, J. (1995). Outwash plains and thermokarst on Mars. *Icarus*, **114**, 93–112.

Costard, F., Aguirre-Puente, J., Greeley, R., and Makhloufi, N. (1999). Martian fluvial-thermal erosion: laboratory simulation. *Journal of Geophysical Research*, **104** (E6), 14091–8.

Costard, F., Dupeyrat, L., Gautier, E., and Carey-Gailhardis, E. (2003). Fluvial thermal erosion investigations along a rapidly eroding river bank: application to the Lena river (central Yakutia). *Earth Surface Processes and Landforms*, **28**, 1349–59.

Fanale, F. P., Salvail, J. R., Zent, A. P., and Postawko, S. E. (1986). Global distribution and migration of subsurface ice on Mars. *Icarus*, **67**, 1–18.

Feldman, W. C., Boynton, W. V., Tokar, R. L. *et al.* (2002). Global distribution of neutrons from Mars: results from Mars Odyssey. *Science*, **297**, 75–8.

French, H. M. (1988). The periglacial phenomena: ancient and modern. *Journal of Quaternary Science*, **3** (1), 110 pp.

Gautier, E. and Costard, F. (2000). Les systèmes fluviaux à chenaux anastomosés en milieu périglaciaire: la Léna et ses principaux affluents en Sibérie Centrale. *Géographie Physique et Quaternaire*, **54** (3), 327–42.

Gautier, E., Brunstein, D., Costard, F., and Lodina, R. (2003). Fluvial dynamics in a deep permafrost zone: the case of the middle Lena River (Central Yakutia). In *Proceedings of the 8th International Conference on Permafrost, Zurich*, ed. M. Phillips. Balkema, pp. 271–5.

Gordeev, V. V. and Sidorov, I. S. (1993). Concentrations of major elements and their outflow into the Laptev Sea by the Lena River. *Marine Chemistry*, **43**, 33–45.

Jahn, A. (1975). *Problems of the Periglacial Zone*. Warsaw: Panstwowe Wydawnictwo Naukowe, reprinted by National Technical Information Service, Springfield, VA 22161, as PB–248 901.

Jakosky, B. M., Mellon, M. T., Kieffer, H. H. *et al.* (2000). The thermal inertia of Mars from the Mars Global Surveyor Thermal Emission Spectrometer. *Journal of Geophysical Research*, **105**, (E4), 9643–52.

Journaux, A. and Dresch, J. (1972). Phénomènes périglaciaires en Sibérie Orientale, définition d'une nouvelle province périglaciaire actuelle. *Bulletin de l'Association des Géographes Français*, 57–73.

Katasonov, E. M. and Solovie, P. A. (1969). Guide to trip in Central Iakoutia. In *Palaeogeography and Periglacial Phenomena*, International Symposium, Iakutsk.

Kieffer, H. H., Chase, S. C., Miner, E. D., Munch, G., and Neugebauer, G. (1973). Preliminary report on infrared radiometric measurements from the Mariner 9 spacecraft. *Journal of Geophysical Research*, **78**, 4291–312.

Knighton, A. D. (1999). Downstream variation in stream power. *Geomorphology*, **29** (3–4), 293–306.

Knighton, A. D. and Nanson, G. C. (1993). Anastomosis and the continuum of channel pattern. *Earth Surface Processes and Landforms*, **18**, 613–25.

Leopold, L. B. and Wolman, M. G. (1957). Rivers channel patterns; braided, meandering and straight. US Geological Survey Professional Paper 282–B, pp. 39–85.

Lopatin, G. V. (1952). Nanosy rek SSSR (obrazovanie i perenos). *Izvestiya Vsesoyuznogo Geograficheskogo Obshchestva*, **14**, 366 pp.

Lucchitta, B. K. (1982). Ice sculpture in the Martian outflow channels. *Journal of Geophysical Research*, **87**, 9951–73.

Lucchitta, B. K. (2001). Antarctic ice streams and outflow channels on Mars. *Geophysical Research Letters*, **28**, 403–6.

Lucchitta, B. K. and Ferguson, H. M. (1983). Chryse basin channels: low gradients and ponded flows. *Journal of Geophysical Research*, **88**, A553–68.

Lvovitch, M. I. (1971). The water balance of the continents of the world and the method of studying it. Moscow: Academy of Science USSR, Instute of Geography.

Makogon, Y. F., Holditch, S. A., and Makogon, T. Y. (2005). Gas hydrate production: 1. Russian field illustrates gas-hydrate production. *Oil and Gas Journal*, **103** (5), 43–7.

Nanson, G. C. and Knighton, A. D. (1996). Anabranching rivers: their cause, character and classification. *Earth Surface Processes and Landforms*, **21**, 217–39.

Nowicki, S. A. and Christensen, P. R. (2001). The thermophysical properties of Ares Vallis from thermal emission spectrometer data. *Abstracts of Papers Submitted to the 32nd Lunar and Planetary Science Conference*. Houston: Lunar and Planetary Institute, CD 32, Abstract 1931.

Péwé, T. L. (1991). Permafrost. In *The heritage of engineering geology; The first hundred years*, ed. G. A. Kiersch. Geological Society of America, Centennial Special Volume 3, pp. 277–97.

Schumm, S. A. (1977). *The Fluvial System*. New York: John Wiley & Sons.

Smith, D. G. (1986). Anastomosing river deposits sedimentation rates and basin susbsidence, Magdalena River, Northwestern Columbia, South America. *Sedimentary Geology*, **46**, 177–96.

Walker, H. J. (1983). E pluribus unum: small landforms and the Arctic landscape. In *Mega-geomorphology; Conference of the British Geomorphological Research Group*, ed. R. Gardner and H. Scoging. Oxford: Oxford University Press, pp. 39–55.

Walker, H. J. and Arnborg, L. (1983). Nature of the Colville River during the late winter and breakup periods, 1962. Geological Society of America Special Paper.

Wallace, D. and Sagan, C. (1979). Evaporation of ice in planetary atmospheres – ice covered rivers on Mars. *Icarus*, **39**, 385–400.

Yamskikh, A. F., Yamskikh, A. A., and Brown, A. G. (1999). Siberian-type Quaternary floodplain sedimentation: the example of the Yenisei river. In *Fluvial Processes and Environmental change*, ed. A. G. Brown and T. A. Quine. New York: John Wiley & Sons, pp. 241–52.

12

Formation of valleys and cataclysmic flood channels on Earth and Mars

Goro Komatsu
International Research School of Planetary Sciences, Universit' d'Annunzio, Pescara

and

Victor R. Baker
Department of Hydrology and Water Resources, University of Arizona

12.1 Introduction

The origin of fluvial valleys has been one of the great problems of science ever since the debates in the eighteenth and early nineteenth centuries over the role of cataclysmic events in the shaping of Earth's valleys. The famous founders of geology, including Hutton, Lyell, and Cuvier all participated in these great debates. It can be argued that geology emerged as a science because of the scientific reasoning that was applied to this problem (Davies, 1969).

It was one of the great surprises of modern planetary science that fluvial valleys were discovered on the planet Mars by study of the vidicon images returned by the Mariner 9 spacecraft (Masursky, 1973). From the much more extensive coverage of the Viking mission we know that the heavily cratered Martian highlands are locally dissected by networks of tributary valleys with widths of 10 km or less and lengths of a few kilometers to nearly 1000 km. These valley networks are one major type of large elongate Martian troughs thought to have fluvial origins (Baker, 1982).

In distinction from fluvial valleys, large outflow channels show extensive evidence of large-scale fluid flow on their floors and walls (Baker *et al.*, 1992a), so technically speaking they are channels, rather than valleys. This is a rather curious circumstance, since Earth experience leads us to suppose that fluvial channels are invariably much smaller than fluvial valleys. However, the Martian outflow channels may arise from the peculiar geological history of Mars and its water endowment (Baker *et al.*, 1991).

The Geology of Mars: Evidence from Earth-based Analogs, ed. Mary Chapman. Published by Cambridge University Press.

Fretted terrain, not considered in this chapter, on the other hand, contains extensive lineated valley fill, which probably reflects the role of ice deformation of overlying debris (Squyres, 1989). New evidence from the Mars Global Surveyor missions suggests that some of this morphology may result from debris-covered glaciers (Baker, 2001).

Manifestations of fluvial valley or channel phenomena occur at least on two planets, Earth and Mars (Komatsu and Baker, 1996a; Baker and Komatsu, 1999). In this chapter, we describe the morphology of valleys and channels on Mars, compare them with terrestrial examples, and discuss implications for the fluvial processes on the two planets.

12.2 Formation of Martian valleys

12.2.1 Comparisons of Martian valleys and terrestrial analogs

Nirgal Vallis (Figure 12.1a) illustrates the major morphological characteristics of many Martian valley networks, including theater-like tributary heads, prominent structural control, low junction angles, quasi-parallel patterns, hanging tributary valleys, irregular widening and narrowing, and indistinct terminal areas. Excellent terrestrial examples of valleys that resemble such Martian valley networks occur in the Colorado Plateau (Laity and Malin, 1985), Hawaii (Kochel and Piper, 1986), and in the uplands of the Sahara (Nicoll and Komatsu, 1999).

For example, extensive linear drainage patterns of Wadi Mareef in southern Egypt (Figure 12.1b) have morphology strikingly similar to that of Nirgal Vallis (Nicoll and Komatsu, 1999). Side valleys are not highly integrated and upland dissection is limited in both systems. Field study indicates that Wadi Mareef was a spring-fed stream that intermittently flowed during the Cenozoic, which was dominantly an arid time period (Kröpelin, 1993). Along Wadi Mareef, carbonates were precipitated via discharge of ambient waters emerging from springs. Petrographic examination of prepared thin sections reveals no clear textural evidence for a hydrothermal origin of the carbonate rocks, suggesting a shallow meteoric origin of the fossil-spring waters. Though the Egyptian landscape has been dominantly hyperarid over the past 2 million years (Nicoll, 1998), the region of Wadi Mareef has effectively preserved the geomorphic evidence of rare rainfall and streamflow events.

Sapping and runoff dissection processes played roles in the geomorphic evolution of drainage networks in Egypt (Luo *et al.*, 1997). The sapping process involves the undermining of hillslopes or cliffs, thereby allowing

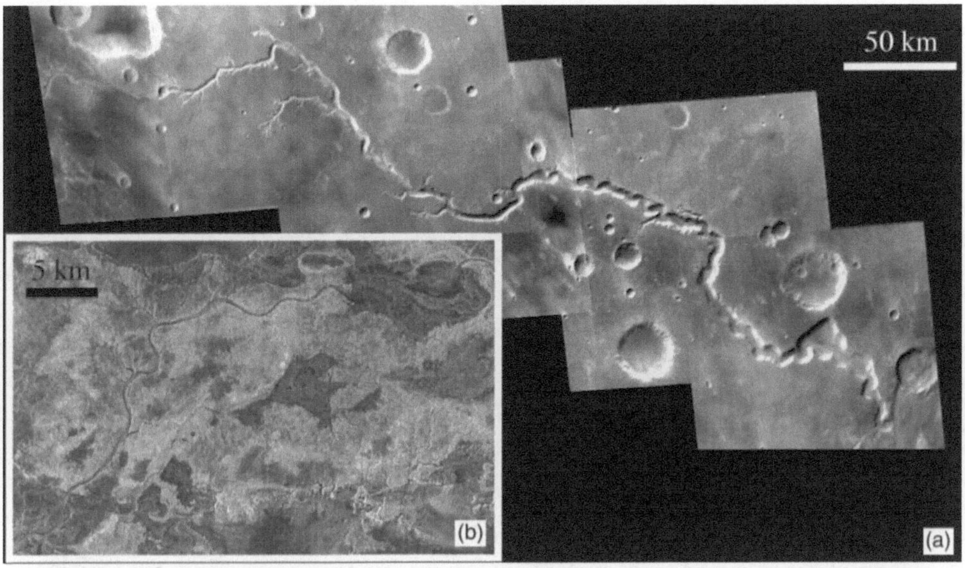

Figure 12.1. Comparison of Nirgal Vallis on Mars ((a) MOC wide angle images, M0705584, M0301844, M0003118, M0400786, M1901954, M0002227, and M0302291, mosaicing courtesy of Hanna Lahtela and Paul Geissler) and Wadi Mareef in southern Egypt ((b) RADARSAT standard mode R111135058G1S003).

upper portions of the slopes to collapse. The term derives from military engineering, in which fortifications are breached by undermining and resultant collapse. A variety of sapping is facilitated by the action of groundwater (springs), generally at the heads of valleys that concentrate groundwater flow, the action which promotes the extension of the valley in a headward direction (Dunne, 1990). This process has been demonstrated experimentally by Howard and McLane (1988).

Nanedi Valles system is one of several Martian valleys that were imaged at high resolution with Mars Orbiter Camera (MOC) onboard the Mars Global Surveyor. The MOC data set together with the Mars Orbiter Laser Altimeter (MOLA) data provide great improvements over the Viking imagery and topographic maps. Image resolution improved to a maximum of ~1.5 m per pixel in MOC data, and vertically accuracy of topographic maps improved to < 10 m with MOLA data. Note that Nanedi's meandering paths (Figure 12.2) display a complex pattern of cutoff meanders and paleochannels (Carr and Malin, 2000). This type of pattern is commonly observed with terrestrial analogs. An excellent example is the Finke River (Pickup *et al.*, 1988) in arid central Australia, where dry-beds exhibit overlapping paleochannels (Figure 12.3). Patterns of paleochannels are often

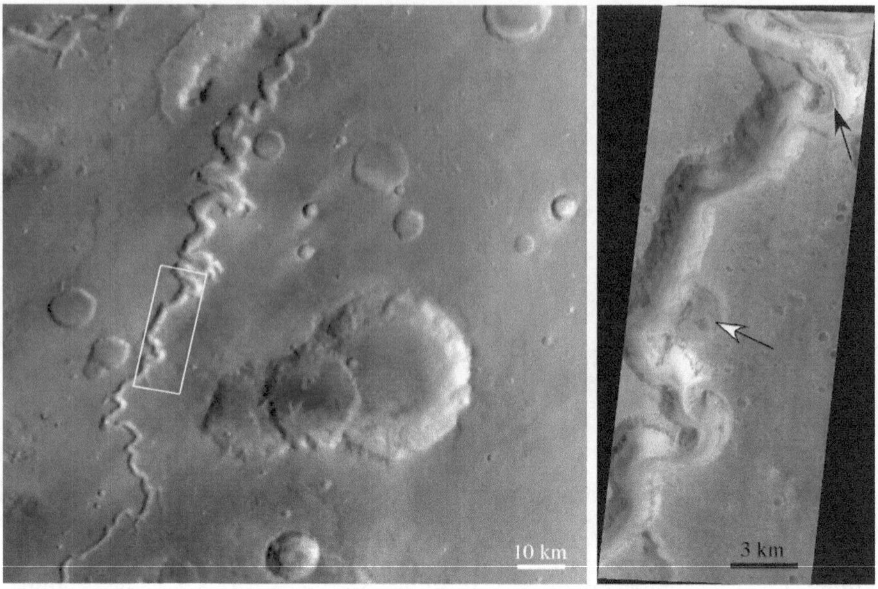

Figure 12.2. At a low resolution, Nanedi Valles exhibit a typical valley network morphology observed in the southern highlands of Mars (wide angle MOC image E0501621). The high-resolution MOC image (narrow angle MOC image AB108704) clearly indicates a complex history of Nanedi Valles (white arrow indicates a paleochannel and black arrow indicates an inner channel).

attributed to changes in discharge rate, climatic shift, and neotectonics, implying that Mars also experienced such changes.

The fine-scale inner channel of Nanedi Valles (shown at the top of the high-resolution frame in Figure 12.2) shows an example of underfitness. On Earth underfitness develops when a valley has on its floor a stream, often meandering, that seems to be small in regard to the meandering valley that it presumably generated (Dury, 1964). This relationship implies a change of flow conditions from the very large flows that generated the meandering valley, to much smaller flows that produced the relatively small inner channel.

Integrated Martian valley networks (Figure 12.4), although relatively limited in their occurrence, are particularly well developed on some geologic units on Mars. On Earth, analogous patterns are observed in both arid and humid regions, and their morphology generally implies surface runoff that was caused by precipitation. Surface runoff is a possibility on Mars although the mode of liquid water flowing on its surface is not well understood (for example, ice- or snow-covered or not). The assumption of precipitation may not easily apply to Mars where drainage density is relatively low and there are wide spaces between drainage networks (Gulick, 2001). However, the

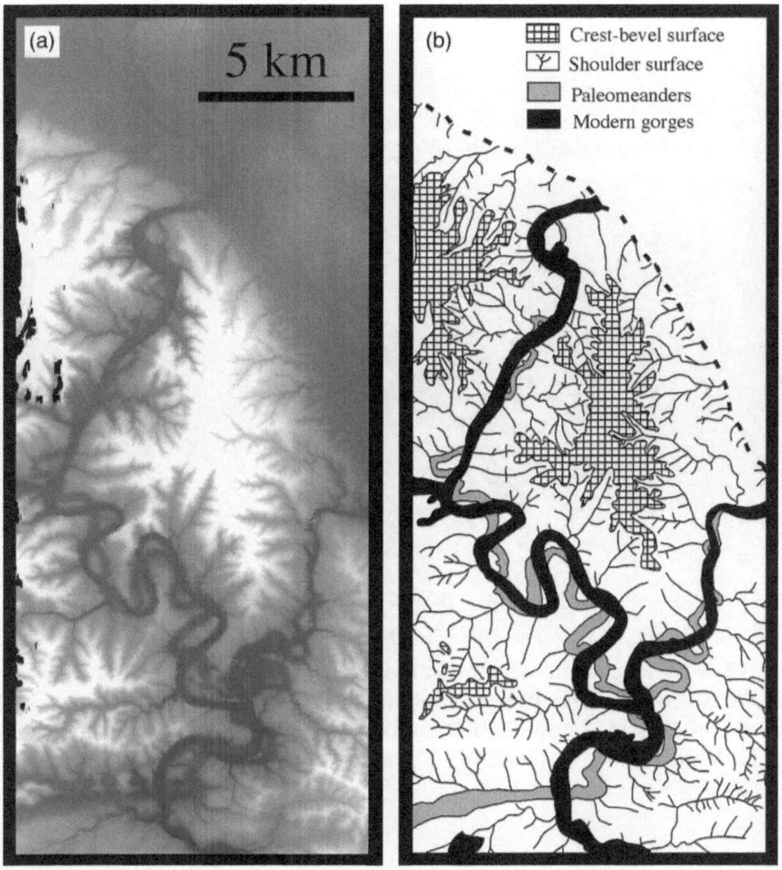

Figure 12.3. (a) The Digital Elevation Model of Finke River in central Australia derived from TOPSAR (Topographic Synthetic Aperture Radar) data by using the radar interferometry technique (Kwoun *et al.*, 1998). Finke River exhibits an excellent example of paleochanncels overprinted by the younger drainages. Black and white spots are areas of no data. (b) Sketch map of Finke River showing ancient and modern drainage systems.

improved image and topographic data set from recent missions are revealing more mature, integrated drainage systems than previously recognized, strengthening the hypothesis of past precipitation on Mars (Hynek and Phillips, 2003).

12.2.2 Proposed origin for Martian valleys

Current thinking on the origin of Martian valley networks is reviewed by Gulick (2001). The general features described for many valley networks

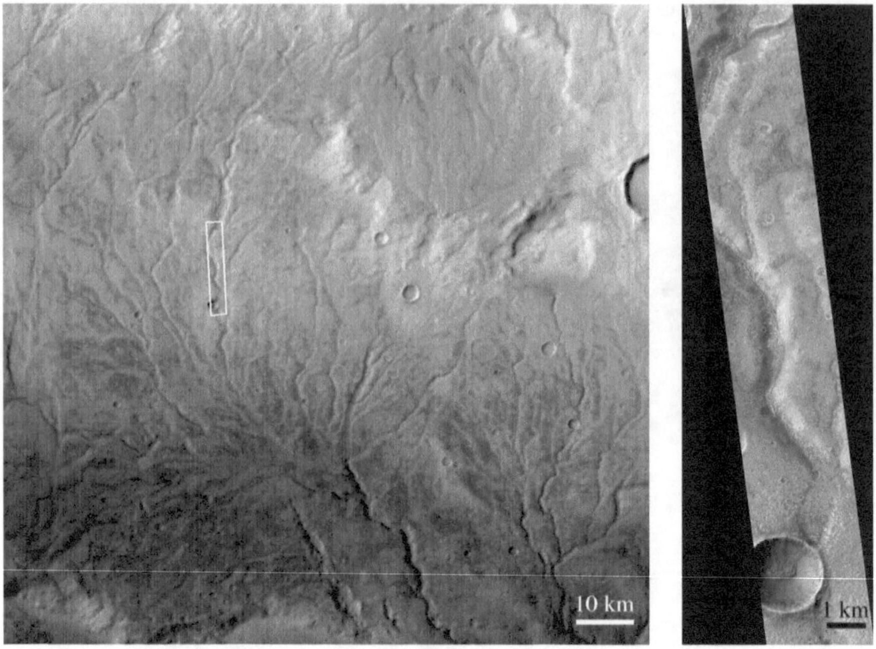

Figure 12.4. An example of integrated valley networks on Mars (Warrego Valles; Wide angle MOC image M0803364). The high-resolution MOC image (narrow angle MOC image M0803363) shows details of a valley segment of Warrego Valles.

including Nirgal Vallis (see Section 12.2.1) were recognized very soon after discovery (Pieri, 1980), and we generally ascribe the whole assemblage to an origin by groundwater sapping processes (Baker *et al.*, 1990). Although groundwater sapping processes are generally accepted as the primary formation mechanism for Martian valleys, surface runoff also appears to have played a role as implied from integrated Martian valley networks. Furthermore, drainage patterns of valleys such as Nanedi Valles indicate complex evolutional paths. Questions regarding the origin of Martian valley networks still remain. The clustered nature of the valleys is associated with incomplete dissection of upland areas and a lack of fine-scale tributaries at the very high-resolution scale of the MOC images (Carr and Malin, 2000). Explanations for these relationships remain rather contentious. Probable high infiltration capacities of the Martian surfaces (Baker and Partridge, 1986), burial of small-scale features beneath mantling deposits, lack of precipitation (Carr and Malin, 2000), and local hydrothermal/snowmelt processes (Gulick, 2001) are all possibilities.

The role of precipitation in Martian valley formation has been particularly controversial. There remain many questions regarding precipitation on Mars: (a) whether or not the precipitation scenario is completely ruled out, (b) if precipitation occurred locally and temporarily or not, and (c) if other factors such as lithology contributed to the low drainage density and sparse distribution of the valleys. The solutions to all these questions require further research, and they should be consistent with various aspects of valleys such as latitudinal dependency of sizes (Williams and Phillips, 2001). The planetary analog question is complicated by the fact that there is no place on Earth where precipitation has not occurred, at least during the last few thousands of years. Even in the hyperarid lands such as the Absolute Desert of the Atacama and the Western Desert in Egypt, relict landforms preserve evidence of a past wetter climate. There is no example of terrestrial valleys formed only by hydrothermally recycled water. This makes assessment of the Martian valleys in comparison to hypothesized non-precipitation environments especially difficult.

Phillips *et al.* (2001) showed that many valley networks have orientations consistent with the effects of surface deformation caused by the rapid volcanic loading of the Tharsis region in late Noachian time (the heavy bombardment period for Mars). The networks also seem to show evidence of extensive erosion during the late Noachian (Hynek and Phillips, 2001). While the association of most networks with the heavily cratered terrains of Mars has been used to argue for great age (Carr and Clow, 1981), there are numerous younger valley networks that postdate the heavy bombardment period (Scott *et al.*, 1995). Gulick and Baker (1989, 1990) documented extensive young valley development on Martian volcanoes, implying hydrological processes on Mars continued to late in Mars history.

Comparisons to Earth valleys suggest that erosion has been rather limited on Mars. The low-order mainstream valleys often have relatively flat longitudinal profiles, in contrast to the pronounced concave up profiles of typical terrestrial river valleys (Aharonson *et al.*, 2001). Such profiles may form by sapping processes, as argued by Luo (2002), but they may also be connected to immature dissection processes, or the spilling of lakes in the headwaters of drainages. The latter may have been significant for some Martian valleys (Irwin *et al.*, 2002), especially given the abundant evidence for crater lakes in the Martian highlands (Cabrol and Grin, 1999).

Despite the limited valley dissection and abundant evidence for sapping morphologies, there are strong arguments for extensive degradation very early in Martian history. There is good reason to believe that the erosion was fluvial, and promoted by rainfall (Craddock and Howard, 2002).

12.2.3 Recent debris-flow gullies

Perhaps the most striking discovery from the very high-resolution MOC images is that of relatively small debris-flow gullies (see Chapter 10), some of which dissect the walls of larger network valleys (see Figure 10.1 in Chapter 10). These were initially explained as the result of groundwater seepage and subsequent surface runoff (Malin and Edgett, 2000). However, they have morphologies that are common in many terrestrial periglacial and desert environments. On Earth such debris flows occur when soil water becomes saturated in the debris fill of hillslope troughs following very heavy rains, snowmelt, or the melting of ground ice in permafrost areas. Remarkably, the Martian debris-flow gullies are superimposed on extensive uncratered surfaces of sand dunes and contraction-crack polygons. The activity seems to have been very recent, within the last few million years. The orbital effect, particularly of high obliquity on the seepage or melting processes, has been proposed by various investigators (e.g., Costard *et al.*, 2002; Mellon and Phillips, 2002). Christensen (2003) hypothesized that the gullies are the result of melting of water-rich snow that has been transported from the poles to mid-latitudes during periods of high obliquity within the past 10^5 to 10^6 years. The implication of very recent water-related activity on Mars (Baker, 2001) forces rethinking of Martian atmospheric models, and, for example, the hypothesis of recent ice ages has emerged (Head *et al.*, 2003).

12.2.4 Reconstruction of Mars paleoclimate using valley landforms: potentials and problems

One of the most interesting and extensively debated topics in Mars geology is the paleoclimate and the role of water. Dry river beds discovered by Mariner 9 images were striking, and many hypotheses regarding the climate of early Mars were proposed. Unlike Earth, however, there are limited types and amounts of data for the study of paleoclimate. Because most proxy data such as oxygen isotopes and tree rings do not exist for Mars, there must be a heavy reliance on landforms for the reconstruction of paleoenvironments. For example, the morphologies of Martian valley networks were studied to infer their origins and hence climatic conditions at the time of formation. However, we caution about the use of this approach since problems exist, including (1) the neglect of various non-climatic factors that contribute to valley formation, (2) oversimplification of individual valley history and regional climatic variability, and (3) poorly known ages for the valleys. The link between landforms and climate is not simple, and Mars is not an exception in this regard. Although some of the above problems can be dealt with by examining a wide range of landforms and their relationships to each other, reality is always more complex. At this stage of Mars research, we have very limited information with which to investigate past climatic conditions. Hence climatic modeling efforts to reproduce precisely the conditions inferred from landforms could be both non-productive and misleading. The modeling can still be insightful, however, if focused on basic climatic processes and on alternative paleoclimatic scenarios for Mars.

12.3 Cataclysmic flood channels on Mars and Earth

12.3.1 Outflow channels on Mars

Martian outflow channels are large-scale complexes of fluid-eroded troughs (Baker and Milton, 1974; Baker, 1978, 1982; Mars Channel Working Group, 1983). The flows which formed these channels appear to have emanated from discrete collapse zones known as chaotic terrain. A wide variety of flood initiation mechanisms have been proposed including outbursts of overpressured aquifers (Carr, 1979), decomposition of clathrate (Milton, 1974), by breakout of a high-pressure gas–water mixture derived from decomposed gas hydrates (Komatsu *et al.*, 2000a), explosive vaporization of liquid CO_2 (Hoffman, 2000), and induced by sub-ice volcanic eruptions (Chapman *et al.*, 2003). The channels are immense by terrestrial standards, as much as 200 km wide and 2000 km in length. Residual uplands separating channels and partially to fully streamlined upland remnants are common in outflow channels (Figure 12.5). The morphology such as extensive grooved channel floors (Figure 12.6) and scale of the outflow channels clearly indicates genesis by fluid flows of immense magnitude. It is most likely that outflow channels were carved by catastrophic water floods (Baker, 1982) although formation by other fluids such as glaciers (Lucchitta, 1982) and mass flows (Tanaka, 1997) has been proposed. Also secondary modifications by periglacial processes (Costard and Kargel, 1995; Costard and Baker, 2001) and wind are possible. Paleodischarges for selected channels have been estimated using the standard Chezy and Manning equations appropriately modified for Mars gravity (e.g., Komar, 1979; Baker, 1982; Robinson and Tanaka, 1990; De Hon and Pani, 1993; Komatsu and Baker, 1997a; Chapman *et al.*, 2003). Results (some estimates are tabulated for comparison in Table 12.1) indicate that paleodischarges may have been ten to

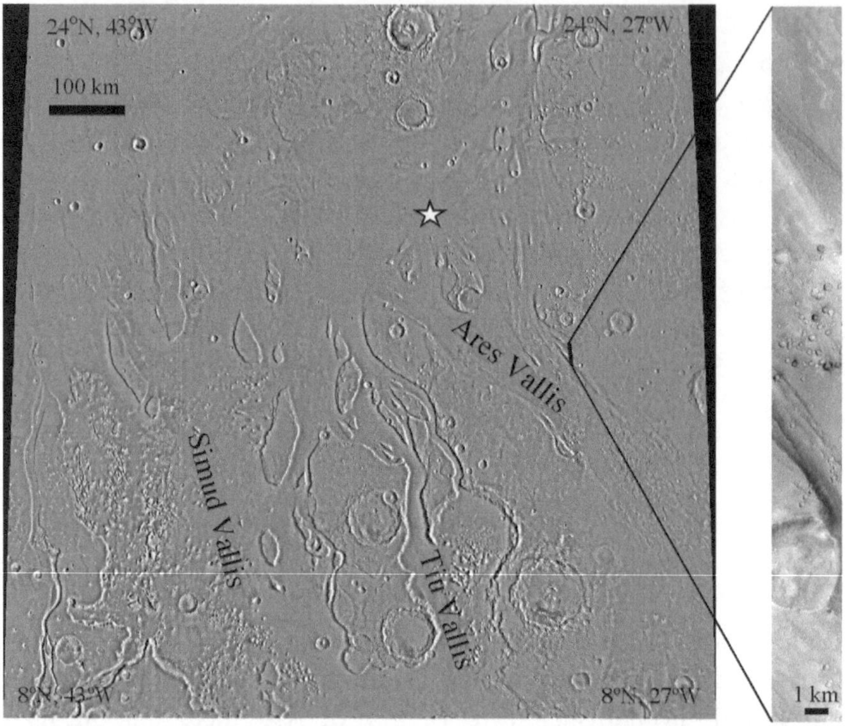

Figure 12.5. MOC image (MGS MOC Release No. MOC2-240) of debris flow gullies in the wall of Nirgal Vallis. NASA/JPL/Malin Space Science Systems. Viking mosaic (Mars Global Digital Image Mosaic 2.0) showing where outflow channels spread out in Chryse Planitia. This area is characterized by numerous erosional uplands, some of which are streamlined and scoured (narrow angle MOC image E1700745). The Mars Pathfinder landing site (star) in the outwash plain of Ares Vallis and Tiu Vallis is also shown.

one hundred times greater than the known biggest prehistoric floods on Earth (Baker, 1982).

Landing of the Mars Pathfinder in Chryse Planitia in 1997 was the first opportunity to see a flood landscape on the ground as the landing site was chosen to be in the outwash plain linked with circum-Chryse outflow channels, Ares Vallis and Tiu Vallis (Figure 12.6). Pre-mission predictions (e.g., Komatsu and Baker, 1997a; Rice and Edgett, 1997) stated flood-carved topography and the presence of large boulders. Despite the fact that accumulation of eolian sediments was prevalent, this situation was clearly visible (Smith *et al.*, 1997). The landing site images show boulders' imbricate (overlapping in a regular pattern) arrangements indicating deposition by high-power flood streams, and hills that may have been shaped also by such floods (Figure 12.7). We caution that alternative hypotheses regarding

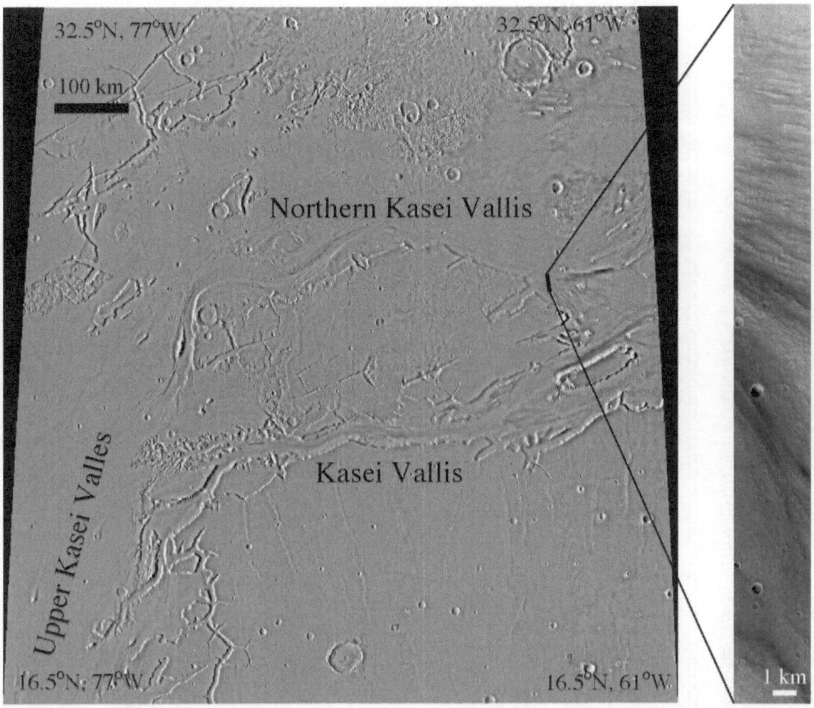

Figure 12.6. Viking mosaic (Mars Global Digital Image Mosaic 2.0) showing northern and southern channels of Kasei Valles. Extensive grooves on the floors of Kasei Valles were formed by flood erosion and/or deposition (narrow angle MOC image E1002538).

the origin of the channel and local view of the Mars Pathfinder landing site exist. These landforms may have resulted from a combination of flood, mass flow, and/or glacial-like, and eolian processes (e.g., Chapman and Kargel, 1999).

Cataclysmic flood channels do not always originate from chaotic terrain. For example, an outflow channel Mangala Vallis originates from Memnonia Fossae, an extensive fracture system of Tharsis. Other fracture-originated outflow channels occur in the Cerberus plains area (Burr *et al.*, 2002). Furthermore, Elaver Vallis south of Gangis Chasma originated as spillover channel from an ancient carter lake basin (Coleman *et al.*, 2003; Komatsu *et al.*, 2003).

Great outflow channels are, in general, Hesperian in age (Carr, 1995). However, outflow channels of very young ages have been identified recently. For example, Athabasca Valles that originates from Cerberus Fossae may have formed geologically very recently (2–8 Ma) based on crater counting (Burr *et al.*, 2002). Therefore, as in the case of Martian valleys,

Table 12.1. *Comparison of maximum discharge rates among terrestrial and Martian cataclysmic flood events*

	Name	Mode of outburst	Estimated discharge rate ($\times 10^6$ m^3/s)	References
Earth	Altai	Failure of glacier-dammed lakes	18	Baker *et al.*, 1993
Earth	Lake Missoula	Failure of glacier-dammed lakes	17	O'Connor and Baker, 1992
Earth	Porcupine River	Failure of glacier-dammed lakes	0.134	Thorson, 1989
Earth	Manych	Spillover of sea water	10	Komatsu and Baker, 1996b
Earth	Lake Bonneville	Spillover of lake water	1	O'Connor, 1993
Mars	Kasei Valles	Outburst of subsurface water	1000–2300	Robinson and Tanaka, 1990[a]
			0.08–20	Williams *et al.*, 2000[b]
Mars	Ares Vallis	Outburst of subsurface water	570	Komatsu and Baker, 1997a[c]
			5000	Smith *et al.*, 1998[b]
Mars	Athabasca Vallis	Outburst from a fracture system	1–2	Burr *et al.*, 2002[b]

Table modified from Komatsu and Baker (1997b).
[a] Based on Viking stereo data.
[b] Based on the improved MOLA topographic data.
[c] Based on the Viking topographic map.
Note that the estimation of discharge rate is strongly dependant on the accuracy of the topographic data, selected channel segments and high-water marks, and choice of the calculation method.

Figure 12.7. A panorama of the Mars Pathfinder landing site. See Figure 12.5 for its location. Twin peaks at the distance may be erosional remnants of preexisting bedrocks, left by a flood of great scale. Some boulders in the foreground show imbricate arrangement, implying flood deposition. We note that other processes such as mass flow and glacier cannot be ruled out as processes responsible for these observations. NASA/JPL.

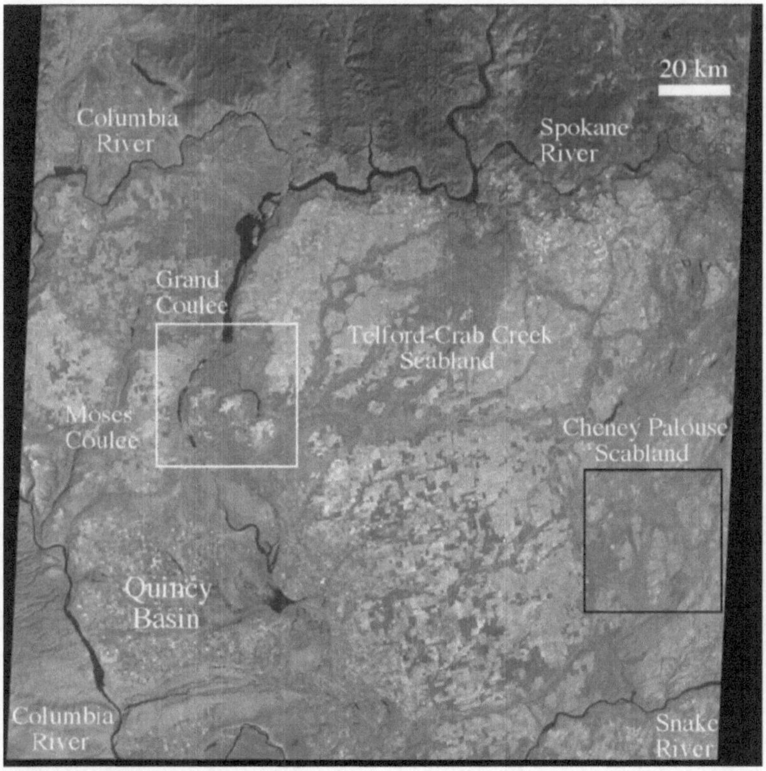

Figure 12.8. This Landsat MSS (Multi-Spectral Scanner) image covers the main reaches of the Channeled Scabland, showing a complex of anastomosing channels. Locations of Figure 12.9a (black frame) and Figure 12.9b (white frame) are indicated.

outflow channels also strongly imply continued hydrological activity late in Mars history.

We note that a model of carbon dioxide processes has emerged (Musselwhite *et al.*, 2001) in order to avoid the environmental change implied by liquid water (Hoffman, 2000). Were this model to be verified, it would be a rather amazing discovery. It would mean that various carbon dioxide processes are capable of producing rather detailed copies of what on Earth is readily recognized as resulting from the long-term effects of water and ice. Specific problems with the model are discussed by Coleman (2003).

12.3.2 Terrestrial analogs for the outflow channels

The presence of terrestrial analogs was essential in revealing formation mechanisms of the outflow channels (e.g., Baker, 1978). The most famous and important analog is the Channeled Scabland (Figure 12.8) of the northwestern United States (Baker and Nummedal, 1978). However, thanks to discoveries of new cataclysmic flood features in northern Eurasia (e.g., Baker *et al.*, 1993; Komatsu *et al.*, 1997c), it is increasingly becoming clear that cataclysmic flood processes were a rather common phenomenon during the Ice Age Earth. And these new examples can also greatly contribute to our understanding of outflow channels on Mars. The origins of terrestrial cataclysmic floods include the failures of glacier-dammed lakes, spillovers of lake water, and possibly subglacial outbursts. The wide range of erosional and depositional landforms produced by such high-energy events (Figures 12.9a and b, 12.10a–d, and 12.11a–d) are useful in identifying Martian features formed by similar catastrophic processes.

The Channeled Scabland consists of a spectacular complex of anastomosing channels (Figure 12.9a), cataracts (Figures 12.9b and 12.10c), streamlined loess "islands" (Figure 12.10a), scabland landscape (Figure 12.10b), longitudinal grooves (Figure 12.10c), and immense gravel bars created by the catastrophic fluvial erosion of the loess and basalt of the Columbia Plateau (Baker, 1973, 1978). Large boulders deposited in the Ephrata Fan testify for the energy of the flood swept through the area (Figure 12.11a). Slackwater deposits (Figure 12.11b) formed in the stagnant water regime during the flooding (Baker and Bunker, 1985), but their rhythmite stratification has been a source of a considerable debate regarding the scale and frequency of the floods (Waitt, 1984; Baker and Bunker, 1985; Shaw *et al.*, 1999; Atwater *et al.*, 2000). The source of the flooding is considered to be the glacier-dammed Lake Missoula and the breakout appears to have occurred repetitively during the late Pleistocene. The resulting peak discharge (Table 12.1) was as great

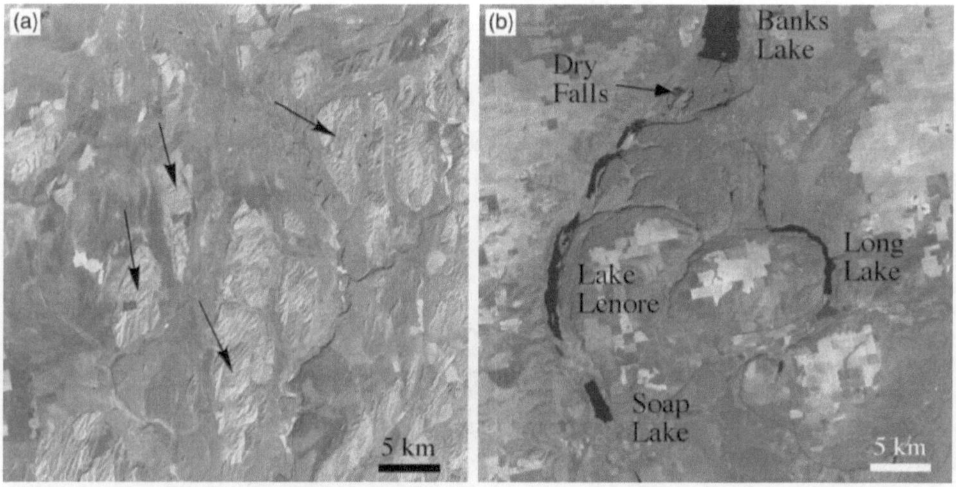

Figure 12.9. (a) Anastomosing channels and streamlined loess "islands" (arrows). Cheney Palouse Scabland tract. (b) Flood-scoured channels and a cataract complex. Grand Coulee. Both images are Landsat MSS (Multi-Spectral Scanner) images.

Figure 12.10. Erosional flood geomorphology. (a) Streamlined loess "island." The Channeled Scabland, Washington, USA. (b) Scabland landscape. Along the Lower Grand Coulee near Lake Lenore (Figure 12.9b), Washington, USA. (c) Longitudinal grooves and a cataract complex, Grand Coulee, Washington, USA. (d) Incision into a bedrock (terrace in the foreground), Porcupine River, Alaska, USA.

Figure 12.11. Depositional flood geomorphology. (a) Boulder field (black dots). Size of these boulders ranges up to about 20 m. Near Rocky Ford Creek Fish Hatchery, Washington, USA. (b) Slackwater deposit at Burlingame Canyon, Washington, USA. (c) "Giant current ripples" (diluvial dunes), Kuray Basin, Altai, Russia. Photo by A.N. Rudoy. (d) Diluvial terrace (giant bar), Katun valley at the mouth of the Little Yeloman River valley, Altai, Russia. Photo by A.N. Rudoy.

as 17×10^6 m^3/s (O'Connor and Baker, 1992). Recently, Shaw *et al.* (1999) proposed subglacial outburst as another mode for the formation of the Channeled Scabland, adding further complexity to the story of cataclysmic flooding (Atwater *et al.*, 2000; Komatsu *et al.*, 2000b).

Smaller, but still significant cataclysmic flood features exist in North America including Lake Bonneville spillover flood landforms (O'Connor, 1993) and flood eroded terrain along the Porcupine River in northern Alaska (Thorson, 1989). The estimated peak discharge rates (Table 12.1) for the Lake Bonneville flood is about 10^6 m^3/s (O'Connor, 1993) and that for the Porcupine flood is approximately 0.134×10^6 m^3/s (Thorson, 1989). The hydraulic forces associated with this range of discharge rates were powerful enough to cause deep incision in bedrock (Figure 12.10d).

Baker *et al.* (1993) studied flood features in the Altai Mountains and concluded that water levels had to exceed 1900 m in order to explain some landforms in some constricted reaches, and the peak discharge may have exceeded 18×10^6 m^3/s, which is the largest known on Earth. In particular, a suite of cataclysmic flood morphologies and sediments distributed in the Altai Mountains (Rudoy and Baker, 1993; Rudoy, 2002) is important for the comparison with those in the Channeled Scabland and also on Mars. The flood sediments include a wide variety, including "giant current ripples" (diluvial dunes) (Figure 12.11c) and diluvial terraces (or giant bars) (Figure 12.11d).

A remarkable paleohydrological pattern is developed in the great system of Pleistocene spillways connecting the Black, Caspian, and Aral Sea basins (Komatsu and Baker, 1996b). Expansion of the paleolakes occupying these basins in the Pleistocene might be attributed to progressive changes in water-balance parameters, including temperature, precipitation, runoff, and evaporation. This explanation is emphasized by Chepalyga (1984). Nevertheless, evidence for cataclysmic flood erosion in the spillways indicates that very rapid, massive influxes of water played a major role in their origin. One mechanism for achieving this is diversion of the north-flowing West Siberian rivers by ice. In one reconstruction of the late Pleistocene Eurasian ice sheets, Grosswald (1980) envisions spectacular damming of the great north-flowing Siberian rivers with associated spillovers and drainage diversions. An alternative reconstruction involves much less ponding and diversion (Velichko *et al.*, 1984), but incomplete field mapping and geochronology precludes scientific preference for either of these hypothetical alternatives.

The great Mansi paleolake was generated south of the Kara dome, inundating much of the West Siberian plain (Grosswald, 1993). The Turgai Hollow leads southward from the Mansi paleolake basin to the Aral Sea basin. Fluvial and lake sediments in the Turgai spillway document the conveyance of flows southward during the latest Pleistocene (Astakhov and Grosswald, 1978). The paleodischarges then entered the Aral Sea basin, which also received cataclysmic flood from the Issyk-Kul region. At its late-glacial highstand, the Aral paleolake had the Uzboi spillway entering the paleo-Caspian. The Manych channel that connects paleo-Caspian Sea and the Black Sea basin displays a suite of landforms characteristic of cataclysmic flooding (Figure 12.12): streamlined hills, elongate depressions (now lakes), and scabland erosion (Komatsu *et al.*, 1997c). The immense channel width (may reach 45 km) further indicates the great scale of the paleoflow. Using a flow velocity of 10 m/s, required for bedrock scabland erosion of this type (Baker and Costa, 1987), the estimated maximum flood discharge (Table 12.1) is 10×10^6 m^3/s (Komatsu and Baker, 1996b), which is comparable to that of ice-dammed lake outbursts like those of glacial Lake Missoula (O'Connor and Baker, 1992).

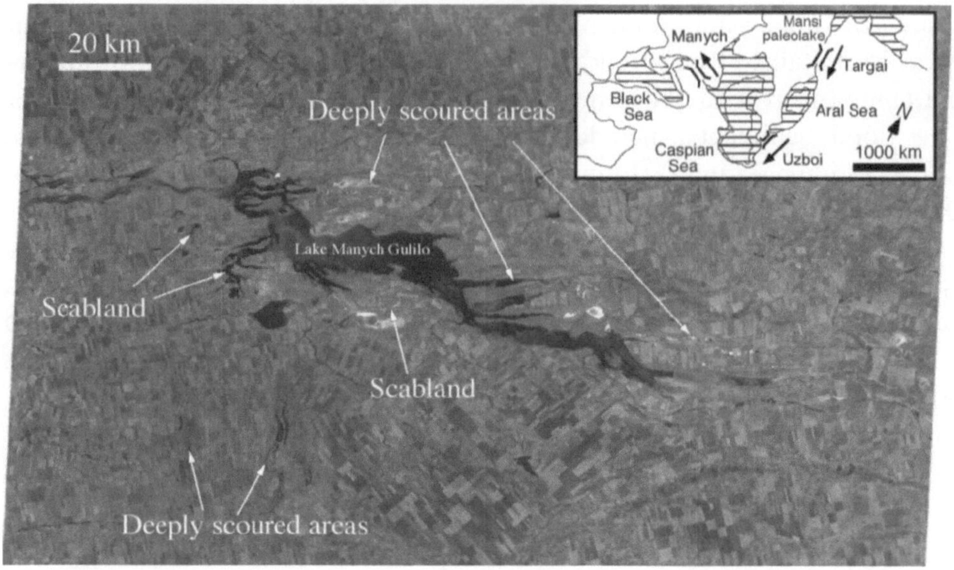

Figure 12.12. Manych spillway that connects paleo-Caspian Sea and the Black Sea basin displays a suite of cataclysmic flood landforms as shown in this Landsat MSS (Multi-Spectral Scanner) image. Inset shows locations of spillways connecting ice age paleoseas (patterned areas) in central Asia.

12.3.3 Scaling of cataclysmic floods and its implications

Peak discharge rates for the cataclysmic floods have been estimated by several researchers, employing different methods (Table 12.1). The estimation of discharge rates is not always straightforward, considering that it depends on the accuracy of the topographic data, selection of the channel segment and high-water marks, and the calculation method used. For example, the peak discharge could have been greater than 10^9 m^3/s for the Kasei Valles, one of the greatest outflow channel system on Mars (Robinson and Tanaka, 1990). However, selection of lower high-water marks and the resulting smaller cross-sectional area leads to a very conservative estimate of 8×10^4–2×10^7 m^3/s (Williams *et al.*, 2000). A discharge estimate for Ares Vallis by using the Viking topographic map with a large error margin (Komatsu and Baker, 1997a) may have led to an order of magnitude lower value (570×10^6 m^3/s) than the number (5000×10^6 m^3/s) calculated from a newly acquired, improved MOLA topographic profile (Smith *et al.*, 1997).

Despite all these problems in flood hydraulic computation, differences in the calculated discharges are attributable predominantly to the scale of the channels. The natural ranges of other factors, such as slope, friction factor, and gravity (terrestrial vs. Martian), are relatively small in comparison, and they do not account for major differences in the calculated discharges. The results indicate that, even at the immense scale of Martian outflow channels, channels are adjusted to the scale of cataclysmic fluid flows in a regular pattern of adjustment. Relatively limited study has been conducted on flood sediments on Mars since at present we lack a definitive method of identifying such sediments by orbital observations. Nonetheless, theoretical studies indicate that sediment transport should be more effective on Mars than on Earth because of the difference in gravity (e.g., Komar, 1980). More importantly, the great scales of the Martian outflow channels strongly imply enormous sediment loads that must have been deposited in vast outwash plains.

In addition to the discharge rate, other factors governing outflow channel development on Mars are important in understanding the evolution of these channels. Such factors include the role of tectonics and magmatism, aquifer characteristics, and how the climatic conditions changed with time. All of them have the capacity to influence the formation of outflow channels, but they are poorly constrained. Therefore they are the topics that warrant further investigation.

Hesperian and Amazonian-age cataclysmic flooding on Mars was a global phenomenon that has been hypothesized to have caused temporal climate change on the planet (Baker *et al.*, 1991). The cataclysmic flood events discussed in our study of Eurasia were more localized in their scales, but they may still have had a tremendous impact on the regional landscape and ecosystem. Comparative studies of Martian and terrestrial cataclysmic floods, therefore, could prompt us to think their great influences on environment.

12.4 Conclusions

Martian valley networks, in general, have characteristics that are generally ascribed to an origin by groundwater sapping processes, but other hypotheses have also been proposed. Excellent terrestrial examples of valley evolution by groundwater sapping are known in the Colorado Plateau, Hawaii, and in the uplands of the Sahara. New high-resolution images acquired by the Mars Global Surveyor show detailed morphologic characteristics never revealed before. They include paleochannels indicative of complex drainage history, and very young relatively small debris-flow gullies. Terrestrial analogs for these features provide insights into processes operating on Mars. Furthermore, because such processes strongly imply certain environmental conditions, their study can deepen our understanding of climatic history of Mars.

Martian outflow channels are gigantic systems of fluid-eroded troughs and they were probably formed by floods of catastrophic scales. Many of them formed by outburst of subsurface water as evidenced by their sources in chaotic terrain and fracture systems. However, spillover of lake basins also appears to have occurred. The outbursts of the Pleistocene Glacial Lake Missoula in the northwestern United States provide good examples of geological processes operating catastrophically. The landforms include streamlined hills, "giant current ripples" (diluvial dunes), and rugged scoured terrains. Recent works on the Eurasian Continent have identified new catastrophic flood features in Siberia and central Asia. These floods originated by catastrophic failure of glacier-dammed lakes, or by spillover of lakes expanded by environmental changes or lakes situated in large proglacial basins. Although the scales of flooding on Mars were, in general, at least an order of magnitude greater than those of terrestrial catastrophic floods, the study of Earth examples is essential in revealing processes of cataclysmic flooding on Mars.

Clearly the origin of valleys and channels continues to be a controversial question in science. We are fortunate in this mystery to have the manifestations of fluvial phenomena on two planets, and possibly fluvial-like valleys and channels on a third, Venus (Baker *et al.*, 1992b; Komatsu *et al.*, 1993, 2001) and in addition on a satellite of Saturn, Titan. Let us hope that reasoning in geology can be similarly advanced in the new controversy over valleys, as it was in the debates of 200 years ago.

Acknowledgments

We thank Ken Tanaka and Devon Burr for their reviews. Our research activities were funded by the Italian Space Agency (ASI) and National Aeronautical and Space Administration (NASA). The authors acknowledge the use of Mars Orbiter Camera images processed by Malin Space Science Systems that are available at http://www.msss.com/moc_gallery/.

References

Aharonson, O., Zuber, M. T., Rothman, D. H., and Whipple, K. X. (2001). Drainage basins and channel incision on Mars. *Proceedings of the National Academy of Sciences, USA*, **99**, 1780–3.

Astakhov, V. I. and Grosswald, M. G. (1978). New data on the age of sediments in the Turgay Valley. *Doklady Akademii Nauk SSSR*, **242**, 891–4 (in Russian).

Atwater, B. F., Smith, G. A., and Waitt, R. B. (2000). Comment on "Shaw et al. 1999, The Channeled Scabland: Back to Bretz ?" *Geology*, **28**, 574–5.

Baker, V. R. (1973). Paleohydrology and sedimentology of Lake Missoula flooding in Eastern Washington. *Geological Society of America Special Paper*, **144**, 73.

Baker, V. R. (1978). The Spokane Flood controversy and the Martian outflow channels. *Science*, **202**, 1249–56.

Baker, V. R. (1982). *The Channels of Mars*. Austin: University of Texas Press.

Baker, V. R. (2001). Water and the Martian landscape. *Nature*, **421**, 228–36.

Baker, V. R. and Bunker, R. C. (1985). Cataclysmic late Pleistocene flooding from Glacial Lake Missoula: a review. *Quaternary Science Reviews*, **4**, 1–41.

Baker, V. R. and Costa, J. E. (1987). Flood power. In *Catastrophic Flooding*, ed. L. Mayer and D. Nash. Boston: Allen and Unwin, pp. 1–21.

Baker, V. R. and Komatsu, G. (1999). Extraterrestrial fluvial forms. In *Varieties of Fluvial Forms*, ed. A. Miller. New York: Wiley, pp. 11–30.

Baker, V. R. and Milton, D. J. (1974). Erosion by catastrophic floods on Mars and Earth. *Icarus*, **23**, 27–41.

Baker, V. R. and Nummedal, D. (1978). *The Channeled Scabland*. National Aeronautics and Space Administration.

Baker, V. R. and Partridge, J. (1986). Small Martian valleys: pristine and degraded morphology. *Journal of Geophysical Research*, **91**, 3561–72.

Baker, V. R., Kochel, R. C., Laity, J. E., and Howard, A. D. (1990). Spring sapping and valley network development. In *Groundwater Sapping*, ed. C. G. Higginsand and D. R. Coates. Geological Society of America Special Paper 252, 235–65.

Baker, V. R., Strom, R. G., Gulick, V. C. *et al.* (1991). Ancient ice sheets and the hydrological cycle on Mars. *Nature*, **352**, 589–94.

Baker, V. R., Carr, M. H., Gulick, V. C., Williams, C. R., and Marley, M. S. (1992a). Channels and valley networks. In *Mars*, ed. H. Kieffer, B. Jakosky, C. Snyder, and M. Matthews. Tucson: University of Arizona Press, pp. 493–522.

Baker, V. R., Komatsu, G., Parker, T. J. *et al.* (1992b). Channels and valleys on Venus: preliminary analysis of Magellan data. *Journal of Geophysical Research*, **97**, 13421–44.

Baker, V. R., Benito, G., and Rudoy, A. N. (1993). Paleohydrology of Late Pleistocene superflooding, Altai Mountains, Siberia. *Science*, **259**, 348–50.

Burr, D. M., Grier, J. A., McEwen, A. S., and Keszthelyi, L. P. (2002). Repeated aqueous flooding from the Cerberus Fossae: evidence for very recently extant, deep groundwater on Mars. *Icarus*, **159**, 53–73.

Cabrol, N. and Grin, E. (1999). Distribution, classification and ages of Martian impact crater lakes. *Icarus*, **142**, 169–72.

Carr, M. H. (1979). Formation of Martian flood features by release of water from confined aquifers. *Journal of Geophysical Research*, **84**, 2995–3007.

Carr, M. H. (1995). *Water on Mars*. Oxford: Oxford University Press.

Carr, M. H. and Clow, G. D. (1981). Martian channels and valleys: their characteristics, distribution, and age. *Icarus*, **48**, 91–117.

Carr, M. H. and Malin, M. C. (2000). Meter-scale characteristics of Martian channels and valleys. *Icarus*, **146**, 366–86.

Chapman, M. G. and Kargel, J. S. (1999). Observations at the Mars Pathfinder Site: do they provide "Unequivocal" evidence of catastrophic flooding? *Journal of Geophysical Research*, **104**, 8671–8.

Chapman, M. G., Gudmundsson, M. T., Russell, A. J., and Hare, T. M. (2003). Possible Juventae Chasma sub-ice volcanic eruptions and Maja Valles ice outburst floods, Mars: implications of MGS crater densities, geomorphology, and topography. *Journal of Geophysical Research*, **108** (E10), doi:10.1029/2002JE002009.

Chepalyga, A. L. (1984). Inland sea basins. In *Late Quaternary Environments of the Soviet Union*, ed. A. A. Velichko. Minneapolis: University of Minnesota Press, pp. 229–47.

Christensen, P. R. (2003). Formation of recent Martian gullies through melting of extensive water-rich snow deposits, *Science*, **299**, 1048–51.

Coleman, N. M. (2003). Aqueous flows carved the outflow channels on Mars. *Journal of Geophysical Research*, **108**, doi:10.1029/2002JE001940.

Coleman, N. M., Dinwiddie, C. L., and Casteel, K. (2003). High channels on Mars indicate Hesperian recharge at low latitudes. In *Sixth International Conference on Mars*. Pasadena: Lunar and Planetary Institute, (CD-ROM), Abstract 3071.

Costard, F. and Baker, V. R. (2001). Thermokarst landforms and processes in the Martian outflow Ares Vallis. *Geomorphology*, **37**, 289–301.

Costard, F. and Kargel, J. S. (1995). Outwash plains and thermokarst on Mars. *Icarus*, **114**, 93–112.

Costard, F., Forget, F., Mangold, N., and Peulvast, J. P. (2002). Formation of recent Martian debris flows by melting of near-surface ground ice at high obliquity. *Science*, **295**, 110–13.

Craddock, R. A. and Howard, A. D. (2002). The case for rainfall on a warm, wet early Mars. *Journal of Geophysical Research*, **107** (E11), doi:10.1029/2001JE001505.

Davies, G. L. (1969). *Earth in Decay: A History of British Geomorphology*. 1578–1878. London: MacDonald.

De Hon, R. A. and Pani, E. A. (1993). Duration and rates of discharge: Maja Valles, Mars. *Journal of Geophysical Research*, **98**, 9129–38.

Dunne, T. (1990). Hydrology, mechanics, and geomorphic implications of erosion by subsurface flow. In *Groundwater Geomorphology*, ed. C. G. Higgins and D. R. Coates. Geological Society of America Special Paper, 252, pp. 1–28.

Dury, G. H. (1964). Principles of underfit streams. *US Geological Survey Professional Paper*, **452-A**, 1–67.

Grosswald, M. G. (1980). Late Weichselian ice sheet of northern Eurasia. *Quaternary Research*, **13**, 1–32 (in Russian).

Grosswald, M. G. (1993). Extent and melting history of the late Weichselian ice sheet, the Barents-Kara continental margin. In *Ice in the Climate System*, ed. W. R. Peltier. Springer-Verlag: Heidelberg, pp. 1–20.

Gulick, V. C. (2001). Origin of the valley networks on Mars: a hydrological perspective. *Geomorphology*, **37**, 241–68.

Gulick, V. C. and Baker, V. R. (1989). Fluvial valleys and Martian paleoclimates. *Nature*, **341**, 514–16.

Gulick, V. C. and Baker, V. R. (1990). Origin and evolution of valleys on Martian volcanoes. *Journal of Geophysical Research*, **95**, 14325–44.

Head, J. W., Mustard, J. F., Kreslavsky, M. A., Milliken, R. E., and Marchant, D. R. (2003). Recent ice ages on Mars. *Nature*, **426**, 797–802.

Hoffman, N. (2000). White Mars: a new model for Mars surface and atmosphere based on CO_2. *Icarus*, **146**, 326–42.

Howard, A. D. and McLane, C. F. III (1988). Erosion of cohesionless sediment by groundwater sapping. *Water Resources Research*, **24**, 1659–74.

Hynek, B. M. and Phillips, R. J. (2001). Evidence for extensive denudation of the Martian highlands. *Geology*, **29**, 407–10.

Hynek, B. M. and Phillips, R. J. (2003). New data reveal mature, integrated drainage systems on Mars indicative of past precipitation. *Geology*, **31**, 757–60.

Irwin, R. P. III, Maxwell, T. A., Howard, A. D., Craddock, R. A., and Leverington, D. W. (2002). A large paleolake basin at the head of Ma'adim Vallis, Mars. *Science*, **296**, 2209–12.

Kochel, R. C. and Piper, J. F. (1986). Morphology of large valleys on Hawaii: evidence for groundwater sapping and comparisons with Martian valleys. *Journal of Geophysical Research*, **91**, 175–92.

Komar, P. D. (1979). Comparison of the hydraulics of water flows in Martian outflow channels with flows of similar scale on Earth. *Icarus*, **37**, 156–81.

Komar, P. D. (1980). Modes of sediment transport in channelized water flows with ramifications to the erosion of Martian outflow channels. *Icarus*, **42**, 317–29.

Komatsu, G. and Baker, V. R. (1996a). Channels in the Solar System. *Planetary and Space Science*, **44**, 801–15.

Komatsu, G. and Baker, V. R. (1996b). Catastrophic flood spillways: comparison of Mars and Earth examples. *Abstracts of Papers Submitted to the 27th Lunar and Planetary Science Conference*. Houston: Lunar and Planetary Institute, pp. 685–6.

Komatsu, G. and Baker, V. R. (1997a). Paleohydrology and flood geomorphology of a Martian outflow channel: Ares Vallis. *Journal of Geophysical Research*, **102**, 4151–60.

Komatsu, G. and Baker, V. R. (1997b). Cataclysmic flooding in the solar system: a new perspective. *Abstracts of Papers Submitted to the 28th Lunar and Planetary Science Conference*. Houston: Lunar and Planetary Institute, pp. 745–6.

Komatsu, G., Clute, S. K., and Baker, V. R. (1997c). Preliminary remote-sensing assessment of Pleistocene cataclysmic floods in central Asia. *Abstracts of Papers Submitted to the 28th Lunar and Planetary Science Conference*. Houston: Lunar and Planetary Institute, pp. 747–8.

Komatsu, G., Baker, V. R., Gulick, V. C., and Parker, T. J. (1993). Venusian channels and valleys: distribution and volcanological implications. *Icarus*, **102**, 1–25.

Komatsu, G., Kargel, J. S., Baker, V. R. *et al.* (2000a). A chaotic terrain formation hypothesis: explosive outgas and outflow by dissociation of clathrate on Mars. *Abstracts of Papers Submitted to the 31st Lunar and Planetary Science Conference*. Houston: Lunar and Planetary Institute, CD 31, Abstract 1434.

Komatsu, G., Miyamoto, H., Ito, K., Tosaka, H., and Tokunaga, T. (2000b). Comment on "Shaw et al. 1999, The Channeled Scabland: Back to Bretz?" *Geology*, **28**, 573–4.

Komatsu, G., Gulick, V. C., and Baker, V. R. (2001). Valley networks on Venus. *Geomorphology*, **37**, 225–40.

Komatsu, G., Rossi, A. P., and Di Lorenzo, S. (2003). An integrated GIS study applied to the Valles Marineris: chaotic terrain transition zone, Mars. In *Proceeding volume for V Convegno di Scienze Planetarie*, Gallipoli (Lecce), Italy, pp. 91–4.

Kröpelin, S. (1993). Geomorphology, landscape evolution and paleoclimates of Southwest Egypt. *Catena*, **26**, 31–65.

Kwoun, O. I., Crawford, M. M., Baker, V. R., and Komatsu, G. (1998). Variable resolution topographic mapping of ancient fluvial landscapes in Australia. In *Proceeding volume for IEEE 1998 IGARSS*, Seattle, WA, 5, pp. 2360–2.

Laity, J. E. and Malin, M. C. (1985). Sapping processes and the development of theater-headed valley networks in the Colorado Plateau. *Geological Society of America Bulletin*, **96**, 203–17.

Lucchitta, B. K. (1982). Ice sculpture in the Martian outflow channels. *Journal of Geophysical Research*, **87**, 9951–73.

Luo, W. (2002). Hypsometric analysis of Margaritifer Sinus and origin of valley networks. *Journal of Geophysical Research*, **107** (E10), doi:10.1029/2001JE001500, 5071.

Luo, W., Arvidson, R. E., Sultan, M. *et al.* (1997). Groundwater sapping processes, Western Desert, Egypt. *Geological Society of America Bulletin*, **109**, 43–62.

Malin, M. C. and Edgett, K. S. (2000). Evidence for recent groundwater seepage and surface runoff on Mars. *Science*, **288**, 2330–5.

Mars Channel Working Group (1983). Channels and valleys on Mars. *Geological Society of America Bulletin*, **94**, 1035–54.

Marsursky, H. (1973). An overview of geologic results from Mariner 9. *Journal of Geophysical Research*, **78**, 4037–47.

Mellon, M. T. and Phillips, R. J. (2002). Recent gullies on Mars and the source of liquid water. *Journal of Geophysical Research*, **106**, 23165–79.

Milton, D. J. (1974). Carbon dioxide hydrate and floods on Mars. *Science*, **183**, 654–6.

Musselwhite, D. S., Swindle, T. D., and Lunine, J. I. (2001). Liquid CO_2 breakout and the formation of recent small gullies on Mars. *Geophysical Research Letters*, **28**, 1283–6.

Nicoll, K. (1998). Holocene playas as sedimentary evidence for recent climate change in the presently hyperarid Western Desert, Egypt. Ph.D. dissertation.

Nicoll, K. and Komatsu, G. (1999). Interpreting Martian paleoclimates from valley network morphologies: insights from terrestrial analogues in Egypt. *Abstracts of Papers Submitted to the 30th Lunar and Planetary Science Conference*. Houston: Lunar and Planetary Institute, CD 30, Abstract 1054.

O'Connor, J. E. (1993). Hydrology, hydraulics, and sediment transport of Pleistocene Lake Bonneville flooding on the Snake River, Idaho. *Geological Society of America Special Paper*, **274**, 83 pp.

O'Connor, J. E. and Baker, V. R. (1992). Magnitudes and implications of peak discharges from Glacial Lake Missoula. *Geological Society of America Bulletin*, **104**, 267–79.

Phillips, R. J., Zuber, M. T., Solomon, S. C. *et al.* (2001). Ancient geodynamics and global scale hydrology on Mars. *Science*, **291**, 2587–91.

Pickup, G., Allan, G., and Baker, V. R. (1988). History, palaeochannels and palaeofloods of the Finke river, central Australia. In *Fluvial Geomorphology of Australia*, ed. R. F. Warner. London: Academic Press, pp.177–200.

Pieri, D. C. (1980). Geomorphology of Martian Valleys. *NASA Technical Report*, TM-81979, Washington, DC.

Rice, J. W. Jr. and Edgett, K. S. (1997). Catastrophic flood sediments in Chryse Basin, Mars, and Quincy Basin, Washington; application of Sandar Facies model. *Journal of Geophysical Research*, **102**, 4185–200.

Robinson, M. S. and Tanaka, K. L. (1990). Magnitude of a catastrophic flood event at Kasei Valles, Mars. *Geology*, **18**, 902–5.

Rudoy, A. N. (2002). Glacier-dammed lakes and geological work of glacial superfloods in the Late Pleistocene, Southern Siberia, Altai Mountains. *Quaternary International*, **87**, 119–40.

Rudoy, A. N. and Baker, V. R. (1993). Sedimentary effects of cataclysmic late Pleistocene glacial outburst flooding, Altai Mountains, Siberia. *Sedimentary Geology*, **85**, 53–62.

Scott, D. H., Dohm, J. M., and Rice, J. W. Jr. (1995). Map showing channels and possible paleolake basins. *US Geological Survey Miscellaneous Investigation Series*. MAP-2461.

Shaw, J., Munro-Stasiuk, M., Sawyer, B. *et al.* (1999). The Channeled Scabland: back to Bretz? *Geology*, **27**, 605–8.

Smith, P. H., Bell, J. F. III, Bridges, N. T. *et al.* (1997). Results from the Mars Pathfinder Camera. *Science*, **278**, 1758–65.

Squyres, S. W. (1989). Urey Prize lecture: Water on Mars. *Icarus*, **79**, 229–88.

Tanaka, K. L. (1997). Sedimentary history and mass flow structures of Chryse and Acidalia Planitiae, Mars. *Journal of Geophysical Research*, **102**, 4131–49.

Thorson, R. M. (1989). Late Quaternary paleofloods along the Porcupine River, Alaska: implications for regional correlation. In *Late Cenozoic History of the Interior Basins of Alaska and the Yukon*, ed. L. D. Carter, T. D. Hamilton and J. P. Galloway. US Geological Survey Circular 1026, pp.51–4.

Velichko, A. A., Isayeva, L. L., Makeyev, V. M., Matishov, G. G., and Faustova, M. A. (1984). Late Plestocene glaciation of the Arctic Shelf, and the reconstruction of Eurasian ice sheets. In *Late Quaternary Environments of the Soviet Union*, ed. A. A. Velichko. Minneapolis: University of Minnesota Press, pp. 35–44.

Waitt, R. B. Jr. (1984). Periodic jökulhlaups from Pleistocene Glacial Lake Missoula: new evidence from varved sediment in northern Idaho and Washington. *Quaternary Research*, **22**, 46–58.

Williams, R. M., Phillips, R. J., and Malin, M. C. (2000). Flow rates and duration within Kasei Valles, Mars: implications for the formation of a Martian ocean. *Geophysical Research Letters*, **27**, 1073–6.

Williams, R. M. E. and Phillips, R. J. (2001). Morphometric measurements of Martian valley networks from Mars Orbiter Laser Altimeter (MOLA) data. *Journal of Geophysical Research*, **106**, 23737–52.

13

Playa environments on Earth: possible analogs for Mars

Goro Komatsu, Gian Gabriele Ori, Lucia Marinangeli
International Research School of Planetary Sciences, Universita' d'Annunzio, Pescara

and

Jeffrey E. Moersch
Dept. of Geological Sciences, University of Tennessee, Knoxville

13.1 Introduction

Playas (dry lakes) are a type of lacustrine system that are dry most of the time, and can be flooded only occasionally. A playa environment, despite its dry conditions, is characterized by an active hydrological cycle. This is evidenced by a wide range of hydrogeological processes operating today or in the recent past. Therefore, playas are a fundamentally different environment from dry desiccated deserts, and identification of playas on Mars has significant implications for the planet's hydroclimatic history.

Mars currently is dominated by a hyperarid environment. Today, water appears to exist abundantly in the Martian polar caps, and also in the surrounding high-latitude regions, but as near-surface ice (Boynton *et al.*, 2002), not liquid water. Whether or not there are localities with recent active hydrogeological processes is uncertain. However, there may have been sites of stable lakes (deep-water lakes) in the past. Such sites would have changed to playa environments, owing to the decline in the water budget, and eventually desiccated completely. Photogeologic surveys have identified possible paleoshorelines in the northern plains (Parker *et al.*, 1989) and crater lakes (Cabrol and Grin, 1999; Ori *et al.*, 2000a; Malin and Edgett, 2003). If these features are in fact paleoshorelines, it would necessarily imply that conditions suitable for stable oceans and lakes must have existed at some point in Mars' history. Ice-covered paleolakes could have also existed, and their shoreline geomorphology could differ from that of paleolakes without ice cover.

The Geology of Mars: Evidence from Earth-based Analogs, ed. Mary Chapman. Published by Cambridge University Press.

In the terrestrial playa environment, a wide range of sedimentary deposits is observed. They include clastic sediments supplied by drainages during flood inundation, and those made of precipitated minerals such as evaporites. A crucial tool in identifying sedimentary deposits on Mars is remote sensing from orbit. Data acquired by remote sensing can help identify candidate sites where future landed missions will study in detail the history of water and sedimentation. Sedimentary deposits (particularly of aqueous origin) on Mars, if they indeed exist, would have great value to the field of exobiology because sedimentary deposits on Earth are often derived from biological processes or preserve biological materials (Komatsu and Ori, 2000). Therefore, identifying sedimentary deposits on Mars is a high priority for future exploration. Regarding exobiological implications, high-altitude lakes in the Andes and their habitats, environments that may be compared to those during periods of enhanced hydrological processes and sedimentation on Mars, are discussed by Cabrol *et al.* (Chapter 14).

In this chapter, we discuss three important elements of playa environment research: geological processes operating in playas, evidence of ancient deep-water lakes, and remote sensing identification of evaporite deposits in playa environments. For this purpose, we introduce three instructive examples of playa environments where we have conducted field studies: Chott el Jerid and Chott el Rharsa in Tunisia; Tsagaan Nuur in the Valley of Lakes, Mongolia; and Badwater Basin, Death Valley, USA.

13.2 Depositional and erosional processes of playa: Chott el Jerid and Chott el Rharsa, Tunisia

13.2.1 Playas of southern Tunisia

Playas of southern Tunisia are ideal sites for comparison with their possible counterparts on Mars. This is because characteristic elements of playa depositional processes including evaporitic and eolian sedimentation are observable there. In addition, hydrological processes of the playas such as flooding and groundwater circulation are still active today, allowing deeper assessment and understanding of the role water might have played in terrestrial and Martian playa environments.

The easternmost portion of the Atlas Mountain verges southward on the Sahara, forming a set of mainly east–west trending gentle folds. Several depressions (sub-basins) lie within the mountain range and many of them are occupied by sebkha systems (*sebkha*: playa in Arabic) (Swezey, 1996). Three major playa systems characterize the border of the Tunisian

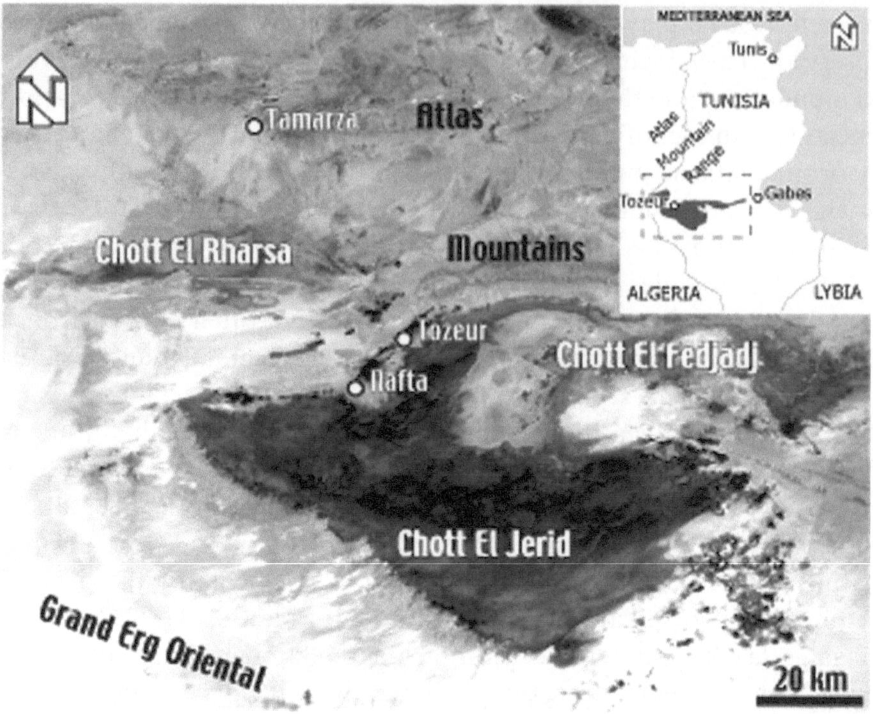

Figure 13.1. Location map of the chotts in southern Tunisia. LANDSAT ETM$^+$ image (panchromatic band).

Sahara: Chott el Jerid, Chott el Rharsa, and Chott el Fedjadj (*chott*: closed depression in Arabic; Figure 13.1). Chott el Rharsa and el Fedjadj rest on top of the most external thrust-sheets of the Atlas range and are limited by partially eroded anticlines. Oued el Melah (*oued*: river in Arabic) debouches into Chott el Rharsa, a large depression between thrusts (Figure 13.2). Chott el Jerid, instead, lies on the proximal portion of the foredeep of the Atlas mountain chain and is bordered by a low-elevation mountain chain, Jebel Zmertene, to the east. Therefore, the chotts are the latest expression of a long tectonic history and they characterize the upper portion of thick sedimentary sequences deposited during the southward migration of the Atlas chain against the Sahara platform. The results of our investigation of the chott system in southern Tunisia are summarized in the field guidebook titled *Exploring Mars Surface and its Terrestrial Analogues* (Ori *et al.*, 2001).

The chotts in southern Tunisia have experienced significant climatic fluctuations, which have affected the Sahara and the adjoining zones

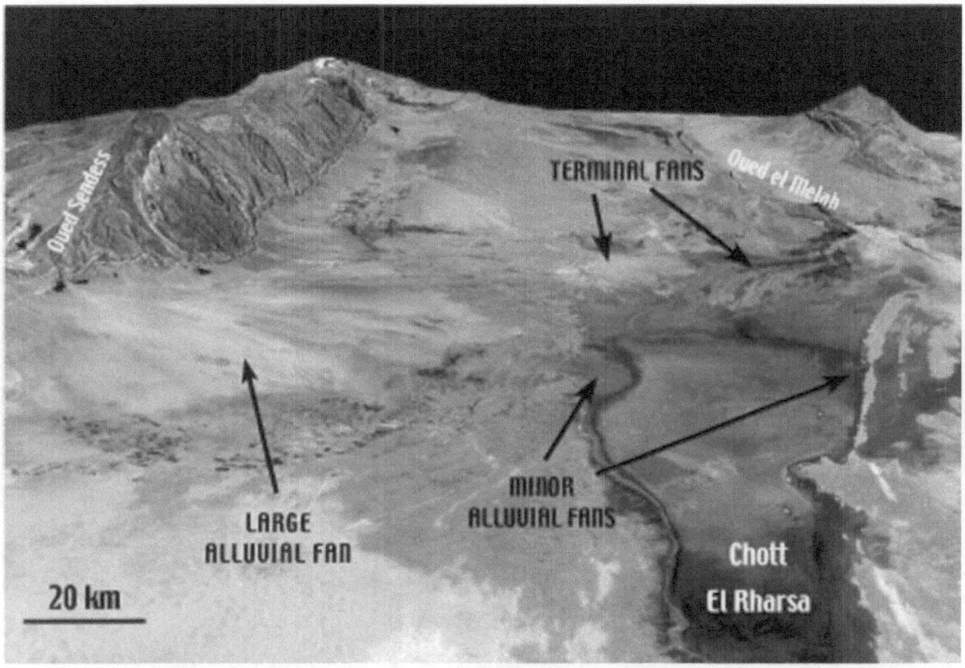

Figure 13.2. Digital Terrain Model (DTM) of the Chott el Rharsa basin obtained by ASTER stereo data and draped by a LANDSAT ETM$^+$ mosaic (panchromatic band).

(Gasse, 2002). The result is a series of intricate patterns of eolian, fluvial, "dry," and "wet" lake deposits. During the late Quaternary, eolian deposition was active from 12 000 to 10 000 years BP, and from 7500 to 5500 years BP (Swezey, 1997). Lacustrine sedimentation linked to perennial or semi-perennial lakes was active before the Holocene and sometimes around 9000 years BP (Gasse and Van Campo, 1994; Swezey, 1997). Evaporitic deposits formed throughout the Holocene, except for during the "wet" lacustrine episode, when mainly clays were deposited. Holocene evaporitic deposits that are exposed at the borders of Chott el Rharsa (Figure 13.3), consist mostly of fragments of selenitic gypsum, microcrystalline gypsum, and microcrystalline carbonate (Barbieri, 2001). These deposits represent, probably, margins of sebkhas during "dry" periods with authigenic (formed or produced in place rather than having been transported and deposited) sedimentation and eolian reworking. Figure 13.3 shows an example of post-flood deposit with evaporitic crust covering Chott el Jerid. Dark spots enriched in organic material can be easily recognized.

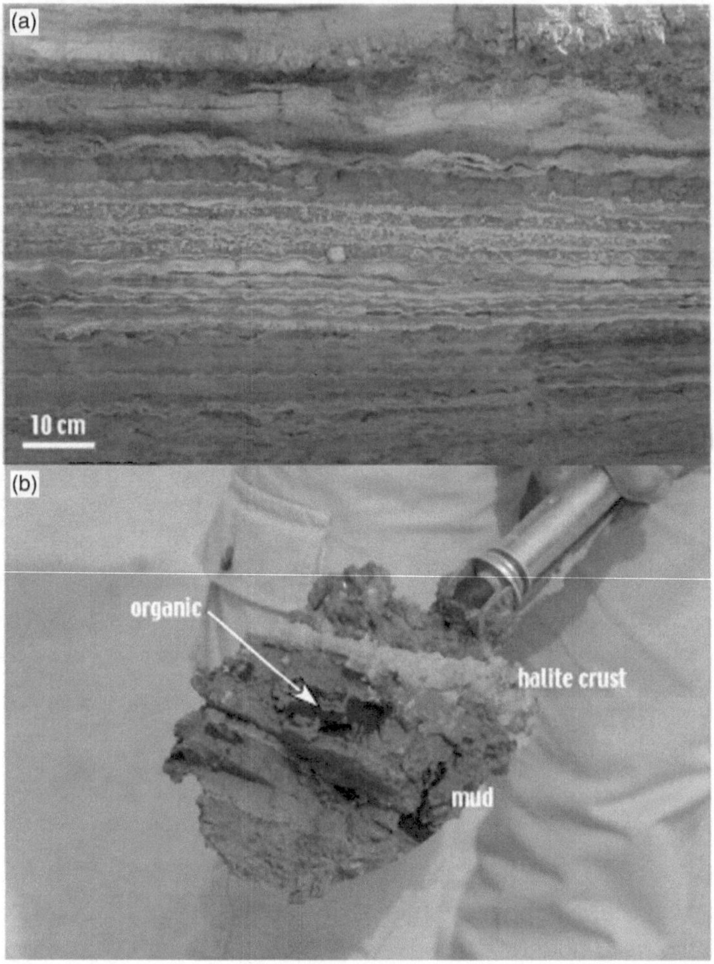

Figure 13.3. Examples of chott's deposits. (a) Sequences of selenitic and microcrystalline gypsum and salt-rich mud forming the top of a low-elevation plateau along the southern border of Chott el Rharsa. (b) Muddy deposit with a salt crust formed after a large flooding event in Chott el Jerid. Enrichment of organic material formed dark and sometimes fetid agglomerates.

13.2.2 Eolian environment

In present times, the Tunisian chotts are bordered by extensive sand sheets. The sand sheet areas mainly consist of flat surfaces with accumulation of sand near obstacles and vegetation (Figure 13.4). Dome-shaped dunes and barchans are scattered throughout the sand sheets. Among the

Figure 13.4. Eolian deposits around Chott el Rharsa. (a) Recent eolian deposits formed against bushes after a 2-day dust storm in Chott el Rharsa (road El Hamma–Chebika). (b) Fossil dunes that are highly eroded along the southwest border of Chott el Rharsa.

sand sheet deposits, a large number of fossil dune deposits occur (Figure 13.4). These fossil dunes have been shaped by wind activity, resulting in yardangs or other eolian erosional landforms. The change of eolian deposit patterns could be due to dryer environmental conditions, but changes in wind regimes can also be the cause. The chotts have been deflation areas. Wind-blown sand is entrapped in the sebkha deposits, but there is not enough detritus to form a sand layer. The processes of eolian deposition and deflation persisted in the same locations respectively throughout the Holocene despite the dramatic climate changes that affected the region.

13.2.3 Fluvial environment

Oued el Melah is an ephemeral stream that flows along the axis of a piggy-back (or thrust sheet top) basin that rests between the main mountain chain and a set of external splays. This stream forms a terminal fan where it debouches into Chott el Rharsa (Figure 13.2). Terminal fans are produced by ephemeral streams, and they are affected by a tremendous loss of water due to evaporation and infiltration. The effect of this process is the loss of sediment transport capacity, with a consequent increase of sedimentary deposition, resulting in channel divergence and an intricate pattern of shallow, distributary channels. Farther downstream, the channels fade out by spreading their water over a large unchanneled flat area (mud flat) that merges into the sebkha. Therefore, ephemeral streams contribute to the sedimentation in the chott by depositing a large amount of mud at the chott border, and in the terminal fan areas.

Alluvial fans are another major component of the fluvial–sebkha systems. In Chott el Rharsa, two types of fluvial apparatus coexist: (1) large fans connected directly with the mountain chain and (2) minor localized fans draining at terraced areas of the chott or at the external anticlines (Figure 13.2). The first case is represented by an alluvial fan formed by Oued Sendess that drains from the Atlas Mountains. The fan is 10 km in radius and consists of poorly to moderately sorted gravels (up to cobble size). The deposition occurs during major flooding events that involve exceptionally large water discharges. According to the gravel facies, the mechanism of deposition ranges from grain flow to highly concentrated turbulent flow. Alluvial fans of the second type are less than 1 km in radius and are made of locally reworked sediments.

13.2.4 The chott surface

The central parts of both Chott el Jerid and Chott el Rharsa are flat and the surfaces are represented by mud entrapping eolian sands and authigenic evaporitic minerals. Figure 13.5 shows this type of surface with large cracks deforming the salt crust. This kind of morphology is usually observed in halite surfaces. The sedimentation in these areas depends on two different processes. The first one occurs in the subsurface and consists

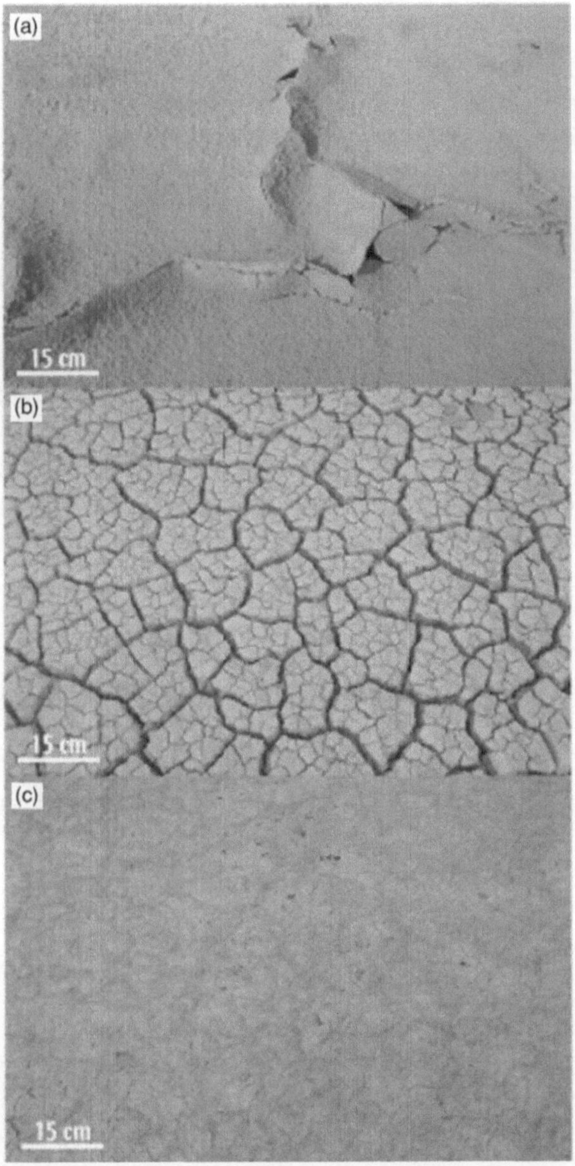

Figure 13.5. Surface textures of chotts. (a) Tepee-shaped cracks in a halite crust (Chott el Jerid). (b) Mud cracks in Chott el Rharsa. (c) Recent mud deposits covering old desiccation cracks in the terminal fan area at the border of Chott el Rharsa.

of the deposition of salts, mostly gypsum, by evaporation of water contained in rock or sediment pores and ascent of brine into the drying layer. This process is called evaporitic pumping. Gypsum is present extensively, and it has its source in Mesozoic to Cenozoic marine evaporite deposits underlying the chott. Halite and carnallite ($KMgCl_3 \cdot 6H_2O$) are other minerals that form in the near surface.

The second type of sedimentation is related to flooding of chotts by rainwater several times per decade. During these pluvial events, the water table moves up and the chott becomes covered by water. Surface crusts of halite and other highly soluble salts are commonly dissolved, and mud is deposited mechanically in this condition. The present mean annual precipitation in the chott area of southern Tunisia is about 80 to 140 mm. But the evaporation rate is about 1500 mm, far exceeding the precipitation. A large quantity of new evaporite minerals and re-precipitated salts is formed after major flooding events. Halite precipitates mainly and gypsum may be a minor component. This dominance of halite is due to the far greater solubility of halite than gypsum. The results of these depositional processes is a flat surface dominated by mud, which is covered at places by halite and, possibly other minerals such as gypsum, usually in the form of selenite. Below the surface, dark and sometimes fetid organic-rich agglomerates are extensively present (Figure 13.3). This is a product of bacterial activity that occurs in the shallow subsurface where moisture is retained.

The nature of chott surfaces at their margins depends on the surrounding environment. In proximity of terminal fans or alluvial fans, the surface is flat or slightly inclined, and composed of mud (Figures 13.5b and 13.5c). In the vicinity of sand sheets, chott deposits are covered by a thin veneer of wind-blown sand.

13.2.5 Spring mounds

In the southeast margin of Chott el Jerid, several mounds are scattered in irregular patches or along lineaments. These mounds are formed by the rise of water from a deep aquifer in primarily Cretaceous rocks, and their formation occurred during the Quaternary pluvial period. They are in the form of ridges, towers, pinnacles, and conical hills with or without central cones. Their heights reach from less than 1 m up to over 30 m and their diameters can be up to 500 m. Most consist of sand and silt size particles cemented by varying amounts of tufa, travertine, and gypsum (Roberts and Mitchell, 1987). Many mounds show a depression where a spring pool was present. Several channels are present around the mounds. They probably acted as drainage channels. Dead palm trees rest on top of the mounds (Figure 13.6). Their presence suggests that these spring mounds are no longer active due to the human exploitation of the water reservoir. Radiocarbon analyses date some mounds as far back as 20 000 years BP

Figure 13.6. A spring mound on the southeast border of Chott el Jerid.

(Swezey, 1997). However, many mounds have been active during historical time.

13.2.6 Implications of chott environment for Mars

The surfaces of chotts in southern Tunisia are remarkably uniform due to the extensive presence of mud flats. Most authigenic minerals remain in the subsurface, but surficial zoning of evaporitic minerals due to sequential crystallization (see Section 13.4.1) can be detected by remote sensing (Drake *et al.*, 1994). On Mars, alternating light- and dark-toned, banded outcrops are observed locally in topographic depressions. For example, in Holden Crater (Malin and Edgett, 2000) the crater floor is interpreted as an evaporite basin (Ori *et al.*, 2000b). If these sequences are indeed evaporites, it is in accordance with the observations made at the Tunisian chotts where most of the evaporitic minerals occur in exposed cliffs. In the Tunisian chotts, the evaporitic minerals are often hidden by the coupled action of stream floods and wind-blown sand. On Mars, this effect can be completely due to the wind-blown detritus that is present in large amounts on the planet. Continuous eolian processes operating on playa surfaces may be akin to those in dry Martian conditions, although there should be differences in the way the processes operate due to thinner atmosphere, smaller gravity, and availability of surface materials.

The spring mounds in Chott el Jerid are unique features that can be used as type examples to differentiate spring mounds (and also mud volcanoes) from other volcanic features. Several features on Mars resemble spring mounds and/or mud volcanoes (e.g., Ori *et al.*, 2000c), and they should be investigated with very high-resolution images and spectroscopy in order to ascertain their origin. The mound features could be indicators of movements of liquids, solids, and gases from the subsurface to the surface.

Exobiology of playa environment deserves a special attention (Farmer and Des Marais, 1999; Komatsu and Ori, 2000). Microbial activity in salt flats is high (e.g., halobacteria) and preservation potential of microbial life forms is also high in evaporite (Norton *et al.*, 1993). These aspects may justify for investigation of Martian playas in future missions. Spring mounds and/or mud volcanoes could serve as localized oases for microbial life, providing additional reason for studies.

On January 4, 2004, a Mars Exploration Rover (MER) named Spirit landed successfully inside Gusev Crater, the 150 km diameter impact basin located near the equator of Mars. Gusev Crater has long been argued to be a site of ancient lakes as summarized in Cabrol *et al.* (2003). The panorama image around the landing site shows a relatively flat landscape with scattered rocks (Figure 13.7). Investigations being conducted with the rover at the time of this writing will hopefully reveal the origin of this landscape, and in particular will determine whether the site bears evidence for ancient "wet" and "dry" lakes.

Three weeks later, on January 25th, another MER Opportunity landed in Sinus Meridiani (Figure 13.8). The area had been known to be covered by the crystalline hematite mineral found by orbital remote sensing (Christensen *et al.*, 2000), but no other clear observation favorable for a playa environment had been reported prior to the landing. Despite this,

Figure 13.7. Mars Exploration Rover "Spirit" landed in Gusev Crater, the hypothesized former crater lake basin. Panoramic Camera (Pancam) image. Courtesy of NASA/JPL/Cornell.

Figure 13.8. Mars Exploration Rover "Opportunity" landed in Sinus Meridiani, and the rover has revealed evidence for abundant water existed in the past. Panoramic Camera (Pancam) image. Courtesy of NASA/JPL/Cornell.

a wide range of evidence for a body of shallow water and its desiccation (e.g., layered rocks of a probable sedimentary origin, salts) in the ancient time was discovered by the rover (NASA/JPL MER web page, 2004). This exciting situation prompts us to postulate that playa environments indeed existed on the red planet.

13.3 Morphology of playa. Paleolake landforms: Tsagaan Nuur, the Valley of Lakes, Mongolia

13.3.1 Paleoshoreline landforms of playa

Modern playas exist in topographic basins. Therefore, under climatic conditions different from today, long-lasting lakes could form in the playa basins. Deep-water lakes have hydrological regimes very different from playas and the processes of lake water and associated drainage systems leave distinctive geomorphological and sedimentological traces in and around the basins. Even after the environment shifts to drier playa conditions, the traces often remain. We can study these traces and reconstruct characteristics of paleolakes, and consequently infer paleoenvironments which existed in the past. This approach is valuable also for Mars.

13.3.2 Tsagaan Nuur topographic depression, the Valley of Lakes

The Valley of Lakes is a region currently occupied by a chain of lakes in central and western Mongolia (Figure 13.9). This valley is bounded by the Hangai Plateau to the north and the Altai Mountain Range to the south. A small closed depression exists in the valley between two stable lakes, Böön Tsagaan Nuur and Orog Nuur (*nuur*: lake in Mongolian). Tsagaan Nuur is a lake occupying a small part of the depression, and larger areas can be inundated occasionally. Tsagaan Nuur is located south of the Hangai Plateau, one of the main sources of water for the lake via the Nariin Gol (*gol*: river in Mongolian). In addition, smaller drainages feed the lake. The majority of the depression is essentially a playa environment today, and it contains well-preserved exposures of paleoshorelines. Komatsu *et al.* (2001) conducted paleolake reconstruction from shoreline landforms utilizing Radarsat imagery in conjunction with field observations. The shoreline geomorphology using Landsat TM imagery is presented in Figure 13.9. This topographic depression exhibits a wide range of shoreline geomorphology commonly observed with well-studied paleolakes, such as glacial Lake Bonneville in the USA (e.g., Gilbert, 1890; Currey, 1980, 1990).

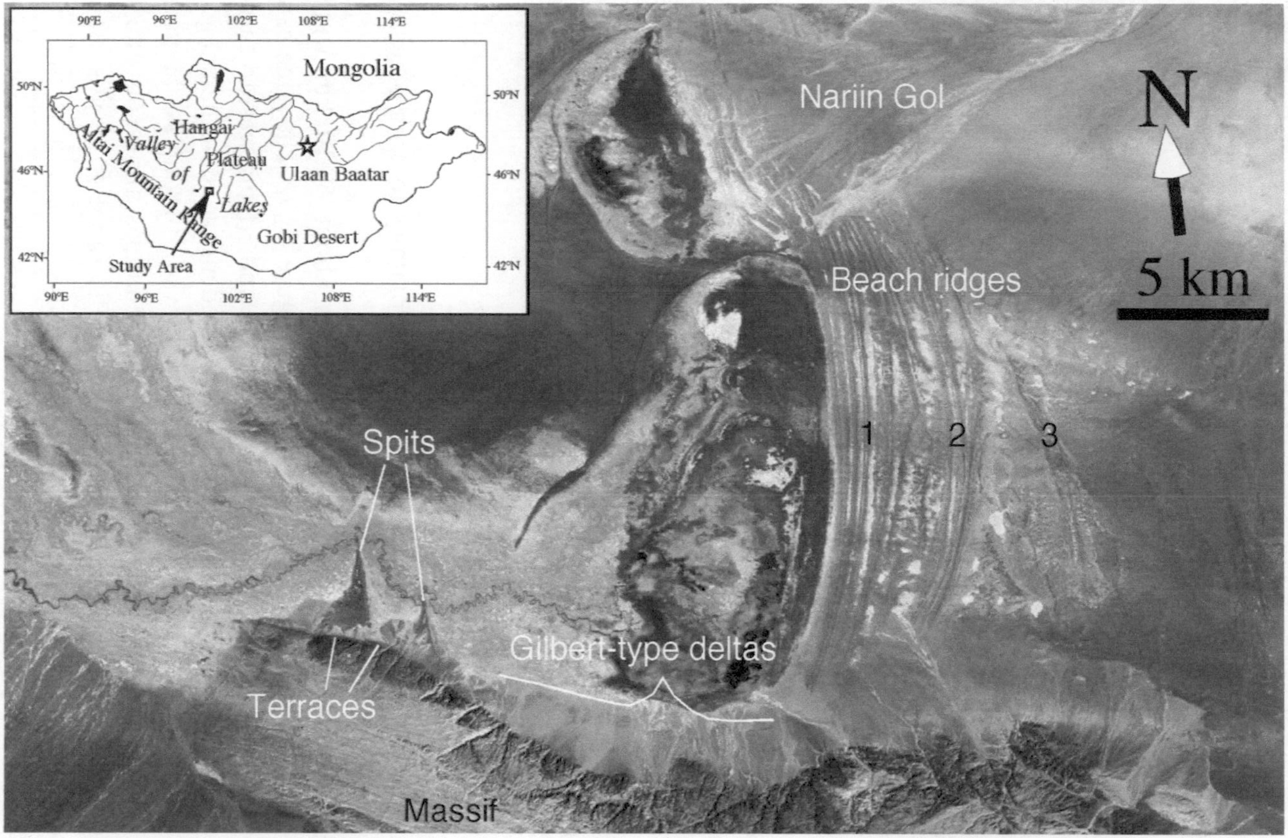

Figure 13.9. Paleoshoreline features in the Tsagaan Nuur topographic depression 60 km east of Böön Tsagaan Nuur. Beach ridge paleostrandline zone can be divided in three sub-groups, designated as 1, 2, and 3. Landsat TM image. The inset shows locality of the study area in Mongolia. Based on Komatsu *et al.* (2001).

Figure 13.10. (a) Sub-group 1 of the beach ridge paleostrandlines. These strandlines are enhanced by the low-angle sunset light. (b) Terrace paleostrandlines bordering a massif. See two people and a truck for scales. Note the valleys that are significantly wider above the strandlines and narrower below. The valley incision was perhaps limited in the lower part during the period when the local water surface was much higher. (c) This gravel-rich ridge is one of two relict spits extending from the massif south of the paleolake. (d) A complex of probable Gilbert-type deltas. Note the flat-topped and steep-sided structure of the fan deltas in front of the mountains. Photos from Komatsu *et al.* (2001).

We can observe these features in relatively close proximity to each other in this topographic depression.

The entire beach ridge paleostrandline zone is about 5–10 km wide east–west, and about 20 km long north–south (Figure 13.9). The zone can be divided into three sub-groups, designated as 1, 2, and 3 from west to east. There are a total of 15 to 20 prominent strandlines identifiable in the image, but not all of them are continuous. Figure 13.10a shows sub-group 1 enhanced by low-angle sunset light. On the ground, the strandlines appear as alternating ridges and troughs. Their amplitudes typically range from less than 1 m to a few meters. The ridge surfaces are covered with coarse-grained gravel a few centimeters in diameter and the intervening troughs are mantled by fine-grained gravel less than a centimeter in diameter. Evaporites occasionally fill the troughs. The preservation of ridges is partly the result of subsurface calcite cementation (caliche).

On the flank of the granitic massif south of the valley there is another type of paleostrandline feature (Figure 13.9). These paleostrandlines have a narrow terrace geometry (typically less than tens of meters wide) and difficult to observe from orbit. The highest of the terrace paleostrandlines is at least 30 m above the valley floor (Figure 13.10b). In the field, they appear to be wave-cut terraces carved on unconsolidated slope materials or on bedrock in some locations. Wave dissipation, and consequently the formation of wave-cut terraces, is controlled by factors including wave geometry, depth of water, and lake floor topography. In reality, many terraces are formed by competing sedimentation and erosion in a complex process. Note that the valleys in Figure 13.10b are significantly wider above the strandlines and narrower below. This may indicate that the valley incision was limited below the strandlines during a period when the local water surface was much higher. Near-shoreline lacustrine deposits may have mantled the lower valleys during this period. Upon desiccation, the lacustrine deposits were incised.

Two prominent relict spits (1 to 3 km long) extend north from the massif at the southern end of the paleolake (Figures 13.9 and 13.10c). These gravel-rich spits were presumably formed by the movement and deposition of materials by longshore currents in the paleolake. The fans placed at the mouths of mountain streams from the massif (Figure 13.9) morphologically show flat tops marking former lake levels (about the same levels as terrace strandlines), and they are steep-sided toward the center of the basin (Figure 13.10d). This geometry is characteristic of Gilbert-type deltas, a type of fan deltas (e.g., Ori and Roveri, 1987; Milligan and Chan, 1998) first described by Gilbert (1890). Gilbert-type deltas are also characterized by a well-defined near horizontal stratification pattern at their tops (topsets) and oblique beds (foresets) in their main bodies underlying the topset facies. The foreset of offlap geometry is formed by incremental addition of detritus on the delta front slope, whereas topset is the result of near shoreline accumulation of coarse-grained bed-load and it corresponds to the local base level.

13.3.3 Shoreline evidence of past water ponding on Mars

Researchers have searched for evidence of past water ponding on Mars. Water pondings at the scales of oceans (Parker *et al.*, 1989, 1993), lakes in large topographic basins (Scott *et al.*, 1992), and crater lakes (Cabrol and Grin, 1999; Ori *et al.*, 2000a) have been inferred. Martian paleolake studies utilize geomorphology and sediments for the recognition of past ponding of water. For example, Cabrol and Grin (1999) identified 179 paleolakes in impact craters based on features such as deltas, terraces, shorelines, and evaporites, all of which were interpreted from Viking imagery. They also determined the age range of the crater lakes to be the Middle Noachian or Late Noachian by employing the crater counting technique.

Using Viking imagery, Ori *et al.* (2000a) conducted a detailed analysis of possible crater lake examples in the Ismenius Lacus and Memnonia regions of Mars. They estimated in some crater lakes, based on a wave modeling on Mars, that almost all wave energy should be dissipated at the coastline and causes terrace formation. They also concluded from a modeling of wave action on terraces and geometries of Gilbert-type deltas, that these crater lakes had been active for thousands of years and developed in several phases during the early to late Hesperian. Assuming a 1 bar atmosphere, wave energy on Mars may be similar to that on Earth (Kraal *et al.*, 2003), and a study of wave propagation models (Parker and Currey, 2001) concluded that conditions on Mars may permit formation of coastal landforms on Mars that are large enough to be visible from orbit.

However, identification of shorelines in orbital images is not an easy task even on Earth. Deep lakes do not necessarily produce well-defined continuous shoreline features and post-formation degradation is always a problem. On Earth, ground truth is essential to confirm the coastal landforms. MOC (Mars Orbiter Camera) onboard the Mars Global Surveyor acquired images with greatly improved spatial resolution (max. ~1.5 m). These images are providing new opportunities for investigating possible evidence of past water ponding on Mars. For example, amazing details of putative shoreline features in the Ismenius Lacus and Memnonia are now revealed (Figure 13.11). Using MOC images, Malin and Edgett (1999) tested the Mars ocean hypothesis by examining the selected localities of paloeshoreline morphology proposed by Parker *et al.* (1989, 1993). They did not find morphologic evidence for water ponding. However, the research of possible paleoshorelines on Mars is far from complete. It needs extensive detailed studies based on experience from terrestrial analogs, and preferably ground truthing in the future.

13.4 Search for playa sediments: spectral analysis of sediments in Badwater Basin, Death Valley, USA

13.4.1 Patterns of mineral deposition in playas

Within terrestrial playas, evaporite minerals are often found in characteristic patterns that result from their order of deposition. Terminal lake basins

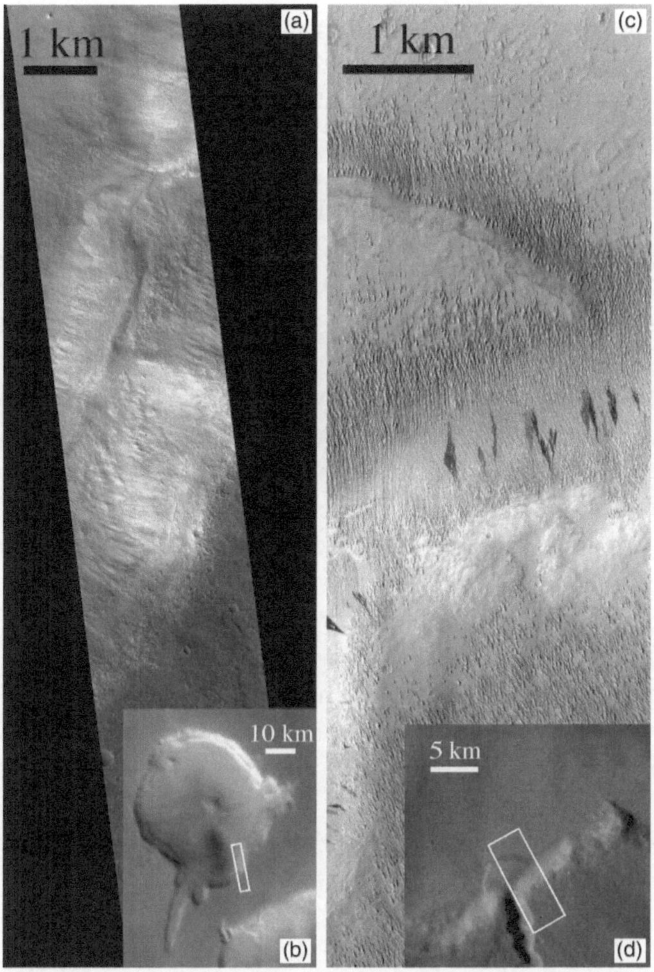

Figure 13.11. Examples of hypothesized shoreline landforms. (a) An example of possible wave-cut terraces. Terrace morphology in the Ismenius Lacus region on Mars. These terraces are built against the crater rim, channel inlet and minor embayments, implying a sedimentary origin. MOC narrow-angle image M0800544. (b) MOC image wide-angle context image M0800545. (c) A probable Gilbert-type delta in the Memnonia region on Mars. MOC narrow-angle image E0500182. (d) Viking context image 456S07.

tend to be particularly rich in evaporite deposits because the concentrations of various salts that flow into the basin increase until the salts precipitate out of solution, having no other outlet. In a receding lacustrine environment, temperature and solubility exercise predominant control over the sequence and products of precipitates. Carbonates, being the least soluable of the salts typically in solution, are deposited in marginal basin areas. Sulfates and halides, having intermediate and high solubility (respectively), precipitate at progressively more basinward locations as concentrations in the brine increase with evaporation (Eugster and Hardie, 1978). The result, as viewed from above, is often a "bull's-eye" or "bathtub ring" pattern of evaporite facies appearing in order of solubility (see Jones, 1965).

Evaporites are not the only precipitated deposits that can form in lacustrine environments. Given proper temperature, pH conditions, and a supply of dissolved atmospheric CO_2, carbonate minerals can precipitate directly out of solution to produce non-evaporitic deposits. Clastic sediments, predominantly clays, are also a major component of lacustrine deposits, and are often found in intimate mixtures with precipitated salts (Hunt *et al.*, 1966). Their specific mineralogy, location, and physical size relate to their mode of deposition. Shoreline and delta environments are often characterized by coarse sediment, while clay and silt-sized particles dominate less energetic regimes, such as the lake bottom. Clastic sediment deposits do not display strikingly characteristic mineralogic zonation patterns like precipitated sediments. The mineralogy of lacustrine clays on Earth is controlled by climate and the source of the clay, with montmorillonite and illite predominating in semi-arid environments. On Mars, palogonites (Allen *et al.*, 1981) or smectite clays (Banin and Margulies, 1983) are considered leading candidate clay minerals, but they would not be confined to paleolake basins. Nevertheless, ancient lakes on Mars would have functioned as traps for clay minerals, and may have left deposits in these locations detectable by remote sensing.

13.4.2 An example of infrared remote sensing of a terrestrial playa: Badwater Basin, Death Valley, California, in the thermal infrared

Death Valley, California, contains a well-studied series of playa deposits with a wide range of evaporite minerals (Hunt *et al.*, 1966). Carbonates, sulfates, and halite deposits are found in the typical basinward sequence along the western margin of the Cottonball, Middle, and Badwater Basins within Death Valley. Because of its easy accessibility and the abundance of previous ground-based work, Death Valley has become a *de facto* type locality for studying evaporite deposits with different remote sensing experiments. In particular, studies of Badwater Basin, Death Valley, in the visible to near-infrared region with data from the Airborne Visible-Infrared Imaging Spectrometer (AVIRIS; Crowley, 1993) and in the thermal infrared region with data from the Thermal Infrared Mapping Spectrometer (TIMS; Crowley and Hook, 1996) and the Modis/ASTER Airborne

Simulator (MASTER; Moersch *et al.*, 2000) have amply demonstrated the utility of remote sensing techniques for mapping playa evaporite deposits on Earth.

MASTER is an airborne "pushbroom imager" that simultaneously acquires data in 50 channels from the visible to the thermal infrared (0.46–12.85 μm) at a spatial resolution of 5–15 m (Hook *et al.*, 2001). Ten of the channels fall within the 7–13 μm region, making MASTER an outstanding analog for the thermal infrared remote sensing experiments NASA has recently sent to Mars on orbiting spacecraft (see Section 13.4.3, below). Figure 13.12a shows a MASTER image of Badwater Basin constructed using the 0.654, 0.542, and 0.460 μm reflectance bands in the red, green, and blue channels (respectively) to approximate natural color.

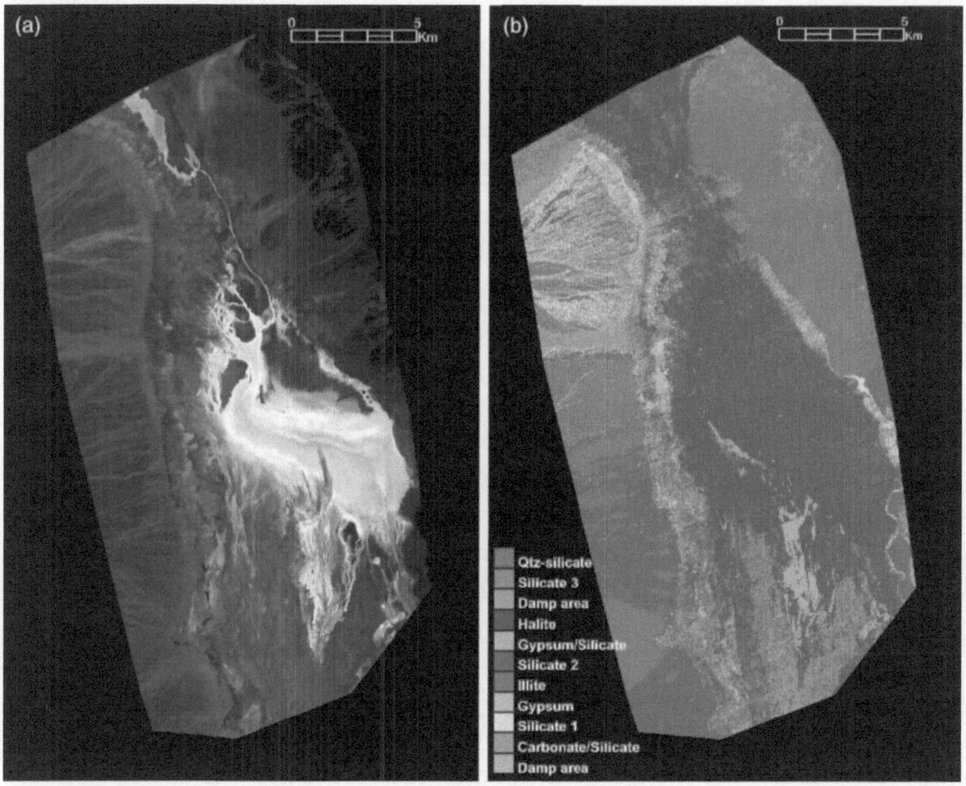

Figure 13.12. MASTER views of Badwater Basin, Death Valley, California. (a) Approximate true color image created using the MASTER 0.654, 0.542, and 0.460 μm reflectance bands. (b) Spectral classification map of MASTER thermal infrared bands. (For a color version of this figure, please refer to color plate section.)

A spectral classification map of Badwater Basin based solely on MASTER's thermal infrared channels appears in Figure 13.12b. Evaporite minerals are seen on the west side of the basin in typical "bathtub rings" of carbonates (shown in medium brown) and sulfates (cyan and aquamarine) surrounding a large expanse composed predominantly of halite (blue). Note that the halite has not been detected directly because it has no significant absorptions in this spectral region. Rather, it is inferred here from its lack of spectral features and from prior knowledge from the field.

In order to assess the accuracy of thermal infrared multispectral mapping of playa deposits, Baldridge *et al.* (2004) completed a comprehensive ground-truthing campaign of the area shown in the MASTER scene presented above. Although the mineral identifications of collected samples were not completely consistent with each other, taken as a set they provided enough constraints for an accurate interpretation that agreed with identifications based solely on the MASTER spectral classification analysis for 14 of the 16 sites.

13.4.3 Remote sensing of putative playa deposits on Mars and lessons learned from terrestrial analogs

In recent years, NASA has made a concerted effort to map the mineralogic composition of the Martian surface using relatively high spatial resolution thermal infrared remote sensing observations from orbit. Since 1997, the Mars Global Surveyor Thermal Emission Spectrometer (TES) has been mapping the planet in the ~6–50 μm spectral range in 143 spectral channels, at a spatial resolution of approximately 3 × 6 km (Christensen *et al.*, 1998). One of the primary goals of this experiment is to locate and map deposits of minerals that form in the presence of liquid water. A major discovery of the TES experiment was the detection of coarse-grained, crystalline hematite (Fe_2O_3) in Meridiani Planum (Christensen *et al.*, 2000) and in smaller deposits in other locations. The favored hypothesis of Christensen *et al.* (2000) for the formation of these deposits is that they were precipitated from iron-rich solutions.

Despite intensive searching, TES has not detected aqueous minerals deposits on Mars. Stockstill *et al.* (2003) have performed a survey of TES data from craters identified by Cabrol and Grin (1999) on the basis of photogeologic observations as putative paleolake basins. Hyperspectral mapping techniques were applied at the limit of spatial resolution to the 80 largest craters (the remainder being too small to resolve any intracrater spectral units). Although discrete, mapable spectral units were found within several craters, deconvolution of their spectra yielded no evidence for any aqueous minerals other than the previously reported hematite. The search for aqueous minerals on Mars is continuing with new data from the Thermal Emission Imaging System (THEMIS) experiment (Christensen *et al.*, 2003), which arrived at Mars aboard the Mars Odyssey orbiter in late 2001 and also from the more recent Visible and Infrared Mineralogical Mapping Spectrometer (OMEGA) experiment onboard the European Mars Express orbiter. These multispectral cameras will image the entire planet in ten thermal infrared bands at a spatial resolution of 100 m/pixel. As with TES, a major goal for these experiments is to identify and map deposits of aqueous minerals. The early calibration and validation of both data are nearly complete, and mineralogic mapping has started. The European OMEGA/Mars Express hyperspectral imager has recently identified hydrated sulfates on Mars, minerals which record direct evidence of aqueous activity (Gendrin *et al.*, 2005). Sulfates could originate from evaporation of playas, but may also form by other means (Gendrin *et al.*, 2005). Much work remains to be done with this growing dataset.

Terrestrial analogs may help shed some light on the question of why TES and THEMIS have not detected aqueous minerals other than hematite on Mars. Thermal infrared remote sensing observations of terrestrial playa environments can be degraded to match the spatial resolutions of data being returned from Mars in an attempt to set thresholds of detection for aqueous minerals in natural settings (Moersch *et al.*, 2001). When the MASTER data used to create the spectral classification map in Section 13.4.2 are re-binned to the 100 m spatial resolution of THEMIS data (Figure 13.13a) and then re-classified, the spatial/spectral patterns characteristic of evaporite basins, such as bands of carbonates and sulfates around the basin's margin (particularly the west side), are still easily discernable. However, when the raw data are further de-resolved to the ~3 km resolution of TES data (Figure 13.13b) and re-classified, all of the spatial patterns are lost. Spectral signatures of evaporites also become extremely subdued as they are washed-out by sub-pixel mixing effects at TES-like spatial resolution.

These results from Badwater Basin suggest that one possible reason TES has not seen evaporite minerals in putative Martian paleolake basins is that it simply lacks sufficient spatial resolution. Further, the 100 m MASTER simulations indicate that THEMIS should be capable of seeing evaporites if they are exposed on the surface of Mars as they are in Badwater Basin. However, Badwater Basin is just one terrestrial evaporite basin and may not be representative of all – analyses of other terrestrial basins are currently underway to address this possibility.

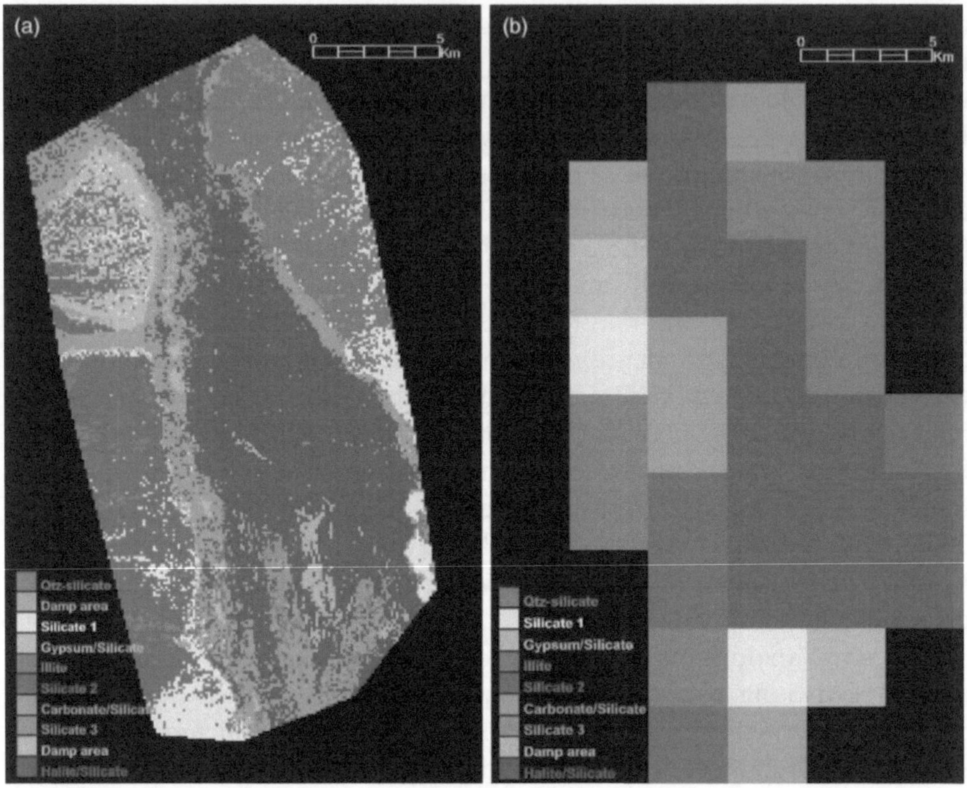

Figure 13.13. The effects of spatial resolution on spectral classification mapping of Badwater Basin. (a) MASTER thermal infrared emissivity data were degraded to the 100 m/pixel spatial resolution of the THEMIS instrument and then spectrally classified as in Figure 13.12b. (b) The same data were degraded to 3 km spatial resolution (approximately equal to TES resolution) and then spectrally classified. Spectral endmembers used to classify the TES-resolution data were taken from THEMIS-resolution data because there were insufficient pixels available at TES resolution to provide useful endmembers. Note that the typical spatial/spectral evaporite patterns on the western margin of the basin are clearly visible at THEMIS resolution, but are lost at TES resolution. (For a color version of this figure, please refer to color plate section.)

It will be interesting to see if THEMIS, given its significantly improved spatial resolution over TES, is able to find spatial/spectral patterns associated with evaporites on Mars. If not, researchers will be forced to consider whether there are environmental factors at play on the surface that have either removed such deposits or obscured them from detection by thermal infrared remote sensing, or if the geochemical models that predict such deposits were present to begin with on Mars are wrong, or if the Martian lakes postulated by photogeologic studies simply never existed. It is interesting to note that the Opportunity rover has detected jarosite and sulfate salts together with the previously known hematite near its landing site in Sinus Meridiani (NASA/JPL MER web page, 2004). In addition, detection of sulfate salts and clay minerals by the OMEGA instrument has been recently reported.

13.5 Conclusions

Playa environments occur in dry regions of Earth. eolian processes dominate modification of playa surfaces for the majority of time. However, in spite of dry conditions prevailing in playas, water plays a major role in geomorphology and sedimentation. There are water-related processes operating over a variety of geological time and spatial scales in playas. For example, evaporitic pumping and spring circulation redistribute sediments on the playa surface, leaving evaporites and mound morphology. Occasional flood inundation is important, bringing new sediments to the playa basin and evaporite deposits form. If inundation becomes more permanent, deep lakes could form. And upon desiccation, such deep lakes could leave a suite of paleoshoreline geomorphology and mineralogy. Detection of playas on Mars depends on identification of water-generated landforms and evaporitic sediments. This can be achieved utilizing orbital remote sensing such as high-resolution optical cameras and infrared spectroscopy, combined with *in situ* analyses.

Acknowledgments

Our research activities were funded by the Italian Space Agency (ASI) and National Aeronautical and Space Administration (NASA). We appreciate Dr. Nabil Guesmi (Faculté des sciences humaines et socials, Tunis, Tunisia) and Dr. Titi Melis (Università di Cagliari, Italy) for their help during our fieldwork in Tunisia. We also thank Nathalie Cabrol and Jeff Kargel for their reviews. The authors acknowledge the use of Mars Orbiter Camera images processed by Malin Space Science Systems that are available at http://www.msss.com/moc_gallery/.

References

Allen, C. C., Gooding, J. L., Jercinovic, M., and Keil, K. (1981). Altered basaltic glass: a terrestrial analog to the soil of Mars. *Icarus*, **45**, 347–69.

Baldridge, A. M., Farmer, J. D., and Moersch, J. E. (2004). Mars remote sensing analog studies in Badwater Basin, Death Valley, California. *Journal of Geophysical Research*, **109**, E12006, doi:10.1029/2004JE002315.

Banin, A. and Margulies, L. (1983). Simulation of Viking biology experiments suggests Smectites not Palagonites, as Martian soil analogs. *Nature*, **305**, 523–5.

Barbieri, R. (2001). Microbial structures from hypersaline playa deposits of the southern Tunisian chotts. In *Exploring Mars Surface and its Terrestrial Analogues, fieldtrip guidebook*, ed. G. G. Ori, G. Komatsu, and L. Marinangeli (http://irsps.sci.unich.it/education/tunisia/index.html).

Boynton, W. V., Feldman, W. C., Squyres, S. W. *et al.* (2002). Distribution of hydrogen in the near surface of Mars: evidence for subsurface ice deposits. *Science*, **297**, 81–5.

Cabrol, N. A. and Grin, E. (1999). Distribution, classification, and ages of Martian impact crater lakes. *Icarus*, **142**, 160–72.

Cabrol, N. A., Grin, E., Carr, M. H. *et al.* (2003). Exploring Gusev Crater with Spirit: review of science objectives and testable hypotheses. *Journal of Geophysical Research*, **108** (E12), 8076, doi:10.1029/2002JE002026.

Christensen, P. R., Anderson, D. L. Chase, S. C. *et al.* (1998). Results from the Mars Global Surveyor Thermal Emission Spectrometer. *Science*, **279**, 1692–8.

Christensen, P. R., Bandfield, J. L., Clark, R. N. *et al.* (2000). Detection of crystalline hematite mineralization on Mars by the Thermal Emission Spectrometer: evidence for near-surface water. *Journal of Geophysical Research*, **105**, 9623–42.

Christensen, P. R., Bandfield, J. L., Bell, J. F. I. *et al.* (2003). Morphology and composition of the surface of Mars: Mars Odyssey THEMIS results. *Science*, **300**, 2056–61, doi:10.1126/science.1080885.

Crowley, J. K. (1993). Mapping playa evaporite minerals with AVIRIS data: a first report from Death Valley, CA. *Remote Sensing of Environment*, **44**, 337–56.

Crowley, J. K. and Hook, S. J. (1996). Mapping playa evaporite minerals and associated sediments in Death Valley, California, with multispectral thermal infrared images. *Journal of Geophysical Research*, **101**, 643–60.

Currey, D. R. (1980). Coastal geomorphology of Great Salt Lake and vicinity. *Utah Geological and Mineral Survey Bulletin*, **116**, 69–82.

Currey, D. R. (1990). Quaternary paleolakes in the evolution of semidesert basins, with special emphais on Lake Bonneville and the Great Basin, USA. *Palaeogeography, Palaeoclimatology, Palaeoecology*, **76**, 189–214.

Drake, N. A., Bryant, R. G., Millington A. C., and Townshend, J. R. G. (1994). Playa sedimentology and geomorphology: mixture modeling applied to Landsat thematic mapper data of Chott el Djerid, Tunisia. In *Sedimentology and Geochemistry of Modern and Ancient Saline Lakes*, SEPM Special Publication 50, pp. 125–31.

Eugster, H. P. and Hardie, L. A. (1978). Saline Lakes. In *Lakes: Chemistry, Geology, Physics*, ed. A. Lerman. New York: Spinger-Verlag, pp. 237–93.

Farmer, J. D. and Des Marais, D. J. (1999). Exploring for a record of ancient Martian life. *Journal of Geophysical Research*, **104**, 26977–95.

Gasse, F. (2002). Diatom-inferred salinity and carbonate oxygen isotopes in Holocene waterbodies of the Western Sahara and Sahel (Africa). *Quaternary Science Reviews*, **21**, 737–67.

Gasse, F. and Van Campo, E. (1994). Abrupt post-glacial climate events in West Asia and North Africa Monsoon domains. *Earth and Planetary Science Letters*, **126**, 435–56.

Gendrin, A., Mangold, N., Bibring, J.-P. *et al.* (2005). Sulfates in Martian layered terrains: the OMEGA/Mars Express view. *Science*, **307**, 1587–91.

Gilbert, G. K. (1890). *Lake Bonneville*. US Geological Survey Monograph 1. Washington, DC, 438 pp.

Hook, S. J., Myers, J. E. J., Thome, K. J., Fitzgerald, M., and Kahle, A. B. (2001). The MODIS/ASTER airborne simulator (MASTER): a new instrument for earth science studies. *Remote Sensing of Environment*, **76**, 93–102.

Hunt, C. B., Robinson, T. W., Bowles, W. A., and Washburn, A. L. (1966). Hydrologic Basin, Death Valley, California, *US Geological Survey Professional Paper*, **494–B**, 138.

Jones, B. F. (1965). The hydrology and mineralogy of Deep Springs Lake, Inyo County, California, *US Geological Survey Professional Paper*. **502**–A, 56.

Kraal, E. R., Asphaug, E. I., Lorenz, R. D. (2003). Wave energy on Mars and Earth: considering lacustrine erosion. *Abstracts of Papers Submitted to the 34th Lunar and Planetary Science Conference*. Houston: Lunar and Planetary Institute, CD 34, Abstract 1725.

Komatsu, G. and Ori, G. G. (2000). Exobiological implications of potential sedimentary deposits on Mars. *Planetary and Space Science*, **48** (11), 1043–52.

Komatsu, G., Brantingham, P. J., Olsen, J. W., and Baker, V. R. (2001). Paleoshoreline geomorphology of Böön Tsagaan Nuur-Tsagaan Nuur system and Orog Nuur: the Valley of Lakes, Mongolia. *Geomorphology*, **39** (3–4), 83–98.

Malin, M. C. and Edgett, K. S. (1999). Oceans or seas in the Martian northern lowlands: high resolution imaging tests of proposed coastlines. *Geophysical Research Letters*, **26**, 3049–52.

Malin, M. C. and Edgett, K. S. (2000). Sedimentary rocks of early Mars. *Science*, **290**, 1927–37.

Malin, M. C. and Edgett, K. S. (2003). Evidence for persistent flow and aqueous sedimentation on early Mars. *Science*, **302**, 1931–4.

Milligan, M. R. and Chan, M. A. (1998). Coarse-grained Gilbert deltas; facies, sequence stratigraphy and relationships to Pleistocene climate at the eastern margin of the Lake Bonneville, northern Utah. In *Relative Role of Eustasy, Climate, and Tectonism in Continental Rocks*, ed. K. W. Shanley and P. J. McCabe. Society for Sedimentary Geology Special Publication 59, pp. 177–89.

Moersch, J., Farmer, J., and Baldridge, A. (2001). Searching for evaporites on Mars: remote sensing of terrestrial analogs for putative Martian paleolake basins. *Geological Society of America Abstracts with Programs*, **33** (6), 403–4.

NASA Jet Propulsion Laboratory, Mars Exploration Rover Mission web page. http://marsrovers.jpl.nasa.gov/.

Norton, C. F., McGenity, T. J., and Grant, W. D. (1993). Archaeal halophiles (halobacteria) from two British salt mines. *Journal of General Microbiology*, **139**, 1077–81.

Ori, G. G. and Roveri, M. (1987). Geometries of Gilbert-type deltas and large channels in the Meteora Conglomerate, Meso-Hellenic basin (Oligo-Miocene), central Greece. *Sedimentology*, **34**, 845–59.

Ori, G. G., Marinangeli, L., and Baliva, A. (2000a). Terraces and Gilbert-type deltas in crater lakes in Ismenius Lacus and Memnonia (Mars). *Journal of Geophysical Research*, **105**, 17629–41.

Ori, G. G., Marinangeli, L., and Komatsu, G. (2000b). Martian paleolacustrine environments and their geological constrains on drilling operations for exobiological research. *Planetary and Space Sciences*, **48** (11), 1027–34.

Ori, G. G., Marinangeli, L., and Komatsu, G. (2000c). Gas (methane)-related features on the surface of Mars and subsurface reservoirs. *Abstracts of Papers Submitted to the 31st Lunar and Planetary Science Conference*. Houston: Lunar and Planetary Institute, CD 31, Abstract 1550.

Ori, G. G., Komatsu, G., and Marinangeli, L., (eds.) (2001). *Exploring Mars Surface and its Terrestrial Analogues*, fieldtrip guidebook compiled by the Italian Space Agency and International Research School of Planetary Sciences, and published by Alenia Spazio, 92 pp., online copy available at http://irsps.sci.unich.it/education/tunisia/index.html.

Parker, T. J. and Currey, D. R. (2001). Extraterrestrial coastal geomorphology, *Geomorphology*, **37**, 303–28.

Parker, T. J., Saunders, R. S., and Schneeberger, D. M. (1989). Transitional morphology in west Deuteronilus Mensae, Mars: implications for modifications of the lowland/upland boundary. *Icarus*, **82**, 111–45.

Parker, T. J., Gorsline, R. S., Saunders, R. S., and Pieri, D. C. (1993). Coastal geomorphology of the Martian northern plains. *Journal of Geophysical Research*, **98**, 11061–78.

Roberts, C. R. and Mitchell, C. W. (1987). Spring mounds in southern Tunisia. In *Desert Sediments: Ancient and Modern*, ed. L. Frostick and I. Reid. Geological Society Special Publication **35**, pp. 321–34.

Scott, D. H., Chapman, M. G., Rice, Jr. J. W., and Dohm, J. M. (1992). New evidence of lacustrine basins on Mars: Amazonis and Utopia Planitiae. *Proceedings of the 22th Lunar and Planetary Science Conference*. Houston: Lunar and Planetary Institute, pp. 53–62.

Stockstill, K. R., Ruff, S. W., Moersch, J., Baldridge, A., and Farmer, J. (2003). TES Hyperspectral mapping of proposed paleolake basins: a search for aqueous minerals. *Abstracts of Papers Submitted to the 32nd Lunar and Planetary Science Conference*. Houston: Lunar and Planetary Institute, CD 31, Abstract 1793.

Swezey, C. S. (1996). Structural controls on Quaternary depocentres within the Chott Trough region of southern Tunisia. *Journal of African Earth Sciences*, **22**, 335–47.

Swezey, C. S. (1997). Climatic and tectonic controls on Quaternary eolian sedimentary sequences of the Chott Rharsa basin, Southern Tunisia. Ph.D. thesis, University of Texas.

14

Signatures of habitats and life in Earth's high-altitude lakes: clues to Noachian aqueous environments on Mars

Nathalie A. Cabrol, Chris P. McKay
Space Science Division, MS 245-3, NASA Ames Research Center, California

Edmond A. Grin
Space Science Division, MS 245-3, NASA Ames Research Center, California

Keve T. Kiss, Éva Ács, Katalin É. Szabó, Bence Tóth
Hungarian Danube Research Station, Institute of Ecology and Botany, Hungarian Academy of Sciences, Göd

István Grigorszky
Debrecen University, Botanical Department, Debrecen

David A. Fike
Department of Earth, Atmospheric and Planetary Sciences, Massachusetts Institute of Technology, Cambridge

Andrew N. Hock
University of California, Los Angeles

Cecilia Demergasso, Lorena Escudero, P. Galleguillos
Laboratorio de Microbiología Técnica, Avda

Guillermo Chong
Departamento de Geología, Universidad Católica del Norte, Avda

Brian H. Grigsby
Schreder Planetarium/ARISE, Redding

Jebner Zambrana Román
Servicio Nacional de Geología y Minería (SERGEOMIN), La Paz

and

Cristian Tambley
Department of Astrophysics, Avda

The Geology of Mars: Evidence from Earth-based Analogs, ed. Mary Chapman. Published by Cambridge University Press.

14.1 Introduction

A series of astrobiological high-altitude expeditions to the South American Andean Mountains were initiated in 2002 to explore the highest perennial lakes on Earth, including several volcanic crater lakes at or above 6000 m in elevation. During the next five years, they will provide the first integrated long-term astrobiological characterization and monitoring of lacustrine environments and their biology at such an altitude. These extreme lakes are natural laboratories that provide the field data, currently missing above 4000 m, to complete our understanding of terrestrial lakes and biota. Research is being performed on the effects of UV in low-altitude lakes and models of UV flux over time have been developed (Cockell, 2000). The lakes showing a high content of dissolved organic material (DOM) shield organisms from UV effects (McKenzie *et al.*, 1999; Rae *et al.*, 2000). DOM acts as a natural sunscreen by influencing water transparency, and therefore is a determinant of photic zone depth (Reche *et al.*, 2000). In sparsely vegetated alpine areas, lakes tend to be clearer and offer less protection from UV to organisms living in the water. Transparent water, combined with high UV irradiance may maximize the penetration and effect of UV radiation as shown for organisms in alpine lakes (e.g., Vincent *et al.*, 1984; Vinebrook and Leavitt, 1996). Shallow-water benthic communities in these lakes are particularly sensitive to UV radiation. Periphyton, which defines communities of microorganisms in bodies of water, can live on various susbtrates. While on rocks, they include immobile species that cannot seek low UV refuges unlike sediment-dwelling periphyton (Happey-Wood, 1988; Vincent *et al.*, 1993) or alpine phytoflagellates (Rott, 1988) which both undergo vertical migration. Inhibition of algal photosynthesis by UV radiation has been documented in the laboratory (Häder, 1993) and it has been shown that phytoplankton production is reduced by formation of nucleic acid lesions (Karentz *et al.*, 1991) or production of peroxides and free oxygen radicals (Cooper *et al.*, 1989). Most of the experiments that have demonstrated *in situ* suppression of algal growth by UV radiation have either used artificially enhanced UV irradiance (Worrest *et al.*, 1978) or shallow systems (< 1 cm) that lack significant natural attenuation of UV radiation (Bothwell *et al.*, 1993, 1994). Our project is providing the field data that are missing from natural laboratories above 4000 m and will complement the postulation of the effects of UV on life and its adaptation modes (or lack thereof).

The exploration of high-altitude lakes could shed light on early Earth's biological evolution as well. For two billion years, Earth's atmosphere lacked an ozone layer and life was subjected to high UV radiation. These lakes represent an opportunity to observe the evolution of microorganisms in shallow waters that do not offer substantial UV protection. Survival strategies in these lakes might prove to be ancient and could provide a rare look into Earth's past. They could also provide critical information for the search for life on other planets. Their unique environmental analogy to Martian paleolakes of the end of the Noachian era 3.5 Ga ago (De Hon, 1992; Cabrol and Grin, 1995, 1999, 2001, 2002; Scott *et al.*, 1995; Wharton *et al.*, 1995; Newsom *et al.*, 1996; Grin and Cabrol, 1997; Malin and Edgett, 2000; Ori *et al.*, 2000; Cabrol *et al.*, 2003) allows for a test of the habitability potential of aqueous oases on early Mars and will help to better prepare for future astrobiological missions.

14.2 Environmental background

The 2002 expedition investigations took place at the Licancabur lake (LC) and several lower saline lakes (lagunas). Located at 22°50' S and 67°53' W in a tectonically active region, the Licancabur volcano (6014 m) borders both Bolivia and Chile (Figure 14.1). Volcanism is ongoing in the region with two active volcanoes 60 km north and 40 km south of the investigation site.

14.2.1 Licancabur summit lake

The lake (5916 m) is nested in the 200 m diameter summit crater of Licancabur. The volcano was formed during the Early-Pliocene and is classified as latent (Marinovic and Lahsen, 1984) with no recorded activity in recent history (600 to 1000 years). A planktonic fauna was mentioned by a previous archeological expedition (Rudolph, 1955). Divers measured the bottom lake temperature at +6 °C (Leach, 1986). Except for these references, the lake's biotic diversity, environment, and survival strategies are undocumented.

We characterized the physical environment (Table 14.1) of LC and focused on elements of the environment that would lend insight to constraints placed on local biology. The stability of LC is of particular interest, since liquid water is uncommon in this section of the Andes. Most lakes above 5200 m on Chile's volcanoes are frozen year-round. At LC, the atmospheric pressure is 480 mb and average air temperature about −12 °C; precipitation is less than 200 mm/yr and the atmospheric 0 °C isotherm lies at 4400 m (Nunez *et al.*, 2002). Still, the summit crater contains a ~100 m wide, and ~4 m deep lake which is ice-covered only part of the year. We measured

Figure 14.1. Simplified geograhical map showing the location of the Licancabur Volcano at the boundary of Chile and Bolivia.

LC surface water pH (~8.5), temperature (+4.9 °C), and total dissolved solids (TDS: 1050 mg/L). The lake waters are two degrees warmer than that predicted by a model of the temperature maximum density for freshwater at this altitude (Eklund, 1983). The temperature and the location of LC atop a young, latent, volcano in an active geothermal setting led us to suspect the existence of a deep magmatic hydrothermal system. We investigated that hypothesis and present new data on the physical and chemical nature of the lake.

The chemical analysis of LC water samples was performed using inductively coupled mass spectrometry (ICP-MS) and ion chromatography (IC). ICP-MS was used to determine elemental abundances across the periodic table to the parts-per-billion (ppb) level. This analysis targeted the elements commonly found in lake fluids near andesitic volcanoes (e.g., Na, Ca, K, Mg, Fe, and Al derived from the dissolution of volcanic glass) and other solutes

Table 14.1. *Physical and environmental data from lakes and springs*

	Thermales hot spring[a]	LB[b] Cold Spring	LB	LV[c]	LC[d]
GPS Coord. Lat.	22°46.96' S	22°48.32' S	22°47.00' S	22°47.32' S	22°50.03' S
GPS Coord. Long.	67°48.15' W	67°46.34' W	67°47.00' W	67°49.16' W	67°53.00' W
Elevation (m)	4328	4340	4340	4332	5916
Length × width (km)	0.010 × 0.005	–	1 × 0.5	7 × 3	0.1 × 0.09
Maximum depth (m)	2	–	0.5	40	4 to 10(4)
UV flux (W/m^2)	81	81	81	81	89
Water temp. (°C)	+36.2	+17.7	+14	+12.9	+4.9
Air temp. (°C, min/max)	−30/+12	−30/+12	−30/+12	−30/+12	−40/+3
Atm. pressure (mb)	600	600	600	600	480
pH	8	7.3	7.2	9	8.5
TDS[e](mg/L)	2120	2740	22400	11 7500	1050

[a] Spring located between LB and LV.
[b] LB: Laguna Blanca.
[c] LV: Laguna Verde.
[d] Still to be determined.
[e] Total dissolved solids.

common to geothermal fluids (Nicholson, 1993; Varekamp *et al.*, 2000). IC was used to determine anion abundances to the parts-per-million (ppm) level. The goal was to quantify Cl, SO_4, and F concentrations, which are elevated in fluids that have been mixed with hot magmatic fluids and gases. Nitrate and phosphate concentrations were also determined using IC to characterize some of the nutrients available to local biology. ICP-MS results show that LC waters are enriched in some major rock-forming elements ([Mg] = 39.6 ppm, [Ca] = 230 ppm, [Al] = 1.75 ppm, [Fe] = 0.902 ppm) with respect to geothermal springs at the base of Licancabur. If it is assumed that the source waters for these features have similar composition, then this enrichment may be representative of increased water–rock interaction as the fluid flowing up to the summit has had more time to react with local lithologies. According to IC, sulfate, fluoride, and chloride concentrations in LC ([SO_4] = 4.37 mg/L, [F] = 0.56 mg/L, and [Cl] = 0.54 mg/L) are also higher than those detected in the lower springs and are over 200 times more concentrated than those detected in local snowmelt. Since solute enrichment is not uniform across the analytes in LC, it is unlikely that this chemistry is a result of evaporative concentration alone.

The bulk properties of the lake, as well as the physical and geochemical analyses, lead us toward the conclusion that LC is still an active volcanic lake (low activity) supported by a small geothermal heating term, and may host a diffuse hydrothermal system. More chemical, isotopic analyses and direct measures of the heat flow in the vicinity of the summit lake will be performed in coming years. LC (Figure 14.2) represents an end-member of the physical environment on Earth where lakes and liquid water are stable. As such, it is of considerable interest for astrobiological studies, as fading geothermal flux and evaporating lakes could have been common environments on Mars at the Noachian/Hesperian transition.

14.2.2 Laguna Verde and Laguna Blanca

Two lower hypersaline lakes, Laguna Verde (LV) and Laguna Blanca (LB), and a small warm hydrothermal pond (Thermales) connecting the two lagunas (see Figures 14.2b–d) were also explored. They are located at the foot of Licancabur at 4340 m. The physical environment of the lagunas combines: high UV radiation (40% higher than at sea level) enhanced by the tropical latitude of the site, as well as low atmospheric pressure, low oxygen (58% of pO_2 at sea level), high temperature amplitude, low humidity (proximity to the Atacama Desert), and low precipitation (< 200 mm/yr), see Table 14.1. Because of its salt content (three times that of seawater) LV can stay free of ice at temperatures reaching −25 °C (Cabrol *et al.*, 2003). LB freezes at −5 °C.

The water supply is from influx of cold and warm hydrothermal springs. Compared to LC, the geothermal fluid input through springs is clearly active at the lagunas, especially in Laguna Blanca. LB and LV are joined today by a small channel and the "Thermales" pond. Dataloggers were placed to monitor yearly air, soil, water temperature, UV flux, and relative humidity. A survey of thermal water input to LB covered approximately half of the lake's shoreline. Fifteen individual springs between +12 °C and +36 °C were observed and documented. Surface manifestation of thermal fluid upwelling there is highly dependent on coastal geology along LB and LV and a clear relationship between the lakes, their biological communities, and the local geothermal environment was established (see Section 14.2.2 and Figure 14.4).

Figure 14.2. (a) Licancabur lake 100 m below the summit rim. Paleoshorelines are visible. The lake currently ~100 × 90 m and possibly up to 10 m deep, may have reached 65 m and ~200 m in diameter at its peak. (b) Laguna Verde with Licancabur in the background. (c) Hydrothermal springs in Laguna Blanca and algal mat. (d) Oxygen producing algae in the "Thermales" hot spring. Algae abound in the +36 °C water. Credit photographs: Brian H. Grigsby and Nathalie A. Cabrol. (For a color version of this figure, please refer to color plate section.)

Most of the brackish lagunas in the Bolivian Altiplano were formed as freshwater lakes and experienced their peak volumes during the early Holocene between 18 000 and 9200 BP, when precipitation reached 400–500 mm/yr (Tauca phase, Messerli *et al.*, 1993; Grosjean *et al.*, 1995; Wirrmann and Mourguiart, 1995; Baucom and Rigsby, 1999; Sylvestre *et al.*, 1999; Vuille *et al.*, 2003). Our topographic survey using Global Positioning System (GPS) and Laser Range Finder shows that the two lagunas were originally one large (~20 × 3 km), deep (~55 m) lake which has receded significantly since its formation. The topography, increasing aridity, and local hydrothermal systems at the lagunas have resulted in their evolution from a single large basin (Cabrol *et al.*, 2003) to extremely different lake environments today (pH, temperature, depth, salinity) generating an exceptional diversity of habitats, sometimes only a few meters apart. LB and LV's most ancient terrace belt is stratigraphically correlated with the highest levels of the large paleolake of the wet Holocene episode as shown by ^{14}C analyses (see Section 14.2.2). The main variable since their formation and the present is the amount of precipitation received per year in this part of the Altiplano.

The environmental conditions shown by these three different lakes might be the closest possible terrestrial analogs to the late Noachian on Mars. Their geosignatures and biosignatures provide detection criteria for the search of sites of astrobiological interest on Mars. Habitats, life diversity, and life survival strategies in environments that provide little protection against UV effects give important information that can test Mars' past habitability potential. Moreover, these lakes provide clues about life's ancient survival strategies on early Earth before the ozone layer was formed. It also provides a test of the limits of life on our own planet.

14.3 Present habitats and life

Water and sediment samples were collected at all lakes to document the diversity of life and habitats and understand the impact of environmental stresses. These analyses also contribute to the understanding of the hydrological and algological state of the Andean high lakes (Hustedt, 1927; Rumrich *et al.*, 2000). Several Bolivian lower lakes (mostly in the Unuyi region) were investigated as tracers of modern climate changes (Sylvestre *et al.*, 1999; Servant Vilardy *et al.*, 2000).

14.3.1 Diversity

Our geophysical and chemical analyses show that the three lakes differ considerably from each other. This is reflected in the significant differences of diatom species numbers in the samples: LB has 113 taxa, LV 50 taxa, and LC 22 taxa (Ács *et al.*, 2003). More analyses are ongoing; however, it comes as an unambiguous result that in spite of the high UV radiation and high altitude, even LC shows a species-rich algal flora. Diatom assemblages seem to adjust to enhanced UV radiation up to a certain point. The species identified in our samples show only slight differences in species composition and species numbers compared to the control group

(Reizopoulou *et al.*, 2000). Plankton samples were collected at both lagunas using 10 and 50 μm plankton nets and water samples (all lakes). Periphyton samples were collected from several macroscopically different littoral biotops at LB, small cold springs, Thermales and the central part of the lake, and along the shoreline of LC. The abundance of phytoplankton and the relative abundance of periphytic diatoms were determined by the Utermöhl method (Paxinos and Michell, 2000).

14.3.2 Life in the summit lake

LC is a small alpine lake. Its water is crystal-clear and transparent to the bottom. Its encased location prevents winds that could stir it up. Total dissolved solid (TDS) results lead to its classification as a mesohalobic (moderately saline with intense growth of salt-loving organisms) lake (see Table 14.1). The dominant salt is $CaSO_4$. The freezing point is ~0 °C. The geothermal heating term supported by our data might contribute to maintaining the lake's temperature. From an algological standpoint, LC's concentration of Si is adequate; however N and P (NO_3, NO_2, PO_4) concentrations are low, leading to the classification of the lake as ultra-oligotrophic (highly deficient in nutrients with abundant dissolved oxygen). In the phytoplankton, only one *Synechococcus* (Cyanobacteria) colony, three Chrysophyceae, one Euglenophyta, and two Chlorophyceae species were found. The individual number of phytoplankton is low (15 ind/ml). We found a dozen specimens of 1 to 2 Ciliata species, and a few cells of a Hyphomyceta species. Both observations support the presence of a food web (primary producers, consumers, decomposers) at the summit lake (Figure 14.3).

14.3.3 High diatom abnormality rate in Laguna Blanca

LB is shallow (deepest point < 1m) and thus belongs to the littoral zone. It is irradiated to the bottom (euphotic zone) in spite of winds that often stir up the sediment. Its dissolved salt content is high (22 400 mg/l). LB freezes at about −5 °C but not to the bottom in spite of the altitude or winter because of the +36 °C spring influx of Thermales and several cooler springs. Limnologically, its high NaCl content defines it as brackish. The SO_4 and HCO_3 content are high; however the concentration of Ca and Mg is not abnormal. N and P are abundant, making the water potentially polytrophic or hypertrophic (large supply of nutrients and low visibility depths). High winds increase the turbidity and the suspended matter content

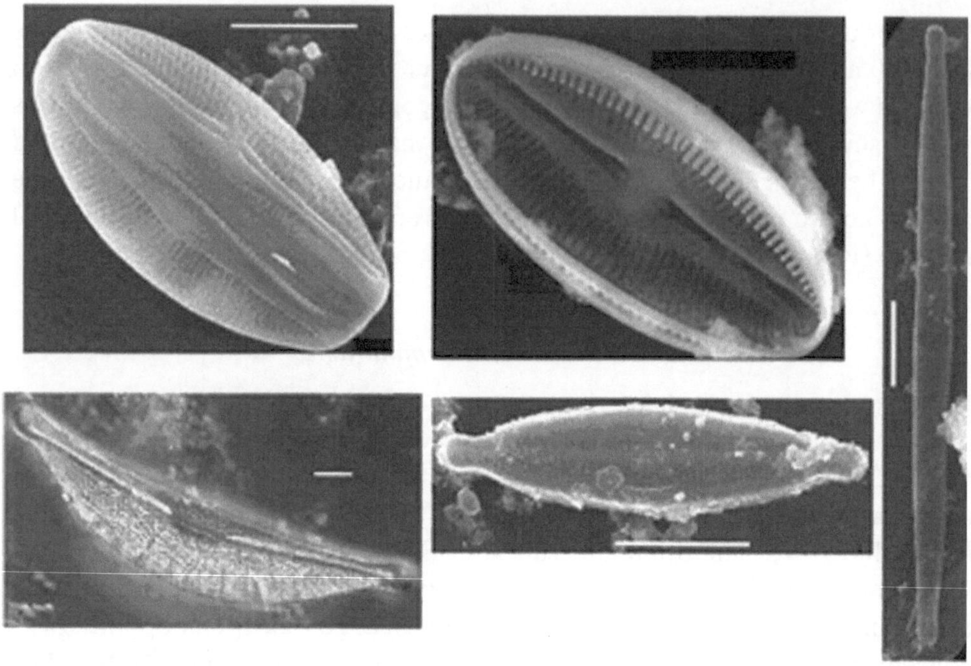

Figure 14.3. Micrographs of diatoms in LC. Upper row, from left to right: *Amphora veneta* (SEM) and *Diploneis chilensis* (SEM). Right vertical: *Fragilaria pulchella* (SEM). Lower row: *Amphora platensis* (LM) and *Craticula buderi* (SEM). Scale bars: 10 μm.

reaches 207 mg/l, while Secchi transparency decreases to 20–40 cm. Most of the suspended matter consists of diatom frustules. The high Si concentration of the water favors the development of the huge diatom mass observed. The phytoplankton in LB is poor in euplanktonic elements, both in species and individual numbers. Diatoms living on the sediment float in the water as tychoplanktonic (adapted to living in still water) elements as a result of wind activity.

The number of living cells is extremely high: 82 000 ind/ml. As the species vary considerably in size, the conversion into true biomass is difficult. Moreover, the size of the chloroplasts and central vacuoles is also variable even among individuals of the same species. The primary production of the lake is considerable, allowing its classification as at least mesoeutrophic (moderate levels of organisms, nutrients, and visibility). Because of the stirring effect of the wind, 98% of the floating cells are frustules of dead diatoms. On the basis of the dominant diatoms in the 11 periphyton samples collected, an obvious micro-heterogenity characterizes the different habitats (Figure 14.4) that are at times only few meters apart and have the same

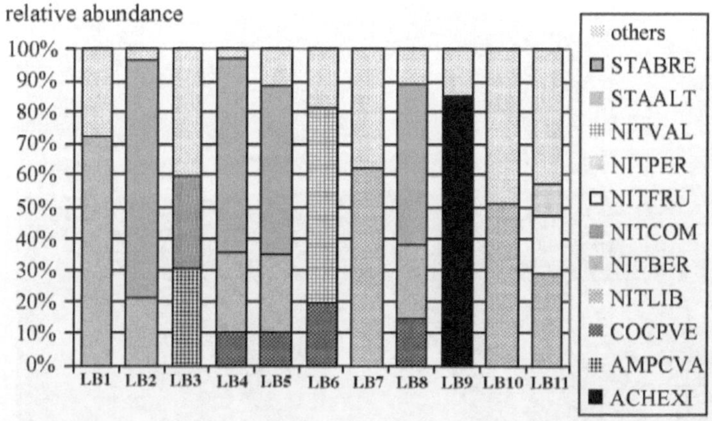

Figure 14.4. Distribution of dominant benthic diatoms in Laguna Blanca (species ≥10% relative abundance). ACHEXI: *Achnanthes exigua*; AMPCVA: *Amphora coffaeaformis* var. *acutiuscula*; COCPVE: *Cocconeis placentula* var. *euglypta*; NITLIB: *Nitzchia libetruthii*; NITBER: *Nitzschia bergii*; NITCOM: *Nitzschia communis*; NITFRU: *Nitzschia frustulum*; NITPER: *Nitzschia perminuta*; NITVAL: *Nitzschia valdecostata*; STAALT: *Staurosira altiplanensis*; STABRE: *Staurosira brevistriata*. Sampling sites: LB 1–4: east shore; LB 5–7: north shore; LB 8, 9: Thermales; LB 10, 11: north shore.

5–20 cm water column topping them. These differences could be attributed to, for instance, the distance from warm or cold springs.

The majority of samples were collected at LB, which may explain why most teratological (abnormal) diatoms were found in this lake (Figures 14.5). At LB, the proportion of teratological diatom frustules was found to be 1–2% (normal: 0.1–0.2%). Because of LB's characteristics, the sampling depth was also shallower than average when compared to that of LC and LV. The probability of developing teratological forms amongst diatoms is enhanced by increased UV irradiation, as shown by *Cocconeis placentula* found in LB, which is a UV-sensitive species and for which a wide range of deformities were observed (Figures 14.5a–d). Finally, the flora and fauna are extremely varied in the different biotops. Close to the shoreline, diverse cyanobacterium colonies were sampled. Several macrozoobenthic organisms (Gammaridae, Mollusca species) also occur and masses of copepods live in the plankton.

14.3.4 Hypersaline Laguna Verde

Though connected to LB, LV has an extremely different environment. It reaches 40 m at its deepest and is surprisingly rough in windy weather,

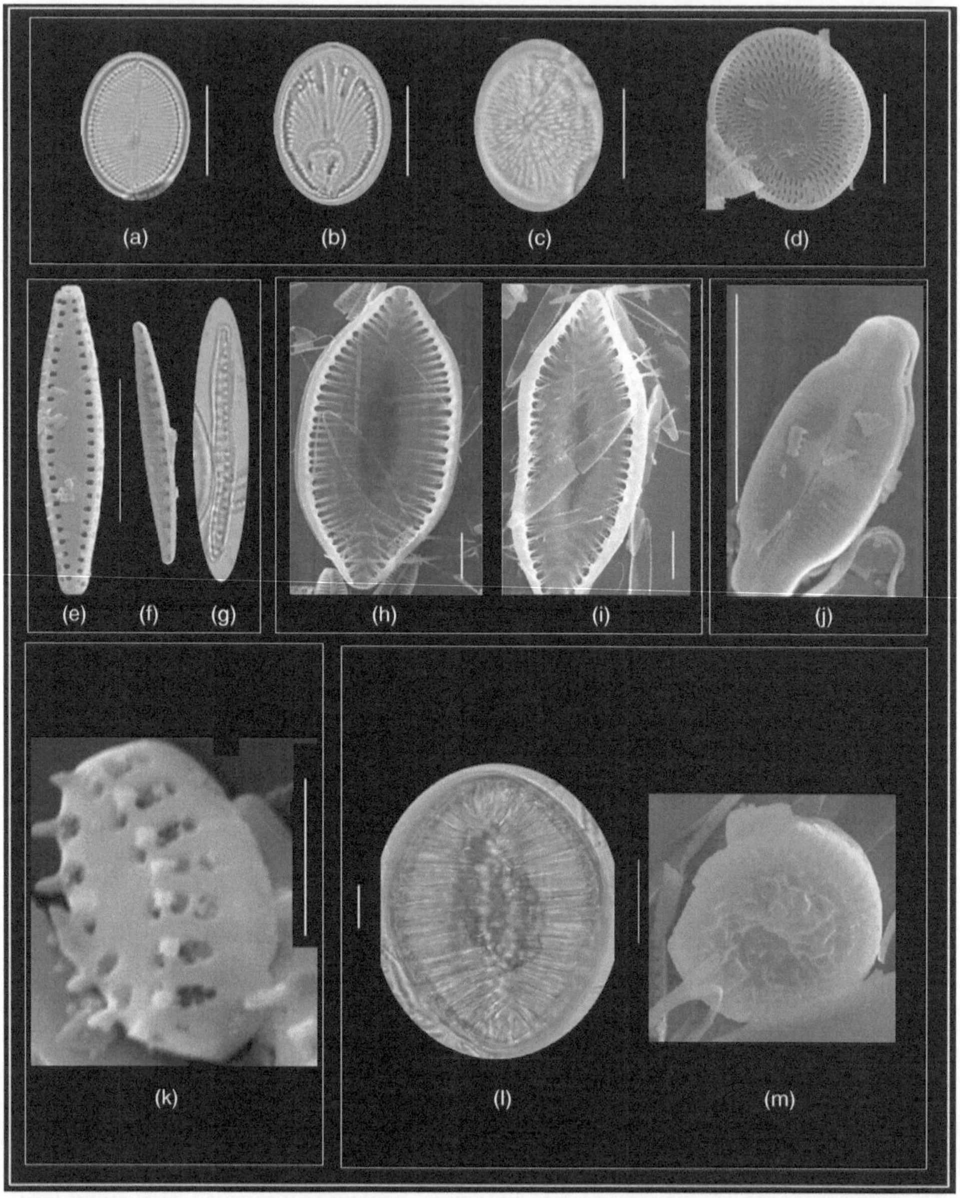

Figure 14.5. (a) *Cocconeis placentula*, normal valve, LM (LB and LV); (b, c) Teratological valves, LM; (d) teratological valve (SEM); (e) *Staurosira brevistriata* (LB); (f, g) Normal and teratological *Nitzchia libertruti* (LB); (h, i) Inside views of normal and teratological forms of *Surirella chilensis* (LB); (j) *Achnanthes exigua*, SEM (LB); (k) *Staurosira altiplanensis*, SEM (LB); (l) *Surirella sella* (LM) living cell with chloroplast found in the central region (LV and LB); (m) *Cyclotella baltica*. First record of the species outside the Baltic Sea and in a lake. All scale bars 10 μm except *Staurosira altiplanensis* (2.5 μm) and *Cyclotella baltica* (5 μm).

with significant waves leading to a decrease in transparency and the color of the water to become turquoise-opaque. In the suspended matter many complete and broken diatom frustules and various size mineral granules were observed. Due to an anomalously high suspended salt content (113 g/l) the freezing point of its water is −25 °C. The water TDS value (117 500 mg/l) is three times higher than that of seawater. Moreover, the concentration of Cl, SO_4, HCO_3, Mg, and Ca ions is considerable. The lake can be classified as continental salt-water. Only one euplanktonic species (*Gymnodinium* sp.) was found in the samples collected with plankton nets. All other species are tychoplanktonic diatoms. The individual number of algae is low (91 ind/ml), classifying the lake as oligotrophic.

Both lagunas yielded unexpected observations and discoveries. We found high numbers of live *Surirella sella* in both LB and LV (Figure 14.5). This species was described mostly from fossil samples, and is rarely mentioned alive (e.g., shallow lake Lipez in Bolivia with high concentration of $NaSO_4$; Hustedt, 1927; Servant Vilardy *et al.*, 2000) and in LV (Servant Vilardy *et al.*, 2000). The species was characterized as an endemic, athalassic (saline) and of a saltwater form. It is abundant in both lagunas.

Only two centric diatom species were found in the lagunas. This is unusual or rare compared with other lakes that usually host 5–10 centric species. One of these species, *Cyclotella baltica*, has only previously been described in the Baltic Sea (Håkansson, 2002), see Figure 14.5. This is the first record of this species both in the southern hemisphere and from a lake.

14.3.5 The effect of UV on biomass

Preliminary examination of surface samples reveals a general decrease in total lipid (waxy oxygenated fats or fatty acides) extracts (TLE) generated per gram dry weight of sediment and per milliliter of water from LC as compared to LB and LV (Table 14.2). UV intensity is ~10% greater at LC but temperature, pH, and alkalinity also vary between these lakes. Further study is thus needed to rigorously establish the limiting effect of each parameter. However, support for the UV hypothesis limiting the biological productivity at the summit lake comes from the analysis of TLE derived from sediments from different depths (with pH, and alkalinity held constant). Sediment taken from under a 50 cm water column had more than three times as much TLE per gram sediment as the sediment taken from under a 5 cm water column (0.800/2.350). The depth difference corresponds to approximately 33% less UV at the sediment–water interface for the deeper sample.

Table 14.2. *Characterization of lipids extracted from 2002 expedition samples*

	Lipids (mg lipid/g)				
Sample location	TLE[(a)]1	TLE2	ATLE[(b)]	Nlip[(c)]	Plip[(d)]
LC – water	0.015		0.015	0.007	0.008
LC – sediment (shallow)	0.800	0.200	0.500	0.350	0.450
LC – sediment (deep)	2.350	1.300	1.825	1.850	0.500
LB – water	0.013		0.013	0.008	0.005
LB – sediment	1.500	10.250	5.875	1.100	0.400
LV – sediment	0.750	0.400	0.575	0.350	0.400
LV – mat	6.500	57.500	32.000	4.000	2.500
Thermales – sediment	2.050	0.100	1.075	0.350	1.700
Thermales – mat	14.000	33.500	23.750	3.500	10.500

[a] Total lipid extract;
[b] average total lipid extract;
[c] neutral lipid;
[d] polar lipid.

14.3.6 Preliminary assessment of genetic diversity

Denaturing gradient gel electrophoresis (DGGE) was selected as a fingerprinting technique and sequence analysis of the resulting 16S rDNA bands to carry out a first-order assessment of the genetic diversity of the samples. Although some polymerase chain reactions (PCRs) and primer biases could emerge during the amplification process (Von Wintzingerode *et al.*, 1997) DGGE offers a rapid means of detecting predominant PCR-targeted populations.

The main difference observed through contrast phases and fluorescent microscopy among samples from LC, LB, LV, and Thermales results from their specific populations of cyanobacteria and diatoms species. PCR products were obtained directly from the original DNA without nested amplification, except for the Thermales, LB-2 and LC-1 samples. Only two samples (LC-1 and Thermales) yield nested PCR amplification product using the universal primers combination for archaeal fragments. A variable number of bacterial fragments bands were obtained in the different systems, the total ranging between 1 and 10. The relative intensities of the bands in each lane were used to calculate Shannon's bacterial diversity indices for each sample. Values range from 0 to 1.99. The average Shannon index for bacterial community in three saline lakes studied in the nearby Salars de Llamará, de Atacama, and de Ascotán is 2.21 and 16 bands of bacterial fragments in DGGE gels (Aiken *et al.*, 2002). The 21 main bands (intensity and frequency) were excised, re-amplified, and sequenced. No satisfactory sequence could be recovered from 12 of these bands. The remaining 9 bands produced sequences that could be used for BLAST (Basic Local Alignment Search Tool) analysis. The only amplified archaeal fragments analyzed by DGGE (LC-1) show the occurrence of 4 bands and a Shannon index of 1.32.

Between 42% and 82% of the total band intensity in each lane could not be assigned to known organisms at this time because either the bands could not be re-amplified (8) or the sequences retrieved were of bad quality (1). Some sequences (4) show between 76% and 98% similarity to environmental clones but not to cultured microorganisms, and therefore their physiology could not be assessed. Only four of the sequences have percent similarity values higher than 95% in the 16S rDNA fragments. This allows the assumption that their ecophysiology is similar to cultured organisms. The major contributor in the total band intensity of all samples is unidentified sequences. The second contributor is the group of sequences belonging to the phylum Proteobacteria. In six of the samples γ-proteobacteria were the most important bands (LB, LB-1, LB-2, Thermales, LV-3, and LC-2), then sequences belonging to β-proteobacteria in three samples (LB-1, LC-1, and LC-2). The smallest contribution to the phylum is α-proteobacteria found in two samples (LB-1 and Thermales). In LV-2 the assemblage was dominated by the Cytophaga–Flavobacterium–Bacteroides (CFB) phylum. Sequences whose closest relative was a member of the genus *Pseudomonas* were recovered in two of the bands (*P. jessenii* and *P.* sp.). They are the major contributors in the total band intensity of the lane for LB, LB-2, LV-3, and contribute in almost the same way with γ-Proteobacteria in LC-2. One sequence reveals the occurrence of microorganisms related to sulfur oxidizers at the summit lake while another shows low similarity value (80%) with *Acidithiobacillus thiooxidans* (another sulfur oxidizer). Analysis is ongoing to confirm yet unidentified and suspected new species in our first-year samples.

14.4 Fossil life

A $\sim$100 km^2 field of fossil stromatolites occupies most of the ancient terraces of the former paleolake that originally included LB and LV. Their distribution is stratigraphically discontinuous: the structures are widespread on the oldest terraces, disappear from some intermediate shoreline levels, and reappear on younger ones. These gaps are often associated with sharper stratigraphic transitions (e.g., larger terraced shorelines and thicker deposits), changes in grain size and the nature of material on the deposits.

^{14}C analysis was performed on samples of the main terrace and indicates an age of 15 330 ± 210 years. Smaller stromatolites located stratigraphically closer to the current shore are dated 11 210 ± 120 years. These results are consistent with the large paleolake being formed during the Holocene wet climate episode (Messerli *et al.*, 1993; Wirrmann and Mourguiart, 1995; Grosjean *et al.*, 1995; Baucom and Rigsby, 1999; Sylvestre *et al.*, 1999; Vuille *et al.*, 2003). Below this stratigraphic level, and closer to the current active shoreline, the density of fossil structures is scarce. However, the survey of Laguna Verde showed the existence of modern cyanobacterial mat on the present-day shore. Several small (10 cm) round cyanobacterial colonies were observed building domes resembling stromatolites and covering them with cyanobacterial mat.

The short spatial and temporal scale variability in distribution of the stromatolites is associated with an exceptional diversity of forms and sizes ranging from 10 cm to 20 m (Figure 14.6). Both stromatolites (laminated) and thrombolites (clotted structures; Kennard and James, 1986) are observed.

Microscopic examination shows that cyanobacterial filaments formed most of the structure within the laminations. Their morphology shares many characteristics of the Precambrian stromatolites. In addition to cyanobacteria, other layers are mixed and have diatoms and ostrocod shell fragments. Mineralogical variations within the lamination of individual structures are indicative of changes in the environment within short geological periods of time. Future detailed sampling and analysis of the stromatolites along stratigraphic transects through the paleoterraces and at the lamination scale will help understand the magnitude of the changes and their implications for the microorganic communities. In upcoming campaigns the discontinuity of this field at such short temporal scale added to sharp stratigraphic relationships will also allow for the testing of the sensitivity to rapid environmental changes of one of the most ancient forms of life. In less than 15 000 years, precipitation rates decreased from 500 to $\leq$200 mm/yr (Messerli *et al.*, 1993). The extent of the original lake gives a water volume close to 5 km^3. Both lagunas now combine only ~0.3 km^3. Most of the volume was lost in the past 10 000 years as deduced by our paleobathymetry and ^{14}C analysis. In spite of these short geological timescale variations and restricted physical environments, cyanobacteria, one of the most ancient terrestrial forms of life, colonized these shores and survived up to the present. Understanding how they survive could provide precious clues for the search for biota on Mars.

Moreover, the textures observed in the samples share many of the features of Precambrian stromatolites, including Mesoproterozoic specimens from

Figure 14.6. (a) 5 m collapsed stromatolite (LV); (b) Cross-shaped structure; (c) 50 cm-high domes (Licancabur in the background); (d–i): Photomicrographs of a stromatolite fragment. (d) Dendrites (dark gray) consist of micrite with cyanobacterial filaments throughout. Calcite cements fills the area and contains abundant cyanobacterial filaments. Occasionally, dendrites penetrate into the calcite cement layers, and sometimes connect successive dendrite layers. (Scale photo, S: 2.5 mm); (e) Close up of dendritic micrite structures revealing cyanobacterial filaments (black) throughout the structure. These filaments are also found throughout the calcite cement (S: 625 μm): (f) Diatom hash at the termination of additional micritic structures (S: 625 μm); (g) (same as E, S: 625 μm); (h) Ostrocod and diatom hash at the termination of micritic structures (S: 1.25 mm); (i) Fine-scale laminations observed at the edges of the sample (S: 2.5 mm).

the Billyakh Group, Siberia (Bartley *et al.*, 2000) which possess dendritic textures similar to those observed in the lagunas as well as micrometer-scale lamination similar to that found on the edge of the examined dendrites. While the Mesoproterozoic stromatolites formed in a peritidal marine environment and the lagunas stromatolites formed in a shallow lacustrine and arid environment, the morphological resemblance between them suggests that the physical parameters that played a role in forming the specimens at the lagunas may be important to understand the depositional environment of the Mesoproterozoic samples.

14.5 Conclusion

A detailed picture of a macro- to microscale diversity in habitats and life is emerging through the study of these extreme lakes, reinforcing the notion of life's ability to colonize every possible niche while under severe environmental stress as revealed by high levels of species deformities. Yearly follow-up expeditions will revisit LC, LV, and LB and other 6000+ m lakes, retrieving environmental and biological data from *in situ* experiments and long-range monitoring. The investigation will focus on the role of UV on productivity, environmental stress, and species malformation shown by our first analyses. UV stations and ELDONET dosimeters will establish clear UV profiles and characterize the impact of UV on biology. Our goal is to better understand if these results point to adaptation-driven mutations (evolution) or to pathways to extinction (limits of life). Ultimately, they will provide new insights into the evolution of early Earth and Mars environments.

The water chemistry at the lagunas and the potential geothermal flux at the summit lake show that water can flow and pond under temperatures that range between −40 °C and +10 °C and reduced atmospheric pressures between 480 and 550 mb). These temperature and pressure ranges are similar to those modeled for the Noachian/Hesperian transition. These lakes harbor abundant microorganic life coping to some extant with high UV radiation. Diatom-like organisms are unlikely to have ever evolved on Mars as they appeared late in Earth's biological history. However, by their rate of mutation, adaptation, and or extinction, they, and other microorganisms, can provide important clues for the search for life on Mars. Both habitat and life can leave characteristic geo- and biosignatures that we might learn to recognize on distant Martian shores.

Acknowledgments

The authors would like to thank the two reviewers, Goro Komatsu and Kevin Mullins, for their comments that helped improved the manuscript. The 2002 Licancabur expedition was funded by the NASA

Ames Research Center (ARC) Directorate Discretionary Funds and supported by the ARC Space Science Division and the SETI Institute. The NASA Ames Astrobiology Institute provided support funding through its Student Research Grant program. SERGEOMIN and SERNAP supported the logistics of the investigation in Bolivia and Universidad Catolicà del Norte (Antofagasta) in Chile. The SEM studies were performed with the Hungarian National Scientific Fund (OTKA M 041686).

References

Ács, É., Cabrol, N., Grigorszky, I. *et al.* (2003). Similarities and dissimilarities in biodiversity of three high-altitude mountain lakes (Andes, Bolivia). *6th Hung. Ecol. Congress*, ed. M. Dombos and G. Lakner. Gödöllő: 305.

Aiken, G., Kaplan, A., and Weishaar, J. (2002). Assessment of relative accuracy in the determination of organic matter concentration in aquatic systems. *Journal of Environmental Monitoring*, **4**, 70–4.

Bartley, J. K., Knoll, A. N., Grotzinger, J. P. *et al.* (2000). Lithification and fabric genesis in precipitated stromatolites and associated peritidal carbonates, Mesoproterozoic Billyakah Group, Siberia: carbonate sedimentation and diagenesis. *The Evolving Precambrian world, Special Publication, Society for Sedimentary Geology*, **67**, 59–73.

Baucom, P. C. and Rigsby, C. A. (1999). Climate and lake-level history of the northern Altiplano, Bolivia, as recorded in Holocene sediments of the Rio Desaguadero. *Journal of Sedimentary Research*, **69**(3), 597–611.

Bothwell, M. L., Sherbot, D., Roberge, A. C., and Daley, R. J. (1993). Influence of natural ultraviolet radiation on lotic periphytic diatom community growth, biomass accrual, and species composition: short-term versus long-term effects. *Journal of Phycology*, **29**, 24–35.

Bothwell, M. L., Sherbot, D. M. J., and Pollock, C. M. (1994). Ecosystem response to solar ultraviolet-B radiation: influence of trophic level interactions. *Science*, **265**, 97–100.

Cabrol, N. A. and Grin, E. A. (1995). A morphological view on potential niches for exobiology on Mars. *Planetary & Space Science*, **43**(1), 179–88.

Cabrol, N. A. and Grin, E. A. (1999). Distribution, classification and ages of Martian impact crater lakes. *Icarus*, **142**, 160–72.

Cabrol, N. A. and Grin, E. A. (2001). The evolution of lacustrine environments on Mars: is Mars only hydrologically dormant? *Icarus*, **149**, 291–328.

Cabrol, N. A. and Grin, E. A. (2002). Overview on the formation of paleolakes and ponds in impact craters on Mars. *Global and Planetary Change*, **35**, 199–219.

Cabrol, N. A., Grin, E. A., Friedmann, R. *et al.* (2003). Licancabur: exploring the limits of life in the highest lake on Earth. *NASA Technical Memorandum*, 2003–211862, pp. 64–7.

Cockell, C. S. (2000). The ultraviolet history of the terrestrial planets: implications for biological evolution. *Planetary and Space Science*, **48**, 203–14.

Cooper, W. J., Lean, D. R. S., and Carey, J. H. (1989). Spatial and temporal patterns of hydrogen peroxide in lake waters. *Candian Journal of Fishery and Aquatic Science*, **46**, 1227–31.

De Hon, R. (1992). Martian lake basins and lacustrine plains. *Earth, Moon, and Planets*, **56**, 95–122.

Eklund, H. (1983). Stability of lakes near the temperature of maximum density. *Science*, **142**, 1457–8.

Grin, E. A. and Cabrol, N. A. (1997). Limnologic analysis of Gusev crater paleolake, Mars. *Icarus*, **130**; 461–74.

Grosjean, M., Messerli, B., Ammann, C. *et al.* (1995). Holocene environmental changes in the Atacama Altiplano and paleoclimatic implications. *Bull. Inst. Français des Etudes Andines*, **24**, 585–94.

Häder, D.-P. (1993). Risks of enhanced solar ultraviolet radiation for aquatic ecosystems. *Progress in Phycological Research*, **9**, 1–45.

Håkansson, H. (2002). A compilation and evaluation of species in the general Stephanodiscus, Cyclostephanos and Cyclotella with a new genus in the family Stephanodiscaceae. *Diatom Research*, **17**, 1–139.

Happey-Wood, C. M. (1988). Vertical migration patterns of flagellates in a community of freshwater benthic algae. *Developmental Hydrobiology*, **45**, 99–123.

Hustedt, F. (1927). Die Diatomeen der interstadialen Seekreide. *International Review of Hydrobiology*, **18**, 317–20.

Karentz, D., Cleaver, J. E., and Mitchell, D. L. (1991). DNA damage in the Antarctic. *Nature*, **350**, 28–30.

Kennard, J. M. and James, N. P. (1986). Thrombolites and stromatolites; two distinct types of microbial structures. *Palaios*, **1**, 492–503.

Leach, J. W. P. (1986). Andean high altitude expedition. *Underwater Technology*, **12**(1), 27–31.

Malin, M. C. and Edgett, K. S. (2000). Sedimentary rocks on early Mars. *Science*, **290**, 1927–37.

Marinovic, N. and Lahsen, A. (1984). Hoja Calama, Region de Antofagasta. Carta geologica de Chile No. 58, 1:250,000. *Servicos Nacional Minero Geologico.*

McKenzie, R., Bodeker, G., and Connor, B. (1999). Increased UV in New Zealand: a cautionary tale. *Water and Atmospheres*, **7**(4), 7–8.

Messerli, B., Grosjean, M., Bonani, G. *et al.* (1993). Climate change and dynamics of natural resources in the Altiplano of northern Chile during Late Glacial and Holocene time. First synthesis. *Mountain Research and Development*, **13**(2), 117–27.

Newsom, H. E., Britelle, G. E., Hibbits, C. A., Crossey, L. J., and Kudo, A. M. (1996). Impact crater lakes on Mars. *Journal of Geophysical Research*, **101**, 14951–5.

Nicholson, K. (1993). *Geothermal Fluids: Chemistry and Exploration Techniques.* Berlin: Springer Verlag.

Nunez, L., Grosjean, M., and Cartajena, I. (2002). Human occupations and climate change in the Puna de Atacama, Chile. *Science*, **298**, 821–4.

Ori, G. G., Marinangeli, L., and Baliva, A. (2000). Terraces in Gilbert-type deltas in crater lakes in Ismenius Lacus and Memnonia (Mars). *Journal of Geophysical Research*, **105**, 17629–43.

Paxinos, R. and Mitchell, J. G. (2000). A rapid Utermöhl-method for estimating algal numbers. *Journal of Plankton Research*, **22**(12), 2255–62.

Rae, R., Howard-Williams, C., and Vincent, W. F. (2000). Temperature dependence of photosynthetic recovery from solar damage in Antarctic phytoplankton. In *Antarctic Ecosystems: Models for Wider Ecological Understanding*, ed. C. Howard-William and P. Broady. Christchurch: New Zealand Natural Sciences.

Reche, E., Pulido-Villena, J., Conde-Porcuna, M., and Carrillo, P. (2000). Photoreactivity of dissolved organic matter from high-mountain lakes of Sierra Nevada, Spain. *Arctic, Antarctic and Alpine Research*, **33**(4), 426–34.

Reizopoulou, S., Santas, P., Danielidis, D., Haeder, D.-P., and Santas, R. (2000). UV effects of invertebrate and diatom assemblages of Greece. *Journal of Photochemistry, Photobiology, & Biology*, **56**, 172–80.

Rott, E. (1988). Some aspects of the seasonal distribution of phytoflagellates in mountain lakes. *Hydrobiology*, **161**, 159–70.

Rudolph, W. (1955). Licancabur: mountain of the Atacameños. *Geographical Review*, **45**(2), 151–71.

Rumrich, U., Lange-Bertalot, H., and Rumrich, M. (2000). Diatomeen der Anden. In *Iconographia Diatomologica* 9 (von Venezuela bis Patagonien/Tierra del Fuego). ed. Lange Bertalot.

Scott, D. H., Dohm, J. M., and Rice, Jr. J. W. (1995). Map of Mars showing channels and possible paleolakes. *US Geological Survey Miscellaneaous Investigations Map*. I-2461.

Servant Vilardy S., Risacher, F., Roux, M., Landre, J., and Cornee A. (2000). Les diatomées des milieux salées (Ouest Lipez, SW de l'Altiplano bolivien). http://mnhn.fr/mnhn/geo/diatoms/.

Sylvestre, F., Servant, M., Servant-Vildary, S., Causse, C., and Fournier, M. (1999). Lake-level chronology on the southern Bolivian Altiplano (18° S–23° S) during late-glacial time and the early Holocene. *Quaternary Research*, **51**, 54–66.

Varekamp, J. C., Pasternack, G. B., and Rowe, G. L. (2000). Volcanic lake systematics II. Chemical constraints. *Journal of Volcanology and Geothermal Research*, **97**, 161–80.

Vinebrook, R. R. and Leavitt, P. R. (1996). Effects of ultraviolet radiation on periphyton in an alpine lake. *Limnology and Oceanography*, **41**(5), 1035–40.

Vincent, W., Wurtsbaugh, W., Vincent, C., and Richerson, P. (1984). Seasonal dynamics of nutrient limitation in a tropical high-altitude lake (Lake Titicaca, Peru-Bolivia): application of physiological bioassays. *Limnology and Oceanography*, **29**, 540–52.

Vincent, W., Castenholz, R., Downes, M., and Howard-Williams, C. (1993). Antarctic cyanobacteria: Light, nutrients and photosysnthsis in the microbial mat environment. *Journal of Phycology*, **29**; 745–55.

von Wintzingerode, F., Göbel, U., and Stackebrandt, E. (1997). Determination of microbial diversity in enviromental samples: pitfalls of PCR-based rRNA analysis. *Federation of European Microbiological Societies Microbiology Review*, **21**, 213–29.

Vuille, M., Bradley, R. S., Werner, M., and Keimig, F. (2003). 20th century climate change in the tropical Andes: observations and model results. *Climatic Change*, **59**(1–2), 75–99.

Wharton, R. A., Crosby, J. M., McKay, C. P., and Rice, Jr. J. W. (1995). Paleolakes on Mars. *Journal of Palaeolimnology*, **13**, 267–83.

Wirrmann, D. and Mourguiart, P. (1995). Late Quaternary spatio-temporal limnological variations in the Altiplano of Bolivia. *Quaternary Research*, **43**, 344–54.

Worrest, R.C., Van Dyke, H., and Thomson, B.E. (1978). Impact of enhanced simulated solar ultraviolet radiation upon a marine community. *Photochemistry and Photobiology*, **27**, 471–8.

15

The Canyonlands model for planetary grabens: revised physical basis and implications

Richard A. Schultz
Department of Geological Sciences, University of Nevada, Reno

Jason M. Moore
William Cotton & Associates, Los Gatos, California

Eric B. Grosfils
Department of Geology, Pomona College, Claremont

Kenneth L. Tanaka
US Geological Survey, Flagstaff

and

Daniel Mège
Laboratoire de planétologie et géodynamique, Université de Nantes

15.1 Introduction

For more than a quarter of a century, the spectacular grabens of Canyonlands National Park, Utah, have provided planetologists with a fundamental analog for understanding what planetary grabens should look like and – more importantly – what may be implied about the depth variation of mechanical properties and horizontal extensional strain.

The seminal work on Canyonlands grabens was done by George McGill and coworkers in support of their investigations of the origin and kinematic significance of lunar and Martian straight rilles (McGill, 1971; McGill and Stromquist, 1975, 1979; Stromquist, 1976; Wise, 1976). McGill and Stromquist (1979) hoped to invert graben widths, assessed on an aerial or orbital image, for the depth of faulting (i.e., fault intersection depth). By equating this depth with stratigraphic layer thickness and assuming a symmetric graben geometry and plausible values of fault dip angles, grabens provided ready and seemingly reliable probes of the near-surface planetary stratigraphy and strain. Interestingly, the analog modeling of brittle-layer extension over a ductile (quasiplastic) substrate, appropriate to Canyonlands

The Geology of Mars: Evidence from Earth-based Analogs, ed. Mary Chapman. Published by Cambridge University Press.

stratigraphy (McGill and Stromquist, 1975, 1979), anticipated the key role of faulting in triggering and mobilizing salt or shale diapirism at depth (Jackson and Vendeville, 1994; Jackson, 1995). Other observations and inferences made in the 1970s, including flexure of rock layers at ramps near graben terminations and incremental growth of fault slip (McGill and Stromquist, 1979), anticipated these fundamentally important ideas by at least a decade (Sibson, 1989; Peacock and Sanderson, 1991; Cowie and Scholz, 1992).

Kinematically, a symmetric graben geometry was thought to limit the type of structure beneath the graben to either a dilatant ("tension") crack or a décollement (for various rationales see Golombek, 1979; Golombek and McGill, 1983; Tanaka and Golombek, 1989; Banerdt *et al.*, 1992; Figure 15.1). These hypotheses are at variance with more recent (as well as some older) field, experimental, and theoretical investigations of developing grabens

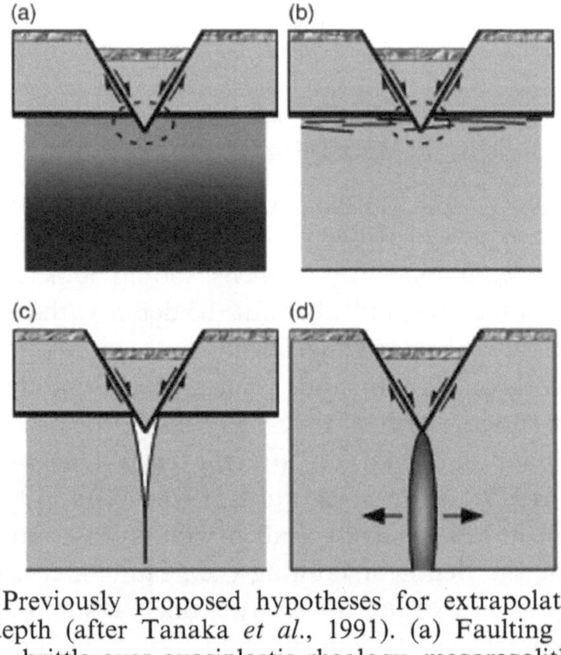

Figure 15.1. Previously proposed hypotheses for extrapolating planetary grabens to depth (after Tanaka *et al.*, 1991). (a) Faulting at bimaterial interface (e.g., brittle over quasiplastic rheology, megaregolith over basalt, dessicated ground over icy ground), (b) Faulted upper layer separated by sills or detachment zones from undeformed substrate, (c) Graben wedge falling into space-accommodating tensile crack in substrate, (d) Graben faults nucleated by dike dilation at depth. Dashed circles in (a) and (b) indicate areas of kinematic incompatibility; extension in (d) due to dike is subequal to that accommodated by superjacent graben. All these ideas imply a thin-skinned upper faulted layer that is uncoupled from subjacent strata.

(symmetric and asymmetric) on the Earth (see reviews by Schultz, 1992, 1999). The idea of "thin-skinned" extension, and the resulting problem of the mechanisms by which strain would then be accommodated at greater depths below the graben wedges, have motivated a variety of formulations of Tharsis tectonics (Tanaka *et al.*, 1991; Banerdt *et al.*, 1992; Thomas and Allemand, 1993).

Although reasonable when proposed, many of the initial ideas from Canyonlands faults, as applied to Mars and other planetary bodies, have not withstood the test of time. Substantial advances in understanding the kinematics of relay-ramps and systematic measurement of fault displacements over the past decades have led to more sophisticated approaches to graben geometry and development being commonplace today (Barnett *et al.*, 1987; Peacock and Sanderson, 1991, 1994; Odonne and Massonnat, 1992; Davison, 1994; Trudgill and Cartwright, 1994; Cartwright *et al.*, 1996; Moore and Schultz, 1999; Schultz, 1999, 2000). Nevertheless, it is still common to encounter papers from planetary researchers who attempt to map subsurface horizons using graben widths. We expect planetary grabens to look like those in Canyonlands and – despite the likely lack of thick, wet, evaporite sequences underlying large regions of post-Noachian Mars – to behave like them.

The purpose of this chapter is to propose a new model for planetary grabens, based on an improved understanding of Canyonlands graben analogs. The new "hourglass" model is derived from current concepts demonstrated by the field relations, kinematics, and mechanics of well-studied terrestrial faults and grabens. First, we review the salient aspects that motivated the original Canyonlands model for planetary grabens in the 1970s. After noting some important planetary applications and expansions to this hypothesis, we briefly review several new observations of Canyonlands grabens that require the present revision of this model. Last, we touch on some of the implications for extensional tectonics on the terrestrial planets and satellites.

15.2 Historical development of the model

The "simple" graben paradigm follows directly from an interpretation of the surface expression of the Canyonlands grabens. These interpretations were based on an understanding of fault kinematics and mechanics available in the 1970s. Because planetary grabens on Mars and elsewhere are, in many cases, still being interpreted by using this paradigm, a brief review of the origin and application of the original Canyonlands model will permit a clearer understanding of the changes that are needed to better interpret planetary grabens.

15.3 Grabens at Canyonlands National Park, Utah

The grabens occur within the Needles District of the Park and define an arcuate, northwest-trending system (McGill and Stromquist, 1979; Trudgill and Cartwright, 1994; Cartwright *et al.*, 1995; Schultz and Moore, 1996; Schultz-Ela and Walsh, 2002). The grabens range from ~100 m to 6 km in length and are generally spaced 700–1000 m apart. Widths at the surface range from < 100 m to >400 m, with normal-fault displacements – defined here by the exposed stratigraphic offsets along the faults – varying from less than 25 m to more than 100 m (McGill and Stromquist, 1975; Cartwright *et al.*, 1995; Moore and Schultz, 1999).

Graben floors are mantled by Quaternary colluvial, eolian, and alluvial sediments (Biggar and Adams, 1987) that produce a smooth, sub-horizontal surface. Previous work suggested that these sediments were negligibly thin (< 10–15 m) in relation to the fault offsets (~100 m), implying that the present geomorphic surface mirrors the configuration of the structural graben floor (i.e., depth and attitude) (Cartwright and Mansfield, 1998).

The grabens deform a ~460 m thick section of clastic sedimentary rocks that overlie gypsum and other evaporites in the Paradox Formation (Lewis and Campbell, 1965; Condon, 1997). Down-dip sliding of the clastic sequence and/or flow of subjacent Paradox evaporites is thought to have initiated graben growth and related extension in the overlying rocks (e.g., Baker, 1933; Lewis and Campbell, 1965; McGill and Stromquist, 1975; Huntoon, 1982; Trudgill and Cartwright, 1994; Cartwright *et al.*, 1995; Schultz-Ela and Walsh, 2002; Walsh and Schultz-Ela, 2003).

Together, the faulted clastic sequence and the subjacent evaporites approximate a brittle-over-quasiplastic rheologic system that has two relevant attributes. First, the layered sequence prohibits fault development in the ductile substrate, leading to a special – and particular – tectonic setting for these structures. Second, the deformable substrate facilitates the accommodation of potential volumetric problems beneath the grabens by flow and growth of reactive salt diapirs (Jackson and Vendeville, 1994; Jackson, 1995; Moore and Schultz, 1999; Schultz-Ela and Walsh, 2002).

The geometric graben model that emerged from this body of work was based on several key observations. Graben walls at the surface were vertical to depths of ~100 m, implying that normal-fault slip occurred along preexisting subvertical joints (McGill and Stromquist, 1974, 1975, 1979). Where observed in crosscutting drainages (such as Lower Red Lake Canyon), graben faults steepened in average dip angle from vertical to ~70° below depths of ~100 m, leading to vertical gaps along graben walls as the graben wedge moved downward (McGill and Stromquist, 1975) along with inward rotation and failure of joint-bounded slabs. Normal faults having shallow and steep segments in particular layers, as a result of down-dip linkage of the individual joint and fault segments, are well documented in the literature (Gudmundsson, 1992; Gross *et al.*, 1997; Peacock, 2002; Wilkins and Gross, 2002; Ferrill and Morris, 2003; Soliva, 2004). The magnitude of vertical offset of Cedar Mesa sandstone caprock onto graben floors was taken to be subequal on both sides of the graben, leading to a symmetric graben configuration (McGill and Stromquist, 1974, 1979). Although asymmetric offsets were locally observed at the time (G.E. McGill, personal communication, May 1997), the symmetric graben model appeared more successful in accounting for the range of observations obtained, and it was consistent with physical experiments that simulated graben development (McGill and Stromquist, 1974, 1979; Stromquist, 1976).

Kinematically, the growth of two equal-displacement, conjugate normal faults that span the thickness of a brittle sequence was most easily interpreted as contemporaneous nucleation of both faults at depth, at the brittle–quasiplastic interface, and simultaneous upward propagation to the

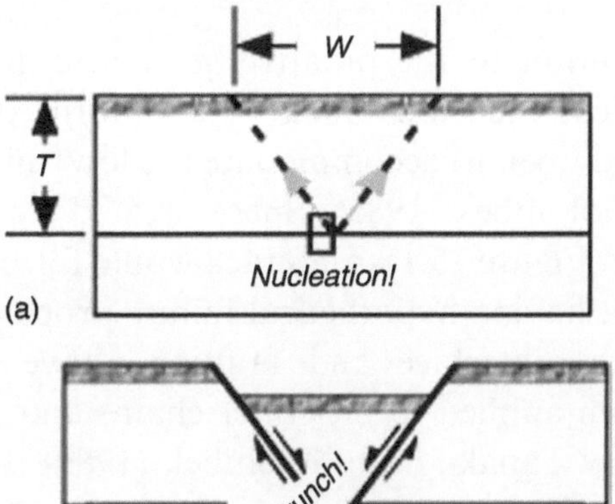

Figure 15.2. Kinematic sequence required by the original Canyonlands model for grabens. In (a), two conjugate fault surfaces nucleate simultaneously at the interface between the upper and lower layers, then propagate upward to the surface. Initial width of incipient graben *W* is taken to be proportional to faulted-layer thickness *T*. (b) Nonzero displacement along

surface (Golombek, 1979; Golombek and McGill, 1983; Golombek, 1985). This sequence, shown in Figure 15.2, implies that the incipient graben width at the surface is proportional, through an assumed (constant) value of fault dip angle, to the faulted-layer thickness (Figure 15.2). Continued extension and vertical offset would, however, necessarily lead to increasingly inaccurate relationships between width and thickness (Figure 15.2).

Horizontal extension of a layer that is accommodated by downward translation of a symmetric-graben wedge implies geometric (space) problems (Golombek, 1979; Golombek and McGill, 1983). In particular, the lower tip of the wedge must either deform internally, punch downward into quasiplastic rocks, or be truncated by a detachment surface at the interface (Golombek, 1979; Tanaka *et al.*, 1991; Figure 15.1) for horizontal extension to occur and accumulate. This geometric problem is circumvented in many terrestrial cases by the development of listric bounding faults and block rotations (Roberts and Yielding, 1994) for large-strain (and sufficiently deep) examples. Large horizontal displacements on planets other than the Earth that would produce listric faults and detachments are not typically observed, however (Okubo and Schultz, 2003, 2004). Alternatively, clay models reported by Cloos (1968) demonstrate how symmetric graben geometries can result from nucleation at depth of asymmetric faults that slip progressively with increasing horizontal strain, leading to internal deformations and a final symmetric graben geometry for relatively low-strain examples.

An alternative solution to the putative geometric incompatibility hypothesizes that a vertical "tension crack" could form directly beneath the symmetric graben and open to accommodate the lowering wedge (Golombek and McGill, 1983; Golombek, 1985). Space created by the crack would be invaded by the wedge (Figure 15.1). The crack would form in a subjacent layer, such as megaregolith, having hypothetical failure properties that would lead to jointing below the interface and faulting above (Golombek, 1985). This scenario has been applied to pit-crater chains and collapse depressions at Valles Marineris by Tanaka and Golombek (1989). However, the idea is at variance with typical rock-mass failure criteria (Hoek and Brown, 1980), that generally predict cracking at shallower depths than faulting (Schultz, 1992; Schultz and Zuber, 1994) and the implied volumetric balance between the wedge and subjacent void volume in the crack (Mège and Masson, 1996). This layer-cake scenario differs from cracking or faulting of individual intercalated units that later link up to form a "dilational fault" (Ferrill and Morris, 2003) having an average dip angle steeper than the optimal one for Coulomb frictional sliding alone (Gudmundsson, 1992; Gross *et al.*, 1997;

Peacock, 2002; Wilkins and Gross, 2002; Ferrill and Morris, 2003; Soliva, 2004). This "crack-below fault" conjecture – for the two-layer sequence (Figure 15.1) – is not considered robust for crustal-scale faults today.

Based on the echelon patterns of Martian pit-crater chains, Schultz (1989) originally suggested that these depressions may be the surface expression of dilatant hydrofractures at shallow depth, pressurized by either magma or groundwater. Other Martian grabens such as those in: (1) Tharsis (Wilson and Head, 2002; see Tanaka *et al.* (1991), Mège and Masson (1996) and Mège *et al.* (2003) for reviews and discussion), (2) the Valles Marineris system (Schonfeld, 1979; recently revived conceptually by McKenzie and Nimmo, 1999), (3) on the Moon (Head and Wilson, 1994), and (4) on Venus (Grosfils and Head, 1994; Ernst *et al.*, 1995) have also been proposed to extend as a result of dike dilation below. Planetary grabens associated with pit-crater chains might not be formed directly by regional crustal extension per se, but instead might nucleate and accumulate displacement in concert with local, inhomogeneous stresses associated with the free-surface interaction of near-surface dikes (Pollard *et al.*, 1983; Mastin and Pollard, 1988; Rubin and Pollard, 1988; Rubin, 1992; Mège and Masson, 1996; Mège *et al.*, 2003). Dike dilation and lateral propagation represent a balance between internal magma pressure and crustal stress state (Rubin, 1995; Fialko and Rubin, 1999), leading to an indirect relationship between crustal stress and graben strain. In this scenario (Figure 15.1), no stratigraphic or mechanical discontinuity is required, and the depth of faulting would be related to the details of dike emplacement (Mège and Masson, 1996; Wilson and Head, 2002). Schultz *et al.* (2004) recently inferred the existence of a dike below a Martian graben from MOLA topography.

15.4 Planetary implications of the symmetric graben model

One prominent and persistent byproduct of the Canyonlands model of graben faulting is the putative correlation of fault-intersection depth with the thickness of the megaregolith, or other mechanically distinct layer, on the Moon (Golombek, 1979; Golombek and McGill, 1983). The megaregolith is thought to be an intercalated sequence of laterally discontinuous ejecta deposits from impact craters and basins, along with lavas and sediments (Melosh, 1989, p. 197). Originally identified for the Moon (Hartmann, 1973; Squyres *et al.*, 1992), the putative Martian megaregolith is thought to attain cumulative thicknesses of perhaps 2 km (Fanale, 1976) and may comprise a substantial part of the older, Noachian terrains as impact-basin ejecta blankets (MacKinnon and Tanaka, 1989).

Patterned after the lunar stratigraphy, a layer-cake model of planets such as Mars was developed (MacKinnon and Tanaka, 1989; Squyres *et al.*, 1992). The idea was that grabens would nucleate at the contact between basalts above and weaker, more plastically deforming megaregolith below, with graben widths directly indicating the thickness of the brittle basaltic layers (Wise, 1976). For Mars, this brittle-over-quasiplastic stratigraphic template (Davis and Golombek, 1990; Plescia, 1991; Banerdt *et al.*, 1992) was expanded to include ice-rich layers (Soderblom and Wenner, 1978; Squyres and Carr, 1986; Squyres *et al.*, 1992) as mechanical "discontinuities" (Golombek, 1985) or plastically deforming substrates that could lead to comparable graben-fault nucleation and geometric interpretations (Davis *et al.*, 1995; Golombek *et al.*, 1996). This general sequence – faulted basement, megaregolith, capping veneer of lava flows, ground ice, and/or sediments – has motivated many interpretations of Martian and planetary tectonics (Tanaka and Golombek, 1989; Allemand and Thomas, 1992; Thomas and Allemand, 1993).

The thin-skinned models discussed by Tanaka *et al.* (1991) all require either a brittle-over-quasiplastic layering (Figure 15.1) or a décollement (detachment) surface (Figure 15.1), both of which serve to decouple the faulted and extending upper carapace from everything below (Jackson and Vendeville, 1994). Thus, this paradigm for Martian extensional tectonics retains a clear heritage from the early ideas for the Moon and those developed for the grabens in Canyonlands. This simple two-layer scenario has been applied with varying success to icy satellites (Golombek and Banerdt, 1986; Pappalardo and Greeley, 1995) and to thermally induced rheologic stratification on Venus (Hansen and Willis, 1998; Ghent and Hansen, 1999; Ghent and Tibuleac, 2002).

Recent high-resolution images of Valles Marineris trough walls and other slope faces on Mars, from the Mars Global Surveyor spacecraft (Albee *et al.*, 1998), have been unable, however, to identify either the megaregolith or its basal contact with the fractured bedrock below (McEwen *et al.*, 1999; Williams *et al.*, 2003). Previous searches using Viking data (Lucchitta *et al.*, 1992) have also been unsuccessful in locating and demonstrating this contact. Ongoing analyses of high-resolution MOC images demonstrate the presence and regional extent of fine-scale layering on Mars (Williams *et al.*, 2003), morphologically similar to terrestrial flood-basalt sequences with sedimentary interbeds. Sequences of individual layers ~5–10 m thick attain aggregate thicknesses exceeding 3–6 km in the Valles Marineris region (McEwen *et al.*, 1999). These new observations of thinly layered sequences of strata in the Martian crust (Malin and Edgett, 2000) provide a firm basis for identifying Martian grabens in cross section by revealing offset markers and

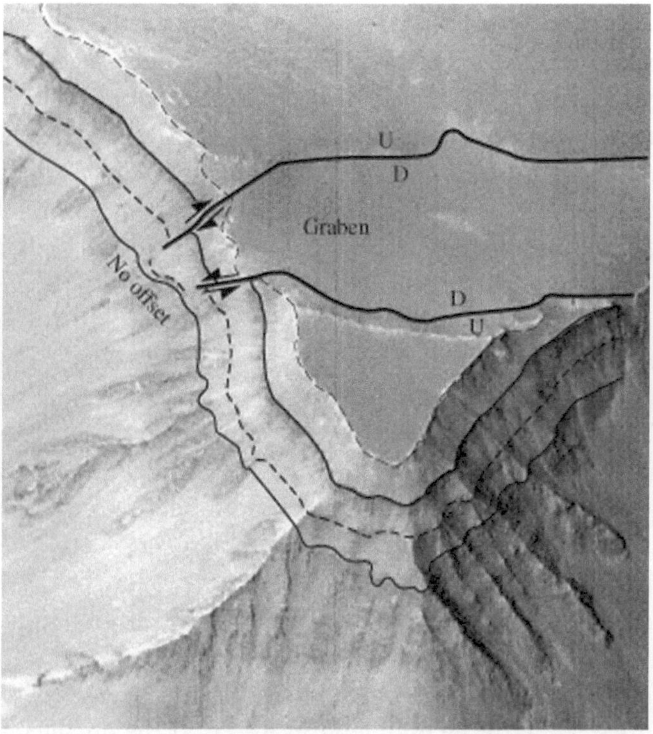

Figure 15.3. Map of a representative Martian graben showing offset of layers along slope and no offset of deeper layers (see McEwen *et al.*, 1999, for discussion of this area and the layers). Normal separation of layers on slope exceeds the dip-slip offset due to the shallow slope angle. Part of MOC image 568174924.8003 P080-03 in eastern Coprates Chasma centered near 14.5° S, 55.8° W; graben width ~ 1 km.

do not support assertions that Martian grabens must sole into a detachment, horizontal discontinuity, or change in stratigraphy, as illustrated in Figure 15.3.

The three-layer scenario is too simplistic to describe much Martian geology and stratigraphy (McEwen *et al.*, 1999; Malin and Edgett, 2000). As a result, the original Canyonlands model may be of limited applicability to describe the three-dimensional deformation and strain in Martian terrains underlain by thick sequences of layered rocks (Hauber and Kronberg, 2001; Wilkins *et al.*, 2002).

15.5 Canyonlands in the 1990s and beyond

Work on the Canyonlands grabens continued into the 1980s and 1990s by several researchers and agencies, mostly in support of candidate sites for high-level radioactive waste disposal (Woodward-Clyde Consultants, 1983) and petroleum exploration (Trudgill and Cartwright, 1994; Fossen, 1995). Drill cores became available that documented the stratigraphy of the Paradox Basin in many key places. Sedimentologic and stratigraphic work refined the interpretation of important units, such as the Cedar Mesa Sandstone (Loope, 1984) that caps several of the grabens (Lewis and Campbell, 1965; McGill and Stromquist, 1979), as well as the regional stratigraphy (Condon, 1997). Although no longer considered suitable (geologically or politically) for radioactive waste isolation, the Grabens area of Canyonlands National Park now provides a template – and scale model – for oil and gas fields associated with crustal rifting and graben-related structural and stratigraphic traps (Fossen, 1995). The Canyonlands grabens now also represent the type example of (normal) fault growth by segment linkage (Trudgill and Cartwright, 1994) as well as an important and widely cited data set for displacement-length scaling (Cartwright *et al.*, 1995).

This continuing work by independent research groups has exploded the legend of symmetric, keystone-collapse grabens in Canyonlands (Table 15.1; Figure 15.4). For example, Trudgill and Cartwright (1994) and Cartwright *et al.* (1995, 1996) have demonstrated that the displacements along graben-bounding faults scale with the map lengths and that the displacement maxima are located near the fault-segment midpoints – just like on other examples of normal of faults (Peacock and Sanderson, 1991, 1994; Dawers *et al.*, 1993; Soliva and Benedicto, 2004). This important and fundamental observation indicates that estimates of graben-floor depths, obtained by averaging several measurements made at arbitrary positions along the graben (Davis *et al.*,

Table 15.1. *Evolution of thought on graben mechanics*

Old	New
Symmetric grabens	Asymmetric grabens
Conjugate faults of equal offset	Opposing faults have unequal offset
Flat floors, shallow, and indefinitely long	Tilted floors, depth $= f$(length, position)
Strain independent of position	Strain dependent on along-strike position
Conjugate faults are synchronous	Sequential master-antithetic fault development
Faults grow upward from interface at depth	Subsurface interface/discontinuity not required
Rigid-block tectonics (wedge indentor)	Deforming blocks, variable slip on faults
Wedge crunch implies a crack, dike, or truncation at depth	Zero slip at wedge tip; fault networks at depth
Thin-skinned tectonics	Thick-skinned tectonics

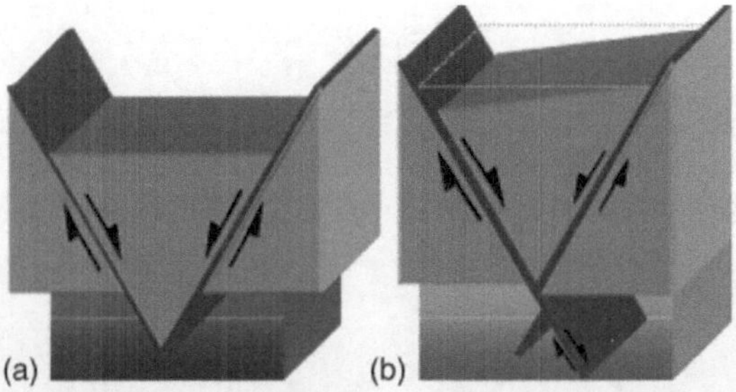

Figure 15.4. Comparison of symmetric "simple" graben (a) with asymmetric graben (b). Note variable offset along graben-bounding normal fault in (b). See text and Table 15.1 for discussion.

1995), are only marginally informative because they ignore the fundamental position-dependence of these depth values (Schultz and Fori, 1996). The work also demonstrates the necessity of identifying the fault-segmentation lengths appropriate to the observed surface displacements (Schultz, 1997; Wilkins *et al.*, 2002), rather than just mapping the aggregate length of an echelon graben array.

Detailed structural and topographic mapping (Schultz and Moore, 1996; Moore and Schultz, 1999; McGill *et al.*, 2000) has demonstrated that the grabens are characteristically asymmetric, rather than the simple, symmetric, keystone-collapse wedges previously thought. Characteristics of asymmetric graben geometry in cross section include (Schultz and Moore, 1996; Moore and Schultz, 1999):

(1) Significant differences in the amount of stratigraphic offset across graben. These differential offsets, indicating master and antithetic faults, are documented in grabens that vary widely in size (Devils Pocket, Devils Lane, Cyclone Canyon, Red Lake Canyon; Moore and Schultz, 1999).
(2) Distinct map pattern of graben bounding faults. The fault having greater stratigraphic offset (master fault) is continuous along the graben's length, whereas that of the facing graben wall (antithetic fault) is discontinuous, segmented, and echelon.
(3) Rollover anticlines formed adjacent to the antithetic fault are related to the translation of strata down the master fault, resulting in local flexure (Higgs *et al.*, 1991). The widths of preexisting joints that parallel the graben (see Figure 15.5) also differ considerably across a graben: greater widths (individual joint openings of perhaps several meters) on the antithetic side are associated with increased surface area and bending of the jointed rocks along the upper, outer surface of the rollover anticline.

Figure 15.5. Aerial view of Devils Pocket graben in Canyonlands National Park (photo by Matt Soby, looking north) showing echelon, segmented geometry of grabens, and stratigraphic offsets of preexisting Needles topography. Jeep trail on graben floor for scale.

(4) Footwall uplift and gentle flexure, perhaps tens of meters in amplitude, occurs adjacent to the master fault; it decreases both along strike toward the fault terminations and across strike away from the fault trace. The pre-existing joints are closed in the footwall area.
(5) Seismic refraction and gravity results (Grosfils *et al.*, 2003) demonstrate substantial floor tilt in northern Devils Lane graben, down toward the master fault and deeper (>65 m) in the center of the graben than near the ends (~15 m). The graben floor is best described as "spoon-shaped," or deepest in the center and tilted down toward the master fault, beneath the overlying sedimentary wedge.

These attributes, either individually or together, are definitive indicators of asymmetric cross-sectional geometry.

Construction of balanced cross sections (Moore and Schultz, 1999; Grosfils *et al.*, 2003), as well as field observation of grabens exposed in cross section (McGill *et al.*, 2000), suggest that the graben faults do not necessarily intersect at the base of the brittle layer – as previously postulated – but at perhaps only two-thirds to three-quarters of the layer thickness (see also parallel conclusions by Schultz-Ela and Walsh, 2002). The intersection depths and fault geometries are sensitive to the layer properties, curvature of the layer in cross section, and other parameters (Schultz-Ela and Walsh, 2002). Thus, graben widths measured at the planetary surface do not provide a reliable measure of the thickness of the faulted layer (see discussion of lunar graben widths below).

Independent mechanical modeling of graben nucleation during down-dip extension of the faulted sequence by Schultz-Ela and Walsh (2002), using a finite-element simulation that incorporates strain-softening and fault nucleation processes (Schultz-Ela *et al.*, 1993), is consistent with the field results reported by Moore and Schultz (1999) and McGill *et al.* (2000). They found that grabens nucleated with asymmetric geometries and had bounding faults that intersected well above the base of the brittle layer. Reactive salt diapirs (Vendeville and Jackson, 1992; Jackson, 1995) continued down from the graben wedge and "hourglass" fault geometries (Figure 15.4) were produced (see also the pioneering analog models of McGill and Stromquist, 1979, and the work of Nicol *et al.*, 1995). Graben formation in this layered mechanical system is more complex than the symmetric-graben model can accommodate, leading to a poor correlation between graben width, depth of faulting, and faulted-layer thickness (Schultz-Ela and Walsh, 2002).

Recent seismic refraction and gravity experiments carried out in Canyonlands grabens demonstrate that a thick sedimentary wedge exceeding 70 m in thickness overlies the structural floor in at least one graben (northern Devils Lanei; Grosfils *et al.*, 2003). This finding indicates that the exposed stratigraphic offsets along the graben-bounding faults are minimum values and that the attitude of the graben floor (strictly speaking, the attitude of the surficial sediments at the ground surface) does not necessarily contain information about the degree of horizontality of the buried structural floor (Moore *et al.*, 1997; Moore and Schultz, 1999; Grosfils *et al.*, 2003; Figure 15.4).

15.6 The new hourglass model for grabens and implications for planetary faulting

Recognition of systematically varying (nonconstant) displacements on graben-bounding normal faults, both along-strike and down-dip, since the original 1970s-era Canyonlands model, as summarized in this chapter, motivates a revised model for planetary grabens. In this section we first redefine a graben, following current usage, highlighting useful observables in planetary data sets. Then we briefly examine some implications of the new "hourglass" model – as viewed in cross section – for the Moon, Venus, and Mars.

A graben can be defined as a pair of parallel, non-coplanar normal faults, that dip toward each other, that completely overlap with small cross-strike separations relative to their lengths. Grabens commonly exhibit a greater amount of vertical stratigraphic offset (throw) on one (master) fault than on the other (antithetic) fault. Such asymmetric grabens (Gibbs, 1984; Rosendahl, 1987; Groshong, 1989; Jackson and White, 1989) can display a rich assemblage of topographic features (Davison, 1994) such as footwall uplift (Weissel and Karner, 1989), hangingwall subsidence (Gudmundsson and Bäckström, 1991), and rollover anticlines (Moore and Schultz, 1999). These topographic elements increase in amplitude from zero at the graben terminations to maximum values near the middle regions of the fault, tracking the shape of the displacement distribution (Dawers *et al.*, 1993; Dawers and Anders, 1995; Davies *et al.*, 1997) and location of maximum offset, D_{max} (Barnett *et al.*, 1987; Pollard and Segall, 1987; Walsh and Watterson, 1987; Bürgmann *et al.*, 1994; Soliva and Benedicto, 2004).

Depocenters along faults and grabens are closely associated with the position of maximum depth, or maximum displacement, along the faults (Gibbs, 1990; Roberts and Yielding, 1994; Gupta *et al.*, 1998; Gawthorpe and Leeder, 2000; Gupta and Cowie, 2000; McLeod *et al.*, 2000; Grosfils *et al.*, 2003). These depocenters normally occur near fault midpoints and, with time, shift toward the stepover of echelon or interacting faults as linkage begins (Morley *et al.*, 1990; Morley, 1999). Each fault in a planetary graben has its own maximum displacement, and depocenter, created prior to linkage and its incorporation into the observed graben (for examples from Mars see Schultz, 1995; Wilkins and Schultz, 2003). Although deposition of dust and other eolian sediments in Martian grabens may partly fill and thus obscure the original structural topography within the graben (i.e., its tilted floor), the uplifted topography outside the graben remains, mirroring in reverse the topography within the graben itself. In addition, the depocenters will serve to collect any hydrothermal (or other) fluids that migrate into the graben, either along its faults or from surrounding rock units. The segmented geometry of planetary normal faults (Schultz, 1991, 1995, 1997, 1999; Schultz and Fori, 1996; Wilkins and Schultz, 2003) provides ready clues to the best locations of depositional sinks within grabens (Figure 15.7).

15.6.1 Lunar grabens revisited

The new thinking about planetary grabens summarized in this chapter provides a renewed impetus to investigate lunar grabens – the place where planetary structural geology largely began. In this section we discuss two main issues: map-view geometry and its implications, and the venerable relationship (or lack of one) between graben width and depth of faulting.

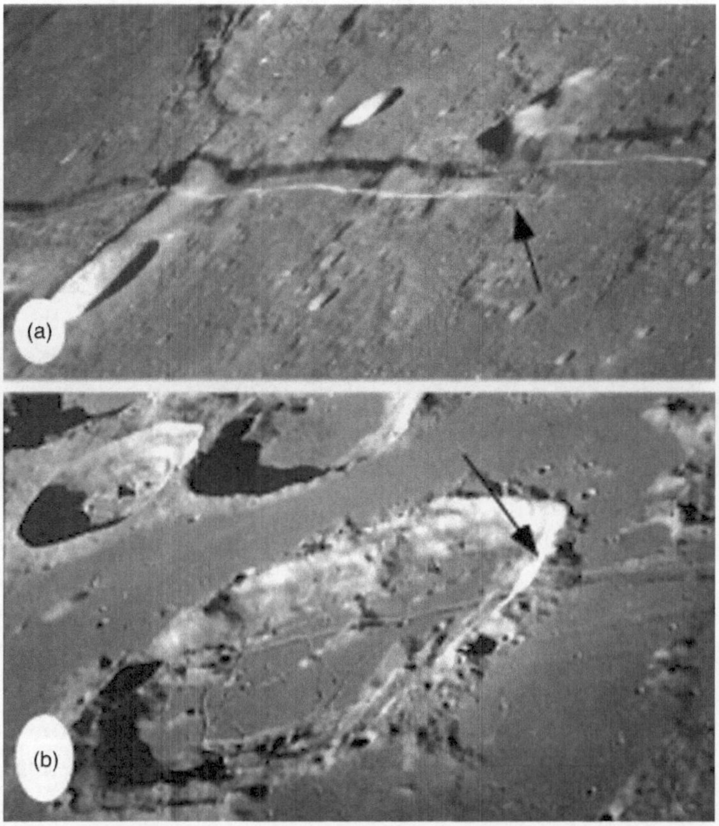

Figure 15.6. Examples of lunar grabens that demonstrate key geometric and mechanical characteristics. (a) Echelon graben (Rima Ariadaeus) traversing lunar plains materials. Note change in graben width over the ridge, and ramp with change of asymmetry sense at stepover (arrow). Part of AS10–31–4645 (H). (b) Grabens traversing basaltic lava flows in Mare Fecunditatus and rim (arrow) and interior fill of crater Goclenius. Note ramps and echelon segmentation of grabens on mare and continuity along strike through all three geologic units. Part of AS8–13–2225 (H).

Several typical lunar grabens are shown in Figure 15.6. These grabens, photographed by Apollo 8 and Apollo 10 astronauts before 1970, provide spectacular and unusually informative examples of normal fault growth and development that, with hindsight, we can readily interpret. A clear echelon stepover is recorded in Figure 15.6 along Rima Ariadaeus. One can see how the central fault changes its sense of dip direction ("polarity") as the graben steps to the left. The graben changes width as it crosses the non-mare ridge near the large crater: in addition to evidence for non-vertical fault dip angles (McGill, 1971), continuous fault traces across basalt, non-mare ridges, and crater wall and floor materials all demonstrate that the graben is not confined to a thin sequence of mare fill material (basalt), but has propagated along-strike into a different lithologic and stratigraphic sequence (Schultz and Zuber, 1994).

A similar paradoxical example is found in Figure 15.6, where grabens transect mare basalts outside of the large central-peak crater Goclenius, the crater rim, and interior fill within the crater. Using the classic Canyonlands model, these spatial relationships are impossible, because: (a) the grabens must be symmetric (violated by the first example, Figure 15.6, where the outer fault extends farther along-strike than the inner one); (b) the grabens must be confined to the "megaregolith" (Golombek, 1979) and thus cannot propagate across different lithologies (Figures 15.6a and 15.6b); and (c) although the graben width in Figure 15.6a is constant at the step, the offset increases from zero at the graben tip, down its ramp, toward a larger value far from the step – an impossible situation if the depth of faulting scales only with the width of the graben (Golombek, 1979; Golombek and McGill, 1983; and many others). These examples suggest that the classic Canyonlands model is inadequate at best at explaining the spatial relationships that we find along typical lunar grabens.

The width of lunar (or other) grabens is not a reliable indicator of the depth of faulting, for several reasons discussed in this section. Theoretical work has suggested that many grabens may grow as their bounding normal faults nucleate at the surface and propagate downward (Melosh and Williams, 1989; Schultz-Ela and Walsh, 2002; Walsh and Schultz-Ela, 2003), rather than up from a common horizon (McGill and Stromquist, 1974; Golombek, 1979; Golombek and McGill, 1983). Slip along the first (usually master) fault changes the stress state in its vicinity, leading to nucleation and growth of a second normal fault nearby; the second fault grows into an antithetic configuration in order to compensate for rotations of the faulted layer due to the first fault (Reches, 1978, 1983; Aydin and Reches, 1982; Krantz, 1988, 1989; Melosh and Williams, 1989), leading to a graben. The flexural rigidity of the faulted layer thus determines the point of nucleation of the second, antithetic fault (Turcotte and Schubert, 1982; Buck, 1988), and therefore, the initial graben width. Because flexural rigidity depends on layer properties (principally Young's modulus) in addition to layer thickness, initial graben spacing is sensitive to the lithology of the faulted layer. The flexural rigidity term is influenced by (and thus dependent on) the degree of homogeneity (i.e., layering) of the faulted crust and by the presence of the fault itself (Buck, 1988); Schultz-Ela and Walsh (2002) found that flow patterns in the subjacent salt layer also influenced initial graben width in Canyonlands.

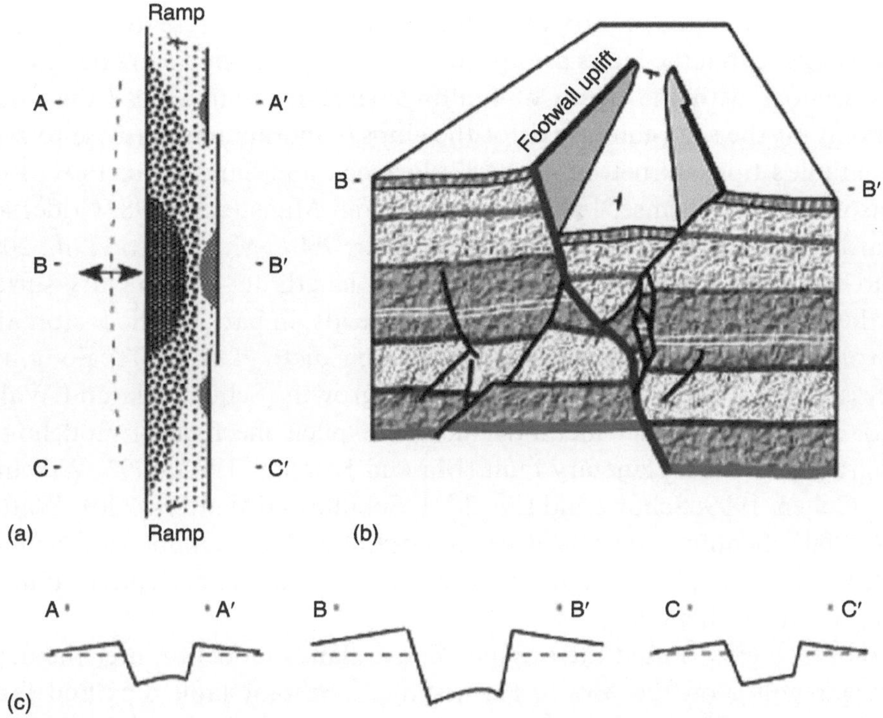

Figure 15.7. (a) Map view sketch of representative planetary graben emphasizing topography of graben floor (shaded); footwall uplift anticline shown by arrow on dashed curve. Locations of section lines A–A′, B–B′, and C–C′ as indicated. (b) Hypothetical cross sectional view of Martian graben along B–B′, showing the "hourglass" model for surface-breaking Martian grabens (constructed after Watterson *et al.* [1998]; the fault sets shown can retrodeform using dip-slip offsets). Note that volumetric changes associated with the fault intersection region are accommodated by nucleation and growth of additional small faults and by block rotations during horizontal extension, along with development of the inverted graben below. (c) Schematic topographic cross sections of the graben shown in (a).

In addition, graben widths change with ongoing extension of the faulted layer. Extension of a layer containing a graben necessarily leads to a progressively widening graben (Figure 15.2; Walsh and Schultz-Ela, 2003). However, continued extension can also lead to growth of new faults outside the graben (Jackson and Vendeville, 1994) as well as new faults within it (as a response to a reactive diapir at depth; Walsh and Schultz-Ela, 2003), so that increasing extensional strains lead to increasing graben widths and/or more complex patterns of graben-bounding faults.

Because the stratigraphic offset within a graben varies with position along the structure (Figure 15.7), a simple relationship between width and depth of faulting cannot, and does not, exist. Field and theoretical research on normal faults suggests that the faults are approximately semi-elliptical or triangular in cross section, with the depth of faulting being maximum near the fault's midpoint (at the semi-minor axis of the ellipse) and minimum (close to zero) at the fault's tips (Barnett *et al.*, 1987; Peacock and Sanderson, 1991, 1994; Davison, 1994; Willemse, 1997; Cartwright and Mansfield, 1998; Crider and Pollard, 1998; Kattenhorn *et al.*, 2000; Crider, 2001; Manighetti *et al.*, 2001; Figure 15.7). The relationship between fault length at the planetary surface and the maximum depth of faulting thus depends, in part, on the factors that control the fault shape (Nicol *et al.*, 1996; Benedicto *et al.*, 2003) along with processes that lead to fault nucleation and growth (Schultz-Ela and Walsh, 2002). As a result, other methods such as explicit mechanical modeling of topography above a planetary fault (Ma and Kusznir, 1992, 1993; Willemse, 1997; Cohen, 1999; Schultz and Lin, 2001; Schultz and Watters, 2001; Watters *et al.*, 2002; Schultz, 2003; Wilkins and Schultz, 2003; Okubo and Schultz, 2004; Schultz *et al.*, 2004) provide more reliable estimates of depth of faulting than graben width alone.

Golombek (1979) used the original Canyonlands model to infer the depth to megaregolith on the Moon. He assumed constant fault dips and faults that meet at (and therefore originated at) the upper surface of the buried megaregolith. Using the revised model in this chapter, the flexural rigidity of the faulted layer and its substrate jointly determine the initial graben width. As a result, the grabens may crudely map out the thickness of the faulted layer as Golombek suggested, but for different physical reasons and with larger uncertainties than previously estimated (dependence of width and fault dip on strain magnitude and layer properties). In particular, the requirement for graben-bounding faults to meet precisely at a point at the megaregolith's upper surface is relaxed. Similarly, layer thicknesses obtained by using this method for Mars (Banerdt *et al.*, 1992; Golombek *et al.*, 1996), Venus (Ghent and Hansen, 1999), and elsewhere (Golombek and Banerdt, 1986) may provide useful estimates (and again, on different physical grounds) but with larger uncertainties as discussed in this section.

15.6.2 Implications for grabens on Venus

On Venus, the current absence of high spatial- and vertical-resolution topography data makes it difficult to estimate the depth of faulting (or D_{max}). However, the general abundance of distributed deformation at the surface, including numerous grabens, provides a rich dataset for assessing graben properties. For example, complex sequences of intersecting graben sets

(Radunitsa Labyrinthus; Kortz *et al.*, 2003), provide interesting new opportunities to study fault set interactions and thus to evaluate the behavior of faulting at depth.

Distributed sets of "tensile-fracture ribbons" and "shear-fracture ribbons" (i.e., grabens), identified as some of the earliest-formed structural elements in many tessera terrains, are interpreted as deformation accommodated within a thin brittle layer overlying a ductile substrate (Hansen and Willis, 1998; Hansen *et al.*, 2000). While "tensile-fracture ribbons" represent a different style of deformation, mechanically, as noted by Hansen *et al.* (2000), "shear-fracture ribbon" formation is thought to be closely analogous to the style of deformation observed in Canyonlands. The thickness of the thin brittle layer is estimated using various methods, for instance using radargrammetric calculations to measure visible fault topography at point locations and by employing layer instability theory (Ghent and Tibuleac, 2002), to relate dominant ribbon spacing to the thickness of the deformed layer. Improved mechanical fault models should, however, provide a powerful way to augment the approaches used to date. For "shear-fracture ribbons" with lengths ranging from tens to ~170 km (Hansen and Willis, 1998), the topographic offset should vary as a function of position along an individual structure, leading to the appropriate "spoon shaped" along-strike topographic profile of the graben floor – the preservation of which is a reasonable possibility in the absence of younger lava embayment because both erosion and sedimentation rates are thought to be extremely low on Venus. As our ability to determine surface topography around the "shear-fracture ribbons" improves for specific structures, mechanical modeling can provide additional insight into the expected extent of subsurface faulting as a function of position along the bounding faults, yielding new estimates for brittle layer thickness at the time the faults were forming.

15.6.3 Martian grabens and Tharsis tectonics

Planetary megaregoliths (consisting of impact-crater ejecta and other poorly consolidated materials) were previously thought to be significantly weaker than an overlying basaltic sequence (Golombek and McGill, 1983; Golombek, 1985; Allemand and Thomas, 1992; Davis *et al.*, 1995; Golombek *et al.*, 1996). However, recent experimental work suggests that this assumption is not correct. Terrestrial deposits considered analogous to Martian megaregolith consist of highly angular clasts (Grant and Schultz, 1993; Urrita-Fucugauchi *et al.*, 1996). Non-indurated materials composed of angular clasts can be as strong as other crustal rocks, such as basaltic lava flows, regardless of clast size and sorting (Mair *et al.*, 2002). The presence of water ice in the Martian subsurface can, however, lead to a mechanical stratigraphy with horizons of reduced strength (as inferred for wrinkle ridges by Mangold *et al.*, 1998, and Okubo and Schultz, 2004), but this phenomenon cannot provide a Canyonlands-type brittle-over-plastic sequence without extraordinarily large (i.e., a kilometer or more), and conjectural, thicknesses of ice. Thus, the Martian megaregolith cannot be considered to be a mechanically weak layer or a potential site for detachment horizons (Okubo and Schultz, 2003) without independent and compelling evidence to the contrary.

The assumption of weak megaregolith beneath a basaltic caprock sequence was central to the formation of narrow, closely spaced Tharsis-radial graben swarms (Wise *et al.*, 1979, 1982). Systematic mapping shows that these grabens developed simultaneously on Tharsis and on its periphery coeval with or younger than wrinkle ridges (Scott and Tanaka, 1986; Banerdt and Golombek, 1990; Tanaka *et al.*, 1991). Okubo and Schultz (2003) showed that the assumption of a weak Martian megaregolith was inconsistent with the vergence directions of the thrust faults that underlie, and control the formation of, wrinkle ridges. Therefore, the presence of a weak megaregolith appears incompatible with the formation of both wrinkle ridges and grabens in Tharsis (and beyond) that require regional shearing along mid-crustal detachments in weak megaregolith. Instead, scenarios such as dike-induced graben formation (Rubin and Pollard, 1988; Rubin, 1992) and normal faulting down to the brittle-quasiplastic transition (Wilkins and Schultz, 2003) are suggested. Thus, Tharsis grabens most likely reflect lithosphere-scale thick-skinned deformation, rather than thin-skinned deformation above a weak Martian megaregolith (Okubo and Schultz, 2003).

15.7 Conclusions

The classic Canyonlands model of planetary grabens, formulated in the 1970s, has been supplanted by new concepts and understanding of fault and graben development. The Canyonlands-based stratigraphic model of "brittle over ductile" is not supported by recent laboratory investigations of megaregolith analogs and by observations of Martian crustal sections. An "hourglass" geometric model for planetary grabens is in better accord with current understanding of these structures than the former symmetric undeformable wedge. Instead, the revised graben model and the current literature on the map-view development of terrestrial normal faults and grabens, provides an improved basis for mapping and interpreting graben arrays on the Moon, Mars, Venus, and elsewhere.

Acknowledgments

We are grateful to the many field participants in the Canyonlands Grabens Initiatives for 1996, 1997, and 1999, including George McGill, who helped to advance some of the ideas developed here. This chapter was conceived during an enjoyable session on Martian tectonics at the 5th International Conference on Mars, held in Pasadena in July 1999. Discussions with Martin Jackson and Dan Schultz-Ela have helped clarify relationships between grabens in Canyonlands and salt tectonics. We thank the National Park Service for facilitating access to the Needles District Grabens and for granting permission for us to pursue structural and seismic research in the Park. Thanks to Marie Dowling of the National Space Science Data Center for providing the Apollo 8 and 10 images of faulted lunar craters. We thank Jim Skinner, Sue Priest, and an anonymous referee for reviews that sharpened the exposition in this chapter. This work was supported in part by NASA's Planetary Geology and Geophysics Program, the Mars Data Analysis Program, and Chevron Petroleum Technology Company, grants to R.A. Schultz.

References

Albee, A. L., Palluconi, F. D., and Arvidson, R. E. (1998). Mars Global Surveyor mission: overview and status, *Science*, **279**, 1671–2.

Allemand, P. and Thomas, P. (1992). Modèle fragile des rides martiennes constraint par la géométrie de surface, *C. R. Acad. Sci. Paris*, **215**, Série II, 1397–402.

Aydin, A. and Reches, Z. (1982). Number and orientation of fault sets in the field and in experiments. *Geology*, **10**, 107–12.

Baker, A. A. (1933). Geology and oil possibilities of the Moab District, Grand and San Juan Counties, Utah. US Geological Survey Bulletin, 841.

Banerdt, W. B. and Golombek, M. P. (1990). The evolution of Tharsis: implications of gravity, topography, and tectonics (abstract), *LPSC* **XXI**, 42–3.

Banerdt, W. B., Golombek, M. P., and Tanaka, K. L. (1992). Stress and tectonics on Mars. In *Mars*, ed. H. H. Kieffer, B. M. Jakosky, C. W. Snyder, and M. S. Matthews. Tucson: Univ. Arizona Press, pp. 249–97.

Barnett, J. A. M., Mortimer, J., Rippon, J. H., Walsh, J. J., and Watterson, J. (1987). Displacement geometry in the volume containing a single normal fault. *American Association of Petroleum Geologists Bulletin*, **71**, 925–37.

Benedicto, A., Schultz, R., and Soliva, R. (2003). Layer thickness and the shape of faults, *Geophysical Research Letters*, **30**, 2076, 10.1029/2003GL018237.

Biggar, N. E. and Adams, J. A. (1987). Dates derived from Quaternary strata in the vicinity of Canyonlands National Park. *Field Symposium, Guidebook of the Four Corners Geological Society*, **10**, 127–36.

Buck, W. R. (1988). Flexural rotation of normal faults, *Tectonics*, **7**, 959–73.

Bürgmann, R., Pollard, D. D., and Martel, S. J. (1994). Slip distributions on faults: effects of stress gradients, inelastic deformation, heterogeneous host-rock stiffness, and fault interaction. *Journal of Structural Geology*, **16**, 1675–90.

Cartwright, J. A. and Mansfield, C. S. (1998). Lateral displacement variation and lateral tip geometry of normal faults in the Canyonlands National Park, Utah. *Journal of Structural Geology*, **20**, 3–19.

Cartwright, J., Mansfield, C., and Trudgill, B. (1995). Fault growth by segment linkage: an explanation for scatter in maximum displacement and trace length data from the Canyonlands Grabens of SE Utah. *Journal of Structural Geology*, **17**, 1319–26.

Cartwright, J., Mansfield, C., and Trudgill, B. (1996). Fault growth by segment linkage. In *Modern Developments in Structural Interpretation, Validation and Modelling*, ed. P. G. Buchanan and D. A. Nieuland, Spec. Publ. Geol. Soc. London, 99, pp. 163–77.

Cloos, E. (1968). Experimental analysis of Gulf Coast fracture patterns. *American Association of Petroleum Geologists Bulletin*, **52**, 420–44.

Cohen, S. C. (1999). Numerical models of crustal deformation in seismic zones. *Advances in Geophysics*, **41**, 133–231.

Condon, S. M. (1997). Geology of the Pennsylvanian and Permian Cutler Group and Permian Kaibab limestone in the Paradox Basin, southeastern Utah and southwestern Colorado. US Geological Survey Bulletin 2000-P, pp. 1–46.

Cowie, P. A. and Scholz, C. H. (1992). Growth of faults by accumulation of seismic slip. *Journal of Geophysical Research*, **97**, 11085–95.

Crider, J. G. (2001). Oblique slip and the geometry of normal-fault linakge: mechanics and a case study from the Basin and Range in Oregon. *Journal of Structural Geology*, **23**, 1997–2009.

Crider, J. G. and Pollard, D. D. (1998). Fault linkage: three-dimensional mechanical interaction between echelon normal faults. *Journal of Geophysical Research*, **103**, 24373–91.

Davies, R. K., Crawford, M., Dula, Jr. W. F., Cole, M. J., and Dorn, G. A. (1997). Outcrop interpretation of seismic-scale normal faults in southern Oregon: description of structural styles and evaluation of subsurface interpretation methods. *Leading Edge*, **16**, 1135–41.

Davis, P. A. and Golombek, M. P. (1990). Discontinuities in the shallow Martian crust at Lunae, Syria, and Sinai Plana. *Journal of Geophysical Research*, **95**, 14231–48.

Davis, P. A., Tanaka, K. L., and Golombek, M. P. (1995). Topography of closed depressions, scarps, and grabens in the north Tharsis region of Mars: implications for shallow crustal discontinuities and graben formation, *Icarus*, **114**, 403–22.

Davison, I. (1994). Linked fault systems; extensional, strike-slip and contractional. In *Continental Deformation*, ed. P. L. Hancock. Pergamon, pp. 121–42.

Dawers, N. H., Anders, M. H., and Scholz, C. H. (1993). Growth of normal faults: displacement-length scaling. *Geology*, **21**, 1107–10.

Dawers, N. H. and Anders, M. H. (1995). Displacement-length scaling and fault linkage. *Journal of Structural Geology*, **17**, 607–14.

Ernst, R. E., Head, J. W., Parfitt, E., Grosfils, E., and Wilson, L. (1995). Giant radiating dike swarms on Earth and Venus. *Earth Science Reviews*, **39**, 1–58.

Fanale, F. P. (1976). Martian volatiles: their degassing history and geochemical fate. *Icarus*, **28**, 179–202.

Ferrill, D. A. and Morris, A. P. (2003). Dilational normal faults. *Journal of Structural Geology*, **25**, 183–96.

Fialko, Y. A. and Rubin, A. M. (1999). Thermal and mechanical aspects of magma emplacement in giant dike swarms. *Journal of Geophysical Research*, **104**, 23033–49.

Fossen, H. (1995). *Lisensekskursjon–Gullfaks–Høsten 1995 Ekskursjonsguide* (in Norwegian), Statoil, Bergen, Norway, 76 pp.

Gawthorpe, R. L. and Leeder, M. R. (2000). Tectono-sedimentary evolution of active extensional basins, *Basin Research*, **12**, 195–218.

Ghent, R. and Hansen, V. (1999). Structural and kinematic analysis of eastern Ovda Regio, Venus: implications for crustal plateau formation. *Icarus*, **139**, 116–36.

Ghent, R. R. and Tibuleac, I. M. (2002). Ribbon spacing in Venusian tesserae: implications for layer thickness and thermal state. *Geophysical Research Letters*, **29**, 2000, doi: 10.1029/2002GL015994.

Gibbs, A. D. (1984). Structural evolution of extensional basin margins. *Journal of the Geological Society of London*, **141**, 609–20.

Gibbs, A. D. (1990). Linked fault families in basin formation. *Journal of Structural Geology*, **12**, 795–803.

Golombek, M. P. (1979). Structural analysis of lunar grabens and the shallow crustal structure of the Moon. *Journal of Geophysical Research*, **84**, 4567–666.

Golombek, M. P. (1985). Fault type predictions from stress distributions on planetary surfaces: importance of fault initiation depth. *Journal of Geophysical Research*, **90**, 3065–74.

Golombek, M. P. and Banerdt, W. B. (1986). Early thermal profiles and lithospheric strength of Ganymede from extensional tectonic features. *Icarus*, **68**, 252–65.

Golombek, M. P. and McGill, G. E. (1983). Grabens, basin tectonics, and the maximum total expansion of the Moon. *Journal of Geophysical Research*, **88**, 3563–78.

Golombek, M. P., Tanaka, K. L., and Franklin, B. J. (1996). Extension across Tempe Terra, Mars, from measurements of fault scarp widths and deformed craters. *Journal of Geophysical Research*, **101**, 26119–30.

Grant, J. A. and Schultz, P. H. (1993). Erosion of ejecta at Meteor Crater, Arizona. *Journal of Geophysical Research*, **98**, 15033–47.

Grosfils, E. and Head, J. W. (1994). The global distribution of giant radiating dike swarms on Venus: implications for the global stress state, *Geophysical Research Letters*, **21**, 701–4.

Grosfils, E. B., Schultz, R. A., and Kroeger, G. (2003). Geophysical exploration within northern Devils Lane graben, Canyonlands National Park, Utah: implications for sediment thickness and tectonic evolution. *Journal of Structural Geology*, **25**, 455–67.

Groshong, Jr. R. H. (1989). Half-graben structures: balanced models of extensional fault-bend folds. *Geological Society of America Bulletin*, **101**, 96–105.

Gross, M. R., Gutiérrez-Alonso, G., Bai, T., Wacker, M. A., Collinsworth, K. B., and Behl, R. J. (1997). Influence of mechanical stratigraphy and kinematics on fault scaling relations. *Journal of Structural Geology*, **19**, 171–83.

Gudmundsson, A. (1992). Formation and growth of normal faults at the divergent plate boundary in Iceland. *Terra Nova*, **4**, 464–71.

Gudmundsson, A. and Bäckström (1991). Structure and development of the Sveinagja graben, Northeast Iceland. *Tectonophysics*, **200**, 111–25.

Gupta, S. and Cowie, P. (2000). Invited editorial: processes and controls in the stratigraphic development of extensional basins. *Basin Research*, **12**, 185–94.

Gupta, S., Cowie, P. A., Dawers, N. H., and Underhill, J. R. (1998). A mechanism to explain rift-basin subsidence and stratigraphic patterns through fault-array evolution. *Geology*, **26**, 595–8.

Hansen, V. L. and Willis, J. J. (1998). Ribbon terrain formation, southwestern Fortuna Tessera, Venus: implications for lithosphere evolution. *Icarus*, **132**, 321–43.

Hansen, V. L., Phillips, R. J., Willis, J. J., and Ghent, R. R. (2000). Structures in tessera terrain: issues and answers. *Journal of Geophysical Research*, **105**, 4135–52.

Hartmann, W. K. (1973). Ancient lunar megaregolith and subsurface structure. *Icarus*, **18**, 634–6.

Hauber, E. and Kronberg, P. (2001). Tempe Fosse, Mars: a planetary analog to a terrestrial continental rift? *Journal of Geophysical Research*, **106**, 20587–602.

Head, J. W. and Wilson, L. (1994). Lunar graben formation due to near-surface deformation accompanying dike emplacement. *Planetary & Space Science*, **41**, 719–27.

Higgs, W. G., Williams, G. D., and Powell, C. M. (1991). Evidence for flexural shear folding associated with extensional faults. *Geological Society of America Bulletin*, **103**, 710–17.

Hoek, E. and Brown, E. T. (1980). Empirical strength criterion for rock masses. *J. Geotech. Div. ASCE*, **106**, 1013–35.

Huntoon, P. W. (1982). The Meander anticline, Canyonlands, Utah: an unloading structure resulting from horizontal gliding on salt. *Geological Society of America Bulletin*, **93**, 941–50.

Jackson, J. A. and White, N. J. (1989). Normal faulting in the upper continental crust: observations from regions of active extension. *Journal of Structural Geology*, **11**, 15–36.

Jackson, M. P. A. (1995). Retrospective salt tectonics. In *Salt Tectonics: A Global Perspective*, ed. M. P. A. Jackson, D. G. Roberts, and S. Snelson. American Association of Petroleum Geologists Mem. 65, pp. 1–28.

Jackson, M. P. A. and Vendeville, B. C. (1994). Regional extension as a geologic trigger for diapirism. *Geological Society of America Bulletin*, **106**, 57–73.

Kattenhorn, S. A., Aydin, A., and Pollard, D. D. (2000). Joints at high angles to normal fault strike: an explanation using 3-D numerical models of fault-perturbed stress fields. *Journal of Structural Geology*, **22**, 1–23.

Kortz, K. M., Grosfils, E. B., and Sakimoto, S. E. H. (2003). Emplacement of long lava flows within a graben network in Radunitsa Labyrinthus, Carson quadrangle, Venus. *Geophysical Research Letters*, **30**, doi:10.029/2003GL017471.

Krantz, R. W. (1988). Multiple fault sets and three-dimensional strain: theory and application. *Journal of Structural Geology*, **10**, 225–37.

Krantz, R. W. (1989). Orthorhombic fault patterns: the odd axis model and slip vector orientations. *Tectonics*, **8**, 483–95.

Lewis, Sr. R. Q. and Campbell, R. H. (1965). Geology and uranium deposits of Elk Ridge and vicinity, San Juan County, Utah. US Geological Survey Professional Paper, 474–B.

Loope, D. B. (1984). Eolian origin of upper Paleozoic sandstones, southeastern Utah. *Journal of Sedimentary Petrology*, **54**, 563–80.

Lucchitta, B. K., McEwen, A. S., Clow, G. D. *et al.* (1992). The canyon system of Mars. In *Mars*, ed. H. H. Kieffer, B. M. Jakosky, C. W. Snyder, and M. S. Matthews. Tucson: Univ. Arizona Press, pp. 453–92.

Ma, X. Q. and Kusznir, N. J. (1992). 3-D subsurface displacement and strain fields for faults and fault arrays in a layered elastic half-space. *Geophysical Journal International*, **111**, 542–58.

Ma, X. Q. and Kusznir, N. J. (1993). Modelling of near-field subsurface displacements for generalized faults and fault arrays. *Journal of Structural Geology*, **15**, 1471–84.

MacKinnon, D. J. and Tanaka, K. L. (1989). The impacted Martian crust: structure, hydrology, and some geologic implications. *Journal of Geophysical Research*, **94**, 17359–70.

Mair, K., Frye, K. M., and Marone, C. (2002). Influence of grain characteristics on the friction of granular shear zones. *Journal of Geophysical Research*, **107**, 2219, doi:10.1029/2001JB000516.

Malin, M. C. and Edgett, K. S. (2000). Sedimentary rocks of Mars. *Science*, **290**, 1927–37.

Mangold, N., Allemand, P., and Thomas, P. G. (1998). Wrinkle ridges of Mars: structural analysis and evidence for shallow deformation controlled by ice-rich décollements. *Planetary & Space Science*, **46**, 345–56.

Manighetti, I., King, G. C. P., Gaudemer, Y., Scholz, C. H., and Doubre, C. (2001). Slip accumulation and lateral propagation of active normal faults in Afar. *Journal of Geophysical Research*, **106**, 13667–96.

Mastin, L. G. and Pollard, D. D. (1988). Surface deformation and shallow dike intrusion at Inyo Craters, Long Valley, California. *Journal of Geophysical Research*, **93**, 13221–36.

Masursky, H., Colton, G. W., and El-Baz, F. (eds.) (1978). *Apollo Over the Moon: A View from Orbit*. NASA Spec. Pap., SP–362, Washington, DC.

McEwen, A. S., Malin, M. C., Carr, M. H., and Hartmann, W. K. (1999). Voluminous volcanism on early Mars revealed in Valles Marineris. *Nature*, **397**, 584–6.

McGill, G. E. (1971). Attitude of fractures bounding straight and arcuate lunar rilles. *Icarus*, **14**, 53–8.

McGill, G. E. and Stromquist, A. W. (1974). A model for graben formation by subsurface flow: Canyonlands National Park, Utah. Department of Geology and Geography Contribution 15. Amherst: University of Massachusetts.

McGill, G. E. and Stromquist, A. W. (1975). Origin of graben in the Needles District, Canyonlands National Park, Utah. In *Canyonlands Country*, ed. J. E. Fassett, 8th Field Conference, Guidebook. Durango: Four Corners Geological Society, pp. 235–43.

McGill, G. E. and Stromquist, A. M. (1979). The grabens of Canyonlands National Park, Utah: geometry, mechanics, and kinematics. *Journal of Geophysical Research*, **84**, 4547–63.

McGill, G. E., Schultz, R. A., and Moore, J. M. (2000). Fault growth by segment linkage: an explanation for scatter in maximum displacement and trace length data from the Canyonlands Grabens of SE Utah: discussion. *Journal of Structural Geology*, **22**, 135–40.

McKenzie, D. and Nimmo, F. (1999). The generation of Martian floods by the melting of ground ice above dykes. *Nature*, **397**, 231–3.

McLeod, A. E., Dawers, N. H., and Underhill, J. R. (2000). The propagation and linkage of normal faults: insights from the Strathsprey-Brent-Statfjord fault array, northern North Sea. *Basin Research*, **12**, 263–84.

Mège, D. and Masson, P. (1996). A plume tectonics model for the Tharsis province, Mars. *Planetary & Space Science*, **44**, 1499–546.

Mège, D., Cook, A. C., Garel, E., Lagabrielle, Y., and Cormier, M.-H. (2003). Volcanic rifting at Martian graben. *Journal of Geophysical Research*, **108**, 5044, doi: 10.1029/2002JE001852.

Melosh, H. J. (1989). *Impact Cratering: A Geologic Process*. New York: Oxford University Press.

Melosh, H. J. and Williams, Jr. C. A. (1989). Mechanics of graben formation in crustal rocks: a finite element analysis. *Journal of Geophysical Research*, **94**, 13961–73.

Moore, J. M. and Schultz, R. A. (1999). Processes of faulting in jointed rocks of Canyonlands National Park, Utah. *Geological Society of America Bulletin*, **111**, 808–22.

Moore, J. M., Schultz, R. A., Grosfils, E. B. *et al.* (1997). The 1996 Canyonlands initiative: field study of small planetary grabens (abstract). *Lunar and Planetary Science*, **XXVIII**, 975–6.

Morley, C. K. (1999). Patterns of displacement along large normal faults: implications for basin evolution and fault propagation, based on examples from East Africa., *American Association of Petroleum Geologists Bulletin*, **83**, 613–34.

Morley, C. K., Nelson, R. A., Patton, T. L., and Munn, S. G. (1990). Transfer zones in the East African rift system and their relevance to hydrocarbon exploration in rifts. *American Association of Petroleum Geologists Bulletin*, **74**, 1234–53.

Nicol, A., Walsh, J. J., Watterson, J., and Bretan, P. G. (1995). Three-dimensional geometry and growth of conjugate normal faults. *Journal of Structural Geology*, **17**, 847–62.

Nicol, A., Watterson, J., Walsh, J. J., and Childs, C. (1996). The shapes, major axis orientations and displacement patterns of fault surfaces. *Journal of Structural Geology*, **18**, 235–48.

Odonne, F. and Massonnat, G. (1992). Volume loss and deformation around conjugate fractures: comparison between a natural example and analogue experiments. *Journal of Structural Geology*, **14**, 963–72.

Okubo, C. H. and Schultz, R. A. (2003). Thrust fault vergence directions on Mars: a foundation for investigating global-scale Tharsis-driven tectonics. *Geophysical Research Letters*, **30**, 2154, doi:10.1029/2003GL018664.

Okubo, C. H. and Schultz, R. A. (2004). Mechanical stratigraphy in the western equatorial region of Mars based on thrust fault-related fold topography and implications for near-surface volatile reservoirs. *Geological Society of America Bulletin*, **116**, 594–605.

Pappalardo, R. T. and Greeley, R. (1995). A review of the origins of subparallel ridges and troughs: generalized morphological predictions from terrestrial models. *Journal of Geophysical Research*, **100**, 18985–19007.

Peacock, D. C. P. (2002). Propagation, interaction and linkage in normal fault systems. *Earth Science Reviews*, **58**, 121–42.

Peacock, D. C. P. and Sanderson, D. (1991). Displacements, segment linkage and relay ramps in normal fault zones. *Journal of Structural Geology*, **13**, 721–33.

Peacock, D. C. P. and Sanderson, D. (1994). Geometry and development of relay ramps in normal fault systems. *American Association of Petroleum Geologists Bulletin*, **78**, 147–65.

Plescia, J. B. (1991). Graben and extension in northern Tharsis, Mars. *Journal of Geophysical Research*, **96**, 18883–95.

Pollard, D. D., Delaney, P. T., Duffield, W. A., Endo, E. T., and Okamura, A. T. (1983). Surface deformation in volcanic rift zones. *Tectonophysics*, **94**, 541–84.

Pollard, D. D. and Segall, P. (1987). Theoretical displacements and stresses near fractures in rock. with applications to faults, joints, dikes, and solution surfaces. In *Fracture Mechanics of Rock*, ed. B. K. Atkinson. New York: Academic Press, pp. 277–349.

Reches, Z. (1978). Analysis of faulting in three-dimensional strain field. *Tectonophysics*, **47**, 109–29.

Reches, Z. (1983). Faulting of rocks in three-dimensional strain fields II. Theoretical analysis. *Tectonophysics*, **95**, 133–56.

Roberts, A. and Yielding, G. (1994). Continental tectonics. In *Continental Deformation*, ed. P. L. Hancock. Pergamon, pp. 223–50.

Rosendahl, B. R. (1987). Architecture of continental rifts with special reference to East Africa. *Annual Review of Earth & Planetary Science*, **15**, 445–503.

Rubin, A. M. (1992). Dike-induced faulting and graben subsidence in volcanic rift zones. *Journal of Geophysical Research*, **97**, 1839–58.

Rubin, A. M. (1995). Propagation of magma-filled cracks. *Annual Review of Earth & Planetary Science*, **23**, 287–336.

Rubin, A. M. and Pollard, D. D. (1988). Dike-induced faulting in rift zones of Iceland and Afar. *Geology*, **16**, 413–17.

Schonfeld, E. (1979). Origin of Valles Marineris. *Proc. 10th Lunar Planetary Science Conference*, **11**, 3031–8.

Scott, D. H. and Tanaka, K. L. (1986). Geologic map of the western equatorial region of Mars. US Geol. Surv. Misc. Invest. Map I-1802-A.

Schultz, R. A. (1989). Do pit-crater chains grow up to be Valles Marineris canyons? (abstract). In *Proceedings of MEVTV Workshop on Tectonic Features on Mars*, ed. T. R. Watters, and M. P. Golombek, Lunar and Planetary Institute Technical Report **89–06**, pp. 49–50.

Schultz, R. A. (1991). Structural development of Coprates Chasm and western Ophir Planum, Valles Marineris, Mars. *Planetary & Space Science*, **43**, 1561–6.

Schultz, R. A. (1992). Studies of faulting on the Earth and other planets. *Trends in Geophysical Research,* **1**, 97–111.

Schultz, R. A. (1995). Gradients in extension and strain at Valles Marineris rift, Mars. *Journal of Geophysical Research*, **96**, 22777–92.

Schultz, R. A. (1997). Displacement-length scaling for terrestrial and Martian faults: implications for Valles Marineris and shallow planetary grabens. *Journal of Geophysical Research*, **102**, 12009–15.

Schultz, R. A. (1999). Understanding the process of faulting: selected challenges and opportunities at the edge of the 21st century. *Journal of Structural Geology*, **21**, 985–93.

Schultz, R. A. (2000). Fault-population statistics at the Valles Marineris Extensional Province, Mars: implications for segment linkage, crustal strains, and its geodynamical development. *Tectonophysics*, **316**, 169–93.

Schultz, R. A. (2003). Seismotectonics of the Amenthes Rupes thrust fault population, Mars. *Geophysical Research Letters*, **30**, 1303, doi:10.1029/2002GL016475.

Schultz, R. A. and Fori, A. N. (1996). Fault-length statistics and implications of graben sets at Candor Mensa, Mars. *Journal of Structural Geology*, **18**, 373–83.

Schultz, R. A. and Lin, J. (2001). Three-dimensional normal faulting models of Valles Marineris, Mars, and geodynamic implications. *Journal of Geophysical Research*, **106**, 16549–66.

Schultz, R. A. and Moore, J. M. (1996). New observations of grabens from the Needles District, Canyonlands National Park, Utah. In *Geology and Resources of the Paradox Basin*, ed. A. C. Huffman, Jr., W. R. Lund, and L. H. Godwin. Utah Geological Association Guidebook 25, pp. 295–302.

Schultz, R. A. and Watters, T. R. (2001). Forward mechanical modeling of the Amenthes Rupes thrust fault on Mars. *Geophysical Research Letters*, **28**, 4659–62.

Schultz, R. A. and Zuber, M. T. (1994). Observations, models, and mechanisms of failure of surface rocks surrounding planetary surface loads. *Journal of Geophysical Research*, **99**, 14691–702.

Schultz, R. A., Okubo, C. H., Goudy, C. L., and Wilkins, S. J. (2004). Igneous dikes on Mars revealed by Mars Orbiter Laser Altimeter topography. *Geology*, **32**, 889–92.

Schultz-Ela, D. D. and Walsh, P. (2002). Modeling of grabens extending above evaporates in Canyonlands National Park, Utah. *Journal of Structural Geology*, **24**, 247–75.

Schultz-Ela, D. D., Jackson, M. P. A., and Vendeville, B. C. (1993). Mechanics of active salt diapirism. *Tectonophysics*, **228**, 275–312.

Sibson, R. H. (1989). Earthquake faulting as a structural process. *Journal of Structural Geology*, **11**, 1–14.

Soderblom, L. A. and Wenner, D. B. (1978). Possible fossil H_2O liquid-ice interfaces in the Martian crust. *Icarus*, **34**, 622–37.

Soliva, R. (2004). Normal fault growth in layered rocks: the role of vertical restriction and linkage on scaling laws and fault spatial distribution. Ph.D. dissertation (in French), Université Paris-Sud, Orsay, France.

Soliva, R. and Benedicto, A. (2004). A linkage criterion for segmented normal faults. *Journal of Structural Geology*, **26**, 2251–67.

Squyres, S. W. and Carr, M. H. (1986). Geomorphic evidence for the distribution of ground ice on Mars. *Science*, **231**, 249–52.

Squyres, S. W., Clifford, S. M., Kuzmin, R. O., Zimbelman, J. R., and Costard, F. M. (1992). Ice in the Martian regolith. In *Mars*, ed. H. H. Kieffer, B. M. Jakosky, C. W. Snyder, and M. S. Matthews. Tucson: Univ. Arizona Press, pp. 523–54.

Stromquist, A. W. (1976). Geometry and growth of grabens, Lower Red Lake Canyon area, Canyonlands National Park, Utah, Department of Geology and Geography Contribution 28. Amherst: University of Massachusetts.

Tanaka, K. L. and Golombek, M. P. (1989). Martian tension fractures and the formation of grabens and collapse depressions in Valles Marineris. *Proc.19th Lunar Planetary Science Conference*, pp. 383–96.

Tanaka, K. L., Golombek, M. P., and Banerdt, W. B. (1991). Reconciliation of stress and structural histories of the Tharsis region of Mars. *Journal of Geophysical Research*, **96**, 15617–33.

Thomas, P. G. and Allemand, P. (1993). Quantitative analysis of the extensional tectonics of the Tharsis bulge, Mars: geodynamic implications. *Journal of Geophysical Research*, **98**, 13097–108.

Trudgill, B. and Cartwright, J. (1994). Relay-ramp forms and normal-fault linkages, Canyonlands National Park, Utah. *Geological Society of America Bulletin*, **106**, 1143–57.

Turcotte, D. L. and Schubert, G. (1982). *Geodynamics: Applications of Continuum Physics to Geological Problems*. Wiley.

Urrita-Fucugauchi, J., Marin, L., and Trejo-Garcia, A. (1996). UNAM scientific drilling program of Chicxulub impact structure: evidence for a 300 kilometer crater diameter. *Geophysical Research Letters*, **23**, 1565– 8.

Vendeville, B. C. and Jackson, M. P. A. (1992). The rise of diapirs during thin-skinned extension. *Marine Petroleum Geology*, **9**, 331–53.

Walsh, J. J. and Watterson, J. (1987). Distributions of cumulative displacement and seismic slip on a single normal fault surface. *Journal of Structural Geology*, **9**, 1039–46.

Walsh, P. and Schultz-Ela, D. D. (2003). Mechanics of graben evolution in Canyonlands National Park, Utah. *Geological Society of America Bulletin*, **115**, 259–70.

Watters, T. R., Schultz, R. A., Robinson, M. S., and Cook, A. C. (2002). The mechanical and thermal structure of Mercury's early lithosphere. *Geophysical Research Letters*, **29**, doi:10.1029/2001GL014308.

Watterson, J., Nicol, A., Walsh, J. J., and Meier, D. (1998). Strains at the intersections of synchronous conjugate normal faults. *Journal of Structural Geology*, **20**, 363–70.

Weissel, J. K. and Karner, G. D. (1989). Flexural uplift of rift flanks due to mechanical unloading of the lithosphere during extension. *Journal of Geophysical Research*, **94**, 13919–50.

Wilkins, S. J. and Gross, M. R. (2002). Normal fault growth in layered rocks at Split Mountain, Utah: influence of mechanical stratigraphy on dip linkage, fault restriction and fault scaling. *Journal of Structural Geology*, **24**, 1413–29.

Wilkins, S. J. and Schultz, R. A. (2003). Cross faults in extensional settings: stress triggering, displacement localization, and implications for the origin of blunt troughs in Valles Marineris, Mars. *Journal of Geophysical Research*, **108**, 5056, doi:10.1029/2002JE001968.

Wilkins, S. J., Schultz, R. A., Anderson, R. C., Dohm, J. M., and Dawers, N. C. (2002). Deformation rates from faulting at the Tempe Terra extensional province, Mars. *Geophysical Research Letters*, **29**, 1884, doi: 10.1029/2002GL015391.

Willemse, E. J. M. (1997). Segmented normal faults: correspondence between three-dimensional mechanical models and field data. *Journal of Geophysical Research*, **102**, 675–92.

Williams, J. K., Paige, D. A., and Manning, C. E. (2003). Layering in the wall rock of Valles Marineris: intrusive and extrusive magmatism. *Geophysical Research Letters*, **30**, 1623, doi: 10.1029/2003GL017662.

Wilson, L. and Head, J. W. (2002). Tharsis-radial graben systems as the surface manifestation of plume-related dike intrusion complexes: models and implications. *Journal of Geophysical Research*, **107**, doi: 10.1029/2001JE001593.

Wise, D. U. (1976). Faulting and stress trajectories near Alba volcano, northern Tharsis ridge of Mars. *Geol. Romana*, **15**, 430–3.

Wise, D. U., Golombek, M. P., and McGill, G. E. (1979). Tharsis province of Mars: geologic sequence, geometry, and a deformation mechanism. *Icarus*, **38**, 456–72.

Wise, D. U., Golombek, M. P., and McGill, G. E. (1982). Tectonic evolution of Mars. *Journal of Geophysical Research*, **84**, 7934–9.

Woodward-Clyde Consultants (1983). Overview of the regional geology of the Paradox Basin study region. Battelle Memorial Institute Office of Nuclear Waste Isolation, ONWI-92.

16

Geochemical analogs and Martian meteorites

Horton E. Newsom
Institute of Meteoritics and Department of Earth and Planetary Sciences, University of New Mexico, Albuquerque

16.1 Introduction

The formation and evolution of Mars involved both physical and chemical processes that are revealed in the chemistry of the Martian meteorites, and in the chemistry of the surface of Mars determined by remote sensing from spacecraft in orbit and on the surface. The interpretation of the chemistry revealed by these studies has been strongly influenced by our knowledge of geochemical processes on the Earth, Moon, and asteroidal parent bodies. In a sense, the entire Earth, Moon, and a number of asteroid parent bodies can be considered Mars analogs! The most studied differentiated body (melted and chemically evolved) from the asteroid belt is the parent body of the Howardite, Eucrite, and Diogenite (HED) igneous meteorite classes, thought to be the asteroid 4 Vesta (Mittlefehldt *et al.*, 1998). These HED meteorites are igneous rocks that are basaltic in nature with slightly different mineral assemblages (McSween, 1999). In this chapter we use data from samples on the Earth including the meteorites from the HED parent body and the Martian meteorites to understand the chemical fractionations that have affected Martian rocks and surface materials. These chemical fractionations are the changes in chemistry due to the different behavior of particular groups of chemical elements according to their properties. We will begin by looking at the evidence for the formation of Mars, the early differentiation of the planet, the later formation of igneous rocks by mantle melting, and end with surface processes leading to formation of the Martian fine-grained regolith OR soils.

Martian meteorites provide our best geochemical information about Mars (Figure 16.1). How do we know that these meteorites are from the same parent

The Geology of Mars: Evidence from Earth-based Analogs, ed. Mary Chapman. Published by Cambridge University Press.

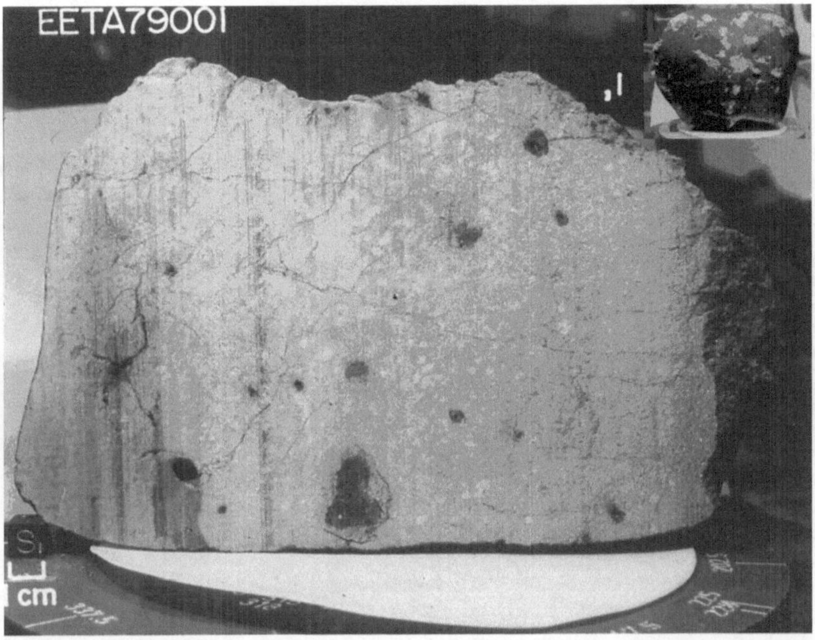

Figure 16.1. The Martian meteorite EETA 79001 was found in Elephant Moraine, Antarctica in 1979. This meteorite is a basaltic rock of the Shergottite class of Martian meteorites. The inset shows the original sample with its fusion crust, while the larger image shows a sawn face and the igneous texture. The dark areas on the cut face are impact melt pockets that contain trapped Martian atmosphere whose composition is the strongest evidence for a Martian origin for this meteorite. (For a color version of this figure, please refer to color plate section.)

body and that this parent body is Mars? These meteorite groups have similar oxygen isotopic compositions linking their origin to the same parent body. The first clue to their possible Martian origin was the young crystallization ages (Rb–Sr and Sm–Nd ages <3.6 Gyr) of most of the rocks compared to samples from asteroids, which requires a relatively large planet such as Mars (Nyquist *et al.*, 2001). The second clue came from measurements of the abundances and isotopic ratios of trace gases in shock veins in some meteorites that match abundances measured by mass spectrometers on the Viking landers. Four different types of Martian meteorites have been found on the Earth so far (McSween and Treiman, 1998). The Shergottite group consists of mantle-derived basaltic or ultramafic rocks and these rock types are important for comparisons between the planets. Recently the Opportunity rover on Mars analyzed the rock "Bounce" (Figure 16.2), which has a composition identical to the ALH 79001 Shergottite (Figure 16.1). Other types of Martian meteorites are the Nakhlites, which are rocks that

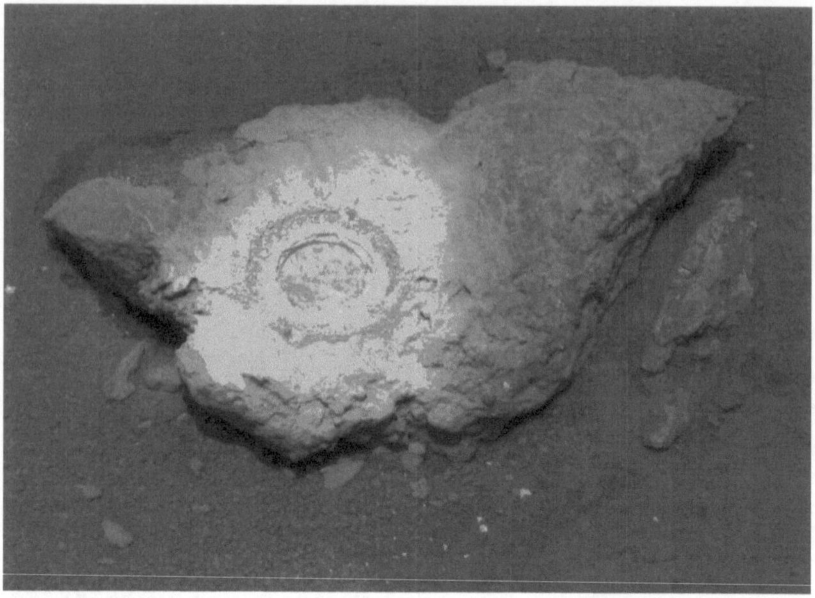

Figure 16.2. The Mars Exploration Rover Opportunity studied a rock dubbed 'Bounce' as shown in this false-color composite taken on sol 68. The 40 cm long rock was drilled to a depth of 7 mm by the rover's rock abrasion tool. The chemical composition of this sample measured by the Rover's Alpha Particle X-ray Spectrometer is nearly identical to the Martian meteorite EETA 79001 illustrated in Figure 16.1. (For a color version of this figure, please refer to color plate section.)

formed by the accumulation of minerals crystallizing from magmas. These samples are important because they have been shocked and then altered by aqueous fluids before ejection from Mars. The Chassignites are cumulate rocks consisting primarily of olivine, which probably formed by the concentration of olivine sinking to the bottom of a magma chamber. The only orthopyroxenite is the ALH 84001 meteorite that formed by accumulation of minerals or by recrystallization or both processes. The name "ALH 84001" is based on the location where the sample was found in Antarctica, Allan Hills, and the numbers indicates that this sample was the first sample studied from the samples found in 1984. This meteorite is important for several reasons: it formed very early in the history of Mars at 4510 million years (Ma) ago (Nyquist *et al.*, 2001); and it also experienced deposition of minerals from fluids and has been intensively studied for evidence of the presence of life.

The similar behavior of different groups of elements during geochemical processes involving the formation of the Earth was first recognized by the geochemist Victor Goldschmidt (Mason, 1992). In 1922 Goldschmidt divided the chemical elements into groups that include the siderophile, chalcophile, lithophile, and atmophile elements. Siderophile elements such as Ni and Co have an affinity for iron-metal, while chalcophile elements such as Cu, Zn, and S have an affinity for sulfide minerals. The lithophile elements are those elements such as Si, Mg, Ca, and Al that have an affinity for silicate liquids (e.g., magma). The atmophile elements, including N, Ne, and Ar, are concentrated in the atmosphere. During planetary formation these groups of elements are largely separated from each other; with siderophile and chalcophile elements in the core, lithophile elements in the crust and mantle, and atmophile elements in the atmosphere. Goldschmidt also recognized the different behavior of chemical elements during igneous fractionation processes that occur as a function of their ionic radii. In general, elements that are enriched in silicate liquids during crystallization are called the incompatible elements, while elements that are concentrated in the common igneous minerals (e.g., pyroxene, plagioclase, and olivine) are called compatible. Incompatible elements, such as potassium, titanium, and the rare earth elements, generally have too a large an ionic size that causes them to concentrate in melts as they are excluded from most minerals. In contrast, the compatible elements such as silicon and magnesium are concentrated in the residue of melting, such as the source regions for basaltic magmas in planetary mantles. In addition to changing the ratios of different groups of elements, chemical fractionations can change parent-to-daughter ratios of radiogenic systems, such as U–Pb, Th–Pb, W–Hf, Rb–Sr, and Sm–Nd. The resulting isotopic shifts that build up over time provide an additional tool to constrain the magnitude and timing of chemical fractionations. Systems involving extinct radionuclides that were only present in the early solar system provide additional constraints on the timing of the fractionations. An example is the ^{182}W–^{182}Hf system, which can constrain the timing of metal silicate fractionation or core formation in the planets. The relative abundances of the different groups of elements in samples from different bodies provide clues to the different types of processes that have occurred on these bodies. In general the standard against which most of the fractionations are measured is the abundances of the different groups of elements in carbonaceous chondrites. The carbonaceous chondrites are assumed to be similar to the original material from which the planets formed. Interestingly, in the available samples from all planetary objects, the volatile elements and siderophile elements are depleted relative to carbonaceous chondrites.

Based on geochemistry of Martian meteorites, we know that Mars formed within a few million years of the formation of the first solid material in the solar nebula. After separation into core and mantle an early crust formed which, like the Moon, was battered by numerous crater-forming asteroids and comets. During the early history of the planet water formed large channels, probably supplied by precipitation in the form of rain or snow. Later, giant floods of groundwater burst forth and created rivers and lakes for brief periods of time. Meanwhile, large and small volcanoes deposited lava flows in many areas of Mars. At present, volcanism may still be somewhat active, and there is evidence for the presence of near-surface water in the form of ice and possibly shallow groundwater in some areas. Much of the surface is covered with a thin regolith or "soil," that is roughly basaltic in nature, that formed by alteration and erosion of the surface rocks on Mars. Martian regolith differs from terrestrial soils in being devoid of organic carbon.

16.2 Accretion of Mars

The geochemical evolution of Mars appears to be closer in nature to smaller-sized bodies, such as the parent bodies of the differentiated meteorites and the Moon than to the Earth. Our information about the bulk composition and early differentiation of Mars comes from the geochemistry of Martian meteorites and geophysical constraints on the mass and density of Mars. The geochemical processes that resulted in the Martian meteorites can be compared with the processes that formed differentiated rocks from asteroidal parent bodies, from lunar samples, and from differentiated terrestrial rocks. A major handicap in studying the bulk composition and geochemical evolution of planetary bodies other than the Earth is the lack of samples of the mantles of these bodies. Mantle samples are important because they contain most of the compatible lithophile elements in a planet. The mantle also represents the starting material for the formation of the crustal rocks by melting and differentiation.

Fractionations in the solar nebula established the differences in the chemical makeup of the material that formed the planets. The solar nebula is the cloud of dust and gas that collapsed into a disk and into the proto-sun forming our Solar System. This collection of dust formed into small millimeter-sized dust balls and bodies called chondrules, which accreted into small meter to kilometer-sized planetesimals that made up the building blocks from which the planets formed. We know from the many different kinds of primitive chondritic meteorites (meteorites from asteroids that did not melt) that there are significant variations in the initial abundances of certain elements in different bodies, for example the ratios of volatile elements to refractory elements. At high temperatures, such as during heating and condensation in the solar nebula, volatile elements, such as Na, K, and Rb, will remain largely in the gas, while refractory elements, such as Al, and the rare earth elements, remain as solids.

One of the ways we compare the nature of Mars with the meteorite parent bodies and the Moon and Earth is by comparison of the abundance of elements in similar kinds of rocks, in particular basalt. For comparison with the basaltic Martian meteorites we have basalts from Earth, from the Moon, and from asteroid parent bodies. The normal way of comparing abundances is to divide the abundances in the planetary material by the abundances in the CI carbonaceous chondrites, which represent the initial abundances in the material from which the solar system formed. Because we are considering planets that differentiated we want to compare the abundances in the silicate portions of the planets; therefore we also need to correct for the loss of metal due to core formation in the planets. This correction is usually done by also normalizing the depletion factors to the silicon (refractory element) abundance in the planetary material compared to the abundance in the chondrites.

An examination of volatile elements measured in Martian meteorites provides a record of volatile loss relative to carbonaceous chondrites that can be attributed to nebular processes (Figure 16.3). Compared to Earth and Vesta, Mars is less depleted in Na, K, Rb, Cs, Br, and Zn.

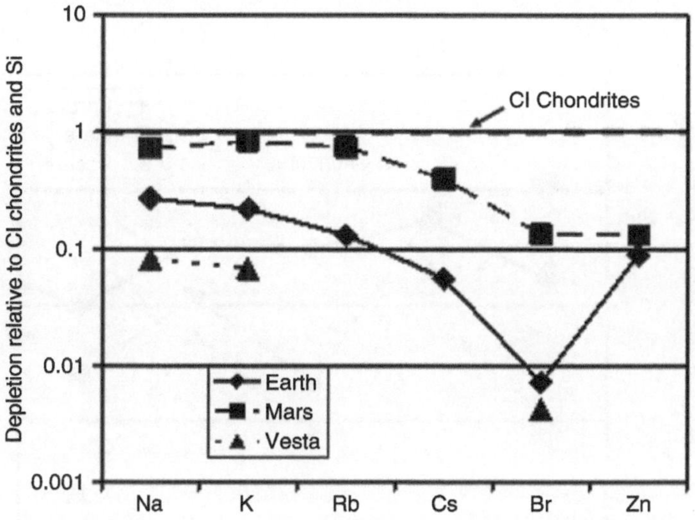

Figure 16.3. Depletion relative to CI chondrites and Si of volatile elements in the Earth, Mars, and Vesta (data compiled in Lodders and Fegley 1998; McDonough and Sun, 1995; Lodders and Fegley, 1997; Dreibus and Wänke, 1980).

Mars apparently accreted from a more volatile-rich material than the other planets. The possibility exists, however, that some compositional differences between planets are due to large impacts very early in solar system history.

16.3 Major fractionations after accretion

16.3.1 Core formation

After accretion, the next important fractionation in the planets was due to the process of core formation, which occurs when the denser metal phases sink to form a core during melting of a planet. The presence of an iron core in Mars can be inferred from density and the depletion of iron-loving siderophile elements in the meteorites. These elements include Fe, Ni, Co, and other siderophile elements which are carried into the core along with the Fe–Ni metal phases during melting of the planets. The depletion of the moderately siderophile elements P, W, Ni, Co, and Mo is shown in Figure 16.4. Mars is less depleted in such elements as P, Mn, Cr, and W than is the Earth, perhaps because of more oxidizing conditions, and a more sulfur-rich metal composition during metal silicate fractionation (Halliday *et al.*, 2001). These abundances also are reflected in the composition of common minerals as well. For example, the Fe/Mn ratio in pyroxenes is 30 for Mars and 32 for the eucrite meteorites from Vesta, but 42 for Earth and 60 for the Moon (Papike *et al.*, 2003). The magnitudes of the siderophile

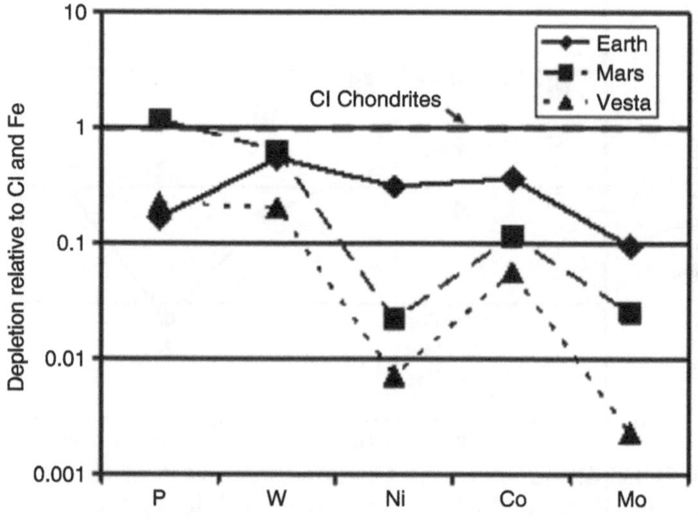

Figure 16.4. Depletion of siderophile elements in the Earth, Mars, and Vesta (data compiled in Lodders and Fegley 1998; Righter and Drake, 1996).

element depletions are controlled by temperature and pressure during equilibration and by the composition of the metallic phase. The large depletions of siderophile elements in samples from Vesta are consistent with metal–silicate equilibration at low pressures in a very reducing (low oxygen fugacity) environment. The relatively high abundances of most siderophile elements in the Earth compared to Mars and Vesta are probably due to equilibration at the base of a magma ocean at high temperatures (Walter *et al.*, 2000). The core of Mars probably contains more sulfur, and pressures and temperatures were lower during core separation, which probably also occurred in a magma ocean (Righter and Drake, 1996).

Some siderophile elements belong to radioactive decay systems that also provide constraints on the nature of core formation. Thanks to the development of a new radiogenic system involving the decay of ^{182}Hf (half-life 9 Myr) to ^{182}W, the age of core formation in the planets can now be measured. The estimated age of the Solar System of ~4575 Myr based on U–Pb dating marks the time when the first high-temperature condensates (Ca–Al-rich meteorite inclusions, CAIs) were starting to form (Nyquist *et al.*, 2003). The W–Hf ages for core formation have Vesta forming 3–4 Myr after CAI formation, Mars at 13 $\pm$ 2 Myr, and the Earth at 33 $\pm$ 2 Myr (Kleine *et al.*, 2002; Yin *et al.*, 2002).

16.3.2 Mantle and crust evolution on Mars

The nature of the Martian mantle is only known by geochemical and geophysical models. Cosmochemical principals based on the compositions of primitive meteorites also constrain the bulk composition of Mars (Dreibus and Wänke, 1982). The basaltic crustal rocks of Mars can be compared to basaltic crustal rocks from the Earth, Moon, and 4 Vesta. The major element geochemistry of the basaltic shergottites suggests that the Martian crust formed by differentiation of the mantle of Mars. The Earth's basaltic oceanic crust and the basaltic rocks (HED meteorites) from 4 Vesta also formed by mantle differentiation. However, different mechanisms resulted in the formation of the Moon's crust by accumulation of feldspar from a magma ocean, and the Earth's continental crust, which formed due to subduction zone processes. On the Moon formation of the crust from a magma ocean resulted in a large contrast between the major element composition of the highlands crust, rich in Ca, and the mare basalts derived from the lunar mantle, rich in Mg and Ti (Heiken *et al.*, 1991). These large compositional differences do not exist on Mars. The chemistry of Martian meteorites also does not reflect a substantial evolution to silicon-rich compositions, as seen in the

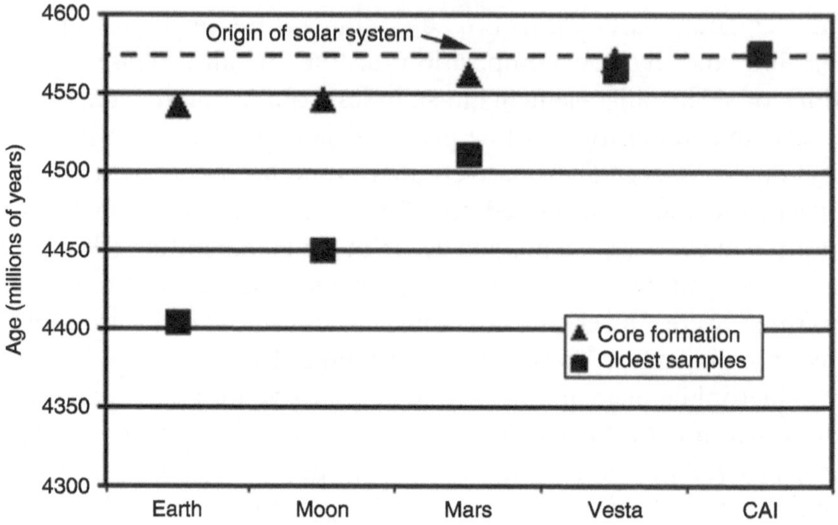

Figure 16.5. Core formation ages from W–Hf constraints and oldest rock ages for Earth, Moon, Mars, and Vesta (see text). The origin of the Solar System is taken as the formation age of the Ca- and Al-rich inclusions in carbonaceous chondrites that formed in the solar system at 4575 Ma (Nyquist *et al.*, 2003).

Earth's continental crust. The controversial identification of "basaltic andesite" rocks in large areas by their spectral signature (Bandfield *et al.*, 2000) is hard to reconcile with any plausible igneous process on Mars, and may represent altered basaltic rocks (Wyatt and McSween, 2003).

Isotopic systems and incompatible trace elements also provide clues to the timing of core formation and crustal formation processes (Figure 16.5). The HED meteorites from 4 Vesta represent the oldest planetary crust, which has an age of 4565 Ma. This age is only about 10 Ma younger than the first solid material (the Ca- and Al-rich inclusions in carbonaceous chondrites) that formed in the solar system at 4575 Ma (Nyquist *et al.*, 2003). The Moon also has an ancient crust, with anorthositic rocks dating back to 4450 Ma (Nyquist and Shih, 1992). Earth's oldest rocks date close to 4000 Ma, but zircon minerals found in Western Australia from the early crust date back to 4404 Ma (Wilde *et al.*, 2001). Earth's present crust presumably began forming shortly after a giant impact created the Moon. Much of the Martian crust is also very old, based on the large crater density (e.g., Frey, 2003), and the ancient age (4510 Ma) of the ALH84001 Martian meteorite (Nyquist *et al.*, 2001).

Although the major elements suggest a similarity to the rocks from Vesta, in the case of the incompatible elements Mars seems to have more similarities to the Moon (Borg *et al.*, 2002, 2003). With the exception of the very old ALH 84001, Martian meteorites are relatively young, with most between 175 and 475 Ma old (Nyquist *et al.*, 2001). Basaltic Martian meteorites, however, retain evidence of the formation of ancient mantle reservoirs in the source regions for the meteorites. The chemical compositions of the basaltic Martian meteorites reflect mixing between magmas from a depleted mantle reservoir, and an incompatible element-enriched reservoir. Isotopic systems in Martian meteorites, including Rb–Sr and Sm–Nd, indicate that distinct reservoirs were created by a differentiation event in the Martian mantle shortly after formation of the planet. However, these mantle differentiation events largely fractionated trace elements, such as the rare earth elements (REEs), rather than the major elements. The basaltic Martian meteorites have a large range of trace element and isotopic compositions reflecting distinct source regions (Borg and Draper, 2003). Meteorites at one end of the continuum, such as QUE 94201, have REE patterns with light REE depletions and radiogenic isotopic compositions suggesting derivation from a source region with long-term depletion in incompatible elements. Meteorites such as Shergotty, however, have flat REE patterns and radiogenic isotopic compositions suggesting derivation from a source with long-term enrichments in incompatible elements. Because these ancient source regions formed very early (~4.52 AE) there is a drastically larger spread in isotopic compositions among the Martian meteorites than observed in rocks on Earth (Borg *et al.*, 2003). This differentiation event may have involved a magma ocean, but crystallization of a magma ocean on Mars did not produce the extreme major element fractionation observed on the Moon, as represented by the contrast between the feldspar-rich highland crust and the Mare basalts. On Earth, mantle convection and recycling of the crust back into the mantle at subduction zones has led to extensive homogenization and resetting of radiogenic systems.

Another feature of the Martian meteorites is apparently inherited from the incompatible element-enriched source regions. There is evidence from geochemical correlations with oxygen fugacity that the enriched component is hydrous in nature (Herd, 2003). This enriched reservoir may be roughly analogous to KREEP components on the Moon. KREEP (potassium, rare earth elements and phosphorus) is a group of incompatible elements that were enriched in the last dregs of the solidifying magma ocean on the Moon. Assimilation or incorporation of rocks containing enriched KREEP-like material into rising magmas on Mars can help explain their geochemical properties (Borg *et al.*, 2002; 2003; Borg and Draper, 2003).

16.4 Chemical alteration processes on Mars and the trapping of water in the Martian crust

Alteration processes involving liquid water are ubiquitous on planets containing water and result in the formation of water-bearing minerals such as clays and zeolites. Some types of hydrothermal minerals on the Earth can retain signatures of biological activity, including silica and iron oxides (Allen *et al.*, 2000). There is even substantial evidence for aqueous processes resulting in hydrous minerals on small bodies, from spectral studies of asteroids such as Ceres, and evidence of aqueous alteration in many chondritic meteorites (Brearley and Jones, 1998). Evidence for aqueous and hydrothermal alteration processes on Mars from Martian meteorites includes carbonate deposits in ALH 84001, dated at 3900 Ma (Borg *et al.*, 1999), and the presence of hydrous alteration minerals, such as the iddingsite in Lafayette (Figure 16.6) and other Nakhlites (Bridges *et al.*, 2001). Possible evidence of hydrothermal activity on Mars obtained by remote sensing studies includes both infrared spectral observations (Geissler *et al.*, 1993; Bandfield, 2002; Bishop *et al.*, 2002), and the detection of the iron oxide mineral hematite by the Mars Global Surveyor (Christensen *et al.*, 2001, 2003). The remarkable observations of large-scale hematite concretions and massive sulfate evaporite deposits in the outcrops at Meridiani Planum observed by the Opportunity rover have no analogs in known Martian meteorites, although these minerals

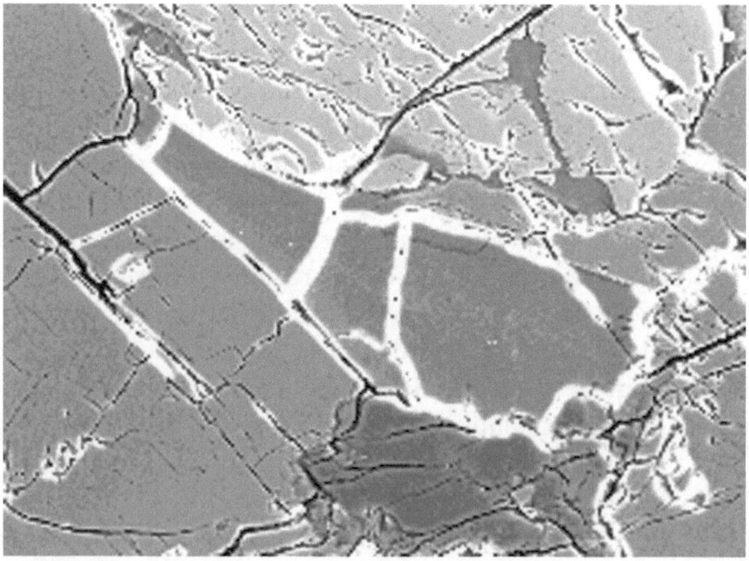

Figure 16.6. Backscattered electron image of alteration material called "iddingsite" in the Lafayette Martian meteorite.

have been observed in Martian meteorites (as summarized in Banin *et al.*, 1992).

Studies of terrestrial analogs for aqueous alteration processes are providing new constraints on Martian processes (Bishop *et al.*, 2002; Hagerty and Newsom, 2003). Volcanism and impacts are generally recognized to have affected most or all Solar System objects, including Mars, and to have resulted in substantial hydrothermal alteration. Hydrothermal systems associated with volcanoes are well known on Earth and could have played a role on Mars (e.g., Gulick, 1998). In the case of large volcanic centers, hydrothermal systems can plausibly explain the presence of channels on some Martian volcanoes (Gulick, 1998). During the later part of Martian history, when cratering rates were very low, volcanism can continue to provide substantial heat needed to drive hydrothermal systems (Wilson and Head, 1997).

The amount of impact melt produced in Martian impact craters has been estimated by Clifford (1993). The heat in a large impact crater available to power hydrothermal processes includes contributions from the impact melt sheet and the central uplift (Newsom *et al.*, 1996). The magnitude of heat available from the uplift in large craters on Mars is approximately equivalent to the amount of heat available from the impact melt (Thorsos *et al.*, 2001). Studies of terrestrial analog impact craters document evidence for crater lakes and impact-induced hydrothermal systems. Terrestrial craters with

Figure 16.7. Tilted basalt flows on the East side of the Lonar crater. The 1.8 km diameter Lonar crater in India is located within the Deccan Trap basalt sequences. Lonar is the only well-studied crater emplaced in basaltic target rocks.

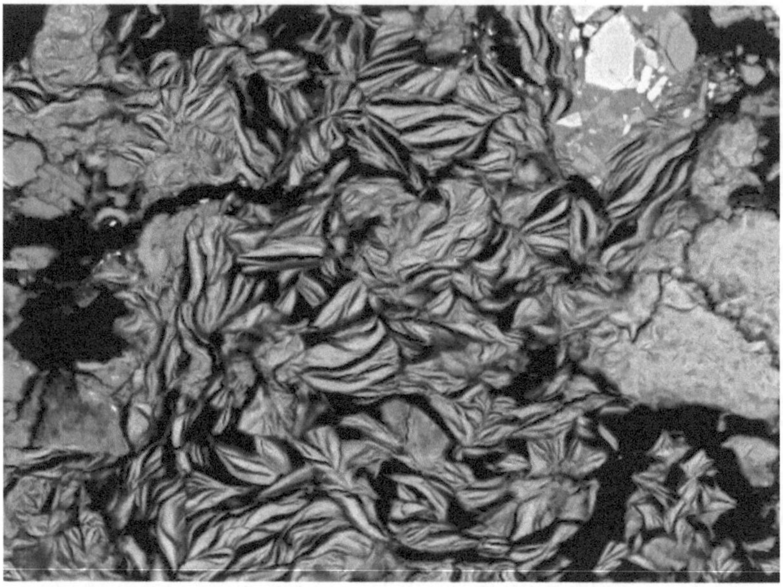

Figure 16.8. Backscattered electron image of "felty" clays in thin section NMNH 116569–13 (drill core LNR-2). Microprobe analyses suggest that this clay is saponite.

evidence for hydrothermal alteration range in diameter from Lonar (2 km), Haughton (20 km), Ries (22 km), and Manson (35 km) craters, up to the very large Sudbury (>140 km) and Vredefort (>250 km) structures. The 1.8 km diameter Lonar crater (Figure 16.7) was emplaced in the Deccan Trap basalt flows that are similar in composition to Martian rocks. At Lonar, rocks below the crater floor experienced impact-induced alteration (Figure 16.8) that led to the production of clay minerals (Fe-saponite, celadonite, glauconite), and carbonate (Hagerty and Newsom, 2003). Basaltic alteration products are of interest because spectral reflectance studies of the Martian surface indicate that the regolith composition is consistent with oxidized iron minerals, palagonite (an amorphous clay-like alteration product), and several members of the montmorillonite clay group (e.g., saponite and nontronite; Morris *et al.*, 2001; Bishop *et al.*, 2002).

One effect that hydrothermal systems may have had on Mars is to trap water in the subsurface in the form of hydrous minerals, such as clays (Griffith and Shock, 1997). Formation of hydrothermal systems during intrusion of the Martian crust by large igneous bodies could also be reflected in the correlation between valley networks and magnetic anomalies (Harrison and Grimm, 2002). The process of aqueous alteration can occur under conditions of either high or low water/rock ratios. Griffith and Shock (1997) estimated that 8% alteration of 10% of the Martian crust could trap 30 m global equivalent of water. This compares to estimates for the total water content on a global basis equivalent to several hundred meters up to 2700 m (Masson *et al.*, 2001; Lunine *et al.*, 2003). Hydrothermal alteration can also occur due to the interaction of water with rock heated by impact events. As an example, the amount of water trapped due to the formation of hydrous alteration minerals in craters on Mars in the size range from 2 to 11 km in diameter was calculated by Hagerty and Newsom (2003). Alteration of these craters over Martian history results in a volume of hydrous alteration phases equivalent to a global layer of 2.8 m (assuming an average depth of alteration of 400 m, and 3 vol.% alteration based on the average of SEM feature scan determinations of alteration in the Lonar crater). Assuming 10 wt% water in the alteration phases (similar to the amount in Lafayette Martian meteorite iddingsite alteration material; Figure 16.6) the 2.8 m of alteration material could trap an amount of water equivalent to a layer of water 0.7 m deep. Impact craters larger than 11 km could result in even a larger amount of trapped water because of their much greater heat sources, which can alter a proportionately larger volume of material.

16.5 Origin of the Martian soil

Among the most interesting comparisons between deserts on the Earth and Mars is the nature of their surface processes, and the presence of a fine-grained surficial regolith or "soil." The justification for using the term soil for the fines on Mars according to Banin *et al.* (1992) lies in the origin of the fines by weathering due to physical (e.g., freeze–thaw) and chemical hydrolytic processes. In contrast, the lunar regolith formed by unique impact processes was never exposed to atmospheric effects and never weathered by processes involving water. Anyone looking at images of the Martian landscapes from successful landers and rovers can see recognizable features, including rocks, pebbles, small dunes, and evidence of windblown dust. Many people living in the desert southwest have these features in their own backyards. Because the focus of this article is on chemical analogs, we will approach the processes responsible for the formation of the fine-grained regolith from a chemical point of view.

Abundant chemical information about the major and minor element compositions of Martian soils was provided by analyses from two Viking landers in widely separated locations (Clark *et al.*, 1982; Clark, 1993) and from analyses by the Pathfinder mission (Economou *et al.*, 2003). The composition of Martian soil is similar to the composition of Shergottite basaltic Mars meteorites, with enrichments in the mobile elements sulfur and chlorine (Clark, 1993; McSween and Keil, 2000; Wänke *et al.*, 2001; Economou *et al.*, 2003). Preliminary results from the MER rover missions suggest that the soils at Gusev and Meridiani are also basaltic in composition, with variations in the abundances of mobile elements such as S and Cl. There is some limited evidence for Martian fines trapped in Martian meteorites by shock effects (Rao *et al.*, 1999). The impact melt glass in EETA 79001 Lithology C is enriched in sulfur but not chlorine, suggesting the presence of 6% Martian soil. Isotopic studies of Martian meteorites also contain evidence for the incorporation of Martian weathering products (e.g., Nyquist *et al.*, 2001).

The Martian soil or regolith is a globally homogenized product of chemical and eolian weathering processes. The composition of Martian soil can be divided into several distinct components, the "basaltic rock" component, the mobile element component (S and Cl), the magnetic component (iron oxides), and probably a small meteoritic component (Fe, Ni, and trace siderophile elements). A number of different theories have been presented over the years to explain the origins of different soil components, but most have focused on the breaking down of surface materials into soil constituents (e.g., Gooding and Keil, 1978; Banin *et al.*, 1997; Blaney, 1998; McSween and Keil, 2000). The mobile element component of the soil is usually assumed to be due to volcanic aerosols, although hydrothermal processes could contribute. The origin of other distinct components in the soil, the meteoritic material, and the magnetic component, are usually considered independently. Unfortunately, our knowledge of the mineralogy (crystalline structure) of the Martian soil components is very poor. We are anxiously awaiting mineralogical information from the Mössbauer and Mini-TES experiments on the MER rovers and near-infrared data from the orbital sensors on Mars Express, and ultimately from studies of samples returned from Mars.

16.5.1 Rock component of the soil

Formation of the Martian soil by alteration of volcanic or impact glass to form palagonite has been discussed over the years by Gooding and Keil (1978), Allen *et al.* (1981), Singer (1982), Bouska and Bell (1993), and Bell *et al.* (2000). Palagonite is an amorphous or microcrystalline gel-like substance produced by a hydrolytic dissolution/precipitation mechanism that forms well-defined surface layers. Gooding and Keil (1978) showed that glass would be thermodynamically unstable on the Martian surface, even under current conditions. Boslough and Cygan (1988) showed how shocked minerals are also susceptible to alteration. Alteration of surface materials also can be accelerated by an acid-fog mechanism, connected with the addition of volcanic aerosols if sufficient water is present to form a thin film (Banin *et al.*, 1997). A number of relevant terrestrial analog studies have looked at palagonite from different localities, especially Hawaii (Morris *et al.*, 2000, 2001; Bishop *et al.*, 2002). A recent analysis of the data from Pathfinder together with data from Viking is consistent with a basaltic rock component affected by palagonitization (McSween and Keil, 2000).

Contributions to the "rock" component of the soil by hydrothermal alteration of the basaltic crust was proposed by Newsom (1980) and further considered by Newsom and Hagerty (1997). However, limited evidence for clay minerals in the soil has lent support for other models. McSween and Keil (2000) investigated the fractionation of major elements between Mars rocks and soils in an effort to distinguish between different processes that contributed to the soil. They showed that the Mars soil is generally depleted in CaO and Na_2O, but not FeO relative to the compositions of basaltic Martian meteorites. Terrestrial ambient temperature chemical weathering trends for basalts are similar to the Martian trends, but such environments are unlikely on Mars at present. They concluded that Martian trends were best explained by palagonitization. Hydrothermal models based on a study of terrestrial ocean floor basalts show strong iron depletion in altered materials and did not match Martian trends. However, the ocean floor environment is not a good analog for Martian processes. Nelson and Newsom (2003) measured the compositions of alteration material formed by impact hydrothermal systems at Lonar and Mistastin craters and obtained data on altered basaltic rocks from other hydrothermal environments. The impact crater materials, particularly those from Lonar, represent an incipient hydrothermal alteration under low water/rock ratio conditions (Hagerty and Newsom, 2003). These processes can also occur when a basalt flow enters a lake or river valley, and in relatively shallow hydrothermal systems above a geothermal heat source. The active Pantelleria volcanic hydrothermal system in Italy, for example, contains altered basalt flows with iron-rich saponite and other alteration phases (Fulignati *et al.*, 1997). Although hydrothermal processes can remove Fe, incipient alteration leads to the formation of Fe-rich clays and retention of Fe in the system. The Fe-rich saponite characteristic of most basalt alteration material

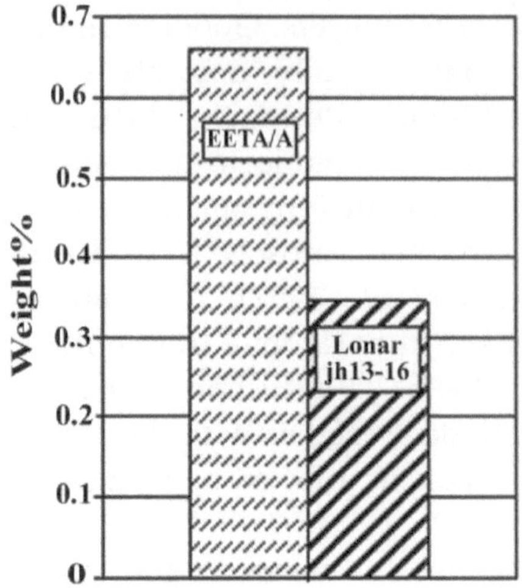

Figure 16.9. The mixing model proposed by Nelson and Newsom (2004) can successfully produce a composition matching the "rock" portion of Martian soil by combining Fe-rich terrestrial impact crater alteration material with basaltic Martian meteorite materials. This example involves a combination of Martian meteorite EETA 79001 lithology A and hydrothermal clays from Lonar crater (Hagerty and Newsom, 2003). The model combines 66% of EETA and 34% of clay to match an average Mars soil composition from Wänke *et al.* (2001).

is chemically similar to palagonite, but can be enriched in Fe, which is desirable as a component of the soil (Hagerty and Newsom, 2003). Nelson and Newsom (2003) showed that the chemical composition of the rock component of the soil could be matched by a mixture of Martian meteorite and the basalt alteration material (Figure 16.9). Although the chemistry of Martian alteration products may be slightly different due to different basalt compositions, this plausibility argument suggests that hydrothermal alteration cannot be ruled out as a possible source or contributor to the soil. Successful models suggest that alteration of either minerals or volcanic glass due to low water/rock ratio hydrothermal alteration in impacts or volcanic settings can produce the depletions in fluid-mobile elements and not deplete iron. Therefore, hydrothermally altered material, eroded by eolian processes on Mars, can yield alteration materials richer in Fe, and contribute part of the Martian soil.

16.5.2 Mobile element component

Since the discovery of the mobile element component, volcanic aerosols have been considered a likely source for enrichments of sulfur and chlorine

(Clark *et al.*, 1982). High-temperature volcanic aerosols and fumarole emissions from volcanoes and lava flows are well known on Earth (e.g., Symonds *et al.*, 1987), and are thought to represent the degassing of magmas. Clark and Baird (1979) suggested that Martian soils might have high concentrations of lead, bromine, antimony, mercury, and arsenic, based directly on the composition of Hawaiian fumarole deposits. Newsom and Hagerty (1997) improved on this approach by using data from many additional sources to establish the range of possible enrichments in the soil. In addition, they corrected the terrestrial compositions for the difference between the composition of the terrestrial mantle and the composition of the Martian mantle (Wänke and Dreibus, 1988).

Extending the volcanic aerosol model, Banin *et al.* (1997) proposed an "acid-fog" model in which volcanic aerosols contributed the mobile element component of the soil. Volatiles deposited on the surface are mobilized by aqueous films and are then involved in the breakdown of surface rocks to form the rock component, explaining the absence of well-crystallized clay minerals in the soil.

Hydrothermal alteration processes can also contribute a mobile element component to the soil through release of fluid-mobile elements that can be transported and deposited at the surface in hot springs and fumarole deposits. The two main types of hydrothermal fluids commonly found on Earth, and probably on Mars (Farmer, 2000), are acid-sulfate and neutral-chloride fluids. The neutral-chloride type is typical of high-temperature systems (usually 200–350 °C) in areas with an abundant supply of water. These systems form as chloride-rich upwelling of geothermal fluids above a magmatic heat source, creating surface hydrothermal features, which include neutral-chloride springs in the up-flow zone and dilute chloride bicarbonate springs in the surrounding outflow zones. The acid-sulfate type of hydrothermal fluid is found in areas of limited water supply (Ellis, 1979; Goff and Janik, 2000). In these systems, deep neutral-chloride waters provide a source of steam which condenses in the near-surface environment (<100 m depth) and forms an acid-sulfate liquid by oxidizing H_2S to H_2SO_4. Steam rising from the boiling reservoir can easily transport sulfur, but not chlorine, compounds resulting in high S/Cl ratios in the condensed vapor. The resulting fluid chemistry also includes high concentrations of volatile elements, such as S, and ammonium ions, and low concentrations of elements such as Cl and Li. Recent data from the Mars Odyssey Gamma Ray experiment indicates that the chlorine content of the soil is variable by a factor of three (Boynton *et al.*, 2004). The observed enrichments west of Tharsis may reflect hydrothermal processes or volcanic aerosol deposition. On Earth acid-mine drainage areas are sulfur-rich and contain sulfuric acid and oxidized iron. Theoretical studies suggest that the sulfur-rich component of the Martian soil may have been produced by similar processes (Newsom *et al.*, 1999; Stoker *et al.*, 2003). The Rio Tinto mining region in Spain contains acid-mine drainage fluids that could be similar to deep environments on Mars. NASA is funding a major drilling project here as part of a search for bacterial life that lives on chemical energy (Stoker *et al.*, 2003).

16.5.3 Meteoritic component

Meteorite debris and cosmic dust are likely to be present in Martian soil. The impact on the chemistry of the soil will be small, however, with the exception of siderophile trace elements, and possibly carbon (Flynn and McKay, 1990). Besides Fe, Ni is the most abundant siderophile element but, unfortunately, the Pathfinder APXS instrument contained nickel in its detector housing and could not measure Ni (T. E. Economou, personal communication, 2001).

16.6 Conclusions

In terms of geochemical processes, Mars has many analogs among the other terrestrial planets. In terms of a simple mantle–crust geochemical evolution, Mars is similar to small bodies in the Solar System, such as the Moon and Vesta. However, the presence of water leads to weathering, alteration, and chemical deposition processes similar to those on the Earth. Although the visual appearance of the surface of Mars is strikingly similar to many sites on Earth, their details may differ significantly. For example, the Martian soil or fine-grained regolith may contain few well-developed clay minerals. The soils may also be loaded with mobile elements, such as arsenic and lead derived from volcanic aerosols, and hydrothermal processes. On Earth rainfall and other processes readily remove these elements. As we continue to study Mars and the terrestrial planets, we will increasingly be able to compare processes on different planets and learn more about our Solar System.

Acknowledgments

Research supported by the NASA Planetary Geology and Geophysics program (NAG 5–10143), H. Newsom P.I. Thanks is extended to Bill Cassidy and John McHone for their insightful suggestions and comments.

References

Allen, C. C., Gooding, J. L., Jercinovic, M., and Keil, K. (1981). Altered basaltic glass: a terrestrial analog to the soil of Mars. *Icarus*, **45**, 347–69.

Allen, C. D., Albert, F. G., Chafetz, H. S., *et al.* (2000). Physical biomarkers in carbonate thermal springs: implications in the search for life on Mars. *Icarus*, **147**, 49–67.

Bandfield, J. L. (2002). Global mineral distributions on Mars. *Journal of Geophysical Research*, **107**(E6), doi:10. 1029/2002JE001510.

Bandfield, J. L., Hamilton, V.E., and Christensen, P. R. (2000). A global view of Martian surface compositions from MGS-TES. *Science* **287**, 1626–30.

Banin, A., Clark, B. C., and Wänke, H. (1992). Surface chemistry and mineralogy. In *Mars*, ed. H. H. Kieffer, B. M. Jakosky, C. W. Snyder, and M. S. Matthews. Tucson: Univ. Arizona Press, pp. 594–625.

Banin, A., Han, F. X., Kan, I, and Cicelsky, A. (1997). Acidic volatiles and the Mars soil. *Journal of Geophysical Research*, **102**, 13341–56.

Bell, III J. F., McSween, H. Y., Crisp, J. A. *et al.* (2000). Mineralogic and compositional properties of Martian soil and dust: results from Mars Pathfinder. *Journal of Geophysical Research*, **105**, 1721–55.

Bishop, J. L., Murchie, S. L., Pieters, C. M., and Zent, A. P. (2002). A model for the formation of dust, soil, and rock coatings on Mars: physical and chemical processes on the Martian surface. *Journal of Geophysical Research*, **107**, E11, 5097, doi:10.1029/2001JE001581.

Blaney, D.L. (1998). Mars dust formation by impact craters into volatile materials and aerosol formation of sulfate duricrust. *Abstracts of Papers Submitted to the 29th Lunar and Planetary Science Conference*. Houston: Lunar and Planetary Institute, CD 29, Abstract 1655.

Borg, L. E. and Draper, D.S. (2003). A petrogenetic model for the origin and compositional variation of the Martian basaltic meteorites. *Meteoritics and Planetary Science*, **38**, 1713–31.

Borg, L. E., Connelly, J. N., Nyquist, L. E. *et al.* (1999). The age of the carbonates in Martian meteorite ALH84001. *Science*, **286**, 90–4.

Borg, L. E., Nyquist, L. E., Reese, Y., and Wiesmann, H. (2003). The age of Dar al Gani 476 and the differentiation history of the Martian meteorities inferred from the Rb-Sr, Sm-Nd, and Lu-Hf isotopic systematics. *Geochimica et Cosmochimica Acta*, **67**, 3519–36.

Borg, L. E., Nyquist, L. E., Wiesmann, H., and Reese, Y. (2002). Constraints on the petrogenesis of Martian meteorites from the Rb-Sr and Sm-Nd isotopic systematics of the lherzolitic shergottites ALH77005 and LEW88516. *Geochimica et Cosmochimica Acta*, **66**(11), 2037–53.

Boslough, M. B. and Cygan, R. T. (1988). Shock-enhanced dissolution of silicate minerals and chemical weathering on planetary surfaces. *Proceedings of the 18th Lunar and Planetary Science Conference*. Houston: Lunar and Planetary Institute, pp. 443–54.

Bouska, V. and Bell III, J. F. (1993). Assumptions about the presence of natural glasses on Mars. *Journal of Geophysical Research*, **98**, 18719–25.

Boynton, W., Janes, D., Kerry, K., *et al.* (2004). The distribution of non-volatile elements on Mars: Mars Odyssey GRS results. *Abstracts of Papers Submitted to the 35th Lunar and Planetary Science Conference*. Houston: Lunar and Planetary Institute, CD 35, Abstract 1950.

Brearley, A. J. and Jones, R. H. (1998). Chondritic meteorites. In *Planetary Materials, Reviews in Mineralogy*, vol. 36. Washington, DC: Mineralogical Society of America.

Bridges, J. C., Catling, D. C., Saxton, J. M. *et al.* (2001). Alteration assemblages in Martian meteorites: implications for near-surface processes. *Space Science Reviews*, **96**, 365–92.

Christensen, P. R., Bandfield, J. L., Hamilton, V. E., *et al.* (2001). Mars Global Surveyor Thermal Emission Spectrometer experiment: investigation description and surface science results. *Journal of Geophysical Research*, **106**, 23823–71.

Christensen, P. R., Bandfield, J. L., Bell, J. F. I., *et al.* (2003). Morphology and composition of the surface of Mars: Mars Odyssey THEMIS results. *Science*, **300**, 2056–61, doi:10.1126/science.1080885.

Clark, B. C. (1993). Geochemical components in Martian soil. *Geochimica et Cosmochimica Acta*, **57**, 4575–81.

Clark, B. C. and Baird, A. K. (1979). Volatiles in the Martian regolith. *Geophysical Research Letters*, **6**, 811–14.

Clark, B. C., Baird, A. K., Weldon, R. J. *et al.* (1982). Chemical composition of Martian fines. *Journal of Geophysical Research*, **87**, 10059–67.

Clifford, S. M. (1993). A model for the hydrologic and climatic behavior of water on Mars. *Journal of Geophysical Research*, **98**, 10973–11016.

Dreibus, G. and Wänke, H. (1980). The bulk composition of the eucrite parent asteroid and its bearing on planetary evolution. *Zeitschrift fur Naturforschung a-a* **35**, 204–16.

Dreibus, G. and Wänke, H. (1982). Parent body of the SNC-meteorites, chemistry, size and formation. *45th Annual Meteoritical Society*, pp. 207–8.

Economou, T.E., Foley, C.N., and Clayton, R.N. (2003). The chemical composition of Martian samples: final results from the pathfinder alpha proton x-ray spectrometer. *Sixth International Conference on Mars*. Lunar and Planetary Institute, Abstract 3155.

Ellis, A. J. (1979). Explored geothermal systems. In *Geochemistry of Hydrothermal Systems*, ed. H. L. Barnes. New York: John Wiley & Sons.

Farmer, J. D. (2000). Hydrothermal systems: doorways to early biosphere evolution. *Geological Society of America Today*, **10**, 1–9.

Fulignati, P., Malfitano, G., and Sbrana, A. (1997) The Pantelleria caldera geothermal system; data from the hydrothermal minerals. *Journal of Volcanology and Geothermal Research*, **75**, 251–70.

Flynn, G. J. and McKay, D. S. (1990). An assessment of the meteoritic contribution to the Martian soil. *Journal of Geophysical Research*, **95**, 14497–509.

Frey, H.V. (2003). Buried impact basins and the earliest history of Mars. *Sixth International Conference on Mars*. Lunar and Planetary Institute, Abstract 3104.

Geissler, P. E., Singer, R. B., Komatsu, G., Murchie, S., and Mustard, J. (1993). An unusual spectral unit in West Candor Chasma: evidence for aqueous or hydrothermal alteration in the Martian canyons. *Icarus*, **106**, 380–91.

Goff, F. and Janik, C. J. (2000). Geothermal systems. In *Encyclopedia of Volcanoes*, ed. H. Sigurdsson *et al.* Academic Press, pp. 817–34.

Gooding, J. L. and Keil, K. (1978). Alteration of glass as a possible source of clay minerals on Mars. *Geophysical Research Letters*, **5**, 727–30.

Griffith, L. L. and Shock, E. L. (1997). Hydrothermal hydration of Martian crust: illustration via geochemical model and calculations. *Journal of Geophysical Research*, **102**, 9135–43.

Griffith, L. L. and Shock, E. L. (2002). A geochemical model for the formation of hydrothermal carbonates on Mars. *Nature*, **377**, 406–8.

Gulick, V. C. (1998). Magmatic intrusions and a hydrothermal origin for fluvial valleys on Mars. *Journal of Geophysical Research*, **103**, 19365–88.

Hagerty, J. J. and Newsom, H.E. (1998). Potential toxicity of the Martian soil. *Lunar and Planetary Science*, **XXIX**, 1697.

Hagerty, J. J. and Newsom, H. E. (2003). Evidence for impact-induced hydrothermal alteration at the Lonar crater, India. *Meteoritics and Planetary Science*, **38**, 365–81.

Halliday, A. N., Wänke, H., Birck, J.-L., and Clayton, R. N. (2001). The accretion, composition, and early differentiation of Mars. *Space Science Reviews*, **96**, 197–230.

Harrison, K. P. and Grimm, R. E. (2002). Controls on Martian hydrothermal systems: application to valley network and magnetic anomaly formation. *Journal of Geophysical Research*, **107**(E5), doi:10. 1029/2001JE001616.

Heiken, G., Vaniman, D., and French, B. M. (1991). *Lunar Sourcebook, A User's guide to the Moon.* Cambridge University Press.

Herd, C. D. K. (2003). The oxygen fugacity of olivine-phyric Martian basalts and the components within the mantle and crust of Mars. *Meteoritics and Planetary Science*, **38**, 1793–805.

Kleine, T., Munker, C., Mezger, K., and Palme, H. (2002). Rapid accretion and early core formation on asteroids and the terrestrial planets from Hf-W chronometry. *Nature*, **418**, 952–5.

Lodders, K. and Fegley, B. (1997). An oxygen isotope model for the composition of Mars. *Icarus*, **126**, 373–94.

Lodders, K. and Fegley, B. (1998). *The Planetary Scientist's Companion.* Oxford University Press.

Lunine, J. I., Chambers, J., Morbidelli, A., and Leshin, L. A. (2003). The origin of water on Mars. *Icarus*, **165**, 1–8.

Mason, B. (1992). *Victor Moritz Goldschmidt: Father of Modern Geochemistry.* The Geochemical Society, Special Publication 4.

Masson, P., Carr, M. H., Costard, F. *et al.* (2001). Geomorphologic evidence for liquid water on Mars. *Space Science Reviews*, **96**, 333–64.

McDonough, W. F. and Sun, S. S. (1995). The composition of the Earth. *Chemical Geology*, **120**, 223–53.

McSween, H. Y. (1999). *Meteorites and their Parent Planets.* Cambridge University Press.

McSween, H. Y. Jr. and Keil, K. (2000). Mixing relationships in the Martian regolith and the composition of globally homogeneous dust. *Geochimica et Cosmochimica Acta*, **64**, 2155–66.

McSween, H. Y. and Treiman, A. H. (1998). Martian meteorites. In *Planetary Materials, Reviews in Mineralogy*, vol. 36. Washington, DC: Mineralogical Society of America.

Mittlefehldt, D. W., McCoy, T. J., Goodrich, C. A., and Kracher, A. (1998). Non-chondritic meteorites from asteroidal bodies. In *Planetary Materials, Reviews in Mineralogy*, vol. 36. Washington, DC: Mineralogical Society of America.

Morris, R. V., Golden, D. C., Bell III, J. F., *et al.* (2000). Mineralogy, composition, and alteration of Mars Pathfinder rocks and soils: evidence from multispectral, elemental, and magnetic data on terrestrial analogue, SNC meteorite, and Pathfinder samples. *Journal of Geophysical Research*, **105**, 1757–817.

Morris, R. V., Golden, D. C., Ming, D. W., *et al.* (2001). Phyllosilicate-poor palagonitic dust from Mauna Kea Volcano (Hawaii): a mineralogical analogue for magnetic Martian dust? *Journal of Geophysical Research*, **106**, 5057–83.

Nelson, M.J. and Newsom, H.E. (2003). Impact hydrothermal alteration of terrestrial basalts: explaining the rock component of the Martian soil. *Third International Conference on Large Meteorite Impacts*. Lunar and Planetary Institute, Abstract 4099.

Newsom, H. E. (1980). Hydrothermal alteration of impact melt sheets with implications for Mars. *Icarus*, **44**, 207–16.

Newsom, H. E. and Hagerty, J. J. (1997). Chemical components of the Martian soil: melt degassing, hydrothermal alteration, and chondritic debris. *Journal of Geophysical Research*, **102**, 19345–56.

Newsom, H. E., Brittelle, G. E., Crossey, L. J., and Kudo, A. M. (1996). Impact cratering and the formation of crater lakes on Mars. *Journal of Geophysical Research*, **101**, 14951–5.

Newsom, H. E., Hagerty, J. J., and Goff, F. (1999). Mixed hydrothermal fluids and the origin of the Martian soil. *Journal of Geophysical Research*, **104**, 8717–28.

Nyquist, L. E., Bogard, D. D., Shih, C. Y. *et al.* (2001a). Ages and geologic histories of Martian meteorites. *Space Science Reviews*, **96**, 105–164.

Nyquist, L. E., Bogard, D. D., Shih, C.-Y. *et al.* (2001b). Ages and histories of Martian meteorites. In *Chronology and Evolution of Mars 96*, ed. R. Kalenback, J. Geiss, and W. K. Hartmann. Kluwer, pp. 105–64.

Nyquist, L. E., Reese, Y., Wiesmann, H., Shih, C.-Y., and Takeda, H. (2003). Fossil 26Al and 53Mn in the Asuka 881394 eucrite: evidence of the earliest crust on asteroid 4 Vesta. *Earth and Planetary Science Letters*, **214**, 11–25.

Nyquist, L. E. and Shih, C. -Y. (1992). The isotopic record of lunar volcanism. *Geochimica et Cosmochimica Acta*, **56**, 2213–34.

Papike, J. J., Karner, J. M., and Shearer, C. K. (2003). Determination of planetary basalt parentage: a simple technique using the electron microprobe. *American Mineralogist*, **88**, 469–72.

Rao, M. N., Borg, L. E., McKay, D. S., and Wentworth, S. J. (1999). Martian soil component in impact glasses in a Martian meteorite. *Geophysical Research Letters*, **26**, 3265–8.

Righter, K. and Drake, M. J. (1996). Core formation in Earth's Moon, Mars, and Vesta. *Icarus*, **124**, 513–29.

Singer, R. B. (1982). Spectral evidence for the mineralogy of high-albedo soils and dust on Mars. *Journal of Geophysical Research*, **87**, 10159–68.

Stoker, C.R., Lemke, L.G., Mandell, H., *et al.* (2003). Mars Analog Research and Technology Experiment (MARTE): a simulated Mars drilling mission to search for subsurface life at the Rio Tinto, Spain. *Abstracts of Papers Submitted to the 34th Lunar and Planetary Science Conference*. Houston: Lunar and Planetary Institute, CD 34, Abstract 1076.

Symonds, R. B., Rose, W. I., Reed, M. H., Lichte, F. E., and Finnegan, D. L. (1987). Volatilization, transport and sublimation of metallic and non-metallic elements in high temperature gases at Merapi Volcano, Indonesia. *Geochimica et Cosmochimica Acta*, **51**, 2083.

Thorsos, I. E., Newsom, H.E., and Davies, A. D. (2001). Availability of heat to drive hydrothermal systems in large Martian impact craters. *Abstracts of Papers Submitted to the 32nd Lunar and Planetary Science Conference*. Houston: Lunar and Planetary Institute, CD 32, Abstract 2011.

Walter, M. J., Newsom, H. E., Erterl, W., and Holzheid, A. (2000). Siderophile elements in the Earth and Moon: metal/silicate partitioning and implications for core formation. In *The Origin of the Earth and Moon*, ed. R. M. Canup, and K. Righter. Tucson: Univ. of Arizona Press, pp. 265–89.

Wänke, H. and Dreibus, G. (1988). Chemical composition and accretion history of terrestrial planets. *Philosophical Transactions of the Royal Society A*, **325**, 545–57.

Wänke, H., Brückner, J., Dreibus, G., Rieder, R., and Ryabchikov, I. (2001). Chemical composition of rocks and soils at the Pathfinder site. *Space Science Reviews*, **96**, 317–30.

Wilde, S. A., Valley, J. W., Peck, W. H., and Graham, C. M. (2001). Evidence from detrital zircons for the existence of continental crust and oceans on the Earth 4.4 Gyr ago. *Nature*, **409**, 175–8.

Wilson, L. and Head III, J. W. (1997). Mars geothermal and volcanic evolution: volcanic intrusions as heat sources to maintain viable ecosystems? In *Conference on Early Mars*. Houston: Lunar and Planetary Institute, pp. 85–6.

Wyatt, M. B. and McSween, Jr. H. Y. (2003). Volcanism or aqueous alteration on Mars? *Nature* (London), **421**, 712–13.

Yin, Q. Z., Jacobsen, S. B., Yamashita, K. *et al.* (2002). A short timescale for terrestrial planet formation from Hf-W chronometry of meteorites. *Nature*, **418**, 949–52.

17

Integrated analog mission design for planetary exploration with humans and robots

Kelly Snook
NASA Johnson Space Center/KX, Houston

Brian Glass, Geoffrey Briggs and Jennifer Jasper
NASA Ames Research Center, Moffett Field

17.1 Introduction

For reasons of cost and risk, planetary exploration since Apollo has been carried out by robots with the human input made from Earth. Given communication time delays and the manifest limitations of robots, the pace and quality of such exploration could be greatly improved if humans were more directly involved. Exploration continues using increasingly advanced robotic technologies including those intended to begin the subsurface exploration of the planets. Before such missions will be undertaken we need assurance that these new technologies work adequately under appropriate terrestrial analog conditions. Eventually, humans will re-enter the picture with in-depth exploration of the Moon and Mars as their principal focus. However, such human explorers will not be able to achieve the global reach needed to answer the many questions scientists pursue for a planet as large and diverse as Mars. So, how should humans and robots work together optimally? Can advanced robots tele-operated by humans at short light distances approach the scientific productivity of a trained, yet suit-encumbered, astronaut? To answer these questions, researchers must define scientific return and find ways to compare the productivity of different human–robot exploration systems. Analogs can be used to develop the full range of possible human and robotic exploration systems using metrics that allow us to quantify the effectiveness of each.

Some important outstanding exploration issues that high-fidelity analog missions can inform include:

- Development, testing, and demonstration of exploration hardware, including surface habitats and extra-vehicular activity (EVA) systems.

The Geology of Mars: Evidence from Earth-based Analogs, ed. Mary Chapman. Published by Cambridge University Press.

- Selection of landing sites that maximize access to resources and scientifically interesting terrain.
- Refinement of planetary scientific exploration techniques and goals (described elsewhere in this book).
- Training of ground and flight crews in areas such as human and robotic planetary field geology and biology; traverse planning; sample collection, processing, and curation; data acquisition, reporting, and communications; new EVA and surface systems; light-time delays; maintenance; and extended mission operations.
- Refinement of realistic and feasible crew task scheduling, timelines, and level of autonomy.
- Design of automated systems for support of exploration personnel.
- Refinement of crew selection criteria for long duration surface exploration missions.
- Improved understanding of the impact of science operations on the design of surface system technologies and infrastructure.
- Controlled investigations and experiments in psychology and human factors.
- Development of education and outreach plans for future human missions.
- Development of new management techniques for international planetary exploration programs.

Analog missions are an obvious way of developing insights we need. This chapter describes some of the limited work that has been done to date in systematizing analog mission studies for both robotic and human planetary exploration.

17.2 Design and evaluation of analog missions

In any a analog activity, at least one characteristic of a planetary destination or mission is simulated. For analog lessons learned to be relevant, three criteria must be met: (1) assumptions regarding the elements and characteristics of future actual missions must be clearly and precisely stated, (2) the objectives, scope, and elements of the analog (simulation) activity must be articulated, and (3) the fidelity of the analog must be qualitatively and/or quantitatively determined. In other words, one must know the actual mission or destination being simulated, how it is being simulated, and how closely the simulation matches the real mission in desired key areas.

17.2.1 Identification of mission to be simulated

Analog missions can be most broadly characterized by the type of mission that is being simulated. The principal types of ongoing and future exploration missions are:

I. Orbital or Transit – Robotic missions

II. Orbital or Transit – Joint human and robotic missions
III. Planetary Surface – Semi-autonomous or tele-operated robotic surface missions
IV. Planetary Surface – Joint human and robotic surface missions.

Examples of space mission elements that may be simulated are:

i Mission destination's physical characteristics (terrain, geology, biology, etc.)
ii Mission purpose, scientific goals, and timeline
iii Overall mission architecture, including level and type of human activity
iv Specific mission segment, if simulating only a part of the mission
v Mission command, control, communication and data flow scenarios
vi Daily mission schedules and anticipated task management
vii Anticipated mission technologies
viii Human factors issues
ix Anticipated support personnel, infrastructure and architecture on Earth
x Anticipated mission public outreach and education plans
xi Other relevant anticipated mission elements.

An analog activity must be traceable to elements of one or more mission type. For example, the analog experiment discussed in Section 17.3 was intended to simulate the mid-mission surface segment of Type III and IV missions. It focused on those functional aspects of elements v, vii, ix, and ii specifically related to EVA exploration in a setting chosen for its Mars-like physical characteristics (i).

17.2.2 Analog missions and fidelity

The word analog is used in a wide variety of contexts to mean different things. In this discussion, "analog" is defined as any activity that physically or functionally mimics any element of a future mission. The analog may be as simple as collecting and analyzing a sample from a scientifically analogous place, or as complex as conducting an end-to-end test of flight hardware and software under anticipated mission constraints and conditions. Analog activities can be in a laboratory, in a specially designed testbed facility, or in the field. Budget, schedule, and physical differences between Earth and the Moon or Mars limit the scope of analog activities that can be successfully carried out. Since no planetary mission can be fully simulated on Earth, it is necessary to qualify and quantify the extent of the similarity between the analog and the actual mission or destination. This measure of similarity is referred to as *fidelity*. Analog simulation activities can be designed and evaluated based on their intended level of fidelity.

Experiments should be designed to capitalize on areas of high fidelity within the analog and minimize deleterious or misleading effects caused by low-fidelity elements. The term Analog Mission describes any analog activity that integrates two or more high-fidelity mission elements.

An analog's fidelity should be defined in terms of its science, science operations, technology development, technology integration, mission operations, training, human factors, human physiology, and its educational and outreach value. In the case of science, fidelity hinges on the physical similarity of the chosen analog site to its space counterpart in terms of morphology, chemistry, geology, or other physical similarity such as microgravity for life science experiments. In the case of technology development, e.g., for rovers and drills, fidelity may depend on topography and physical parameters such as temperature, precipitation, rock abundance, and engineering properties of soils and rocks. Technology development might also require high fidelity in the areas of pressurized habitable volumes and closed loop life support systems. In the case of operations, human factors, and training, fidelity may require situational and logistical similarities such as remoteness, isolation, confinement, and limitations in data rate or in pathways of communications. Ideally analog experiments would include high degrees of fidelity in more than one of these categories since, in such cases, unanticipated issues emerge, making them more valuable to space mission planners than a more focused analog experiment.

Analog categories and their fidelity parameters

A. **Science** Fidelity of the science analog category is measured by scientific output in the form of peer-reviewed papers, articles, etc., and by relevance to planetary science goals as outlined by the scientific community, e.g., the report of the international Mars Exploration Program Advisory Group (MEPAG Committee, 2003). Real science and exploration introduce constraints and requirements on a mission that are difficult or impossible to simulate with artificially contrived science tasks. Fidelity is further increased when it includes state-of-the-art technology and multiple, possibly conflicting, scientific tasks and agendas. Geology, astrobiology, astronomy, chemistry, biology, physics, human physiology, life sciences, and psychology are examples of relevant analog sciences.

B. **Science Operations** is a broad category defined as the processes by which science and exploration are carried out by both field and ground crews. The fidelity of analog missions in this category is distinct from the science fidelity discussed above because issues in this category can be tested independently from the science itself. An analog can have high science operations fidelity without high science fidelity. Snook (2005a) provides a comprehensive discussion of this analog category. A few examples of science operations subcategories are:

(1) *Types of science tasks, investigations, and traverses.* The types of tasks, investigations, and traverses carried out in the field on Earth may not be possible to perform on the Moon and Mars because of mission-imposed limitations such as EVA range, mobility, duration, tools, and communications. Examining and characterizing types of tasks required by field science and exploration on Earth will improve the ability to design tools and systems to support equivalent or substitute tasks on the Moon or Mars.

(2) *Science support, remote science, and the "science backroom."* Our experience in planetary surface exploration by humans is limited to the Apollo Program, where the astronauts were supported by a science officer in the mission control room who, in turn, was supported by "backrooms" of geoscientists. This may or may not be the way in which future lunar and Martian exploration is carried out. In the Martian case the communication problem resulting from light-time delay and available bandwidth will effectively eliminate the real-time authority of an Earth-bound science back room. On the other hand, advances in information technology, including the ability to create virtual worlds for the backroom scientists, will add to their (non-real-time) authority. Some of the new issues raised in this area have been explored and simulated (Clancey, 2004; Stoker *et al.*, 2001; Glass *et al.*, 2003; Snook, 2005b), one example of which is described as a case study in this chapter. Analog missions are particularly attractive for investigating the range of possibilities for science support because multiple mission scenarios can be tested at low cost, even within the same analog activity, by strategically changing communications and data flow constraints.

(3) *Scheduling and planning of exploration tasks and traverses.* Any EVA requires significant pre- and post-traverse planning and debriefing. Experiments in this category might include the development of schedules for traverse planning and execution, identification of who does the planning and analysis (e.g., crew on Mars or science teams on Earth), scheduled frequency, duration, and scope of traverses, extent of sample collection, choice of *in situ* analysis vs. sample return, relative emphasis on science activities vs. maintenance and other mission duties, degree of crew and/or robotic autonomy, the allotment of time and resources for plans and results to be documented and communicated, and many other elements beyond the scope of this chapter (see Snook, 2005b for more discussion).

(4) *Functions of technologies used for scientific tasks and traverses.* "Functionality" is more important than "actual" technology in assessing fidelity. If key technologies anticipated or projected for the actual mission

(e.g., for a human mission 15 years from now) are not presently available, the fidelity of the analog may still be maintained by simulating the needed functionality. For example, a 2020 mobile robot may well be capable of transmitting real time panoramic video while avoiding hazards and traversing at a moderate speed. Such functionality can be simulated by a human driver on an ATV – provided that the operator strictly limits the use of his/her cognitive capabilities. Another example might be the use of a mock spacesuit to simulate some of the functionality and encumbrances of a Mars spacesuit.

(5) *Value of human astronauts.* Little quantitative data are available to answer the basic question of how much more productive *in situ* humans are than robots controlled at a distance. Given the fundamental importance of light time delay, we might expect that productivity would be a function of distance. By simulating where the human is located in the exploration architecture (on Earth, in orbit about the planet in question, or on its surface) we can begin to quantify metrics in this subcategory. For example, a given site can be explored using tele-operated robots with different light time delays and the results compared with those obtained by scientists at the field site, as in the detailed case study below (Section 17.3).

C. **Technology Development** Space flight technologies being developed or tested for actual missions are the measure of fidelity in this category. A good metric for this category might be something similar to the Technology Readiness Levels, TRL-1 through TRL-9, used by NASA to quantify a given technology's readiness for flight (NASA, 1996). Examples of especially relevant technologies include, but are not limited to, habitats, closed loop life support systems, spacesuits, rovers and aircraft, scientific instruments, subsurface access equipment, power systems, equipment for medical countermeasures of reduced gravity, biotechnology and advanced horticulture, greenhouses, inflatable structures, radiation shielding, wireless and satellite communications, and intelligent systems.

D. **Technology Integration** Countless new technologies must be integrated together in any successful future robotic or human mission. This category provides a way of assessing an analog's degree of fidelity in integrating new technologies. An analog activity in which new hardware and software interfaces are required will be higher fidelity than one in which technologies are being developed separately.

E. **Mission Operations** The degree of support, control, and communication from a ground station simulating all the functions of an anticipated mission control defines fidelity in this category (Hoffman, 2001). Subcategories to be considered are similar to those discussed for science operations, but are extended to concern all elements of the mission being simulated, including monitoring and troubleshooting all hardware and software systems and subsystems, maintaining crew health and safety, planning and scheduling crew activities, communicating with crew, providing expertise and assistance to the crew, and a host of other mission-critical functions. Few analog activities since the "full up simulations" of Apollo have had high fidelity in this category because a full simulation of mission operations is ambitious and resource-intensive. It must not interfere with actual ongoing flight operations.

F. **Crew/Ground Team Training** Analogs with a high degree of fidelity in this area are those which provide explicit training, practice, or direct preparation for future missions. An example of high fidelity in this category is the geology training that was done with the Apollo astronauts at desert analogs prior to flight. Geological field trips were set up to train the astronaut teams in specific tasks necessary to document the landing site. The fidelity of the exercises increased as the launch date approached, with EVA technology that more closely mimicked the Lunar Rover and suits (Muehlberger, 2003).

G. **Human Physiology and Medicine** High fidelity in this category indicates that the analog provides useful data on the long-term effects of reduced gravity on the human body. The International Space Station is the only current high-fidelity analog in this category.

H. **Human Factors** is often used as a catch-all phrase for an array of important mission elements. For this reason, fidelity in this area may be harder to quantify. In general, however, fidelity can be achieved if the impact of mission conditions on humans, and vice versa, can be studied. The psychology of human performance under pressure or stress from long work hours or living in a confined space is a human factors concern that can be addressed in many analogs (Morphew *et al.*, 1999).

I. **Outreach and Education** High fidelity in this area requires conducting education and outreach activities either as if the mission were an actual mission, or under the constraints of an actual mission. Educational and outreach events about the analog event itself do not contribute to high fidelity in this category.

J. **Overall Integration** The purpose of this category is to provide a way of assessing the degree to which all the previous categories (A–H) are integrated together as if it were a real mission. Some analog activities, such as Case Study #1, involve many diverse projects co-located at a single analog site, whereas others, such as Case Study #9, center on a focused mission simulation and are primarily concerned with the integration of mission elements. Arguably the analog with the highest fidelity to date in this category is the last exercise for Apollo 17 with the full-up simulation in Mission Control supporting astronauts near Sunset Crater. The site was a physical scientific analog for the "dark mantle" that covered the Apollo 17 landing site, and the surrounding volcanic domes substituted for the massif walls at the landing site.

Sample acquisition, sample handling, communications, technology, operations, and human factors were among the mission elements simultaneously integrated and tested (Muehlberger, 2003).

Case studies and metrics

A brief survey of selected analogs is provided in Table 17.1 for the purpose of illustrating how the types of missions and categories of analogs may be useful. The analogs selected are not intended to represent a comprehensive survey of all analog activities, but instead are limited to those with which the authors or their institutions have been involved.

Figure 17.1 gives a graphical qualitative estimate of the fidelity of the ten Case Study analog activities according to the metrics provided in Table 17.2. It should be noted that the information in Figure 17.1 is subjective, and it is included here only as an illustration of how qualitative metrics might be used. Using this approach, one can quickly distinguish the main focus and strong points of each activity. One can begin to appreciate the diversity of information and lessons they represent so that a larger picture can be synthesized. The authors wish to acknowledge that a potential pitfall of this methodology is comparing one-shot tests with whole series of tests that may include different components year-to-year.

All ten examples are high fidelity in Outreach and Education, which illuminates the remarkable role of analog activities in engaging the public. In an extended program of human exploration, analogs can provide tangible shorter-term milestones demonstrating capabilities and generating excitement and support for exploration goals.

Chapter 14 in this book provides another interesting case study and exemplifies important work targeted to specific rover operations and technology preparedness. The remainder of this chapter is devoted to an integrated analog mission, led and summarized here by Dr. Brian Glass, that focused on the issue of productivity of humans and robots engaged in surface exploration.

17.3 A detailed case study of the Haughton remote science experiment (horse): evaluation of human vs. teleoperated robotic performance in field geology tasks at a Mars analog site

This section will describe an initial Type III and Type IV (as defined in Section 17.2.1) analog experiment during the summers of 2002 and 2003 to measure the science productivity of different future exploration approaches

Table 17.1. *Summary of case studies with their Analog Type (Section 17.3.1) and the Analog Fidelity Category (Section 17.3.2) each emphasized*

	Title	Dates and location	Type and Category	Description
1	NASA Haughton-Mars Project (HMP)	On-going, June–Aug. 1997–present Arctic Desert	I – IV **A***, B, **C**, D, E, F, G, **H***, I*, J, K	An annual field campaign led by Dr. Pascal Lee composed of loosely related analog activities selected and carried out at the Haughton Crater on Devon Island in the Canadian High Arctic [Lee *et al.*, 2004]. Science value is very high because published peer-reviewed science is a primary focus of this activity. It is not focused on science operations, but useful SO lessons learned have come from some of the HMP's individual projects. New technologies are frequently developed and tested, including rovers, airplanes, spacesuits, balloons, greenhouses, habitats, and communications systems, although not in a highly integrated way. The project is largely autonomous in the arctic, with no simulated mission control or science operations ground support (with the exception of certain individual projects within the HMP – see Case Study 2). The project has been funded through individual research grants and other small grants, at a small cost overall relative to the number of projects supported and their scientific return.
2	NASA Haughton Remote Science Experiment (HoRSE) at the NASA HMP	July 2001, 2002, 2003 Arctic Desert NASA Ames	I –IV A, **B***, **C***, D, F, G, I, **J***	A three-part exercise in robotic exploration simulating a tele-operated Mars rover mission with a group of scientists controlling the rover from Earth [Glass *et al.*, 2003]. This project is described further in Section II. This activity is high-fidelity in all categories except science value, mission operations, and human factors. The science that was done at the site, while done by actual scientists authentically representing the science that would be done on Mars, was not

Table 17.1 (cont.)

				publishable science geology or biology. There were no mission operations simulations besides the robotic mission operations simulation. In the simulated robotic mission, the relevant human factors were related to the situational awareness and experience of ground crews while remotely exploring and controlling the rover. Simulated space suits were not pressurized, so lower fidelity.
3	NASA Oceanographic Analog Missions Project (NOAMA)	Jul. – Oct. 2003 Atlantic and Pacific Mid – ocean ridges, ships and subs	III, IV **A***, **B***, **C***, **D***, E, **F***, **G***, H, **I***, **J***	A project undertaken to study the exploration of the deep sea hydrothermal vents using oceanographic surface and submersible vessels as an analogue for planetary exploration [Snook, 2005] as part of a joint scientific and filmmaking project led by James Cameron. The expedition's primary goal was the production of a 3D Imax documentary film. Using the ship as an analog for a Mars habitat, and the submersibles as pressurized rovers, high fidelity was achieved in science, science operations, crew/team training, and human factors. The technologies being used were highly integrated with each other and with the science and exploration activities, and some were relevant to space exploration. An instrument was developed by JPL researchers for the explicit purpose of applications to astrobiology space missions. Operations support through NASA JSC's Exploration Planning and Operations (ExPOC) facility were primarily focused on science aspects of the mission. Most general mission operations functions were performed by the film production crew. Analog researchers "piggybacked" on this mission at very low cost to NASA. The filming process and the film itself provide opportunities for experimenting with how Mars missions might be documented and presented to the public.
4				

Table 17.1. (cont.)

	Title	Dates and location	Type and Category	Description
	Independent Crew 6 Rotation in the Mars Society's Mars Arctic Research Station (MDRS)	Apr. – May 2002 Hanksville, Utah US SW Desert	III, IV **A***, **B***, D, E, **F***, **G***, **I***, **J***	Overall, the intent of the mission was to conduct the highest fidelity integrated simulation possible given the technology level, using simulated technology functionality. Maximum possible mission constraints were placed on the mission. The crew donned simulated space suits, whenever leaving the habitat. A macrobiotic food experiment was conducted to test proposed Mars mission planning and scheduling documents. Scientific geological and biological EVAs were planned and conducted autonomously by the crew. The microbiology experiments conducted during the EVAs by the scientists have yielded publishable results; however, the technologies used during the analog merely simulated the functions of those technologies that will be needed on Mars, including the habitat, the rovers, and the field instruments. No new technologies were used or tested. Limited mission operations functions were simulated. Press were treated according to the constraints of a Mars mission with the exception of one open-house day.
5	NASA Desert Research and Technology Study (Desert RATS) project	On-going Sept. 2001–present Meteor Crater, Cinder Lakes, Arizona US SW Desert	III, IV A, B, **C***, **D***, E, **F***, I, J	An ongoing annual multidisciplinary, multi-agency field exercise in the desert Southwest US focused on developing and integrating technologies for scientific exploration. Technologies and capabilities included in the project have included spacesuits, human unpressurized rovers, robotic rover assistants, voice recognition systems, wireless communication systems, and automated exploration modeling. Satellite links to a simulated science backroom at JSC's ExPOC demonstrate the flow of science data back to a distributed science team. This project is focused on fidilty in

Table 17.1 (cont.)

				the categories of technology and tech. integration. Science tasks are simulated, but useful qualitative science operations lessons are learned from the exercise of returning science data to the ExPOC. Mission operations are largely autonomous in the field, with only some mission operations functions being simulated through the ExPOC. Both suited test subjects and ExPOC teams derive useful training from the activity. Overall the costs are higher than some of the other activities rated here, but the lessons learned scale accordingly.
6	NASA Mobilel Agents Project	On- going Apr. 1997–present Various field sites	I – IV **A***, **B***, **C***, **D***, **E***, **F***, G, I, **J***	An ongoing annual multidisciplinary, multi-agency field exercise in the desert Southwest US led by Dr. William Clancey [Clancey, 2004], focused on developing and integrating information, robotic, and sensing technologies for scientific exploration. This effort is related to the Desert RATS activity in its objectives and in some of its technologies, but it also makes use of the Mars Desert Research Station, emphasizes end-to-end planning and data access, and has a more developed science component, including authentic exploration and a distributed remote science team. The primary focus is on automating the functions of Apollo's CapCom and flight controllers in the areas of navigation, science data logging, telemetry, health monitoring, and dynamic EVA rescheduling. Relay of science data to Earth (voice annotations, photographs, and named associations) is included in the simulation via e-mail links to a shared semantic database. Some mission support functions are provided by the Mars Society, but the crew has a high degree of autonomy. The fidelity of this activity is similar to those of Case Study #5 because it has used some of the same technology components, with increased scientific fidelity and increased human factors resulting from the use of the habitat and selected geologists with specific scientific agendas as subjects. Emphasis is on

Table 17.1. (cont.)

	Title	Dates and location	Type and Category	Description
				empirical discovery of requirements for voice commanding, roles for robots in complex terrain, and end-to-end communications (including tools for collaborative EVA planning involving the EVA crew, habitat crew, remote scientists, and mission support).
7	NASA Apollo 17 simulations	1971 NASA JSC	IV ALL	NASA's integrated simulations leading up to Apollo 17 [Muehlberger, 2002]. Astronauts in the field and ground teams in JSC's Mission Control conducted simulated science activities with high fidelity mission operations, science operations, and technology demonstrations. The intent of this activity was both to train flight and ground crews, and to flesh out any potential problems with the lunar rover lunar spacesuits, science equipment.
8	NASA Extreme Environments Mission Operations (NEEMO) project	On-going 2001–present Florida Keys (underwater)	III, IV **A***, **B***, C, D, **E***, **F***, **G***, **H***, **I***, J	NASA's ongoing involvement in the University of North Carolina's Aquarius facility underwater off the coast of Florida [Williams *et al.*, 2004]. Small crews of scientists and engineers are selected and trained for multiple-week marine biology and exploration research missions. This is overall one of the highest fidelity integrated human exploration analogs at NASA. Life support, communications, and habitat technologies are directly relevant to the design of future surface system design, with the exception of the spacesuit, which is represented here functionally by SCUBA gear donned by the crew in their simulated EVAs. Psychological and human factors fidelity of the environment is high, partially due to the submersion if the small habitat in water inhospitable to humans without proper gear. Longer missions will increase the fidelity in this area.

Table 17.1 (cont.)

9	NASA Mars Analog Research and Technology (MARTE) project	2003–2005 Rio Tinto river bed, Spain	II, III **A***, **B***, **C***, **D***, E, **F***, G, **I***, **J***	A simulated robotic Mars drilling project led by Dr. Carol Stoker to search for subsurface chemolithoautotrophic microorganism biosignatures, using the Rio Tinto field site as an analog for possible microbial life on Mars. Ground truth drilling is being conducted to provide information for assessing the success of the simulated Mars drill in its search for life. A team of scientists on Earth will receive the information from the drill as it is being acquired and will make inputs to the sample handling system requesting specific subsampling to be done by the robotic system. Humans at the site are simulating the functions of the projected robotic drilling system. This analog is medium to high fidelity in all categories except (H).
10	The International Space Station (ISS)	On-going 1998–2010 Earth orbit	I – IV A, B, C, D, **E, F, G, H***, I, J	An international orbiting scientific laboratory, mostly useful in the context of analog studies only if explicit analog experiments are conducted there. If no directed planetary analog experiments are done, it serves naturally as a Type II analog activity. There is a potential for Type I–IV experiments using the space station, provided dedicated crew and ground time are available for this research. The lack of gravity and scientifically explorable terrain, however, limit its usefulness as a full Type IV analog for surface planetary exploration. The primary benefit of the ISS for a Type IV analog would be its fidelity in mission operations and crew training.

Bold category letters indicate higher fidelity. Asterisks indicate explicit focus.

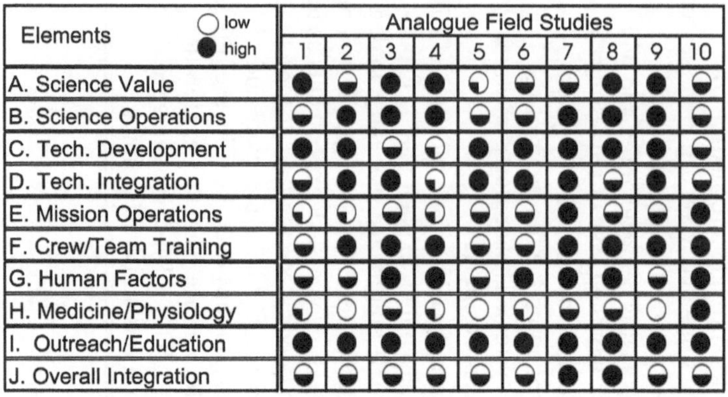

Figure 17.1. Analog Fidelity parameters for ten case studies. (1) HMP, (2) HoRSE, (3) NOAMA, (4) MDRS Crew 6, (5) Desert RATS, (6) Mobile Agents, (7) AIM, (8) NEEMO, (9) MARTE, (10) ISS.

through field simulations. Observations made by humans in simulated EVA were compared with those made by simulated future robots controlled by a (real) remote science team in a special facility on Earth. After defining figures of merit, the section will discuss the expected capabilities of 2015-era surface exploration robots. The field experiment will then be described, including controls, sites, and initial conditions. The section will conclude with a discussion of the initial results and conclusions.

17.3.1 Case study goals

Few data exist on the relative science return produced in missions using different approaches to the *in situ* exploration of a planetary surface. In planning the most effective way of exploring the Moon and Mars, cost–benefit analyses require a means of comparing the relative field science productivity of spacesuited humans vs. a mobile robot controlled using "tele-presence" techniques from a nearby habitat or orbital station, vs. traditional rover operations controlled from Earth. The goals of this quick turn-around study were first to define appropriate science-return metrics for such field science comparisons and then to begin comparing productivity between tele-operated rovers and spacesuited humans. Tests with a *simulated* 2015-class rover and with partially encumbered geologists were conducted at Haughton Crater in the Canadian Arctic in July–August 2002 and 2003. Although these were somewhat rough, first-order experiments, early results imply that humans could be 1–2 orders of magnitude more productive per unit time in exploration than future terrestrially controlled robots.

Table 17.2. *Metrics for analog exploration fidelity parameters*

	○		◒	●
Science Value	None/Not appropriate	Simulated science tasks or science tasks not relevant to planetary science. Little/no publishable results.	Valid scientific objectives/tasks relevant to future planetary exploration. No intent to publish science results or publishible science results not directly relevant to planetary science.	Planetary science tasks lead to publishable (peer-reviewed) science results.
Science Operations	Science operations not relevant to future missions.	Low fidelity relevant science operations, but not focused on operations lessons learned.	Medium fidelity to actual projected science procedures for planetary surface missions. Qualitative lessons learned.	High fidelity of science planning, procedures, communications, and reporting to planetary surface missions. Quantitative metrics.
Technology Development	Little/no technology development	Relevant technology used but not developed. Primarily application of existing technology.	Relevant technology developed but not dependent on analog environment.	New technology tested by taking full advantage of analog environment.
Technology Integration	None/Not appropriate	Different systems used simultaneously but not integrated.	Only a small number of applicable technologies used in an integrated fashion.	Multiple technologies used in an integrated fashion as proposed for actual mission.
Mission Operations	Mission operations not relevant to future missions.	Low fidelity relevant mission operations, but not focused on operations lessons learned.	Medium fidelity to actual projected mission operations for planetary surface missions. Qualitative lessons learned.	High fidelity of mission planning, procedures, communications, and reporting to planetary surface missions. Quantitative metrics.

Table 17.2. (cont.)

	○		◒	●
Crew/Team Training	None/Not appropriate	Tasks developed to meet immediate needs of test. Some applicability to flight or ground crew training.	Tasks are representative of space mission. Alternative procedures are tested and compared.	Tasks are directly applicable to mission preparation for flight or ground crews, e.g. motion flight simulators.
Human Factors	None/Not appropriate	Human crews involved, but low fidelity to planetary habitats or surface activities.	Medium fidelity habitat or surface simulation.	High fidelity habitat or surface simulation.
Medicine/ Physiology	None/Not appropriate	Some studies relevant to future long-term human missions	Studies relevant to maintaining crew medical and health for orbital, transit, and/or surface human missions	Direct medical and physiological experiments on humans in long-duration space flight conditions
Outreach/ Education	None/Not appropriate	Low level of activity. Low visual content/difficult to explain. Not directly relevant to mission.	Moderate level of activity. Moderate visual content/relatively easy to explain. Partially relevant to mission	High level of activity. High visual content/easy to explain. Directly relevant to mission.
Overall Integration	None/Not appropriate	Low level of overall coordination among analog element (science value, science operations, etc.)	Moderate level of overall coordination among analog element (science value, science operations, etc.)	High level of overall coordination among analog elements (science value, science operations, etc.)

17.3.2 Introduction

Strong differences of opinion are held regarding the merits of exploring planets using robots or humans. It is often argued that robotic exploration is much more cost effective – and this may well be the case for relatively simple or particularly hostile planetary bodies (like the Moon and Venus, respectively). For a distant, complex planet like Mars (which may still be geologically active and has a surface area comparable to the Earth's continents) it is not obvious that robots controlled from Earth will be capable of answering the most important science questions (especially the question of whether or not life developed independently elsewhere in our Solar System) in a reasonable amount of time (e.g., a human lifetime) unless true artificial intelligence soon emerges from advanced information technology research. We anticipate that the "right" answer will be to combine the unique cognitive strengths of *in situ* human explorers with the unique strengths of robots (cost, dispensability, endurance, etc.). The purpose of this research is to test our intuitive anticipation by developing a quantitative understanding of the capabilities of humans and robots – separately and together.

Robotic spacecraft carry advanced instrumentation that adds a broader set of sensory inputs and observes more of the electromagnetic spectrum than can unaided humans (Coates, 1999); this is helpful in mineral and chemical identification and in other scientific measurements. As a result, remote-sensing orbital robots cannot be challenged by astronauts. In the case of surface robots, safety issues are relatively minimal, life support needs are eliminated, costs are greatly reduced, and improved strength and endurance are possible. These facts have led some to question sending humans into space at all. However, even with the expected advances in robotics over the next decade or two, robots will still lack human cognitive capabilities in real-time perception, planning and recovery from, or adaptation to unexpected or adverse circumstances (Pedersen *et al.*, 1992). As regards instrumentation, this can be miniaturized and made available to astronauts operating on a planetary surface. Spudis and Taylor (1992) note that increasing levels of detail in geological or biological field work require real-time decisions about choosing subunits to measure or sample, without a priori knowledge. Any kind of field work requires human cognitive and perceptual capabilities and as such research moves to more detailed levels those capabilities become dominant. Thus, unfortunately, if the human is tens or hundreds of millions of miles from the field site, the pace of the effort will be correspondingly slow.

17.3.3 Formulation issues

Definitions of science productivity in exploration

How can one usefully quantify "science return?" Generally, scientific insight is built up piece by piece like a jigsaw puzzle, but sometimes a single dataset or sample may prove to be the key to understanding a site, confirming a given theory, or changing a paradigm. A set of several possible metrics for assessing field exploration was compiled after interviewing current practicing field geologists and planetary scientists. The initial set, as shown in Table 17.3, became a starting point for consideration.

Metrics A through C (Table 17.3) were dropped because, in selecting figures of merit, the issue of significance looms: there may be only one interesting image in a collection or one key subunit exposure in a kilometer-long traverse; sometimes highly valuable fieldwork may be done in a comparatively limited area, while some long-distance traverses return little data of interest. At the risk of introducing subjectivity, those features deemed worth describing or reporting by experienced practitioners are used, as in proposed metrics F–H. These are much higher-level measures of scientific merit compared with simple bit-counts or traverse lengths. Proposed metric E was discarded as it could just as easily be the result of engineering or lighting issues as to the productivity of a given site. The figures of merit then used in this study were metrics D and F–H.

Effects of prior experience, initial conditions

Issues in conducting comparative field studies include the relative expertise of the human participants (the simulated astronauts and those making up the

Table 17.3. *Proposed metrics for human/robotic traverses*

	Metric	How measured
A	Total distance	GPS waypoints recorded by test director; video footage
B	Number of stops	Video footage, notebook entries, number of rover panoramas
C	Number of sites visited	Same as (B) without revisits (alternate metric)
D	Number of samples kept	Counted post-test
E	Number of image retakes	Number of closeup image files
F	Number of site descriptive phrases	Text analysis of descriptive report
G	Number of hypotheses	Text analysis of descriptive report
H	Number of external references	Text analysis of descriptive report

remote science team) and their relative experience with respect to the geological units under study, to the processes that formed the region, and to the specific field test sites. Variations in experience affect the depth of questions asked by the scientific participants and their ability to interpret what they see – and hence strongly influence the scientific return. While this problem is difficult to eliminate, it can be ameliorated by using judgment in selecting the experiment participants based on their backgrounds and by ensuring that none is already familiar with the chosen test sites.

To avoid bias, the remote robotic and on-site human science teams must be given the same briefing and starting data. In the summer of 2001, ten potential Mars analog sites at Haughton Crater had been surveyed, collecting remote sensing and aerial imaging for each. A science panel formulated a list of questions for field investigation based on these surveys and reconnaissance, thereby replicating the likely starting point of either a robotic or human EVA investigation (Remote Science Team Report, 2002).

Shirtsleeve geologist as a control

Even with these precautions ("fresh" participants given the same initial data), there is no absolute yardstick against which to judge scientists' respective surveys. For example, ground truthing was difficult for the chosen test sites because we had not carried out sufficiently painstaking prior study and analysis. Therefore, to provide improved control for our experiment, a third parallel set of surveys was made. Unrestricted human geologists "in shirtsleeves" were given no time limit to complete their field analysis. The time taken, however, was recorded as an important reference.

Need to compare modes of exploration

Humans in EVA and remote-controlled robotic missions operate on much different operational schedules. Human endurance and life support consumables limit EVAs to a few hours while robots may operate in the field for months with periodic command and data relay cycles. Robots controlled from Earth, of course, effectively waste a large portion of mission/experiment time while awaiting ground instructions (and correcting misinterpreted or off-target instructions), while humans are capable of independent, real-time re-prioritization and re-planning in the field (e.g., Apollo Astronaut Harrison Schmitt's lunar vitreous "orange soil"). It is, of course, this time wasted because of extreme latency that robs the efficiency of robotic operations on Mars (given that all the cognitive capability is on distant Earth) and provides the human explorer with his/her performance advantage.

Practical field experiment protocol design requires that some time constraints be placed on both human and robotic sorties (except for the "shirt-sleeved" control sorties). In this study, the remote science team operating the simulated 2015 robot could have one or two successive three-hour shifts at a given test site, while humans in simulated EVA were limited to one three-hour shift. Both the rover and suited tests were pre-positioned at a given starting point at each test site. However, just as the "control" geologists were given time flexibility, the space-suited humans were not required to spend the full three hours allowed in the field if their curiosity and objectives had been satisfied or if suit safety or health concerns arose.

17.3.4 Procedure

Assume advanced rover capabilities

It was assumed that human exploration of Mars is still one or two decades away. In assessing the potential performance of humans and robots (controlled from Earth or from Mars) the relevant timeframe is 10–20 years in the future. Comparison field testing with a current-generation rover such as Carnegie-Mellon's *Nomad* (Wettergreen *et al.*, 1999) or Ames' *K9* (Bresina, 1999) would obviously not be a fair representation of 2015-class robotic exploration systems. Therefore, in this study, it was decided to remotely control a 2015-class *rover-equivalent*, which was simulated by an all-terrain vehicle (ATV), shown in Figure 17.2, piloted by a student and instrumented with panoramic and other imaging capabilities.

In a recent study of future space robotic capabilities conducted by Pedersen (2002), the expected capabilities of 2015-class surface exploration rovers are defined. These include the following nominal capabilities:

- Scientists interact directly, at mission level (high-level commands)
- Autonomous obstacle avoidance and target tracking
- Drivability on rough or soft flat surfaces, not steep slopes or boulder fields
- Active rebalancing and/or center-of-gravity control
- Autonomous grasping of samples, or can blast/break off a sample
- Self-diagnostics, with preset recovery procedures
- No self-righting or repair
- No auto-map-building or global self-navigation (must be pre-positioned)
- Lack of human-level cognitive and perceptual capabilities
- Onboard "distillation" of some science and status data
- Some virtual presence of ground team (visual, not tactile).

Figure 17.2. Human-operated simulated 2015-class rover at Haughton Crater, Devon Island, Canada.

In this experiment the "rover" (ATV) operator communicated with the remote science team via mission-level text messages and commands sent via satellite and local network to an ATV-mounted display. The operator autonomously handled obstacle avoidance, tracking, and rebalancing. As directed by the remote science team, the operator could park the "rover" to acquire samples, use a rock hammer, and take targeted images. The operator was instructed to use initiative only in the case of basic safety and recovery procedures (e.g., backing away from a canyon edge or retracing a path out of a loss-of-signal zone), tasks that could reasonably be expected of a 2015-class rover.

Mars-analog field site in the Arctic

The site chosen for the July 2002 tests was the Haughton impact structure (Jessburger, 1988), centered at 75°23' N, 89°39' W on Devon Island in the Canadian Arctic. Haughton Crater formed about 23 million years ago and is a well-preserved structure with an original rim diameter of ~24 km (Scott and Hajnal, 1988). It is an excellent polar-desert Mars analog that has been shaped by post-impact surface glaciation and periglacial effects; it is nearly devoid of multi-celled life. Two researchers were familiar with this area from past geophysical studies conducted there (Glass and Lee, 2001). An existing

NASA-run summer field camp adjacent to the crater provided necessary logistical and scientific support.

17.3.5 Remote science experiment design

Local and trunk wireless networks

As shown in Figure 17.3, the simulated rover tests required that the ATV operator in the Arctic remain in continuous data communication with a remote science team in California. Building on previous work in wireless exploration networks (Gilbaugh *et al.*, 2001), field communications with the rover/ATV were maintained via a tactical network of 802.11b repeaters, which were in turn connected to base camp by high-speed point-to-point digital spread-spectrum trunk radio links (Braham, 1999). A commercially leased 768 Kbps satellite link provided connectivity from the field camp to NASA-Ames Research Center. Including transmission, error-checking, and buffering effects, the typical data transmission times from the rover back to NASA-Ames ranged from a few seconds for still images of rock specimens to 70 to 90 s or more for 3 MB three-color panoramic images.

The rover tests were run with no added delays, as though they were being operated locally (i.e., from a surface habitat or Mars orbit). This arrangement was consistent with the goal of assessing tele-presence performance

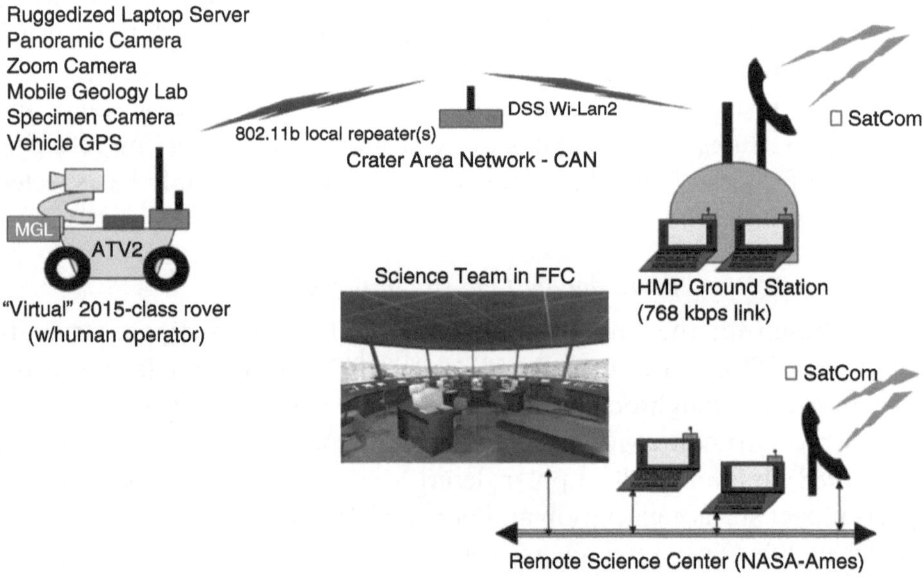

Figure 17.3. Local and trunk wireless datalinks from the FFC backroom to the rover.

(also, it was not feasible to recruit scientists with the patience to voluntarily operate with inserted two-way 20-minute delays to/from the rover). Transmissions to/from the rover were therefore logged with timestamps and a separate post-facto analysis was conducted to construct a similar timeline with Earth–Mars delays inserted.

Virtual presence capability during rover tests

One of the expected capabilities of a 2015-class rover and operations facility is some degree of virtual visual "presence" for the remote science team (whether on Earth or Mars). The Future Flight Central (FFC) facility at NASA-Ames is a full-scale (8 m diameter) virtual air traffic control tower with computer-generated projected 360° out-the-window visuals. For this study, the FFC consoles were used to provide image displays (panoramas, targeted higher resolution images, close-ups of samples) to the science team, as well as compose commands for the rover. Panoramic images from the rover were displayed on the FFC "windows," creating a sense of visual immersion for the science team. (Note: the returned imaging data can be processed to create a virtual environment through which a remote scientist can "move" and, in doing so, achieve a startling level of immersion: this capability was not available to the team.)

Responsibilities for data capture (rover tests, maximum 3-hour runs):

- Science team: command traverses, image acquisition, and sample acquisition. Maintain personal notes to be combined and turned into a 1–2 page written report afterwards.
- Support staff: open and log communications between science team and "rover," archive images by test run, by type, daily
- Rover operator: acquire images as commanded, dump local image files daily to CDs, clear the onboard storage daily, acquire samples as commanded
- Test director: note stop/start/locations, take GPS waypoints, count total number of samples at the end of each run.

Tests using HS prototype spacesuit

Prototype spacesuits were provided through the Hamilton-Sundstrand Company (HS). Geologist test subjects were organized by the SETI Institute. The HS suit, shown in Figure 17.4, was an *unpressurized* "engineering prototype." Retired Shuttle suit gloves were used with the rigid torso assembly, but there were no leggings. HS provided a crew of two to monitor the safety and health of the geologist test subjects and to transport and maintain the suit. A real pressurized spacesuit would introduce significantly greater restrictions on the suit subject which were not quantifiable in this experiment.

Figure 17.4. Hamilton-Sundstrand prototype at Site T11.

Responsibilities for data capture (human in suit, maximum 3-hour runs):

- Geologist in suit: audio recording of personal observations (in suit or external via RF), turned later into a 1–2 page summary; close-up images on camera; choose samples to retain (up to 5 kg limit)
- Suit assistant: carry specimen bag, camera, and hammer
- Support staff: monitor geologist health and safety, download and archive camera images after each deployment, capture handheld video in the field
- Test director: note stop/start/locations, take GPS waypoints, count total number of samples at the end of each run.

"Shirtsleeved"-geologist surveys

Ground-truth surveys by unencumbered geologists were conducted beginning at identical site starting points as in the other test series. An assistant carried cameras, tools, and samples.

Responsibility for data capture (unencumbered human, unconstrained test time):

- "Shirt-sleeve" geologist: written report
- Assistant: download and archive camera images.

Given the figures of merit, tallies of observations, conclusions, and hypotheses were compiled from each report. Figure 17.5 shows a paragraph from a "raw" report and a corresponding breakout.

Line	Observation	Type	Conclusion	Hypothesis
31	on the inner side of the middle ring/rim side of a hill covered in brownish scree	Relational		
32	scree or clasts	Simple		
32	Most ... is covered clasts	Relational		
32	a broad valley	Simple		
33	standing water	Simple		
32	gently descended ... at the bottom	Relational		
33	polygonal freeze-thaw features	Simple		
34	clasts in the ... boundaries	Relational		
34	quik-clay type material	Simple		
34	material inside the polygons	Relational		

The general geologic setting is that of a 23Ma impact into Ordovician-Silurian carbonates. This event gave rise to a concentric series of ridges, valleys, and faults. The test area was on the side of a hill, on the inner side of the middle ring/rim that was covered in brownish scree. Most of the area is covered in scree or clasts. The hill gently descended into a broad valley with standing water a the bottom. There are polygonal freeze-thaw features (larger clasts are concentrated in the polygonal boundaries). There is quick-clay type material inside the polygons. The pebble-to-boulder size scree (or brecciated carbonates) showed signs of weathering. There was amber-streaking at least down to 10 cm depth in some of the boulders.

Figure 17.5. Survey report and observation tally sheet.

17.3.6 Case study results

Four remote science team traverses, three simulated EVA traverses, two unencumbered geologist traverses.

The three-person science team assembled at NASA-Ames at the FFC facility during the week of 22 July 2002. A total of four tele-operated rover test runs were made over a week's time at three separate sites (see Table 17.4). Weather conditions at the Haughton Crater site were poor throughout: this limited the number of sites visited, time spent at each site, and also reduced the contrast of the images returned from the four remote runs. Better weather prevailed in the subsequent traverses to these same three locations by space-suited (Figure 17.4) and by unencumbered geologists. However, no descriptive report was turned in by the suited participants at Site T9 or the "shirt-sleeved" visit to T11 (Table 17.4).

17.3.7 Discussion

Although a comparative assessment of the scientific productivity of the several approaches to field exploration was carried out, this assessment should be viewed as highly preliminary because of the limited duration of the tests resulting from poor weather, technical difficulties, the lack of remote team

Table 17.4. *Field tests in July 2002 with teleoperated rover*

ID	Date	Location	UTM-E	UTM-N	Stops	Comments
H024A-1	2002.07.22	Site T4	16 417855	8374807	7	Rover commanded into comms hole at 2.5 h "safing procedure"
H029A-1	2002.07.24	Site T9a	16 421089	8378712	4	Science team wanted multiple pans at each stop
H029A-2	2002.07.25	Site T9b	16 421125	8379050	6	Found water
H0211A-1	2002.07.26	Site T11	16 420136	8370769	3*	Surprise starting point of landing ellipse. *Lost comms at 1.5 h

training, and the choice of scientifically less interesting sites within the Haughton Impact Crater. Technical improvements include a reduction in the still-substantial latency of communications between the remote science team and the ATV, increasing the panoramic camera's coverage to a full 360°, and providing the team with some imaging coverage during the traverses to maintain situational awareness. The remote science team's limited time (four days) operating the simulated rover without prior training meant that the tests were primarily a learning experience for the team rather than a real operational test.

Table 17.5 lists the science return (as defined above in Section) from each test. The elapsed times for the rover tests do not include Mars–Earth equivalent inserted time delays. Assuming that commands would be grouped whenever possible and that communications were continuous between Earth and Mars, post-test analysis indicates that inserting delays into the transmission transcripts increased the typical test duration by a factor of five. If team deliberation time were less for a Mars-based science team (likely of no more than two astronauts) compared to that typical for a large Earth-based science team then this factor might be more like twenty. And, if Earth access to a rover were further constrained (i.e., only to typical Deep Space Network twice/day access periods), it is estimated that that would reduce productivity by roughly another factor of five.

Among the pre-selected metrics, the number-of-hypotheses metric proved not to be useful. This was perhaps because during the short tests there were relatively few observations made by the remote science team from which

Table 17.5. *Summary of 2002 field results*

Site	Type	Duration (min)	Observations (number)	Conclusions (number)	Hypotheses (number)
T4	Remote	150	22	4	7
T4	Suited	92	28	4	1
T4	Free	30	41	2	1
T9	Remote	335	18	6	3
T9	Free	7	32	8	3
T11	Remote	90	2	0	0
T11	Suited	63	24	3	2
	Averages by type				
	Remote	192	14	3	3
	Suited	78	26	4	2
	Free	19	37	5	2

hypotheses could be drawn. The most differentiation between the three test types proved to be in the total observation count. The observation averages by test type show, as expected, that human geologists in shirtsleeves are far more productive than either spacesuited humans or tele-operated robots. And, not surprisingly, simulated EVA crew were more productive than the 2015-simulated rovers even if operated from a Mars base with reduced latency compared to Earth-based operations. Figure 17.6 shows the relative observational rates – observations per unit time, normalized.

Preliminary results imply that a local spacesuited human would achieve about 25 times the science productivity of a 2015-class Earth-controlled rover. Given that 2015-class rovers are expected to be 10 to 20 times as productive as the current state-of-the-art (Mars Exploration Rover), this further implies a relative productivity advantage per unit time of around 300–400 times for local EVA vs. current capability.

As sketched in Figure 17.7, rather than the rover becoming an extension of the remote human science team, the science team became more mechanistic in their planning and execution during the rover tele-operated tests. Targets of opportunity were bypassed if they were not on the original traverse plan, or would significantly slow the arrival of the rover at its next waypoint. Conversely, both the simulated suit subjects and shirt-sleeved humans diverted their traverses to cover nearby targets of interest (such as hydrothermal vents, or a large ejecta block containing macrofossils at site T11). We anticipate that, with more training and experience, the remote science team would rapidly learn to function more as they would in person in the field.

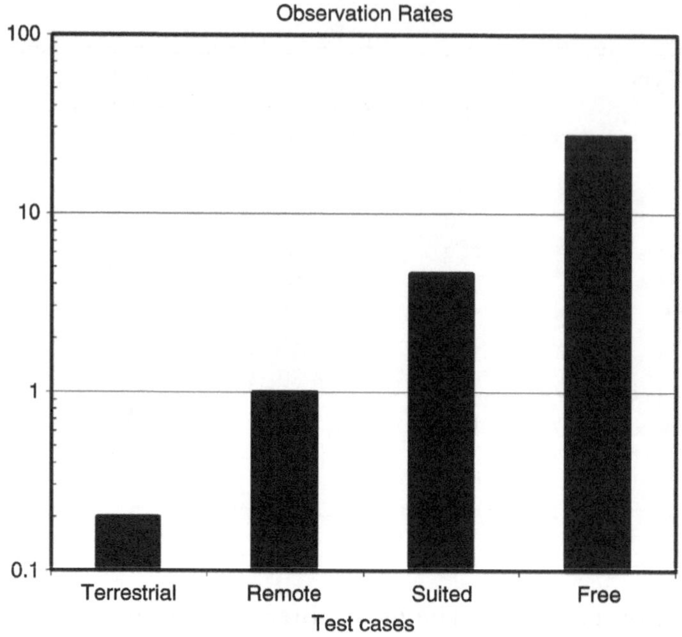

Figure 17.6. Observations per unit time for the given test cases.

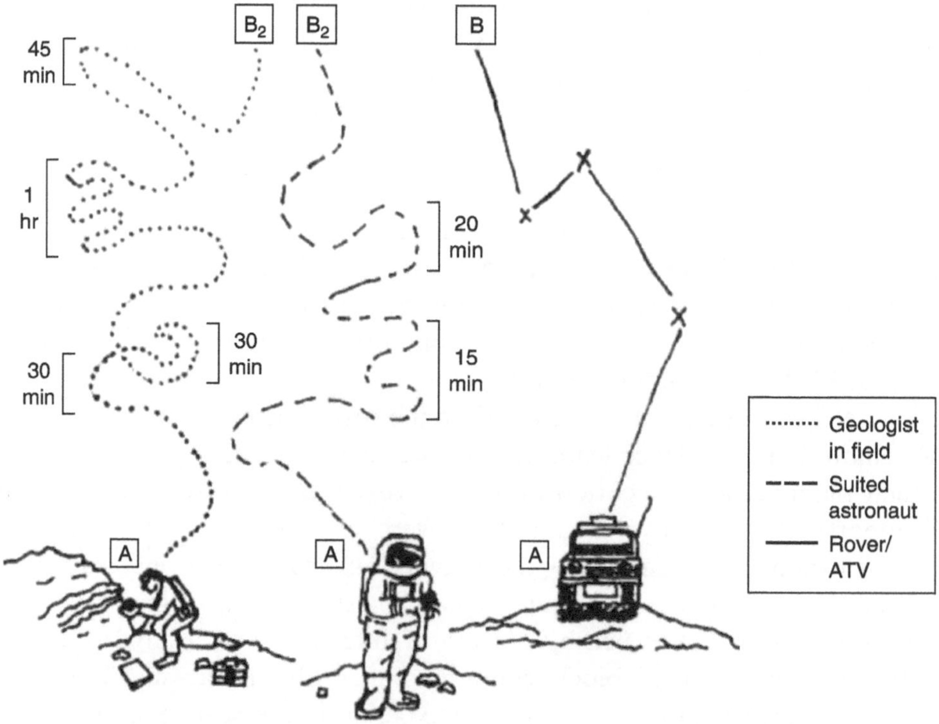

Figure 17.7. Sketch of rover and human paths in field exploration.

17.3.8 HoRSE case study conclusions

The authors re-emphasize the need to avoid inferring too much from this first experiment. This study compares human exploration performance to that of a hypothetical 2015 rover at only three test sites in a single geological setting (Haughton Crater). Small changes in mission profile or in the rate of robotic technology maturation can easily skew results in either direction. Given that three test points are far from enough to be significant, additional field tests of this type are needed.

Follow-on studies must address at least two needs: first, to incorporate all the lessons learned regarding technical implementation, team training, and metrics into an improved system. Second, to add more data points to the field science evaluations begun in 2002, probably by finding an appropriate Mars analog site in the US desert southwest where weather conditions are better than at Haughton Crater. A total of 8–10 three-way science site productivity comparisons between humans and robots would be the minimum needed to provide results that could be used with some confidence in future cost–benefit tradeoffs and other mission planning activities.

17.4 Conclusions

Case studies like those discussed in this chapter illustrate a need for a more comprehensive Analog Missions Program. Work to date has been driven by the needs or interests of specific researchers or projects, has generally been under-funded, and has therefore been limited in scope and level of integration. Current plans for the human exploration of the Moon and Mars require the use of analogs for the development, testing, and integration of new exploration technologies, scientific methodologies, and operational protocols. Analogs are also needed for training the astronauts and ground crews, researching the effects of planetary exploration on humans, and demonstrating readiness for flight. Thousands of testbeds and independent scientific investigations refine each system and subsystem, and Analog Missions provide the opportunity to integrate systems together in higher fidelity simulations. Some of these simulations will go beyond even the "full-up sims" of Apollo 17, using advanced technologies, automation, intelligent systems, new management techniques, new models of international cooperation, and new platforms in space. Using the Earth and Moon as analogs for Mars will be one of the keys to meeting the needs in the next era of human exploration.

Acknowledgments

The authors wish to thank Bill Muehlberger, Dean Eppler, and Chris McKay for their thoughtful reviews and contributions to this manuscript. The authors also wish to thank Pascal Lee and the SETI Institute for their field support of this study through the NASA Haughton-Mars Project. Carol Stoker, Samantha Domville, Victor Rundquist, Richard Alena, Lori Haven, Badi Abad, and Shamim Samadi of NASA Ames, Stephen Hoffman of NASA Johnson Space Center, Jeffrey Moersch of University of Tennessee, and Melissa Lane of the Planetary Science Institute, all made critical contributions to the success to the HoRSE field seasons. The Hamilton Sundstrand contribution was essential and much appreciated. This study was funded in 2002 by the Human-Robotics Working Group of the NASA Exploration Team (NExT).

References

Akin, D. (2001). Robotic capabilities for complex space operations. American Institute of Aeronautics Adminsistration Space 2001 Conference, Albuquerque, New Mexico, August 2001, (abstract) 4538.

Braham, S. (1999). Canada and analog sites for Mars exploration. Second Canadian Space Exploration Workshop (abstract).

Bresina, J. (1999). Increased flexibility and robustness of Mars rovers. 5th International Symposium on Artificial Intelligence, Robotics and Automation in Space (i-SAIRAS), Noordwijk, The Netherlands, June 1999 (abstract).

Clancey, W. J. (2004). Automating CapCom: pragmatic operations and technology research for human exploration of Mars. In *Martian Expedition Planning*, ed. C. Cockell. AAS Science and Technology Series, vol. 107, pp. 411–30.

Coates, A. (1999). Limited by cost: the case against humans in the scientific exploration of space. *Earth, Moon, and Planets*, **87**, 213.

Friedman, L. (2000). Connecting robots and humans in Mars exploration. Concepts and Approaches for Mars Exploration, July 2000, p. 118 (abstract).

Gilbaugh, B., Glass, B., and Alena, R. (2001). Mobile network field testing at HMP-2000. 2001 IEEE Aerospace Conference, Big Sky, Montana, March 2001 (abstract).

Glass, B. J. and Lee, P. (2001). Airborne geomagnetic investigations at the Haughton impact structure. *Abstracts of Papers Submitted to the 32nd Lunar and Planetary Science Conference*. Houston: Lunar and Planetary Institute, CD 32, Abstract 2155.

Hoffman, S. (2001). The human exploration of Mars: the reference mission of the human exploration study team. NASA Johnson Space Center, online document at: http://ares.jsc.nasa.gov/HumanExplore/Exploration/EXLibrary/EXdocuments.htm

Jessburger, E. K. (1988). 40Ar-39Ar dating of the Haughton impact structure. *Meteoritics*, **23**, 233–4.

MEPAG Committee (2004). *Scientific Goals, Objectives, Investigations, and Priorities for Mars Exploration*, ongoing, ed. J. Taylor *et al.* Published online: http://mepag.jpl.nasa.gov/reports/index.html.

Muehlberger, W. R. (2003). Geological training of astronaut during the Apollo Era. Workshop on Analog Sites and Facilities for the Human Exploration of the Moon and Mars, Colorado School of Mines, Golden, CO, May 21–23, 2003 (abstract).

Neal, C. (2001). Geological investigations of Mars: the human factor. Workshop on Science and the Human Exploration of Mars, Lunar and Planetary Institute, January 2001, Lunar and Planetary Institute Contribution 1089, p. 154 (abstract).

Pedersen, L. (2002). NASA Exploration Team (NEXT) Space Robotics Technology Assessment Report, Computational Sciences Division, NASA-Ames Research Center, Moffett Field, California.

Scott, D. and Hajnal, Z. (1988). Seismic signature of the Haughton structure. *Meteoritics*, **23**, 239–47.

Remote Science Team Report (2002). Haughton Remote Science Experiment 2002, NASA-Johnson Space Center, November 2002, ed. K. Snook.

Snook, K. (2005a). Technical Report of the NASA Oceanographic Analog Missions Project (NOAMA). *National Aernautics, & Space Administration Technical Memorandum* (submitted).

Snook, K. (2005b). A review of analog studies and lessons learned, 1960–2004. *National Aernautics & Space Administration Technical Memorandum* (submitted).

Spudis, P. and Taylor, G. J. (1992). The roles of humans and robots as field geologists on the Moon. 1992 Symposium on Lunar Bases and Space Activities, NASA Johnson Space Center, 1992, p. 307 (abstract).

Wettergreen, D., Bapna, D., Maimone, M., and Thomas, G. (1999). Developing Nomad for robotic exploration of the Atacama Desert. *Robotics and Autonomous Systems*, **26**, 127–48.

Index